# Vascular Transport
# in Plants

This is a volume in the

PHYSIOLOGICAL ECOLOGY Series
Edited by Harold A. Mooney
Stanford University, Stanford, California

*A complete list of books in this series appears at the end of the volume.*

# Vascular Transport in Plants

## N. Michele Holbrook

Organismic and Evolutionary Biology
Harvard University
Cambridge, Massachusetts, USA

## Maciej A. Zwieniecki

The Arnold Arboretum
Harvard University
Cambridge, Massachusetts, USA

ELSEVIER
ACADEMIC
PRESS

Amsterdam • Boston • Heidelberg • London
New York • Oxford • Paris • San Diego
San Francisco • Singapore • Sydney • Tokyo

Elsevier Academic Press
30 Corporate Drive, Suite 400, Burlington, MA 01803, USA
525 B Street, Suite 1900, San Diego, California 92101-4495, USA
84 Theobald's Road, London WC1X 8RR, UK

This book is printed on acid-free paper.

**Library of Congress Cataloging-in-Publication Data**
Holbrook, N. Michele (Noel Michele)
  Vascular transport in plants / N. Michele Holbrook, Maciej A. Zwieniecki.
    p. cm.
  Includes bibliographical references and index.
  ISBN 0-12-088457-7 (alk. paper)
  1. Vascular system of plants. I. Zwieniecki, Maciej A. II. Title.
  QK725.H58 2005
  575.7′–dc22

                                                        2005003325

**British Library Cataloguing in Publication Data**
A catalogue record for this book is available from the British Library
ISBN-13: 978-0-1208-8457-5
ISBN-10: 0-12-088457-7

For all information on all Elsevier Academic Press publications
visit our Web site at www.books.elsevier.com

Printed and bound by CPI Group (UK) Ltd, Croydon, CR0 4YY

Transferred to Digital Print 2011

# Contents

v

## Part II   Transport Attributes of Leaves, Roots, and Fruits

# Part III    Integration of Xylem and Phloem

# Part IV   Development, Structure, and Function

# Part V   Limits to Long Distance Transport

## Part VII    Synthesis

# Contributors

*Number in Parentheses after each name indicates the chapter number for the author's contribution*

**Brian G. Ayre** (3)
Department of Biological Sciences
University of North Texas
Denton, TX 76203-5220
USA

**Benjamin Babst** (17)
Department of Biology
Tufts University
Medford, MA 02155
USA

**Peter Barlow** (14)
School of Biological Sciences
University of Bristol
Woodland Road
Bristol BS8 1UG
UNITED KINGDOM

**Arnold Bloom** (12)
Department of Vegetable Crops
University of California
One Shields Avenue
Davis, CA 95616

**C.E.J. Botha** (6)
Department of Botany
Schonland Botanical Laboratories
Rhodes University
Grahamstown 1640
SOUTH AFRICA

**Timothy J. Bowen** (20)
Natural Science Division
Pepperdine University
Malibu, CA 90263-4321
USA

**C. Kevin Boyce** (23)
Department of Geophysical
Sciences and the College
The University of Chicago
Chicago, IL 60637
USA

**Tim Brodribb** (4)
School of Plant Science
University of Tasmania
Hobart 7001
TASMANIA

**Pamela L. Brown** (20)
Natural Science Division
Pepperdine University
Malibu, CA 90263-4321
USA

**Zoë G. Cardon** (13)
Department of Ecology and
Evolutionary Biology
University of Connecticut, U-43
Storrs, CT 06269
USA

**Jeannine Cavender-Bares** (19)
Department of Ecology,
Evolution, and Behavior
University of Minnesota
St. Paul, MN 55108
USA

**Michael Clearwater** (18)
HortResearch Te Puke
Te Puke Research Centre
412 No 1 Road RD2, Te Puke
NEW ZEALAND

**Stephen D. Davis** (20)
Natural Science Division
Pepperdine University
Malibu, CA 90263-4321
USA

**Frank W. Ewers** (20)
Department of Plant Biology
Michigan State University
East Lansing, MI 48824-1312
USA

**Taylor S. Feild** (24)
Department of Botany
University of Toronto
Toronto, Ontario M5S 3B2
CANADA

**Brian Ford-Lloyd** (8)
School of Biosciences
The University of Birmingham
Edgbaston
Birmingham B 15 2TT
UNITED KINGDOM

**Peter Franks** (4)
School of Tropical Biology
James Cook University
Cairns, Cairns
Queensland 4870
AUSTRALIA

**Arthur L. Fredeen** (21)
Natural Resources and
Environmental Studies
University of Northern
British Columbia
British Columbia
CANADA

**Barbara L. Gartner** (15)
Department of Wood
Science and Engineering
Oregon State University
Corvallis, OR 97331
USA

**Guillermo Goldstein** (18)
Tropical Plant Ecophysiology and
Functional Ecology
Department of Biology
University of Miani
Coral Gables, FL 33124-0421
USA

**Nick Gould** (10)
HortResearch Ruakura
Hamilton
NEW ZEALAND

**Uwe G. Hacke** (16)
Department of Biology
University of Utah
Salt Lake City, UT 84112
USA

**Jens B. Hafke** (2)
Institute of General Botany
Justus-Liebig-University
Senckenbergstrasse 17
D-35390 Giessen
GERMANY

**Patrick M. Herron** (13)
Department of Ecology and
Evolutionary Biology
75 N. Eagleville Road, U-43
University of Connecticut
Storrs, CT 06269
USA

**N. Michele Holbrook** (26)
Department of Organismic and
Evolutionary Biology
Harvard University
16 Divinity Avenue
Cambridge, MA 02138
USA

**Ferit Kocacinar** (25)
Department of Botany
University of Toronto
Toronto, Ontario M5S 3B2
CANADA

**George W. Koch** (21)
Biological Sciences
Northern Arizona University
Flagstaff, AZ 86011-5640

**Mark Matthews** (9)
Department of Viticulture &
Enology
University of California
Davis, CA 05717
USA

**Joanna McQueen** (10)
University of Waikato
Hamilton
NEW ZEALAND

**Frederick C. Meinzer** (15)
USDA Forestry Service
Forestry Sciences Laboratory
Corvallis, OR 97333
USA

**Peter Melcher** (1)
Department of Biology
Ithaca College
Center for Natural Sciences
Ithaca, New York 14850
USA

**Peter Minchin** (10)
ICG-III Phytosphaere
Forschungszentrum Juelich
D-52425 Juelich
GERMANY

**John Newbury** (8)
School of Biosciences
The University of Birmingham
Birmingham B15 2%%
UNITED KINGDOM

**Gretchen B. North** (7)
Department of Biology
Occidental College
Los Angeles, CA 90041
USA

**Colin M. Orians** (17)
Department of Biology
Tufts University
Medford, MA 02155
USA

**Carol A. Peterson** (7)
Department of Biology
Univesity of Waterloo
Ontario N2L 3G1
CANADA

**William Pickard** (1)
Department of Electrical and
Systems Engineering
Washington University
Saint Louis, MO 63130
USA

**Jarmila Pittermann** (16)
Department of Biology
University of Utah
Salt Lake City, UT 84112
USA

**Brandon Pratt** (20)
Natural Science Division
Pepperdine University
Malibu, CA 90263-4321
USA

**Jeremy Pritchard** (8)
School of Biosciences
The University of Birmingham
Edgbaston
Birmingham B 15 2TT
UNITED KINGDOM

**Lawren Sack** (5)
Department of Botany
University of Hawaii
Honolulu, HI 96822
USA

**Rowan F. Sage** (25)
Department of Botany
University of Toronto
Toronto, Ontario M5S 3B2
CANADA

**Ken A. Shackel** (9)
Department of Pomology
University of California
Davis, CA 95616-8683
USA

**John S. Sperry** (16)
Department of Biology
University of Utah
Salt Lake City, UT 84112
USA

**Rachel Spicer** (22)
Department of Organismic and
Evolutionary Biology
Harvard University
Cambridge, MA 02138
USA

**Matthew V. Thompson** (11)
Department of Agronomy
University of Kentucky
Lexington, KY 40546-0312
USA

**Michael Thorpe** (10)
Chemistry Department
Brookhaven National Laboratory
Upton, NY 1197
USA

**Robert Turgeon** (3)
Section of Plant Biology
Cornell University
Ithaca, NY 14853
USA

**Melvin T. Tyree** (5)
USDA Forest Service
Northeastern Research Station
Vermont
USA

**Aart J.E. van Bel** (2)
Institüt für Allgemeine Botanik
und Pflanzenphysiologie
Justus-Liebig-Universität Giessen
35390 Giessen
GERMANY

**Amy E. Zanne** (17)
Department of Biology
120 Dana Building
163 Packard Avenue
Tufts University
Medford, MA 02155
USA

**Maciej A. Zwieniecki** (11, 26)
The Arnold Arboretum
Harvard University
16 Divinity Avenue
Cambridge, MA 02138
USA

# Preface

A major challenge in plant biology is to understand how fundamentally decentralized organisms function in a coordinated fashion. Leaves and roots form the major sites of energy and material exchange with the environment, but their sustained function requires that they be interconnected in a manner that allows for physiological integration at the whole plant level. The tissues that connect these often quite distant regions of growth and acquisition are likely to play a key role in achieving such coordination, suggesting that the xylem and phloem function not only as conduits for mass transport, but as a system for communication and integration. The goal of this book is to review and synthesize new ideas and insights in the area of long distance transport in plants, emphasizing the biophysics and physiology of transport through the xylem and phloem as well as the emerging area exploring interactions between the two vascular tissues.

Fundamental differences in the nature of transport in xylem and phloem, as well as differences in tools and approaches, have led to the development of increasingly separate disciplines dedicated to their study. The field of xylem transport is dominated by individuals whose backgrounds are primarily in physics; while studies of phloem transport increasingly fall within the realm of molecular biology. Yet the two transport tissues are intimately connected, not only in terms of their proximity and derivation from the same layer of meristematic cells, but also through the lateral exchange of water, ions, carbohydrates, and hormones. Because water flow through the xylem is largely responsible for determining gradients in water potential, the xylem exerts a marked effect on transport through the phloem. However, the phloem is also capable of influencing the xylem both through the recirculation of potassium ions from phloem, which influences the hydraulic resistance of the xylem, and by supplying energy and solutes used in the repair of cavitated vessels.

Over the past two decades there have been substantial advances in our understanding of transport processes in the xylem and the phloem. In the area of xylem transport, these include greater understanding of factors related to cavitation and embolism repair, as well as the lateral exchanges that occur in roots and leaf tissues. In the phloem, application of molecular approaches and a variety of sophisticated imaging techniques has amplified our understanding of companion cell-sieve tube interactions and demonstrated the complexity of macromolecular substances that move along this pathway. In addition, there is now greater appreciation for the role of living cells and biochemical processes in the xylem, as well as the need to understand the hydraulic components involved in carbohydrate movement. Thus, the time is ripe to bring together the two converging areas and to explore the ways in which these two pathways influence each other.

The organization of this book reflects an underlying philosophy that long distance transport in plants is more than the sum of its parts. Thus, there are no sections devoted entirely to either the xylem or the phloem and each author has been challenged to consider how the specific issues discussed in their chapter may influence the integrated function of the vascular system. The chapters are grouped into six sections beginning with a review of the fundamentals of transport in both the xylem and phloem, and ending with a discussion of the evolution of plant vascular systems. The intervening sections focus on the transport attributes of the major exchange surfaces and sinks; interactions between xylem and phloem, development and structure of the transport tissues, and factors that limit long distance transport in plants. It is our hope that this book will stimulate further research into the integrated function of the vascular system as a whole, as well as deeper understanding of the important role played by these transport tissues in coordinating developmental and physiological activities at the whole plant level.

**N. Michele Holbrook**
**Maciej A. Zwieniecki**

# Acknowledgment

The ideas presented in this book benefited greatly from discussions associated with the workshop on Long Distance Transport Processes in Plants held at the Harvard Forest, Petersham, Massachusetts, in October 2002. We wish to thank the National Science Foundation (IBN-0211683), the USDA (NRICGP 2002-35100-12288), and the David Rockefeller Center for Latin American Studies for their support. The staff of the Harvard Forest provided invaluable logistic support as well as a beautiful venue for the workshop.

We are grateful to the many individuals who contributed to the success of this project. Harold Mooney provided the initial inspiration for this book, as well as ongoing advice and discussion. Peter Melcher was instrumental in developing the overall vision for the workshop, as well as in helping transform this into reality. Kevin Boyce, Saharah Moon Chapotin, Lawren Sack, Rachel Spicer, Matthew Thompson, and Randol Villalobos helped with all manner of logistics, while Wendy Heywood provided administrative leadership throughout the entire project. Finally, we wish to thank the workshop participants themselves for their enthusiasm, excitement, and commitment, all of which were essential in seeing this project to completion.

# Part I

## Fundamentals of Transport

# 1

# Perspectives on the Biophysics of Xylem Transport

*William F. Pickard and Peter J. Melcher*

> *"I'm sorry to say that the subject I most disliked*
> *was mathematics. I have thought about it. I think the reason*
> *was that mathematics leaves no room for argument.*
> *If you made a mistake, that was all there was to it."*
> *(X, 1992, p. 35)*

The fundamentals of the cohesion-tension theory of sap ascent are now well covered in textbooks (Taiz and Zeiger, 2002; Fisher, 2000) with which most in our potential readership are familiar. Moreover, many deeper questions of xylem biophysics were covered in now hoary reviews (Pickard, 1981; Zimmermann, 1983) and have been updated by Tyree in a substantial monograph (Tyree and Zimmermann, 2002); there is no need to be encyclopedic. Nevertheless, any review of xylem biophysics must include at least some discussion of the cohesion-tension theory. This will be followed by a rather more extensive discussion of how embolisms within the transpiration stream might be formed and, more important, resorbed. Since cohesion-tension theory has only recently survived a significant assault on its hegemony, we present a brief discussion of this controversy, including what has been learned. Finally, we focus on factors that affect hydraulic resistance in plants. In each of these endeavors, some effort will be expended to define each concept with clarity and to frame each explanation rigorously.[*]

---

[*]Martin Zimmermann's observation that discussions of *"sap ascent have become so mathematical that they are not read by many plant anatomists"* (1983, p. 1) may be no less apt today than it was twenty years ago. But it misses the take-home lesson of Malcolm X's dictum quoted at the beginning of this paper that mathematical rigor does tend to squelch wrangling. And if it can't quite do that, it may at least pinpoint the root causes of the dispute.

## The Biophysics of Sap Ascent in the Xylem

### Fundamentals of Cohesion-Tension Theory

In broad outline, the cohesion-tension theory is familiar to virtually every worker in plant biology. We shall therefore treat it as an exercise in dogmatics requiring but little exposition.

Dogma I (Hales, 1727, expt. XXXV):[*] The rise of the xylem sap during transpiration is due to the transpiration itself, the "capillary . . . orifices" in the leaves being able "as any sap is evaporated off" to "supply the great quantities of sap drawn off by perspiration" by "their strong attraction (assisted by the genial warmth of the sun)." In modern scientific terminology[†]: *the capillary menisci of the cell walls within the substomatal cavities, being evaporatively depleted by solar heating, contract and draw the sap upwards.* Ascribing this core dogma of cohesion-tension theory to Hales is uncommon (but *cf.* Floto, 1999), it being more usual (e.g., Dixon, 1914; Pickard, 1981) to ascribe it instead to Dixon and Joly (1895) or (e.g., Steudle, 2001) to Böhm (1893).

Dogma II (*cf.* Dixon, 1914): Water possesses considerable tensile strength, which allows it to be drawn upward by capillary menisci in the leaves. Evidence for this has been reviewed, for example, by Dixon (1914), Pickard (1981), and Steudle (2001).

Dogma III (*cf.* Haberlandt, 1914): Nevertheless bubbles can form in tracheary elements and interrupt sap flow. This was reported early on (Dixon, 1914; Haberlandt, 1914) and has been much commented upon since (Tyree and Sperry, 1989; McCully, 1999; Canny *et al.*, 2001; Facette *et al.*, 2001).

Dogma IV: Fortunately, plants have developed ways of coping with the challenges posed by emboli. These include (1) anatomical adaptations such as cavitation-resistant conduits and redundancy of hydraulic capacity, (2) new growth, (3) and global pressurization of the xylem as by root pressure. Within the past decade, evidence has appeared for embolus resorption despite the existence of nearby water putatively under tension (e.g., Canny, 1997; McCully *et al.*, 2000; Cochard *et al.*, 2000; Facette *et al.*, 2001; Pickard, 2001; Tyree and Zimmermann, 2002); however, there is as yet no consensus on the precise mechanism(s) by which this might occur (e.g., Hacke and Sperry, 2003; Konrad and Roth-Nebelsick, 2003; Pickard, 2003a; Vesala *et al.*, 2003; Chapter 18, this volume).

### The Etiology of an Embolism

Despite the long awareness of tracheary bubbles, there has always been some uncertainty about the mechanism by which they got there, especially

---

[*]Where appropriate, pointers will be given to page (p.), section (s.), chapter (ch.), equation (eq.), figure (fig.), appendix (app.), table (tab.), or experiment (expt.) of the pertinent reference.
[†]Hales' use of the term *perspiration* was correct usage for his era.

after the discovery within the stem of acoustic (Ritman and Milburn, 1988) and electric (Pickard, 2001) transients, which were assumed to arise from putative* cavitation events within the tracheae.

There are four obvious ways by which tension within a liquid-filled pipe could produce an embolism (Pickard, 1981; Zimmermann, 1983; Tyree, 1997).

***Homogeneous Nucleation***  Consider a spherical bubble of radius r [m] in a liquid of pressure P [Pa] and surface tension $\gamma$ [N·m$^{-1}$]; and let the amount of *ideal* vapor in the bubble be N [mol]. It then follows from the Young-Laplace equation (Pickard, 1981) and the perfect gas law that the pressure differential across the interface is

$$2\,\gamma/r + P - \tfrac{3}{4}\,NRT/(\pi r^3) \tag{1}$$

where R [= 8.314 J·mol$^{-1}$·K$^{-1}$] is the gas constant and T [K] is the absolute temperature. If the net pressure begins positive, then the bubble will shrink to zero as liquid vapor condenses on its inner surface and other gas is pressurized into solution. If it begins negative and P < 0, then the bubble should expand and fill the available space. What is required is an initial bubble radius such that the net pressure across the interface is negative; and we do not know why that might occur. Detailed theoretical treatments have traditionally yielded tensile limits for water greater than 50 MPa (Pickard, 1981) and continue to do so (Xiao and Heyes, 2002). Similar results continue to be obtained experimentally (Zheng *et al.*, 1991; Williams *et al.*, 1999).

***Heterogeneous Nucleation from a Hydrophobic Surface***  Water, being less firmly bonded to a hydrophobic surface than a hydrophilic one (or to itself), should more easily be pulled loose from a hydrophobic patch to nucleate an embolism. Various theories of this process exist (e.g., Fisher, 1948; Blander and Katz, 1975; Brereton *et al.*, 1998), of which the least inaccessible seems to be the first. All predict that the most important determinant of the bubble generation rate at a surface will be an exponential term of the form $\exp(-W_{max}/kT)$, where $W_{max}$ [J] is the energy required to expand the bubble to critical radius, $k_B$ [1.381 × 10$^{-23}$ J·K$^{-1}$] is the Boltzmann constant, and T [K] is the absolute temperature. The expansion energy is given as:

$$W_{max} = \frac{16\pi\gamma^3\,\psi}{3\,(P_V - P)^2}, \tag{2}$$

where $\gamma$ [~70 mN·m$^{-1}$] is the surface tension of the sap-embolus interface, $\psi$ [dimensionless] is a geometric factor thought to lie between 0 and 1 and

---

*The modifier *putative* has been used because unresolved doubts persist concerning the validity of the indirect methods (*cf.*, McCully 1999, p. 1005).

to be smaller for less hydrophilic surfaces, $P_V$ [Pa] is the vapor pressure of the sap, and P [Pa] is the pressure within the sap. That is, the nucleation rate should rise appreciably as a vessel's walls become less hydrophilic. On the other hand, compared to the bulk of the fluid, there must be many orders of magnitude fewer potential sites of heterogeneous nucleation at the sap-wall interface than there are potential sites of homogeneous nucleation in the bulk sap. How these two factors balance out is unclear as we know of no relevant experimental data or molecular dynamics simulations.

***Heterogeneous Nucleation from a Preexistent Bubble***    Because the interior surface of the xylem element is assumed to be hydrophilic, it is presumed that no such bubbles would have been formed during xylem development (Pickard, 1981; Steudle, 2001). Furthermore, if one had commenced to form, it should have been reabsorbed by the same processes that are thought to defeat homogeneous nucleation.

***Heterogeneous Nucleation from an Air-Water Interface (Air-Seeding)***    This and the other three mechanisms all predict embolus formation under sufficient sap tension. But this alone predicts that the determining variable is $(P - P_A)$, where $P_A$ [Pa] is the pressure of *ambient* air. Thus, it would appear that the key to embolization in this scenario is making $(P - P_A)$ sufficiently negative, and this is done experimentally either by inducing tension within the sap or by pressurizing the air at the interface. Experiments validating this prediction have been reviewed elsewhere (Tyree, 1997).

It would seem therefore that air-seeding, the currently hegemonic explanation of embolism formation (Tyree, 1997; Steudle, 2001), might be satisfactory if only one could finesse the anatomical observation that the location of such interfaces could well be the pores of the boundary pit membranes (Shane *et al.*, 2000b). Such pores have been observed to be ~3.5 nm in diameter (Shane *et al.*, 2000a) or < 20 nm (Choat *et al.*, 2003, 2004); hence meniscal failure should require a sap tension T [Pa], where (Pickard, 1981, Eq. 4)

$$T \sim \frac{4}{\delta}\gamma\cos\theta \tag{3}$$

and $\delta$ [m] is the effective pore diameter. For an air-water interface, this implies a failure tension well above 10 MPa and also above the upper limit of what is observed experimentally (Tyree, 1997; Sperry and Hacke, 2004). One obvious way of reconciling these observations is to postulate either (1) rare much larger pores (Choat *et al.*, 2004) or (2) contact angle modulation of the bounding surfaces of the pit pores such that $|\cos\theta|$ is reduced from unity by a factor of 10 or more, making the pore boundary quasihydrophobic. At present, such hypotheses are speculative, although there is no firm evidence against either.

### The Healing of an Embolism

The possibility of bubble resorption within tracheary elements has been the subject of considerable experimentation and speculation. As described by Holbrook and Zwieniecki (1999), a complete model of refilling an embolized vessel while the still functioning xylem is under transpiration-induced tension must at the least explain:

- How sap is extruded into the empty tracheary element even though the free energy gradient from soil to xylem seems often to be in the wrong direction.
- How the element is hydraulically isolated during refilling to avoid its new contents being sucked away in the transpiration stream.
- How a refilled vessel ultimately reestablishes hydraulic contact with neighboring conduits in such fashion as to become once again useful in water transport.

These issues will now be addressed sequentially, but the reader is warned that the answers provided, even if ultimately accorded enthusiastic acceptance, are more prolegomena to the desired explanation than the explanation itself.

***Extruding Sap into an Embolized Vessel***    How can a plant extrude water into an embolized xylem vessel from regions of apparently *lower* free energy (tissue and/or soil)? This question is presently the focus of much theorizing, all of which predicts that such extrusion can occur. The relevant literature has recently been reviewed by Clearwater and Goldstein (see Chapter 18) who identified no winning candidate but helpfully subdivided the various models into four distinct categories. We shall not recapitulate their efforts here but rather point out that the question itself may infelicitously direct the reader's thinking. That is, the focus on free energy may betoken a covert assumption that the process can be explained using an exclusively thermodynamic model, and this has not been proven. For example, extrusion through a pipe is commonly considered to be a hydraulic/hydrodynamic problem and is treated using the Navier-Stokes equations. Theoretically this leads to the Hagen-Poiseuille equation, which is much used by modelers of flow in the plant vasculature (e.g., Nonweiler, 1975), and which has been subject to extensive experimental tests. Moreover, molecular dynamics simulations uniformly show that the Navier-Stokes equations should work acceptably until the pipe diameter is less than a few diameters of the solvent molecule itself (e.g., Koplik and Banavar, 1995; Fan *et al.*, 2002); over this entire size range, the flow down a pipe should be hydraulic, with a net solvent flux proportional to the hydraulic pressure difference between the ends of the pipe. Conversely, the passage of solvent (e.g., water) through a pipe is bounded from below by the diameter of the solvent molecule for water in the range 0.2 to 0.3 nm; this is the aquaporin realm where the flow is osmotic and proportional to the chemical potential difference between the ends of the pore (e.g., Fujiyoshi *et al.*, 2002; Zhu *et al.*, 2004). The transition regimen between the hydraulic and osmotic limits seems to be largely uninvestigated.

This distinction between hydraulic and osmotic flows has been discussed at greater length elsewhere (Pickard, 2003a), where it was essential to a quantitative model of root pressure. This model has been (1) specifically advanced as a mechanism for extrusion of sap into embolized vessels (Pickard, 2003a) and (2) shown to predict recent novel data on the stem exudation of detopped plants (Pickard, 2003b). Our belief is that a distinction between hydraulic and osmotic flows must be made in any successful biophysical explanation of vessel refilling and that models that do not carefully treat this distinction (e.g., Hacke and Sperry, 2003) may be incomplete.

***Hydraulic Isolation of an Embolized Vessel***    Hydraulic isolation of an embolized vessel to permit luminal refilling has been hypothesized to be enabled by the surface chemistry of the vessel lumen and the pit channels connecting to it (Holbrook and Zwieniecki, 1999). Their idea was that water extruded into a lumen would preferentially wet the hydrophilic walls, forcing the gas (1) into a pressurized luminal bubble from which it resorbed or (2) into pit channels that somehow did not at once refill with water. At issue is the mechanism by which a pit channel might contrive to function like a valve, opposing the entry of water (but not gas), thereby meeting the challenge for the plant to fill the vessel lumen completely *before* filling pathways to adjacent vessels under tension. How this might be accomplished is illustrated in Fig. 1.1A. For an arbitrary liquid-gas interface in a cylinder of radius $\rho_\lambda$ [m], it is easy to show that the equilibrium pressure difference is,

$$P_{gas} - P_{liq} = \frac{2\gamma\cos\theta_\lambda}{\rho_\lambda}; \tag{4}$$

for water and a cylinder radius of 1 μm, $2\gamma/\rho_\lambda$ is roughly 150 kPa. As water continues to be extruded into the lumen, the luminal gas will be compressed and go gradually into (supersaturated) solution. If the pit channels are even mildly hydrophobic ($\cos\theta_{ch} < 0$), their capillary *repulsion* (associated with their negative $\cos\theta_{ch}$ and their relatively small radii) will resist filling and inhibit pit bubble resorption until the vessel lumen is full and continuing extrusion of water into it can increase $P_{liq}$ appreciably. The postulated resistance of an embolized pit to refilling would of course be augmented by (1) the energy required to increase the sap/gas interfacial area once the interface reenters the chamber and (2) the compression of the intrachamber embolus. That is, successful refilling to this stage will depend critically on the microstructure of the cell wall and especially on local variations in its surface chemistry. Hydrophobic inhibition of pit refilling might even be advantageous in that it could help to synchronize pit refilling and pit sap reconnection to the transpiration stream.

***Reestablishment of Hydraulic Connections***    Once the vessel lumen has been refilled and sap forced past a postulated hydrophobic band in the pit channel, refilling of the pit chamber should proceed swiftly as the pressurized sap flows up the presumedly hydrophilic walls of the pit chamber; this is shown schematically in Fig. 1.1B. After the initial capillary repulsion of the advancing air/sap interface has been overcome, hydraulic flow through a narrow hydrophobic pipe is actually *easier* than through a hydrophilic pipe (Vinogradova, 1999). Within the pit chamber, it would be desirable for the sap to make an optimal contact with the presumedly hydrophilic pit membrane (Zwieniecki and Holbrook, 2000), and this we define to occur when the sap surface and the pit membrane are as parallel as possible at first contact because this will leave the smallest volume of residual embolus (Holbrook and Zwieniecki, 1999). Referring to the idealization of Fig. 1.1B, parallelism will be achieved when

$$\beta = \theta_{chamb} + \alpha - 90. \tag{5}$$

Zwieniecki and Holbrook (2000) provided typical experimental values of $\alpha$ ($\sim 75°$), of $\theta$ ($\sim 50°$) for the walls of vessels (but none for the walls of pit chambers), and none for $\beta$; if we assume pit chamber symmetry and a taut pit membrane, then $\beta \sim 0°$ and we estimate $\theta_{chamb} \sim 15°$, roughly approximating that of cellulose (Luner and Sandel, 1969).

## Challenges to the Cohesion-Tension Theory

*Anomalous Results from the Xylem Pressure Probe*    The pressure probe, broadly defined, is a tool for measuring the internal pressure of a tissue or a single cell; reviews have been provided for tissue by Milburn and Ranasinghe (1996) and for single cells by Tomos and Leigh (1999) and Wei *et al.* (2001). The challenge to cohesion-tension arose when, from intravessel pressure probe studies, it became apparent that the negative hydraulic pressures actually experimentally detected within xylem elements often failed to support the predictions of the cohesion-tension theory (Zimmermann *et al.*, 1993, 1995, 2002, 2004). These data have generated persisting uncertainty (Tyree, 1997; Meinzer *et al.*, 2001; Steudle, 2001; Zimmermann *et al.*, 2004).

Because experimental data have primacy in science, an obvious first step in resolving the conundrum is to examine the reliability of the data themselves. First, Holbrook *et al.* (1995) applied tension to the xylem water in a leaf by spinning its branch in a centrifuge and found that the calculated centrifugal tension agreed with pressure bomb measurements on the leaf

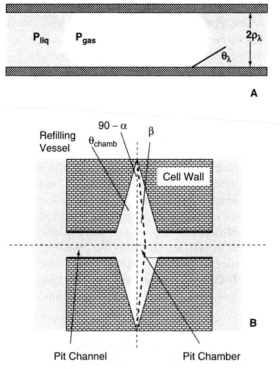

**Figure 1.1**    Schematic representations of (A) the final stage of refilling a vessel lumen and (B) an intermediate stage of refilling a pit; in each case, cylindrical symmetry is assumed. (A) Surface tension forces in a refilling lumen of radius $\rho_\lambda$ [m] and contact angle $\theta_\lambda$ [deg] will suck fluid along the lumen with a force $[2\pi\rho_\lambda][\gamma\cos\theta_\lambda]$ that must be balanced by the liquid pressure $P_{liq}$ [Pa] and the gas pressure $P_{gas}$ [Pa] on the gas-liquid interface, which give an axial force resisting compressing of $[P_{gas} - P_{liq}][\pi\rho\lambda^2]$. (B) Schematic (after Zwieniecki and Holbrook, 2000) of a pit; the diagram is rotationally symmetrical about the horizontal axis. The angle between the vertical axis and the sloping wall of the pit chamber is $90 - \alpha$; that between the vertical and the pit membrane is $\beta$. The curved surfaces of the refilling sap and of the pit membrane have been idealized as conical. Notation: [⬚] embolus; [⬚] sap; [▨] cell wall; ⎯⎯ hydrophilic surface; ▬▬ hydrophobic surface; ‑ ‑ ‑ ‑ pit membrane.

after centrifugation. Second, Pockman *et al.* (1995) found "using centrifugal force to induce negative pressure between −0.5 MPa and −3.5 MPa in intact stems . . . that xylem conduits remained water-filled and conductive." Third, negative pressures down to about −2 MPa have been demonstrated directly using mercury penetrometry, implying "that serious limitations arise from the pressure probe method itself" (Milburn, 1996, p. 399). Fourth, Sperry *et al.* (1996) have offered a harsh critique of the xylem pressure probe at pressures more negative than a few hundred kilopascals: (1) "the xylem conduit wall will not necessarily seal tightly around the glass probe," a problem that should not occur at positive pressures if the seal behaves in the manner of a flapper valve like the tricuspid, and (2) the internal surfaces of the probe itself may provide heterogeneous nucleation sites more failure prone than those of the xylem conduit wall. Fifth, "there is an obvious need for groups other than Zimmermann's to make measurements with the xylem pressure probe" (Tomos and Leigh, 1999); in general, it is scientifically desirable for controversial results to be confirmed by workers outside the orbit of their originator.

***Simultaneous Comparison Between the Xylem Pressure Probe and the Pressure Chamber*** The discrepancy between the pressure probe data reported by U. Zimmermann's group and the pressure chamber measurements obtained by others was resolved by Melcher *et al.* (1997, 1998) and Wei *et al.* (1999). They demonstrated that the two techniques agreed when measurements were made on nontranspiring leaves (i.e., when water potential gradients within the leaf were eliminated).

***The "Compensating Pressure Theory" of Martin Canny*** The existence of U. Zimmermann's disquieting pressure probe measurements prompted Canny (1995, 1998) to reconsider other possibly contrary data and to develop the "compensating pressure theory" of sap ascent, which would replace cohesion-tension with "cohesion sustained by tissue pressure" (Canny, 1995, p. 348), where "tissue pressure" is the "mutual pressure of cells" (Canny, 1998, p. 897). These proposals have generated significant controversy but little support (Milburn, 1996; Tyree, 1997; Wei *et al.*, 1999). In the absence of detailed mathematical realizations of Canny's proposals, we find it hard to give an informed evaluation. Similar difficulties are encountered in trying to assess U. Zimmermann's "multi-force theory" (Zimmermann *et al.*, 2004).

***What Good Came Out of the Debate?*** Although the cohesion-tension theory still stands as the hegemonic model for understanding long-distance sap transport in plants, the controversy propelled the development of new ideas and possible alternative paradigms to replace the postulates of cohesion-tension theory had it been abandoned (Canny, 1998; Zimmermann *et al.*, 2002). Such challenges forced overdue reexamination of the biophysical bases of the cohesion-tension theory (e.g., Tyree, 1997). Nevertheless, significant unknowns in long-distance water transport in plants still remain.

## Factors That Affect Hydraulic Conductance

For xylem sap to flow through the plant, the negative hydrostatic pressure gradient must be large enough to overcome the "resistive" force of sap viscosity and the force of gravity. Because the rate of xylem sap flow through plants is relatively slow, the laminar flow pattern within an individual vessel consists of layers of sap that flow past each other in a sheetlike fashion, with the result that neither eddies nor vortices occur and turbulence is absent.* However, in both laminar and turbulent flow, viscosity leads to internal friction, which produces energy dissipation. If both the viscosity and the vorticity of the fluid are nonzero, then viscous dissipation occurs. For laminar flow at zero gravity through cylindrical pipe, the flow rate is given by the Hagen-Poiseuille relation:

$$J_v = \frac{\Delta V}{\Delta t} = \frac{\Delta P \pi \, r^4}{8 \eta \, l}, \tag{6}$$

where $J_v$ [$m^3 \cdot s^{-1}$] is the volumetric flow rate, V [$m^3$] is the volume, t [s] is time, $\Delta P$ [Pa] is the difference in pressure between the upstream and downstream end of the pipe; r [m] is the radius of the pipe; and $\eta$ [Pa·s] is the dynamic viscosity of the solution in the pipe and $l$ [m] is the length of the pipe. Small changes in the radius of a pipe can greatly affect flow rate through the pipe, an increase of 19% being sufficient to double the flow. Evaporation from the surfaces of leaf menisci provides both the gravitational potential energy stored within the elevated sap and the energy lost to viscous dissipation.

The Hagen-Poiseuille equation works well for idealized systems. For example, recent measurements made using individual vessels in short sections of *Fraxinus americana* stems were in agreement with the Hagen-Poiseuille predictions (Zwieniecki *et al.*, 2001a). Nevertheless, the Hagen-Poiseuille equation is unlikely to hold without taking into account some of the anatomical deviations of the xylem from idealized tubes: (1) xylem radii are seldom constant for the length of a conduit, (2) the sap must pass through bordered pit junctions that are separated by fibrous membranes, (3) the distances between these bordered pit membranes vary greatly (Zimmermann, 1983), (4) vestures and warts (protrusions into the xylem and in bordered pit channels) occur, (5) sap concentration and temperature may change along the flow path and changes in the ionic strength of the xylem sap alter flow rates (Zimmermann, 1978; van Ieperen *et al.*, 2000), thus raising the possibility of a regulatory role for nonliving cells in the xylem (Zwieniecki *et al.*, 2001b), and (6) molecular dynamics simulations show that unexpected flow patterns can, on the nanoscale, occur near boundary irregularities (Fan *et al.*, 2002). Our conclusion is that simplified macroscopic hydrodynamics, artfully applied, enables useful qualitative estimates of behavior and that real anatomical variations within the xylem virtually preclude exact quantitative predictions.

*The Reynolds number in subcellular biological flows has been discussed by Pickard (2003c), whose calculation can be extended to this case to show that it should not exceed 1, even for sap moving at 1 mm·s⁻¹ in a vessel 1 mm in diameter. But the transition to eddy formation is typically well above 10 (Ormières and Provansal, 1999).

# Discussion

Plants, being sessile, are generally plastic in response to the changing environment around them, implying that continuous opportunistic growth occurs to optimize capture of local resources. Plants are always growing and always overcoming the challenge of connecting new growth with extant structures while maintaining performance. Long-distance transport of both mass and messages must occur in sap-filled conduits, driven either by pressure (phloem) or tension (xylem) but not by a mechanical pump (heart). Thus, it is imperative that we strive for understanding how plants connect up their "hardware" as they add new growth while maintaining distributed control of their internal resources, both local and nonlocal. That is, how is the design of the plant's water transport subsystem influenced by (1) the constraints of biophysics, (2) the egregious pure delay in information-transfer imposed by the lack of a nervous system, (3) the limitations of a plant's algorithms for growth, and (4) its typical longevity? The best water transport system in the universe will not guarantee a taxon's reproductive success over multiple generations. Nature awards survival to those taxa that function best as integrated systems, and the cost of the water transport subsystem had better not be exorbitant. Thus, understanding the flow path of water through a plant's intricate plumbing connections from the root to leaf is of great importance.

A major obstacle to deep understanding of plant water transport is the lack of a detailed "wiring diagram" of the xylem. Together with up-to-date information on the hydraulic resistances of the conducting elements, this would greatly add to our understanding. Classic measurements of conduit length (Handley, 1936; Skene and Balodis, 1968) and simple diagrams (Huber, 1935; Vité and Rudinsky, 1959, Kozlowski and Winget, 1963, Zimmermann and Brown, 1971) deserve updating. Nevertheless, the motion pictures created by Martin Zimmermann and Mattmüller (1982) provide a fundamental appreciation of how xylem elements/tracheids grow into each other and how plants ensure reliability of their water transport capacity by overbuilding connections along the path. However, developing a *useful* three-dimensional "wiring diagram" of the xylem from extant literature and movies is problematic because (1) it seems unwise to assume either linearity or time-invariance, (2) the resistance is located at bordered pit connections between xylem conduits (Lancashire and Ennos, 2002), and (3) the degree to which plants use xylem located in older tissue is understudied, thus making it unclear whether measurements that include older xylem yield underestimates of hydraulic resistance (Melcher *et al.*, 2003). This may lead to real problems in interpreting resistance determinations, especially how one maps them onto the observed xylem microanatomy.

Nevertheless, the future of plant water relations seems bright as we profit from the burgeoning development of new tools and methods for studying the transport of both water and minerals within the transpiration stream such as the following:

- Nano-scale observational techniques, such as scanning electron microscopy with EDAX
- Confocal microscopy
- Atomic force microscopy, potentially applicable to the study of the luminal surface chemistry within conducting elements
- Magnetic resonance imaging, which should enable three-dimensional maps of the hydraulic connections in wood
- New infrared imaging techniques, which enable better mapping of surface temperatures and promote insight into water distribution, evaporation, ice formation, and even sap flow
- Abundant computing power, which facilitates the exploration of more realistic models of sap movement

We assert that opportunities now confronting the biology of mass transport in plants can best be exploited by interdisciplinary collaborations of people who are judged by the novelty of the questions they ask and by ingenuity of the techniques they bring to bear rather than the orthodoxy of their training or viewpoint. We conclude with a list of areas where major work is needed.

- The interdependencies and interactions of mass transport in xylem and phloem
- The nature and phenomenology of the distributed control under pure time delay, which seemingly characterizes the integration and regulation of plant processes
- The phenomenology and biophysics of embolism repair across the plant kingdom
- The detailed understanding of hydraulic architecture, both anatomically and mechanistically
- The ontogeny, ultrastructure, and biophysics of plasmodesmata, pits, and perforation plates
- The fate of the transpiration stream, the pathways by which its component water moves from the leaf xylem to the substomatal chamber, and the way in which the leaf handles the stream's nonvolatile solutes
- The biology and biophysics of living wood cells
- The phenomenology and biophysics of lesser studied pipelike structures in plants such as laticifers and resin ducts
- The transport and exchange of metabolically important gases in apex, stem, and root

# References

Blander, M. and Katz, J. L. (1975) Bubble nucleation in liquids. *AIChE J* **21**: 833-848.

Böhm, J. (1893) Capillarität un saftsteigen. *Berichte der Deutschen Botanische Gesellschaft* **11**: 203-212.

Brereton, G. J., Crilly, R. J. and Spears, J. R. (1998) Nucleation in small capillary tubes. *Chem Physics* **230**: 253-265.

Canny, M. J. (1995) A new theory for the ascent of sap: Cohesion supported by tissue pressure. *Ann Bot* **75**: 343-357.

Canny, M. J. (1997) Vessel contents during transpiration: Embolisms and refilling. *Am J Bot* **84**: 1223-1230.

Canny, M. J. (1998) Applications of the compensating pressure theory of water transport. *Am J Bot* **85**: 897-909.

Canny, M. J., McCully, M. E. and Huang, C. X. (2001) Cryo-scanning electron microscopy observations of vessel content during transpiration in walnut petioles. Facts or artefacts? *Plant Physiol Biochem* **39**: 555-563.

Choat, B., Ball, M., Luly, J. and Holtum, J. (2003) Pit membrane porosity and water stress-induced cavitation in four co-existing dry rainforest tree species. *Plant Physiol* **131**: 41-48.

Choat, B., Jansen, S., Zwieniecki, M. A., Smets, E. and Holbrook, N. M. (2004) Changes in pit membrane porosity due to deflection and stretching: The role of vestured pits. *J Exp Bot* **55**: 1569-1575.

Cochard, H., Bodet, C., Thierry, A. and Cruiziat, P. (2000) Cryo-scanning electron microscopy observations of vessel content during transpiration in walnut petioles. Facts or artifacts? *Plant Physiol* **124**: 1191-1202.

Dixon, H. H. (1914). *Transpiration and the Ascent of Sap in Plants*. Macmillan, London.

Dixon, H. H. and Joly J. (1895) On the ascent of sap. *Philos Trans R Soc Lond* **B186**: 563-576.

Facette, M. R., McCully, M. E., Shane, M. W. and Canny, M.J. (2001) Measurement of time to refill embolized vessels. *Plant Physiol Biochem* **39**: 59-66.

Fan, X. J., Phan-Thien, N., Yong, N. T. and Diao, X. (2002) Molecular dynamics simulation of a liquid in a complex nano channel flow. *Physics Fluids* **14**: 1146-1153.

Fisher, D. B. (2000) Long distance transport. In *Biochemistry & Molecular Biology of Plants* (B. B Buchanan, W. Gruissem and R. L. Jones, eds.) American Society of Plant Physiologists, Rockville, MD.

Fisher, J. C. (1948) The fracture of liquids. *J Appl Phys* **19**: 1062-1067.

Floto, F. (1999) Stephen Hales and the cohesion theory. *Trends Plant Sci* **4**: 209.

Fujiyoshi, Y., Mitsuoka, K., de Groot, B. L., Philippsen, A., Grubmüller, H., Agre, P. and Engel, A. (2002) Structure and function of water channels. *Curr Opin Struct Biol* **12**: 509-515.

Haberlandt, G. (1914) *Physiological Plant Anatomy*. Macmillan, London.

Hacke, U. G. and Sperry, J. S. (2003) Limits to xylem refilling under negative pressure in *Lauris nobilis* and *Acer negundo*. *Plant Cell Environ* **28**: 303-311.

Hales, S. (1727) *Vegetable Staticks*. London, UK: W. & J. Innys and T. Woodward. [Reprinted by Scientific Book Guild, London, UK].

Handley, W. R. C. (1936) Some observations on the problem of vessel length determination in woody dicotyledons. *New Phytologist* **35**: 456-471.

Holbrook, N. M., Burns, M. J. and Field, C. B. (1995) Negative xylem pressures in plants: A test of the balancing pressure technique. *Science* **270**: 1193-1194.

Holbrook, N. M. and Zwieniecki, M. A. (1999) Embolism repair and xylem tension: Do we need a miracle? *Plant Physiol* **120**: 7-10.

Huber, B. (1935) Die physiologische bedeutung der ring-und zerstreutporigkeit. *Berlin Deutsche Botanical Gazette* **53**: 711-719.

Konrad, W. and Roth-Nebelsick, A. (2003) The dynamics of gas bubbles in conduits of vascular plants and implications for embolism repair. *J Theoret Bot* **224**: 43-61.

Koplik, J. and Banavar, J. R. (1995) Continuum deductions from molecular hydrodynamics. *Annu Rev Fluid Mechan* **27:** 257-292.

Kozlowski, T. T. and Winget, C. H. (1963) Patterns of water movement in forest trees. *Botan Gazette* **124:** 301-311.

Lancashire, J. R. and Ennos, A. R. (2002) Modelling the hydrodynamic resistance of bordered pits. *J Exp Bot* **53:** 1485-1493.

Luner, P. and Sandel, M. (1969) The wetting of cellulose and wood hemicellulose. *J Polymer Sci* Part C **28:** 115-142.

McCully, M. E. (1999) Root xylem embolisms and refilling. Relation to water potentials of soil, roots, and leaves, and osmotic potentials of the root xylem sap. *Plant Physiol* **119:** 1001-1008.

McCully, M. E., Shane, M. W., Baker, A. N., Huang, C. X., Ling, L. E. C. and Canny, M. J. (2000) The reliability of cryoSEM for the observation and quantification of xylem embolisms and quantitative analysis of xylem sap *in situ. J Microsc* **198:** 24-33.

Meinzer, F. C., Clearwater, M. J. and Goldstein, G. (2001) Water transport in trees: Current perspectives, new insights and some controversies. *Environ Exp Bot* **45:** 239-262.

Melcher P. J., Meinzer, F. C., Yount, D. E. and Goldstein, G. (1997) Measuring high tensions in plants. In *High Pressure Biology and Medicine.* Papers presented at the Vth International Meeting on High Pressure Biology (P. B. Bennett, I. Demchenko and R. E. Marquis, eds.) pp. 93-101. University of Rochester Press, New York.

Melcher, P. J., Meinzer, F. C., Yount, D. E., Goldstein, G. and Zimmermann, U. (1998) Comparative measurements of xylem pressure in transpiring and non-transpiring leaves by means of the pressure chamber and the xylem pressure probe. *J Exp Bot* **49:** 1757-1760.

Melcher, P. J., Zwieniecki, M. Z. and Holbrook, N. M. (2003) Vulnerability of xylem vessels to cavitation in sugar maple. Scaling from individual vessels to whole branches. *Plant Physiol* **131:** 1775-1780.

Milburn, J. A. (1996) Sap ascent in vascular plants: Challengers to the cohesion theory ignore the significance of immature xylem and the recycling of Münch water. *J Exp Bot* **78:** 399-407.

Milburn, J. A. and Ranasinghe, M. S. (1996) A comparison of methods for studying pressure and solute potentials in xylem and also in phloem laticifers of *Hevea brasiliensis. J Exp Bot* **47:** 135-143.

Nonweiler, T. R. F. (1975) Flow of biological fluids through non-ideal capillaries. *Encyclopedia Plant Physiol* (N.S.) **1:** 474-477.

Ormières, D. and Provansal, M. (1999) Transition to turbulence in the wake of a sphere. *Phys Rev Lett* **83:** 80-83.

Pickard, W. F. (1981) The ascent of sap in plants. *Prog Biophys Mol Biol* **37:** 181-229.

Pickard, W. F. (2001) A novel class of fast electrical events recorded by electrodes implanted in tomato shoots. *Aust J Plant Physiol* **28:** 121-129.

Pickard, W. F. (2003a) The riddle of root pressure. I. Putting Maxwell's demon to rest. *Functional Plant Biol* **30:** 121-134.

Pickard, W. F. (2003b) The riddle of root pressure. II. Root exudation at extreme osmolalities. *Functional Plant Biol* **30:** 135-141.

Pickard, W. F. (2003c) The role of cytoplasmic streaming in symplastic transport. *Plant Cell Environ* **26:** 1-15.

Pockman, W. T., Sperry, J. S. and O'Leary, J. W. (1995) Sustained and significant negative water pressure in xylem. *Nature* **378:** 715-716.

Ritman, K. T. and Milburn, J. A. (1988) Acoustic emissions from plants: Ultrasonic and audible compared. *J Exp Bot* **39:** 1237-1248.

Shane, M. W., McCully, M. E. and Canny, M. J. (2000a) Architecture of branch root junctions in corn: Structure of the connecting xylem and porosity of boundary pit membranes. *Ann Bot* **85:** 613-624.

Shane, M. W., McCully, M. E. and Canny, M. J. (2000b) The vascular system of maize stems revisited: Implications for water transport and xylem safety. *Ann Bot* **86:** 245-258.

Skene, D. S. and Balodis, V. (1968) A study of vessel length in *Eucalyptus obliqua* L'Hérit. *J Exp Bot* **19:** 825-830.

Sperry, J. S. and Hacke, U. G. (2004) Analysis of circular bordered pit function. I. Angiosperm vessels with homogenous pit membranes. *Am J Bot* **91:** 369-385.

Sperry, J. S., Saliendra, N. Z., Pockman, W. T., Cochard, H., Cruiziat, P., Davis, S. D., Ewers, F. W. and Tyree, M. T. (1996) New evidence for large negative xylem pressures and their measurement by the pressure chamber method. *Plant Cell Environ* **19:** 427-436.

Steudle, E. (2001) The cohesion-tension mechanism and the acquisition of water by plant roots. *Annu Rev Plant Physiol Plant Mol Biol* **52:** 847-875.

Taiz, L. and Zeiger, E. (2002) *Plant Physiology* (3rd ed.). Sinauer Associates, Sunderland, MA.

Tomos, A. D. and Leigh, R. A. (1999) The pressure probe: A versatile tool in plant cell physiology. *Annu Rev Plant Physiol Plant Mol Biol* **50:** 447-472.

Tyree, M. T. (1997) The cohesion-tension theory of sap ascent: Current controversies. *J Exp Bot* **48:** 1753-1765.

Tyree, M. T. and Sperry, J. S. (1989) Vulnerability of xylem to cavitation and embolism. *Annu Rev Plant Physiol Mol Biol* **40:** 19-38.

Tyree, M. T. and Zimmermann, M. H. (2002) Xylem structure and the ascent of sap. *Springer Series in Wood Science,* 2nd Ed. Springer-Verlag, New York.

van Ieperen, W., van Meeteren, U. and van Gelder, H. (2000) Fluid composition influences hydraulic conductance of xylem conduits. *J Exp Bot* **51:** 769-776.

Vesala, T., Höltta, T., Perämäki, M. and Nikinmaa, E. (2003) Refilling of a hydraulically isolated embolized xylem vessel: Model calculations. *Ann Bot* **91:** 419-428.

Vinogradova, O. L. (1999) Slippage of water over hydrophobic surfaces. *Int J Mineral Processing* **56:** 31-60.

Vité, J. P. and Rudinsky, J. A. (1959) The water-conducting systems in conifers and their importance to the distribution of trunk-injected chemicals. *Contributions Boyce Thompson Institute* **20:** 27-38.

Wei, C., Steudle, E. and Tyree, M. T. (1999) Water ascent in plants: Do ongoing controversies have a sound basis? *Trends Plant Sci* **4:** 372-375.

Wei, C., Steudle, E., Tyree, M. T. and Lintilhac, P. M. (2001) The essentials of direct xylem pressure measurement. *Plant Cell Environ* **24:** 549-555.

Williams, P. R., Williams, P. M., Brown, S. W. J. and Temperley, H. N. V. (1999) On the tensile strength of water under pulsed dynamic stressing. *Proc R Soc London* A **455:** 3311-3323.

X, M. (1992) *The Autobiography of Malcolm X.* Ballantine Books, New York.

Xiao, C. and Heyes, D. M. (2002) Cavitation in stretched liquids. *Proc R Soc London* A **458:** 889-910.

Zheng, Q., Durben, D. J., Wolf, G. H. and Agnell, C. A. (1991) Liquids at large negative pressures: Water at the homogeneous nucleation limit. *Science* **254:** 829-832.

Zhu, F., Tajkhorshid, E. and Schulten, K. (2004) Theory and simulation of water permeation in aquaporin-1. *Biophys J* **86:** 50-57.

Zimmermann, M. H. (1978) Hydraulic architecture of some diffuse-porous tress. *Can J Bot* **56:** 2287-2295.

Zimmermann, M. H. (1983) *Xylem Structure and the Ascent of Sap.* Springer-Verlag, Berlin.

Zimmermann, M. H. and Brown, C. L. (1971) *Trees Structure and Function,* pp. 169-220. Springer-Verlag, New York..

Zimmermann, M. H., and Mattmüller, M. (1982) The vascular pattern in the stem of the palm Rhapis excelsa I. The mature stem. II. The growing tip. Publications of the Institute for Scientific Films Series 15 Nos. 22 and 23. (Films C1404 and D1418). Gottingen.

Zimmermann, U., Haase, A., Langbein, D., and Meinzer, F. (1993) Mechanisms of long-distance water transport in plants: A re-examination of some paradigms in the light of new evidence. *Philos Trans R Soc London,* ser. B **341:** 19-31.

Zimmermann, U., Meinzer, F. and Bentrup, F-W. (1995) How does water ascend in tall trees and other vascular plants? *Ann Bot* **76:** 545-551.

Zimmermann, U., Schneider, H., Wegner, L. H., and Haase, A. (2004) Water ascent in tall trees: Does evolution of land plants rely on a highly metastable state? *New Phytologist* **162**: 575-615.

Zimmermann, U., Schneider, H., Wegner, L. H., Wagner, H-J., Szimtenings, M., Haase, A. and Bentrup, F-W. (2002) What are the driving forces for water lifting in the xylem conduit? *Physiol Plantarum* **114**: 327-335.

Zwieniecki, M. A. and Holbrook, N. M. (2000) Bordered pit structure and vessel wall surface properties. Implications for embolism repair. *Plant Physiol* **123**: 1015-1020.

Zwieniecki, M. A., Melcher, P. J. and Holbrook, N.M. (2001a) Hydraulic properties of individual xylem vessels of *Fraxinus Americana*. *J Exp Bot* **52**: 257-264.

Zwieniecki, M. A., Melcher, P. J. and Holbrook, N. M. (2001b) Hydrogel control of xylem hydraulic resistance in plants. *Science* **291**: 1059-1062.

# 2

# Physiochemical Determinants of Phloem Transport

*Aart J. E. van Bel and Jens B. Hafke*

## Structure-Functional Basics of Phloem Transport

### Driving Forces of Mass Flow in Sieve Tubes

In 1930, Ernst Münch published the first coherent concept of phloem transport in his brilliant book *Die Stoffbewegungen in der Pflanze*. According to his ideas, mass flow ($J_v$) in the phloem was essentially driven by a pressure gradient ($\Delta P$), the difference in turgor pressure (= hydraulic pressure) between source and sink tissues. As has been demonstrated later, turgor pressure differences between source ($P_{source}$) and sink ends ($P_{sink}$) of the sieve tubes are actually responsible for the pressure gradient. In sources and sinks, local turgor pressures are established by differences in chemical water potential (mainly osmopotential) between the sieve tube content and the apoplast. The hydraulic pressure gradient ($P_{source}$ exceeds $P_{sink}$) thus essentially depends on physiological determinants of the sieve-tube osmotic potentials at either end of the phloem system (Fig. 2.1). These physiological activities mainly include metabolic processing of carbohydrates in the tissues bordering the phloem transport channels and the activities of plasma membrane-bound carriers, pumps, and channels lining the conducting channels.

Furthermore, the mass flow is directly proportional to the radius (r) of the sieve tubes and inversely proportional to the distance (L) between source and sink, and the viscosity of the phloem sap ($\eta$) as formulated in the Hagen-Poiseuille equation:

$$J_v = \frac{\Delta P \pi r^4}{8\eta L}$$

The Münch concept encountered fierce opposition with the advent of electron microscopy. Deposits residing on the sieve pores were thought to be incompatible with mass flow. As a number of investigators (Kollmann, 1973; Johnson *et al.*, 1976; Sjolund and Shih, 1983) showed that the

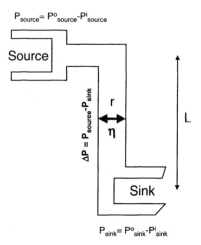

$$P_{source} = P^o_{source} - P^i_{source}$$

$$P_{sink} = P^o_{sink} - P^i_{sink}$$

**Figure 2.1**   The driving force for pressure flow depends on the turgor pressure difference between the sieve tube ends in source and sinks. The difference in turgor pressure ($\Delta P$) between the source ($P_{source}$) and sink ($P_{sink}$) results from the difference in water potential between the local apoplasm ($P^o$) and the adjacent sieve element ($P^i$). Mass flow is further determined by the radius (r) of the tube (in this instance the functional radii of the sieve pores) and the viscosity ($\eta$) of the solute (mainly the viscosity of the sugars translocated) and length of the path (L).

deposits were virtually absent in several plants after careful fixation, the resistance crumbled, and physiological logic regained dominance.

Mathematical calculations using translocation data mainly obtained with radiolabeled photoassimilates built a strong platform supportive of mass flow (e.g., Goeschl and Magnuson, 1986; Magnuson *et al.*, 1986; Murphy, 1989a–d; Thompson and Holbrook, 2003a, b). Moreover, nuclear magnetic resonance (NMR) applied to intact plants (Köckenberger *et al.*, 1997) and confocal laser scanning microscopy (CLSM) imaging of real-time displacement of phloem-mobile fluorochromes (Knoblauch and van Bel, 1998) provided conclusive optical evidence in favor of mass flow through the sieve tubes. Even clots of proteins covering the sieve pores in intact sieve tubes did not prohibit mass flow owing to the presence of narrow corridors in the protein deposits (Knoblauch and van Bel, 1998; Ehlers *et al.*, 2000). Despite all this, fine-regulation between source supply and sink demand (Bancal and Soltani, 2002; Minchin *et al.*, 2002) and quantification of physiochemical determinants of mass flow are still a matter of debate (Henton *et al.*, 2002; Thompson and Holbrook, 2003a,b).

Having adopted mass flow as the mode of translocation, we will deal here with the structural and physiological origin of its physiochemical determinants. Significance and size of these variables associated with various types of phloem "software and hardware" (van Bel, 1996a) are discussed. Major emphasis is laid on $\Delta P$, since this parameter is subject to complex physiological regulation.

## Phloem Zones and Generation of a Turgor Pressure Gradient

Photoassimilates are transported through the sieve tubes, the conducting channels of angiosperm phloem, which are arrays of sieve element/companion cell (SE/CC) modules. Phloem consists of three successive functional zones (or phloem sections), each with a specific task (Fig. 2.2; see also Color Plate section) (e.g., van Bel, 1996a). In the first phloem section (collection phloem), photoassimilates are loaded into the SE/CCs of leaf minor veins after production in the leaf mesophyll. From the minor vein sieve tubes, photoassimilates are translocated via the second phloem section (transport phloem), which is located in leaf major veins, petioles, stems, and roots, to the sinks. There, photoassimilates are unloaded from the SEs in the third phloem section (release phloem) into growing or storage cells, where they are metabolized or sequestered.

Shifting physiological properties of SE/CCs along the phloem pathway, also inferred from the decreasing volume ratio between SEs and CCs towards the sinks (Fig. 2.2), have important consequences for their functional output. The SE/CCs in sources and sinks mostly generate a one-way traffic into and out of the sieve tubes, respectively. In contrast, the SE/CCs of the transport phloem have a dual function (Fig. 2.2) (van Bel, 2003a). Transport phloem carries photoassimilates to terminal sinks such as root/shoot tips. Concurrently, transport phloem is responsible for maintenance and growth (e.g., cambium) of heterotrophic tissues in the plant axis (axial sinks). This is reflected by a continuous leak and retrieval along the phloem pathway (Aloni *et al.*, 1986) as has been demonstrated in bean plants, in which 6% of [11]C-labeled photoassimilates was lost and 3.4% retrieved every centimeter along the phloem pathway (Minchin and Thorpe, 1987). This solute exchange, which requires some degree of symplasmic isolation of SE/CCs (Kempers *et al.*, 1998), goes along with appreciable water fluxes dependent on the dominance of release or retrieval (Fig. 2.2). We next sketch the structural and physiological factors involved in turgor pressure generation in the successive zones of the phloem system.

# Generation of a Hydraulic Pressure Gradient in Collection Phloem

## Modes of Phloem Loading; Consequences for Generation of Hydraulic Pressure?

Particularly in collection phloem, the ultrastructure of CCs is diverse (Fig. 2.3; see also Color Plate section). Three classes of CCs, all having a dense cytoplasm, have been distinguished: intermediary cells (ICs) with numerous vesicles of unknown nature, transfer cells (TCs) with numerous cell wall invaginations, and simple companion cells (SCs) without special features

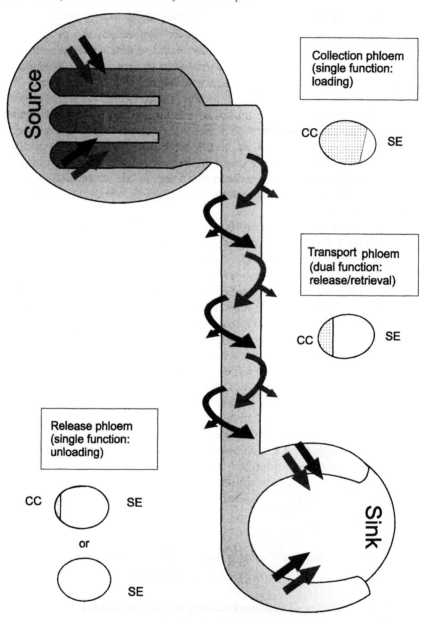

**Figure 2.2**    A dynamic version of the Münch pressure flow model, the local fluxes of photoassimilates (violet arrows) and water (blue arrows) and the relative proportion of sieve elements (SE) and companion cells (CC) in the respective phloem zones. Photoassimilates are translocated via the phloem through essentially leaky instead of hermetically sealed pipes (sieve tubes). The solute content and implicitly the turgor are controlled by release/retrieval

(Gamalei, 1989; Turgeon, 1996a). Plasmodesmal densities between CCs and adjacent parenchyma vary by a factor 1000. The SE/IC complexes (with many plasmodesmata (PD)s towards the mesophyll are mostly associated with symplasmic phloem loading against a sugar gradient (e.g., Turgeon, 1996a) and the SE/TC complexes (very few PDs) with apoplasmic phloem loading against a sugar gradient (Turgeon and Wimmers, 1988; van Bel *et al.*, 1992, 1994). The latter mode of phloem loading is associated with apoplasmic transfer of sucrose; the production of RFOs (raffinose family oligosaccharides) is an integral part of the symplasmic phloem loading mechanism (Turgeon, 1996a).

Turgeon and collaborators (Turgeon *et al.*, 1993; Turgeon and Medville, 1998; Haritatos *et al.*, 2000; Goggin *et al.*, 2001) have drawn ample attention to phloem loading features inconsistent with the conventional subdivision between symplasmic and apoplasmic loaders (see Chapter 3, this volume). In willow, photoassimilates did not seem to concentrate in minor veins with unspecialized CCs (Turgeon and Medville, 1998). A similar mode of phloem supply probably occurs in *Populus, Quercus, Sambucus* (Schrier, 2001) and *Aucuba* (Hoffmann-Thoma *et al.*, 2001). By contrast, other species with unspecialized CCs having either a high (Goggin *et al.*, 2001; van Bel and Wiesmeier, unpublished data) or a low PD frequency (van Bel *et al.*, 1992, 1994; Haritatos *et al.*, 2000) load apoplasmically against a concentration gradient. Based on recent data sets, we designed a tentative classification of phloem loading types and associated CCs (Table 2.1).

When considering the extremes of the vein spectrum (Fig. 2.4, *cf.* van Bel, 2003a), one wonders if differences between the modes of phloem loading could have consequences for generation of hydraulic pressure. In apoplasmic phloem loading, osmotic equivalents are amassed in the sieve tubes due to a wealth of proton pumps (Bouché-Pillon *et al.*, 1994) and carbohydrate carriers (e.g., Gottwald *et al.*, 2000) on the SE/TC plasma membrane. The expected steep osmotic gradient, which requires a commensurate influx of

---

systems located on the plasma membrane of the sieve element/companion cell complexes (SE/CC-complexes). The retrieval mechanisms are energized by the proton-motive force. Differential release/retrieval balances along the pathway control the influx/efflux of sugars and water in the respective phloem zones. In collection phloem (where phloem loading takes place), the retrieval or uptake dominates; in the release phloem (where phloem unloading takes place) the release dominates. In transport phloem having a dual task (nourishment of axial sinks along the pathway and terminal sinks), the balance between release and retrieval varies with the requirements of the plant. The gradual loss of solute and the commensurate amounts of water have been ascribed to a slightly decreasing proton-motive gradient along the phloem pathway. Alternatively, the relative size reduction of companion cells along the source-to-sink path may explain a decreasing retrieval capacity of the SE/CC-complexes in the direction of the sink. The massive photoassimilate delivery in the sinks is assigned to symplasmic and/or apoplasmic loss of photoassimilates from the SEs driven by the high consumption and/or storage rates in the sink tissues. (See also Color Plate section.)

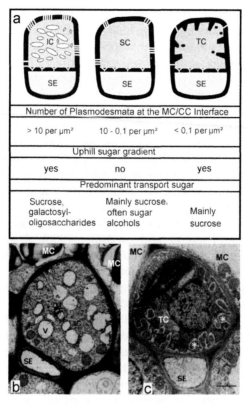

**Figure 2.3** Relationship between ultrastructure of the companion cell (CC) in the minor veins and the mode of phloem loading in dicotyledons. Three types of companion cells are presented here. (a) Intermediary cells (ICs) characterized by numerous cytoplasmic vesicles and many plasmodesmata at the wall interface between mesophyll and companion cells (MC/CC interface). Simple companion cells (SCs), without particular properties, have a moderate plasmodesmal density at the MC/CC-interface. Transfer cells (TC) possess cell wall invaginations to increase plasma membrane surface and have virtually no plasmodesmata at the MC/CC-interface. The structural differences between ICs (b) and TCs (c) are clearly visible in electron-micrographs (courtesy of S. Dimitrovska). V vesicles, * wall invaginations. (See also Color Plate section.)

water, will generate a high local SE turgor. The apoplasmic mode of phloem loading which is driven by proton-motive force (PMF) (TCs, Fig. 2.4) is therefore suspected to be more effective in setting up a hydraulic pressure gradient than the symplasmic phloem loading mechanism (ICs, Fig. 2.4).

In addition there is some support for the suspicion that the high viscosity of highly concentrated raffinose solutions may present an appreciable hindrance to mass flow (Lang, 1978). Data collected by Gamalei (1991) indeed seem to demonstrate that the linear velocity of phloem transport is lower in symplasmically loading species. However, electron microscopy (EM) studies in which the iso-osmolarity of mesophyll cells and SE/CCs

**Table 2.1** Tentative classification of phloem loading mechanisms in dicot minor veins (Schrier 2001, van Bel and Schrier, unpublished)

| Minor vein type[1] | PD density[1] [μm$^{-2}$] | New classification | Functional denomination | Transport sugar | Mode of phloem loading | Chemical loading gradient | Representatives |
|---|---|---|---|---|---|---|---|
| 1 / 1-2a | 1-60 | class 0 | NCC | sucrose | symplasmic | along concentration gradient | Salix,[2] Populus,[3] Cornus,[3] Quercus,[3] Sambucus,[3] Ancuba[4] |
| | | **class 1** | | | | | |
| 1 / 1-2a | 1-60 | class 1A | OCC | mainly sucrose | mainly apoplasmic | against concentration gradient | Liriodendron,[5] Alnus,[6] Fagus[6] |
| 1 | 10-60 | class 1B | IC (cytosolic vesicles) | mainly RFOs | mainly symplasmic | against concentration gradient | Coleus,[7] Epilobium,[8] Origanum,[8] Cuphea,[9] Fraxinus,[6] Syringa[6] |
| | | **class 2** | | | | | |
| 2a | 0.1-1 | class 2A | SCC | sucrose | apoplasmic | against concentration gradient | Pelargonium,[8] Exacum,[9] Ranunculus,[9] Arabidopsis[10] |

*Continued*

**Table 2.1** Tentative Classification of Phloem Loading Mechanisms in Dicot Minor Veins (Schrier 2001, van Bel and Schrier, unpublished)—*cont'd*

| Minor vein type[1] | PD density[1] [$\mu m^{-2}$] | New classification | Functional denomination | Transport sugar | Mode of phloem loading | Chemical loading gradient | Representatives |
|---|---|---|---|---|---|---|---|
| 2b | 0.01 – 0.1 | class 2B | TC (wall invaginations) | sucrose | apoplasmic | against concentration gradient | *Pisum,*[7] *Centaurea,*[8] *Impatiens,*[8] *Symphytum,*[8] *Bellis,*[9] *Anthirrinum,*[9] *Centranthus,*[9] *Linum,*[9] *Valerianella*[9] |

[1]Gamalai 1989, [2]Turgeon and Medville 1998, [3]Schrier 2001, [4]Hoffmann-Thoma *et al.* 2001, [5]Goggin *et al.* 2001, [6]van Bel and Wiesmeier, unpublished, [7]Turgeon and Wimmers 1988, [8]van Bel et al. 1992, [9]van Bel *et al.* 1994, [10]Haritatos *et al.* 2000.

In contrast to the vein typology of Gamalei (1989) where much value is laid on PD frequencies between CC and adjacent parenchyma cells, this classification also uses CC ultrastructure in the minor veins, transport sugars and PCMBS inhibition as decisive indicators for assessment of the loading mechanism. The CCs without specific characteristics fall into three functional categories. The first category called NCCs (normal companion cells) are reservoirs collecting photosynthate diffusing from mesophyll to SEs. The second category OCCs (ordinary companion cells) accumulate sucrose along with RFOs against a concentration gradient. The mechanism of sucrose and RFO loading into OCCs may differ. The third category SCCs (smooth walled companion cells) load only sucrose against a concentration gradient. Structural distinction between OCCs and SSCs can be made, since OCCs generally possess much higher PD densities and a fragmental vacuole. ICs have many more smaller vesicles than OCCs and produce mainly RFOs. SCCs differ from TCs by absence of cell wall invaginations and a higher PD-connectivity with vascular parenchyma. Structurally OCCs and NCCs are similar and can only be distinguished using PCMBS treatment.

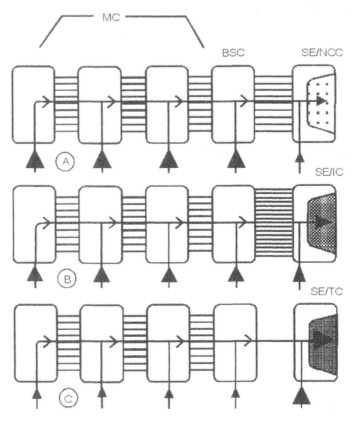

**Figure 2.4** Speculations on the relationship between phloem loading mode and leaf water economy (drawn after Schrier, 2001). (A) In phloem loading occurring along a sugar gradient, water to sustain phloem transport is to be mainly collected by the mesophyll cells. (B, C) In phloem loading against a concentration gradient, the high osmotic potential of sieve tubes implies that the water needed to drive mass flow is mainly taken up directly by the SE/CCs. This short-cut of water uptake (B, C) as compared to species with down-hill (A) sugar gradients from mesophyll to SEs with NCCs, makes water uptake more efficient and phloem transport less vulnerable to water shortage. It is speculated that species with TCs (C) build up a higher osmotic potential in the SEs and a correspondingly higher hydraulic pressure than those with ICs (B). The size of the arrowheads are indicative of the water flux rates (see Table 2.1 for terminology of the companion cells).

was determined by bathing leaf tissues in solutions of increasing osmolarity provide no support for a lower hydrostatic pressure generated in symplasmically loading species (Fisher and Evert, 1982; Fisher, 1986). In the symplasmic loader *Coleus* (Fisher, 1986), the measured jump in osmotic potential between mesophyll (600 mol m$^{-3}$) and minor vein CCs (1400 mol m$^{-3}$) was just as big as in the apoplasmic loader *Amaranthus* (Fisher and

Evert, 1982). It is questionable, however, whether these tissues provide a reliable basis for assessment of cellular osmotic potentials after having gone through an EM-fixation procedure. More sophisticated methods are needed to measure and compare the osmotic potentials in the loading zones before speculations on a relationship between the mode of phloem loading (IC or TC) and hydrostatic pressure can be substantiated.

It looks safe to predict that the lowest hydrostatic pressure in source sieve tubes will be found in veins with normal companion cells (NCCs) (Table 2.1, Fig. 2.4). In keeping with the diffusion model, sugar concentration in the SEs may be even slightly lower than in the mesophyll as shown for *Populus* (Russin and Evert, 1985). Hydrostatic pressure through the sieve tubes is primarily generated here by the mesophyll cells (Fig. 2.4) as in Münch's original model (1930). In such a constellation, water for mass flow propulsion is presumed to be mainly withdrawn from the mesophyll apoplast (NCCs, Fig. 2.4). In other modes of phloem loading (SE/TCs, SE/ICs, Fig. 2.4), photosynthate accumulation by SE/CCs renders sieve tubes more competitive in withdrawing water from the minor vein apoplast including the minor vein xylem vessels (van Bel, 2003a). The accomplished shortcut of water supply to the phloem seems highly beneficial at times of water shortage in the mesophyll (transpiration!) and may have been an evolutionary advantage giving rise to the evolutionary emergence of species that accumulate photoassimilates in the minor vein ends (Turgeon *et al.*, 2001). Unfortunately, an experimental approach to the postulate is difficult, as studies on microflows of water are minefields of methodological errors (Canny, 1990).

### Intercellular Pathways of Water Transport in the Mesophyll

It is common opinion that photoassimilates move by diffusion through the mesophyll symplasm via plasmodesmal connections toward the SE/CCs. Yet, it is worth considering whether, in addition to diffusion, cell-cell solute transport through the mesophyll is dependent on a pressure gradient. In EM studies on cell osmolarity, an osmotic potential gradient through the mesophyll symplasm was not recognized (Russin and Evert, 1985). Using single-cell extracts of barley leaves (Koroleva *et al.*, 1998), however, a sugar gradient through the mesophyll toward the veins has been found. Such a solute gradient may be able to drive mass flow through the mesophyll symplast (van Bel, 1996b). Similarly, pressure gradients have been invoked to explain discrepancies between calculated phloem unloading rates and diffusional transport through plasmodesmata (Patrick *et al.*, 2001; Lalonde *et al.*, 2003). For the same reason, pressure-driven bulk flow of water through plasmodesmata from SEs to the surrounding sink tissues has been proposed for phloem unloading in wheat grains (Fisher and Cash-Clark, 2000). Macroscopic hydrodynamic flow should be established by the time the channel diameter approaches 2 nm (Din and Michaelides, 1997), which is the minimal estimated functional diameter of PDs (Terry and Robards, 1987).

### Deployment of Plasma Membrane Proteins in Collection Phloem

It is the differential deployment of carriers, pumps, channels, and aquaporins along the source-to-sink pathway that enables the plant to build up a hydraulic pressure gradient through the sieve tubes (e.g., Patrick *et al.*, 2001; Patrick and Offler, 2001; Lalonde *et al.*, 2003; van Bel, 2003a). Given the fact that molecular research on transport proteins in phloem has been limited to species with SE/TCs or SE/SCCs in the minor veins, it should be noted that our knowledge on the localization of membrane transport proteins in collection phloem is restricted to sucrose concentrating species.

Diverse molecular strategies have revealed a set of sucrose /H$^+$ symporters involved in phloem loading and photoassimilate export (Lalonde *et al.*, 2003). Blockage of sucrose carriers led to a massive increase of carbohydrates in the mesophyll (e.g., Riesmeier *et al.*, 1994; Kühn *et al.*, 1996; Lemoine *et al.*, 1996; Schulz *et al.*, 1998; Gottwald *et al.*, 2000), which suggests that sucrose is the only sugar loaded in these species. Sucrose carriers are mainly localized in the veins without further distinction between SE and CCs (Riesmeier *et al.*, 1993; Truernit and Sauer, 1995; Noiraud *et al.*, 2000). A higher spatial resolution was reached for proton pumps fueling sucrose carriers by use of EM-immunocytochemistry (Bouché-Pillon *et al.*, 1994; Moriau *et al.*, 1999; Zhao *et al.*, 2000). Proton pump density is much higher on the plasma membrane of CCs than on the plasma membrane of SEs.

The expectedly steep osmotic gradients at the border between mesophyll and SE/TC or SE/SCC require very high permeability constants for water influx, which can be accommodated only by high densities of aquaporins (Tyerman *et al.*, 1999). Aquaporins have actually been found to be associated with minor veins in *Arabidopsis* (Robinson *et al.*, 1996; Schäffner, 1998).

Presence of potassium channels in the loading zone is of major importance since potassium ions may play a compensatory role as osmotic equivalents for maintenance of hydraulic pressure (Fig. 2.5; see also Color Plate section). Thus far, two types of potassium channels have been demonstrated to be associated with the phloem in minor veins of *Arabidopsis* (Marten *et al.*, 1999; Deeken *et al.*, 2000; Lacombe *et al.*, 2000; Pilot *et al.*, 2001). One is an AKT2/3 type K$^+$ channel, which is weakly voltage-dependent (Marten *et al.*, 1999; Deeken *et al.*, 2000; Lacombe *et al.*, 2000), blocked by extracellular calcium (Marten *et al.*, 1999), and affected by apoplasmic acidification (Marten *et al.*, 1999; Lacombe *et al.*, 2000). Its expression is under the control of photosynthate supply by adjacent mesophyll cells (Deeken *et al.*, 2000).

The second is a voltage-gated inward-rectifying KAT2 K$^+$ channel (Pilot *et al.*, 2001). In the phloem of small longitudinal veins of *Zea mays*, similar inward-rectifying KZM1 K[+] channels have been identified (Philippar *et al.*, 2003). A ZMK2 K[+] channel, which is a potential functional homolog to AKT2/3, was associated with functionally homologous vein orders in *Zea* coleoptiles (Bauer *et al.*, 2000). Another AKT2/3 homolog, VFK1, was detected in the stem phloem of *Vicia faba* (Ache *et al.*, 2001). VFK1 is potentially involved in control of the membrane potential as well as in sugar unloading and potassium retrieval (Ache *et al.*, 2001).

### Control of Supply of Photoassimilates to the Collection Phloem

Regulation of the hydraulic pressure gradient through the sieve tubes is a crucial task of the phloem. Because sugars are major components of the sieve tube sap, sugar supply to SEs in minor veins is essential for controlling the osmotic potential of sieve tubes at the source ends. Plants make use of a range of feedback mechanisms to regulate the photoassimilate flux toward the SEs (Kingston-Smith *et al.*, 1998; Hellmann *et al.*, 2000; Komor, 2000; Lewis *et al.*, 2000). High sugar concentrations in the mesophyll cytosol reduce activity of the triose/phosphate shuttle (Stitt, 1997). The resulting accumulation of triose-units within the chloroplast stroma diverts

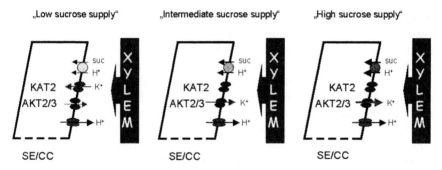

**Figure 2.5**   Schematic drawing of the interaction between sucrose uptake and potassium channel gating in the plasma membrane of SE/CCs in the collection phloem of apoplasmically loading species (collective data of Ache *et al.* 2001, Deeken *et al.* 2002, Philippar *et al.* 2003). Sucrose loading is mediated by sucrose/H[+]-symport and energized by plasma membrane H[+]-ATPase. The voltage drop as a result of sucrose symport is counteracted by a commensurate potassium efflux from the SE/CCs through the weakly voltage-dependent shaker-like AKT2/3 channel (Ache *et al.* 2001, Deeken *et al.* 2002). Thus, AKT2/3 stabilizes the membrane potential, an important feature for maintenance of driving forces. As the sucrose supply increases (left to right) membrane depolarization will increase. Gating of the AKT2/3 channels therefore increases with the supply of sucrose. At low sucrose supply (left), voltage-dependent KAT2 channels open up when the membrane potential drops below a certain negative level. The putative role of KAT2 is the involvement in osmotic homeostasis of SEs. KAT2 channels are probably also engaged in "potassium recycling" from the xylem and act as a functional link between xylem and phloem. (See also Color Plate section.)

the C3-units to the starch synthesis pathway and decreases electron flow through the light-harvesting systems thereby reducing rates of photosynthesis (Stitt, 1997). Photosynthesis is further depressed by effects on high sugar concentrations on the expression of proteins involved in photosynthesis (Koch, 1996; Rolland *et al.*, 2002). Collectively, these mechanisms mostly prevent overproduction of photoassimilates, thus keeping supply to the SEs within appropriate limits. The feedback on photosynthesis sometimes falls short and carbohydrates then amass in the mesophyll (Riesmeier *et al.*, 1993, 1994; Kühn *et al.*, 1996; Lemoine *et al.*, 1996; Schulz *et al.*, 1998; Gottwald *et al.*, 2000).

## Phloem Loading Upregulation/Downregulation

Membrane passages between production and transport compartments at the interface between mesophyll and SE/TCs or SE/OCCs may present other checkpoints for photosynthate supply to SEs. Hardly anything is known about the mechanisms of sugar release from the mesophyll and, hence, nothing about their potential regulation. As for the entry into the SEs, a number of feedback mechanisms on sugar uptake have been identified. High concentrations of SE sucrose exert transinhibitory effects on sucrose carrier activity (Komor, 1977; Wilson and Lucas, 1987). Furthermore, phloem loading is inversely related to turgor-pressure (Smith and Milburn, 1980a). This seems to explain why a drought-induced decrease in SE turgor pressure has a positive effect on phloem loading (Cernusak *et al.*, 2003), resulting in the maintenance of the hydraulic pressure gradient through the phloem. Enhanced sucrose phloem loading is also expected in response to increased supply of sucrose to the apoplast due to its effect on SE turgor pressure. Owing to a resultant increase in the SE sucrose concentration, its turgor-pressure should be backregulated to the previous level after some time. In addition to the metabolic feedback, high sucrose concentrations in SEs suppress the gene expression of proteins involved in sucrose uptake (Chiou and Bush, 1998; Vaughn *et al.*, 2002). This set of devices collectively serves to regulate source supply and to match source supply and sink demand.

## Compensatory Ion Accumulation in Collection Phloem

Photoassimilate transport to the sinks may become insufficient in case of a low photoassimilate production level or a reduced driving force. These phenomena are mostly linked, as sugars act as fuel and cargo simultaneously. If photosynthate supply falls short despite maximal upregulation of the physiological processes in the sources (photosynthesis rate, optimal compartmentation for phloem loading in the mesophyll, phloem loading rate), ion accumulation by SEs in collection phloem is used to compensate for insufficient hydraulic pressure. As early as 1980, Smith and Milburn demonstrated that potassium could function as a supplementary osmoregulator. In *Ricinus,* a decrease in phloem-sap sucrose level over 24 hours in darkness was accompanied by an increased influx of potassium (Smith and Milburn, 1980b). Following this logic, ion accumulation and corresponding channel gating are expected to be induced by low turgor pressure in SEs of collection phloem in accordance with the increased expression of *akt2/3* under drought stress conditions (Lacombe *et al.,* 2000).

## Maintenance of Hydraulic Pressure Gradient in Transport Phloem

### Retention/Release of Osmotic Equivalents in Transport Phloem; the Role of Potassium

Transport phloem has a dual function in that it serves in the nourishment of both axial and terminal sinks (Minchin and Thorpe, 1987; van Bel, 1996a), which requires dynamic release/retrieval mechanism along the pathway. Such a flexible mechanism also readily amends any osmotic disturbance at the local level and hence buffers irregularities in the hydraulic pressure gradient along the pathway.

We speculate that distributed leakage/retrieval plays a role in maintaining the hydraulic pressure gradient. Calculations on mass transport virtually ignore the presumptively open sieve pores as potential bottlenecks in sieve tube transport. However, electronmicrographs (Ehlers *et al.,* 2000) and CLSM pictures of sieve tubes (Knoblauch and van Bel, 1998) indicate that sieve pores are partly covered with proteins, at least in *Vicia* plants. Therefore, the transport corridors in the sieve plates may be considerably narrower than the diameter of the sievepores. Therefore the pressure drop over the sieve plates may be even more substantial than that adopted in modeling (Thompson and Holbrook, 2003a). To counteract appreciable losses in hydraulic pressure, we suppose that a part of the solutes is released in the sieve tube apoplast at the proximal side of the sieve plate and retrieved at the distal side. In this concept, release and retrieval are regulated by elasticity-dependent membrane channels, which react to a slight fluid-induced compression and relaxation at the proximal and distal side of the sieve plate, respectively. Such a circumvention would lead to a regeneration of hydraulic pressure behind every sieve plate. Our proposal is a variation on the relay hypothesis of phloem transport advanced by Lang (1979) and Aikman (1980). The latter concept appears to be in agreement with the release/retrieval events reported for bean (Minchin and Thorpe, 1987), however not with results obtained with *Ricinus* (Murphy and Aikman, 1989).

Potassium channels have been invoked to explain regulation of sucrose concentration and osmolarity along the phloem pathway of *Arabidopsis thaliana* (Deeken *et al.*, 2002) and *Vicia faba* (Ache *et al.*, 2001). According to this concept, AKT2/3 K$^+$-channels, localized on the plasma membrane of SE/CCs in minor veins, compensate sucrose/H$^+$ symport-induced depolarization of the plasma membrane by K$^+$ efflux, thus maintaining protonmotive force for sucrose uptake (Fig. 2.5). Interestingly, expression of *akt2/3* is under the control of photosynthate supply by neighboring mesophyll cells (Deeken *et al.*, 2000). In addition, inward-rectifying voltage-gated KAT2 K$^+$ channels on the SE/CC membranes (Fig. 2.5) are held responsible for potassium uptake (Pilot *et al.*, 2001; Philipppar *et al.*, 2003), in agreement with compensatory potassium uptake when sucrose supply is low (Smith and Milburn, 1980a,b).

In *Ricinus* plants, sucrose concentration and pH decrease as one moves from source to sink (Vreugdenhil, 1985; Vreugdenhil and Koot-Gronsveld, 1989). Rough calculations indicate that loss of sucrose from the sieve tubes correlates with a proton-motive force gradient along the pathway (van Bel, 1993). The decline in pmf along the pathway explains diffusional, perhaps carrier-facilitated loss of sucrose resulting in declining sucrose gradient along the sieve tubes. Dissipation of pmf in the absence of charge compensation by K$^+$ efflux would also lead to sucrose escape from the sieve tubes. This expectation conforms with massive loss of sucrose along the pathway in *akt2/3 Arabidopsis* mutant plants, which are unable to compensate for charge losses in the absence of functional AKT2/3 channels. According to Ache *et al.* (2001) more positive membrane potentials of SEs in the sink region would facilitate potassium release through the AKT2/3 channel homolog VFK1, because the Nernst potential for potassium would become more negative than the local membrane potential.

### Deployment of Plasma Membrane Proteins in Transport Phloem

Most of the knowledge on the deployment of proteins involved in membrane transfer has been gained using SE/CCs of transport phloem. Their large size makes them amenable for experimental manipulation and allows a better visual cell discrimination in techniques such as immunocytochemistry or *in situ* hybridization.

Using different molecular strategies, contrasting observations have been obtained with regard to the deployment of sucrose carriers in SE/CCs of transport phloem. Sucrose transporters appear to be located on the CC-membrane in *Arabidopsis thaliana* and *Plantago major*. In solanaceous species such as potato and tomato, however, sucrose carriers are positioned on the SE plasma membrane (Kühn *et al.*, 1997; Barker *et al.*, 2000; Weise *et al.*, 2000). It appears that the deployment of sucrose carriers on SEs or CCs is indeed species- and most likely even organ-specific (van Bel, 2003b).

Initially, immunolocalization suggested that H⁺/ATPases were exclusively located on the CC plasma membrane (DeWitt and Sussman, 1995; DeWitt *et al.*, 1996). Later, Langhans *et al.* (2001) demonstrated that diverse H⁺-ATPase isoforms occur on the plasma membrane of both CCs and SEs.

Potassium channels, present in transport phloem without detailed localization to either CCs or SEs (Marten *et al.*, 1999; Deeken *et al.*, 2000, 2002; Lacombe *et al.*, 2000; Ache *et al.*, 2001), may play a principal role in osmotic regulation. Dihydropyridine-sensitive calcium channel-like proteins, reported to reside on the SE plasma membrane of *Nicotiana tabacum* and *Pistia stratiotes* (Volk and Franceschi, 2000), seem to be involved in $Ca^{2+}$ processing.

## Manipulation of the Hydraulic Pressure Gradient in Release Phloem

### Modes of Phloem Unloading

At the sink end of the phloem system, symplasmic and apoplasmic phloem unloading may operate, either single, in series, or in parallel (Lalonde *et al.*, 2003). In many sink types, SE unloading occurs symplasmically (Patrick, 1997; Patrick and Offler, 2001; Patrick *et al.*, 2001; Lalonde *et al.*, 2003), most likely along a steep sucrose gradient through an array of cells (Fisher and Wang, 1995; Winch and Pritchard, 1999). Gating of plasmodesmata, the key corridors in symplasmic transport, may therefore regulate symplasmic phloem unloading (Lalonde *et al.*, 2003). However, modulation of phloem unloading rates by manipulation of apoplasmic osmolarity is also conceivable. A high apoplasmic osmolarity would create a surge through the entire sink system, resulting in an enhanced flux of nutrients away from the SEs (Dick and ap Rees, 1975; Schulz, 1994). Both mechanisms may operate in a concerted fashion; bathing pea root tips in high-osmolarity media was shown to result in a transient widening of plasmodesmata (Schulz, 1995) and a coincident, enhanced sugar flux (Schulz, 1994). Supportive of this idea may be the observation that phloem unloading in roots is stimulated by high apoplasmic concentrations of slowly permeant nonmetabolizable mannitol and to a lesser extent by high concentrations of sucrose (Schulz, 1994).

A different situation exists in seeds, where unloading SEs and sink tissues are separated by an apoplasmic interface (Patrick, 1997; Patrick and Offler, 2001; Patrick *et al.*, 2001; Lalonde *et al.*, 2003). In legume seed coats, cells are unloaded along a symplasmic route until the apoplasmic gap between maternal and filial tissues is reached (Patrick and Offler, 2001; Lalonde *et al.*, 2003). As soon as the embryonic withdrawal of nutrients exceeds supply by the seed coat phloem, osmolarity in the small apoplasmic space between seed coat and embryo readily decreases (Patrick and Offler, 2001). This provokes an immediate increase of turgor pressure in the seed coat parenchyma, since the osmotic difference between apoplast and seed coat cells is small (0.1 to 0.2 MPa). Increasing of turgor pressure above a certain turgor set point causes an error signal leading to a compensatory increase in photoassimilate efflux from the seed coat (Patrick, 1997). Turgor-sensitive sugar transporters probably play a crucial role in seed nutrition, as the ability to keep the turgor pressure constant ensures a continuous and sufficient photoassimilate supply (Patrick, 1997). All sink systems may have in common that alterations in osmotic potential difference between symplast and apoplast brought about by changes in phloem import are sensed as turgor pressure changes and adjusted by turgor-sensitive transporters and aquaporins (Chrispeels *et al.*, 1999).

### Deployment of Plasma Membrane Proteins in Release Phloem

In release phloem, transmembrane proteins exert control on turgor pressure as in other phloem zones. The rate at which solutes escape from SEs is decisive for the loss of water from the SEs and, hence, for local turgor pressure (Patrick *et al.*, 2001). At first sight, functioning of carriers seems to be irrelevant in symplasmic SE unloading. At second thought, however, carriers on the plasma membranes must operate in such a collective fashion that solute concentration in the parenchyma cells adjacent to the SEs is kept high to maintain a steep solute gradient through the symplast (van Bel *et al.*, 2003a). Helpful for quick unloading is a high apoplasmic solute concentration around the cells of the sink symplast, particularly those distant from the SE. The water surge created may impose mass flow of water through the symplast. In this context, apoplasmic invertase (Roitsch *et al.*, 2000; Lalonde *et al.*, 2003) could play a prominent part.

A high rate of solute release is required when a major sink lies at the opposite side of the apoplast (e.g., at the interface between maternal tissues and the embryo in seeds) (Patrick *et al.*, 2001; Patrick and Offler, 2001; Lalonde *et al.*, 2003). Transmembrane proteins at the edge of the maternal tissue mediate the delivery of solutes, while those in the epidermal tissue of the embryo must be designed to accumulate substances. We know much about the physiological properties of the transmembrane proteins at this interface, as an experimental access to both surfaces is simple (Lalonde *et al.*, 2003). Given the minute size of the cells, however, exact deployment of proteins in the sink area at the maternal side is difficult. SUT-homologs were found to co-localize with $H^+$-ATPases in seed coat release cells of broad bean (Harrington *et al.*, 1997) and pea (Tegeder *et al.*, 1999). These SUT-proteins could function as antiporters to release sucrose into the apoplasmic space (Lalonde *et al.*, 2003).

## Radius of the Sieve Tubes

Because mass flow is related to the fourth power of the radius, small changes in radius have large effects on flow calculations in phloem conduits. Therefore, the SE radius should be used with caution given the impact of sieve plates on the mass flow rate. Sieve tubes are partitioned by sieve plates with perforations of various sizes. Despite the introduction of a so-called Sampson factor into flow calculations (Thompson and Holbrook, 2003a), sieve pores still remain bottlenecks of unknown impact on flow profiles and mass flow rate. Moreover, partial protein sealing of sieve pores (e.g., Ehlers *et al.*, 2000) makes it difficult to assess their functional diameter. Some degree of physiological control of sieve plate permeability may be achieved by temporary calcium-regulated plugging or gating of sieve pores as has been found for crystalline protein bodies (forisomes) in the sieve tubes of Fabacaea (Knoblauch *et al.*, 2001). Thus, we must be aware of the limitations when the Hagen-Poiseuille equation is applied in such narrow canals with a myctoplasmic layer of unknown thickness and miniscule corridors of undefined size in transverse walls.

To further complicate considerations with regard to sieve tube radius, the diameter of sieve elements varies with the phloem zone. For instance, the sieve tube diameter in collection phloem is generally much smaller than that in transport phloem. How phloem transport capacity is matched between phloem zones has hardly been investigated to the best of our knowledge. In view of their smaller diameters, the aggregate volume of sieve tubes in collection phloem must be appreciably larger than that of the sieve tubes in transport phloem to meet the prerequisites of the Hagen-Poiseuille equation and to guarantee steady solute flow throughout the sieve tube system.

## Viscosity, Sugar Species, and Concentrations in Sieve Tubes

Viscosity is a major determinant of mass flow, which has hardly been investigated, as phloem sap is hard to collect in a natural state. Phloem sap is a complex mixture of inorganic ions (e.g., $K^+$, $Na^+$, $Mg^{2+}$, $Cl^-$, $PO_4^{3-}$, $NO_3^-$), low-molecular (sugars, amino acids, organic acids, small polypeptides) and high-molecular substances (structural proteins, enzymes, RNA) in all forms of complexation (Ziegler, 1975). As soluble carbohydrates make up the major part of the sieve tube content, calculations on sap viscosity are based on the sugar content. Some authors advanced that the viscosity of phloem sap strongly depends on the nature of sugars translocated (Lang, 1978; Turgeon, 1996b). In contrast to raffinose solutions, sucrose solutions stay low-viscous at concentrations up to 400 mM. Translocation efficiency plots, which represent the optimal concentration for bulk C-translocation, show that maximal amounts of raffinose are expected to be transported at 180 mol m$^{-3}$, whereas a maximal flux of sucrose takes place at 700 mol m$^{-3}$ (Lang, 1978). Little attention has been devoted to the potential effect of macromolecules on sieve tube sap viscosity. In particular in species with a high soluble protein content such as *Cucurbita* (Kollmann *et al.*, 1970; Alosi *et al.*, 1988) or *Ricinus* (Hall and Baker, 1972), proteins are expected to contribute significantly to viscosity. Furthermore, it should be noted that viscosity of macromolecules strongly depends on the molecular shape; rod-shaped macromolecular complexes increase sap viscosity more than disc-shaped ones (Atkins, 1990). As a final note, the low oxygen concentrations in sieve tubes (van Dongen *et al.*, 2003) potentially enhance the fluidity of phloem sap, as phloem-specific proteins coagulate when exposed to oxygen (Alosi *et al.*, 1988).

## Physiochemical Relationship Between Xylem and Phloem Pathway

As the sieve tubes are struggling for water supply to maintain their hydraulic pressure gradient, the physiochemical relationship between xylem and phloem pathway is obvious. The osmotic potential in sieve tubes should keep pace with the physical water potential in xylem vessels. Water potential changes in xylem vessels have effects on the apoplasmic water potential of sieve tubes (e.g., in leaf minor veins), with major consequences for accumulation of osmotic equivalents by SE/CCs. To maintain the propulsion in sieve tubes, drops in water potential in the xylem vessels demand commensurate rises of phloem sap concentration. The elegance of the vascular transport system as a whole is the spatial correspondence between low and high osmotic potentials in the sieve tubes and low and high water potentials in the xylem vessels, respectively. In source leaves, where the hydraulic potential in the xylem vessels is the most negative, the competitive strength for water of the sieve tubes is the highest due to the highly negative osmotic potential.

## Concluding Remarks

We have identified several open questions as to the physiochemical determinants of phloem transport. There is a general quantitative unawareness with regard to physiochemical parameters. Moreover, we must be thoughtful of potentially large differences between plant species due to disparate structural/functional conditions.

The steepness of the hydraulic gradient depends on the physiological activities along the whole sieve tube stretch. Thus, diversity in modes of phloem loading, release/retrieval along the transport pathway, and in modes of phloem unloading, may have a strong impact on generation of the hydraulic pressure gradient. This recognition requires detailed studies on deployment and functioning of proteins involved in shuttling osmotic equivalents and water through the SE/CC plasma membrane of the successive phloem zones and the structural frame of the SE/CCs in various plant groups. Furthermore, the reciprocal feedback regulation of sources and sinks and the compensatory and buffering activities of transport phloem must be investigated on the cell-biological level.

Viscosity of the sieve tube sap must be related more precisely to the concentration of sugar mixtures in highly complex chemical environments such as sieve tube contents. The design of a method for entrapment of phloem sap in its natural state is therefore of prime interest for phloem physiology.

As for the diameter of the sieve tubes, the amount of sieve tubes and their diameters must be determined at any point in the pathway to calculate if the flow capacity remains similar throughout the phloem system. It should be computed, for instance, if the flow capacity of a higher number of narrow-diameter sieve tubes in the sources matches that of a lower number of broad-diameter sieve tubes in the transport phloem. It is imperative to further consider the role of sieve pores as bottlenecks in the sieve tube system and the changes in pressure before and after the sieve plate.

## Notations

| | |
|---|---|
| *CC* | companion cell |
| *BSC* | bundle sheath cell |
| *CLSM* | confocal laser scanning microscopy |
| *EM* | electron microscopy |
| $\eta$ | viscosity |
| *IC* | intermediary cell |
| *L* | length of the conducting channel i.e., the length a sieve tube or total length of the phloem path |
| *MC* | mesophyll cell |
| *NCC* | normal companion cell |

| | |
|---|---|
| *NMR* | nuclear magnetic resonance |
| *OCC* | ordinary companion cell |
| $P^i_{source}$ | intracellular water potential of a sieve element in the source region |
| $P^o_{source}$ | extracellular water potential around the sieve element (local apoplast) in the source region |
| $P_{source}$ | turgor pressure of the sieve elements in the source region |
| $P^i_{sink}$ | intracellular water potential of a sieve element in the sink region |
| $P^o_{sink}$ | extracellular water potential around the sieve element (local apoplast) in the sink region |
| $P_{sink}$ | turgor pressure of the sieve elements in the sink region |
| *PDs* | plasmodesmata |
| *r* | radius |
| *RFO* | raffinose family oligosasccharide |
| *SE* | sieve element |
| *SE/CCs* | sieve element/companion cell complex |
| *SC* | simple companion cell |
| *SCC* | smooth walled companion cell |
| *TC* | transfer cell |

# References

Ache, P., Becker, D., Deeken, R., Weber, H., Fromm, J. and Hedrich, R. (2001) VFK1, a Vicia faba K⁺ channel involved in phloem unloading. *Plant J* **27:** 571-580.

Aikman, D. P. (1980) Contractile proteins and hypotheses concerning their role in phloem transport. *Can J Bot* **58:** 826-832.

Aloni, R., Wyse, R. E., and Griffith, S. (1986) Sucrose transport and phloem unloading in the stem of *Vicia faba*: Possible involvement of a sucrose carrier and osmotic regulation. *Plant Physiol* **81:** 482-486.

Alosi M. C., Melroy, D. L. and Park, R. B. (1988) The regulation of gelation of exudate from *Cucurbita* by dilution, glutathione, and glutathione reductase. *Plant Physiol* **86:** 1089-1094.

Atkins, P. W. (1990) *Physical Chemistry*, 4th Ed. Oxford University Press, Oxford.

Bancal, P. and Soltani, F. (2002) Source-sink partitioning. Do we need Münch? *J Exp Bot* **53:** 1919-1928.

Barker, L., Kühn, C., Weise, A., Schulz, A., Gebhardt, C., Hirner, B., Hellmann, H., Schulze, W., Ward, J.M. and Frommer, W.B. (2000) SUT2, a putative sucrose sensor in sieve elements. *Plant Cell* **12:** 1153-1164.

Bauer, C. S., Hoth, S., Haga, K., Philippar, K., Aoki, N. and Hedrich, R. (2000) Differential expression and regulation of K⁺ channels in the maize coleoptile: Molecular and biophysical analysis of cells isolated from cortex and vasculature. *Plant J* **24:** 139-145.

Bouché-Pillon, S., Fleurat-Lessard, P., Fromont, J.-C., Serrano, R. and Bonnemain, J.-L. (1994) Immunolocalisation of the plasma membrane H⁺-ATPase in minor veins of *Vicia faba* in relation to phloem loading. *Plant Physiol* **105:** 691-697.

Canny, M. J. (1990) What becomes of the transpiration stream? *New Phytol* **114:** 341-368.

Cernusak, L. A., Arthur, D. J., Pate, J. S. and Farquhar, G. D. (2003) Water relations link carbon and oxygen isotope discrimination to phloem sap sugar concentration in *Eucalyptus globulus*. *Plant Physiol* **131:** 1544-1554.

Chiou, T. J. and Bush, D. R. (1998) Sucrose is a signal molecule in assimilate partitioning. *Proc Natl Acad Sci U S A* **95:** 4784-4788.

Chrispeels, M. J., Crawford, N. M. and Schroeder, J. I. (1999) Proteins for transport of water and mineral nutrients across the membranes of plant cells. *Plant Cell* **11:** 661-675.

Deeken, R., Geiger, D., Fromm, J., Koroleva, O., Ache, P., Langenfeld-Heyser, R., Sauer, N., May, S. T. and Hedrich, R. (2002) Loss of AKT2/3 potassium channel affects sugar loading into the phloem of *Arabidopsis. Planta* **216:** 334-344.

Deeken, R., Sanders, C., Ache, P. and Hedrich, R. (2000) Development and light-dependent regulation of a phloem-localised K+ channel of *Arabidopsis thaliana. Plant J* **23:** 285-290.

DeWitt, N. D., Hong, B., Sussman, M. R. and Harper, J. F. (1996) Targeting of two *Arabidopsis* H+-ATPase isoforms to the plasma membrane. *Plant Physiol* **112:** 833-844.

DeWitt, N. D. and Sussman, M. R. (1995) Immunocytological localisation of an epitope-tagged plasma membrane proton pump (H+-ATPase) in phloem companion cells. *Plant Cell* **7:** 2053-2067.

Dick, P. S., and ap Rees, T. (1975) The pathway of sugar transport in roots of *Pisum sativum. J Exp Bot* **26:** 305-314.

Din, X.-D. and Michaelides, E. E. (1997) Kinetic theory and molecular dynamics simulations of microscopic flows. *Physics Fluids* **9:** 3915-3925.

Ehlers, K., Knoblauch, M. and van Bel, A. J. E. (2000) Ultrastructural features of well-preserved and injured sieve elements: Minute clamps keep the phloem transport conduits free for mass flow. *Protoplasma* **214:** 80-92.

Fisher, D. B. and Cash-Clark, C. E. (2000) Gradients in water potential and turgor pressure along the translocation pathway during grain filling in normally watered and water-stressed wheat plants. *Plant Physiol* **123:** 139-147.

Fisher, D. B. and Wang, N. (1995) Sucrose concentration gradients along the post-phloem transport pathway in the maternal tissues of developing wheat grains. *Plant Physiol* **109:** 587-592.

Fisher D. G. (1986) Ultrastructure, plasmodesmatal frequency, and solute concentration in green areas of variegated *Coleus blumei* Benth. leaves. *Planta* **169:** 141-152.

Fisher D. G. and Evert R. F. (1982) Studies on the leaf of *Amaranthus retroflexus* (Amaranthaceae): Ultrastructure, plasmodesmatal frequency, and solute concentration in relation to phloem loading. *Planta* **155:** 377-387.

Gamalei, Y. V. (1989). Structure and function of leaf minor veins in trees and herbs: A taxonomic review. *Trees* **3:** 96-110.

Gamalei, Y. V. (1991) Phloem loading and its development related to plant evolution from trees to herbs. *Trees* **5:** 50-64.

Goeschl, J. D. and Magnuson, C. E. (1986) Physiological implications of the Münch-Horwitz theory of phloem loading: Effect of loading rates. *Plant Cell Environ* **9:** 95-102.

Goggin F. L., Medville R. and Turgeon R. (2001) Phloem loading in the tulip tree. Mechanisms and evolutionary implications. *Plant Physiol* **124:** 891-899.

Gottwald, J. R., Krysan, P. J., Young, J. C., Evert, R. F., and Sussman, M. R. (2000). Genetic evidence for the *in planta* role of phloem-specific plasma membrane sucrose transporters. *Proc. Natl Acad Sci U S A* **97:** 13979-13984.

Hall, J. L. and Baker, D. A. (1972). The chemical composition of *Ricinus* phloem exudate. *Planta* **106:** 131-140.

Haritatos E., Ayre B. G. and Turgeon, R. (2000) Identification of phloem involved in assimilate loading in leaves by the activity of the galactinol synthase promotor. *Plant Physiol.* **123:** 929-937.

Harrington, G. N., Franceschi, V. R., Offler, C. E., Patrick, J. W., Tegeder, M., Frommer, W. B., Harper, J. F. and Hitz, W. D. (1997). Cell specific expression of three genes involved in plasma membrane sucrose transport in developing *Vicia faba* seed. *Protoplasma* **197**: 160-173.

Hellmann, H., Barker, L., Funck, D. and Frommer, W. B. (2000) The regulation of assimilate allocation and transport. *Austr J Plant Physiol* **27**: 583-594.

Henton, S. M., Greaves, A. J., Piller, G. J. and Minchin, P. E. H. (2002) Revisiting the Münch pressure-flow hypothesis for long-distance transport of carbohydrates: Modelling the dynamics of solute transport inside a semipermeable tube. *J Exp Bot* **53**: 1411-1419.

Hoffmann-Thoma, G., van Bel, A. J. E. and Ehlers, K. (2001) Ultrastructure of minor-vein phloem and assimilate export in summer and winter leaves of the symplasmically loading evergreens *Ajuga reptans* L., *Acuba japonica* Thumb., and *Hedera helix* L. *Planta* **212**: 231-242.

Johnson, R. P. C., Freundlich, A. and Barclay, G. F. (1976) Transcellular strands in sieve tubes: What are they? *J Exp Bot* **101**: 1127-1136.

Kempers, R., Ammerlaan, A. and van Bel, A. J. E. (1998) Symplasmic constriction and ultra-structural features of the sieve element/companion cell complex in the transport phloem of apoplasmically and symplasmically phloem-loading species. *Plant Physiol* **116**: 271-278.

Kingston-Smith, A. H., Galtier, N., Pollock, C. J. and Foyer, C.H. (1998) Soluble acid invertase activity in leaves is independent of species differences in leaf carbohydrates, diurnal sugar profiles, and paths of phloem loading. *New Phytol* **139**: 283-292.

Knoblauch, M., Peters, W. S., Ehlers, K. and van Bel, A. J. E. (2001) Reversible calcium-regulated stopcocks in legume sieve tubes. *Plant Cell* **13**: 1221-1230.

Knoblauch, M. and van Bel, A. J. E. (1998) Sieve tubes in action. *Plant Cell* **10**: 35-50.

Koch, K. (1996) Carbohydrate-modulated gene expression in plants. *Annu Rev Plant Physiol Plant Mol Biol* **47**: 509-540.

Köckenberger, W., Pope, W. J., Xia, Y., Jeffrey, K. R., Komor, E. and Callaghan, P. T. (1997) A non-invasive measurement of phloem and xylem water flow in castor bean seedlings by nuclear magnetic resonance microimaging. *Planta* **201**: 53-63.

Kollmann, R. (1973) Cytologie des Phloems. In *Grundlagen der Cytologie* (H. Ruska and P. Sitte, eds.) pp. 479-505. Gustav Fischer, Jena.

Kollmann, R., Dörr, I. and Kleinig, H. (1970) Protein filaments: Structural components of the phloem exudate. 1. Observations with *Cucurbita* and *Nicotiana*. *Planta* **95**: 86-94.

Komor, E. (1977) Sucrose uptake by cotyledons of *Ricinus communis* L.: Characteristics, mech-anism and regulation. *Planta* **137**: 119-131.

Komor, E. (2000) Source physiology and assimilate transport: The interaction of sucrose metabolism, starch storage and phloem export in source leaves and the effects on sugar sta-tus in phloem. *Austr J Plant Physiol* **27**: 497-505.

Koroleva, O. A., Farrar, J. F., Tomos, A. D. and Pollock, C. J. (1998) Carbohydrates in individ-ual cells of epidermis, mesophyll, and bundle sheath in barley leaves with changed export or photosynthetic rate. *Plant Physiol* **118**: 1525-1532.

Kühn, C., Franceschi, V. R., Schulz, A., Lemoine, R. and Frommer, W. B. (1997) Macromolecular trafficking indicated by localization and turnover of sucrose transporters in enucleate sieve elements. *Science* **275**: 1298-1300.

Kühn, C., Quick, W. P., Schulz, A., Riesmeier, J. W., Sonnewald, U. and Frommer, W. B. (1996) Companion cell-specific inhibition of the potato sucrose transporter SUT1. *Plant Cell Environ* **19**: 1115-1123.

Lacombe, B., Pilot, G., Michard, E., Gaymard, F., Sentenac, H. and Thibaud, J.-B. (2000) A shaker-like K+ channel with weak rectification is expressed in both source and sink phloem tissues of *Arabidopsis*. *Plant Cell* **12**: 837-851.

Lalonde, M., Tegeder, M., Throne-Holst, M., Frommer, W. B. and Patrick, J. W. (2003) Phloem loading and unloading of sugars and amino acids. *Plant Cell Environ* **26**: 37-56.

Lang, A. (1978) A model of mass flow in the phloem. *Austr J Plant Physiol* **5**: 535-546.

Lang, A. (1979) A relay mechanism for phloem translocation. *Ann Bot* **44**: 141-145.

Langhans, M., Ratajczak, R., Lützelschwab, M., Michalke, W., Wächter, R., Fischer-Schliebs, E. and Ullrich, C.E. (2001) Immunolocalization of plasma membrane $H^+$-ATPase and tonoplast-type pyrophosphatase of the sieve element-companion cell complex in the stem of *Ricinus communis* L. *Planta* **213**: 11-19.

Lemoine, R., Kühn, C., Thiele, N., Delrot, S. and Frommer, W. B. (1996) Antisense inhibition of the sucrose transporter in potato: Effects on amount and activity. *Plant Cell Environ* **19**: 1124-1131.

Lewis, C. E., Noctor, G., Causton, D. and Foyer, C. H. (2000) Regulation of assimilate partitioning in leaves. *Austr J Plant Physiol* **27**: 509-517.

Magnuson, C. E., Goeschl, J. D. and Fares, Y. (1986) Experimental tests of the Münch-Horwitz theory of phloem transport: Effect of loading rates. *Plant Cell Environ* **9**: 103-109.

Marten, I., Hoth, S., Deeken, R., Ketchum, K. A., Hoshi, T. and Hedrich, R. (1999) *AKT3*, a phloem-localised $K^+$ channel is blocked by protons. *Proc Natl Acad Sci U S A* **96**: 7581-7586.

Minchin, P. E. H. and Thorpe, M. R. (1987) Measurement of unloading and reloading of photo-assimilate within the stem of bean. *J Exp Bot* **38**: 211-220.

Minchin, P. E. H., Thorpe, M. R., Farrar, J. F. and Koroleva, O. A. (2002) Source-sink coupling in young barley plants and control of phloem loading. *J Exp Bot* **53**: 1671-1676.

Moriau, L., Michelet, B., Bogaerts, P., Lambert, L. and Michel, A. (1999) Expression analysis of two gene subfamilies encoding the plasma membrane $H^+$-ATPase in *Nicotiana plumbaginifolia* reveals the major transport functions of this enzyme. *Plant J* **19**: 31-41.

Münch, E. (1930) *Die Stoffbewegungen in der Pflanze*. Gustav Fischer, Jena.

Murphy, R. (1989a) Water flow across the sieve-tube boundary: Estimating turgor and some implications for phloem loading and unloading. I. Theory. *Ann Bot* **63**: 541-549.

Murphy, R. (1989b) Water flow across the sieve-tube boundary: Estimating turgor and some implications for phloem loading and unloading. II. Phloem in the stem. *Ann Bot* **63**: 551-559.

Murphy, R. (1989c) Water flow across the sieve-tube boundary: Estimating turgor and some implications for phloem loading and unloading. III. Phloem in the leaf. *Ann Bot* **63**: 561-570.

Murphy, R. (1989d) Water flow across the sieve-tube boundary: Estimating turgor and some implications for phloem loading and unloading. IV. Root tips and seed coats. *Ann Bot* **63**: 571-579.

Murphy, R. and Aikman, D. P. (1989). An investigation of the relay hypothesis of phloem transport in *Ricinus communis* L. *J Exp Bot* **40**: 1079-1088.

Noiraud, N., Delrot, S. and Lemoine, R. (2000). The sucrose transporter of celery. Identification and expression during salt stress. *Plant Physiol* **122**: 1447-1455.

Patrick, J. W. (1997) Phloem unloading: Sieve element unloading and post-sieve element transport. *Annu Rev Plant Physiol Plant Mol Biol* **28**: 165-190.

Patrick, J. W. and Offler, C. E. (2001) Compartmentation of transport and transfer events in developing seeds. *J Exp Bot* **52**: 551-564.

Patrick, J. W., Zhang, W., Tyerman, S. D., Offler, C. E. and Walker, N. A. (2001) Role of membrane transport in phloem translocation of assimilates and water. *Austr J Plant Physiol* **28**: 695-707.

Philippar, K., Büchsenschütz, K., Abshagen, M., Fuchs, I., Geiger, D., Lacombe, B. and Hedrich, R. (2003) The $K^+$ channel KZM1 mediates potassium uptake into the phloem and guard cells of the $C_4$ grass *Zea mays*. *J Biol Chem* **278**: 16973-16981.

Pilot, G., Lacombe, B., Gaymard, F., Chérel, I., Boucherez, J., Thibaud, J.-B. and Sentenac, H. (2001) Guard cell inward $K^+$ channel activity in *Arabidopsis* involves expression of the twin channel units KAT1 and KAT2. *J Biol Chem* **276**: 3215-3221.

Riesmeier, J. W., Hirner, B. and Frommer, W. B. (1993) Potato sucrose transporter expression in minor veins indicates a role in phloem loading. *Plant Cell* **5**: 1591-1598.

Riesmeier, J. W., Willmitzer, L. and Frommer, W. B. (1994) Evidence for an essential role of the sucrose transporter in phloem loading and assimilate partitioning. *EMBO J* **13**: 1-8.

Robinson, D. G., Sieber, H., Kammerloher, W. and Schäffner, A. R. (1996) PIP1 aquaporins are concentrated in plasmalemmasomes of *Arabidopsis thaliana* mesophyll. *Plant Physiol* **111**: 645-649.

Roitsch, T., Ehneß, R., Goetz, M., Hause, B., Hofmann, M. and Sinha, A.K. (2000) Regulation and function of extracellular invertase from higher plants in relation to assimilate partitioning, stress responses and sugar signalling. *Austr J Plant Physiol* **27**: 815-825.

Rolland, F., Moore, B. and Sheen, J. (2002). Sugar sensing and signalling in plants. *Plant Cell* **14**: S185-S205.

Russin, W. A. and Evert, R. F. (1985) Studies on the leaf of *Populus deltoides* (Salicaceae): ultrastructure, plasmodesmatal frequency and solute concentrations. *Am J Bot* **72**: 1232-1247.

Schäffner, A. R. (1998) Aquaporin function, structure, and expression: Are there more surprises to surface in water relations? *Planta* **204**: 131-139.

Schrier, A. A. (2001) *Der Einflua der Temperatur auf das Funktionieren, die Evolution und die Verbreitung apoplasmatischer und symplasmatischer Phloembeladung.* Thesis, University of Giessen, Germany.

Schulz, A. (1994) Phloem transport and differential unloading in pea seedlings after source and sink manipulations. *Planta* **192**: 239-248.

Schulz, A. (1995) Plasmodesmal widening accompanies the short-term increase in symplasmic phloem unloading in root tips under osmotic stress. *Protoplasma* **118**: 22-37

Schulz, A., Kühn, C., Riesmeier, J. W. and Frommer, W. B. (1998) Ultrastructural effects in potato leaves due to anti-sense inhibition of the sucrose transporter indicate an apoplasmic mode of phloem loading. *Planta* **206**: 533-543.

Sjolund, R. D. and Shih, C. Y. (1983) Freeze-fracture analysis of phloem structure in plant tissue cultures. I. The sieve element reticulum. *J Ultrastr Res* **82**: 11-121.

Smith, J. A. C. and Milburn, J. A. (1980a) Osmoregulation and the control of phloem-sap composition in *Ricinus communis* L. *Planta* **148**, 28-34.

Smith, J. A. C. , and Milburn, J. A. (1980b). Phloem turgor and the regulation of sucrose loading in *Ricinus communis* L. *Planta* **148**: 42-48.

Stitt, M. (1997) The flux of carbon between the chloroplast and cytoplasm. In *Plant Metabolism*, (D. T. Dennis, D. H. Turpin, D. D. Lefebvre and D. B. Layzell, eds.), 2nd Ed., pp. 382-400. Longman Press, Harlow, England.

Tegeder, M., Wang, X.-D., Frommer, W. B., Offler, C. E. and Patrick, J. W. (1999) Sucrose transport into developing seeds of *Pisum sativum* L.. *Plant J* **18**: 151-161.

Terry, B. R. and Robards, A. W. (1987) Hydrodynamic radius alone governs the mobility of molecules through plasmodesmata. *Planta* **171**: 145-157.

Thompson, M. V. and Holbrook, N. M. (2003a) Application of a single-solute non-steady-state phloem model to the study of long-distance assimilate transport. *J Theor Biol* **220**: 419-455.

Thompson, M. V. and Holbrook, N. M. (2003b) Scaling phloem transport: Water potential equilibrium and osmoregulatory flow. *Plant Cell Environ* **26**: 1561-1577.

Truernit, E. and Sauer, N. (1995) The promoter of the *Arabidopsis thaliana* SUC2 sucrose-H+ symporter gene directs expression of ß-glucuronidase to the phloem: Evidence for phloem loading by SUC2. *Planta* **196**: 564-570.

Turgeon, R. (1996a) Phloem loading and plasmodesmata. *Trends Plant Sci* **1**: 418-423.

Turgeon, R. (1996b) The selection of raffinose family oligosaccharides as translocates in higher plants. In *Compartmentation and Source-Sink Interactions in Plants* (M.A. Madore and W.J. Lucas, eds.), pp. 195-203. American Society of Plant Physiologists, Rockville, MD.

Turgeon, R., Beebe, D. U. and Gowan, E. (1993) The intermediary cell: Minor-vein anatomy and raffinose oligosaccharide synthesis in the Scrophulariaceae. *Planta* **191**: 446-456.

Turgeon, R. and Medville, R. (1998) The absence of phloem loading in willow leaves. *Proc Natl Acad Sci U S A* **95**: 12055-12060.

Turgeon, R., Medville, R. and Nixon, K. C. (2001) The evolution of minor vein phloem and phloem loading. *Am J Bot* **88:** 1331-1339.

Turgeon, R. and Wimmers, L. E. (1988) Different patterns of vein loading of exogenous [$^{14}$C]-sucrose in leaves of *Pisum sativum* and *Coleus blumei. Plant Physiol* **87:** 179-182.

Tyerman, S. D., Bohnert, H. J., Maurel, C., Steudle, E. and Smith, J. A. C. (1999) Plant aquaporins: Their molecular biology, biophysics and significance for plant water relations. *J Exp Bot* **50:** 1055-1071.

van Bel, A. J. E. (1993) The transport phloem. Specifics of its functioning. *Prog Bot* **54:** 134-150.

van Bel, A. J. E. (1996a) Interaction between sieve element and companion cell and the consequences for photoassimilate distribution. Two structural hardware frames with associated software packages in dicotyledons? *J Exp Bot* **47:** 1129-1140.

van Bel, A. J. E. (1996b) Carbohydrate processing in the mesophyll trajectory in symplasmic and apoplasmic phloem loading. *Prog Bot* **57:** 140-167.

van Bel, A. J. E. (2003a) The phloem, a miracle of ingenuity. *Plant Cell Environ* **26:** 125-150.

van Bel, A. J. E. (2003b) Transport phloem: low profile, high impact. *Plant Physiol* **131:** 1509-1510.

van Bel, A. J. E., Ammerlaan, A. and van Dijk, A. A. (1994) A three-step screening procedure to identify the mode of phloem loading in intact leaves. Evidence for symplasmic and apoplasmic phloem loading associated with the type of companion cell. *Planta* **192:** 31-39.

van Bel, A. J. E., Gamalei, Y. V., Ammerlaan, A. and Bik, L. P. M. (1992) Dissimilar phloem loading in leaves with symplastic and apoplastic minor vein configurations. *Planta* **186:** 518-525.

van Dongen J. T., Schurr U., Pfister M. and Geigenberger P. (2003) Phloem metabolism and function have to cope with low internal oxygen. *Plant Physiol* **131:** 1529-1543.

Vaughn, M. W., Harrington, G. N. and Bush, D. R. (2002) Sucrose-mediated transcriptional regulation of sucrose symporter activity in the phloem. *Proc Natl Acad Sci U S A* **99:** 10876-10880.

Volk, G. and Franceschi, V. R. (2000) Localization of a calcium-channel-like protein in the sieve element plasma membrane. *Austr J Plant Physiol* **27:** 779-786.

Vreugdenhil, D. (1985) Source-to-sink gradient of potassium in the phloem. *Planta* **163:** 238-240.

Vreugdenhil, D. and Koot-Gronsveld, E. A. M. (1989) Measurements of pH, sucrose and potassium ions in the phloem of castor bean (*Ricinus communis*) plants. *Physiol Plant* **77:** 385-388.

Weise, A., Barker, L., Kühn, C., Lalonde, S., Buschmann, H., Frommer, W. B. and Ward, J. M. (2000) A new subfamily of sucrose transporters, SUT4, with low affinity/high capacity localized in enucleate sieve elements of plants. *Plant Cell* **12:** 1345-1355.

Wilson, C. and Lucas, W. J. (1987) Influence of internal sugar levels on apoplasmic retrieval of exogenous sucrose in leaf tissue. *Plant Physiol* **84:** 1083-1095.

Winch, S. and Pritchard, J. (1999) Acid-induced cell-wall loosening is confined to the accelerating region of the root growing zone. *J Exp Bot* **50:** 1481-1488.

Zhao, R., Dielen, V., Kinet, J.-M. and Boutry, M. (2000) Cosuppression of a plasma membrane H$^+$-ATPase isoform impairs sucrose translocation, stomatal opening, plant growth, and male fertility. *Plant Cell* **12:** 535-546.

Ziegler, H. (1975) Nature of transported substances. In *Transport in Plants I. Phloem Transport* (M. H. Zimmermann and J. A. Milburn, eds.) New series, Vol. 1, pp. 59-100. Springer-Verlag, Berlin.

# 3

# Pathways and Mechanisms
# of Phloem Loading

*Robert Turgeon and Brian G. Ayre*

Phloem loading is the starting point for long-distance nutrient transport. By this process, solute is actively concentrated in the phloem of source organs, primarily mature leaves, thus generating hydrostatic pressure. As solute unloads in sink tissues, hydrostatic pressure is reduced in the sink phloem, and a pressure gradient is established that drives bulk flow of solution. It has been argued that these local processes of loading and unloading control the rate of phloem transport, not the magnitude of the pressure drop itself (Thompson and Holbrook, 2003). The primary mechanisms by which loading occurs are still debated, and many details are not well understood (see reviews by Komor, 2000; Turgeon, 2000; Noiraud *et al.*, 2001b; Lalonde *et. al.*, 2003). In particular, the regulation of phloem loading has only recently come under scrutiny (Deeken *et al.*, 2002; Vaughn *et al.*, 2002).

Phloem loading can only be understood by detailed examination of the routes of sugar movement from mesophyll cells to sieve elements. Mapping these routes and determining the mode of transport at different interfaces are challenging, as the pathways are complex, the distances short, and the transport compounds soluble in the solvents normally used to prepare tissues for microscopy. To compound the problems, it is possible that photoassimilate migrates to sieve elements via multiple, parallel pathways. Notwithstanding these difficulties, a great deal of progress has been made in the study of phloem loading by using a variety of physiological, microscopic, and molecular approaches.

Since the accumulation process generally takes place against a concentration gradient, it requires metabolic energy. Although it is useful to think of "loading" as the entire transport process, from mesophyll cells to sieve elements (Oparka and van Bel, 1992), it is important to remember that the term *loading* implies the expenditure of energy. It was first described as an "endothermic pumping action" (Loomis, 1955). Therefore, in specific terms phloem loading is the final, active step or mechanism that carries molecules against their thermodynamic gradient into the phloem.

Plasmodesmata are present at every interface in a leaf along the photoassimilate transport pathway, from mesophyll cells to sieve elements. Although transport routes are often referred to as symplastic (through the plasmodesmata-connected cytosol of cells) or apoplastic (involving at least one extracellular step), we should perhaps think of these as quantitative, rather than strictly qualitative, differences. It is possible that both pathways contribute to photoassimilate flux, more so at some interfaces than at others. It is also possible that the relative distribution of photoassimilate from one pathway to another is subject to regulation in response to nutrient demands or environmental changes.

In this chapter, the potential for symplastic and apoplastic transport across each interface is discussed to provide a framework for further exploration of the mechanisms and regulation of phloem loading. Beebe and Russin (1999) have described the details of plasmodesmatal structure along the loading route. Dicots are emphasized here since monocots are dealt with by Botha (see Chapter 6) in this volume.

## Minor Veins

Three cell types are most commonly found in minor vein phloem: sieve elements, companion cells, and phloem parenchyma cells. Since sieve elements and companion cells are ontogenetically related (Behnke and Sjolund, 1990) and symplastically coupled (see later), they are frequently considered together as the sieve element/companion cell complex (SE/CCC). Companion cells can often be distinguished from phloem parenchyma cells on the basis of their greater cytoplasmic density. However, in some plants the distinction is difficult to make, even in electron micrographs. Minor veins are bounded by a sheath of cells that may be more or less mesophyll-like in structure. This bundle sheath separates the air spaces of the mesophyll from the very limited air spaces within the veins. In some plants the bundle sheath is connected, at the level of the phloem, to a specialized layer of mesophyll cells called the paraveinal mesophyll that is specialized for delivery of photoassimilates to the veins (Lansing and Franceschi, 2000).

A comment on the nomenclature of parenchyma cells is appropriate here. Parenchymatic elements within veins are sometimes referred to as phloem or xylem parenchyma, based on proximity to one of the two types of transport cells, or as vascular parenchyma to avoid this sometimes arbitrary assignment. The way the terms are used is a matter of personal preference. The cells we are specifically concerned with here are within, or closely associated with, the phloem, hence our use of the term *phloem parenchyma.*

Arrangement of the cell types in minor veins varies. There are certain common features, such as the location of the sieve elements toward the inside of the vein, where they have very limited contact with the bundle sheath (Beebe and Russin, 1999). However, the way in which phloem parenchyma cells and companion cells are arranged with respect to one another differs considerably. These patterns of intercellular interaction undoubtedly provide hints to the various pathways of solute flux, but have not been analyzed in a comprehensive way. Four examples are given next.

The simplest known architecture of minor vein phloem is in species that translocate the raffinose family of oligosaccharides (RFOs). In the Cucurbitaceae, the abaxial phloem consists of a pair of very large companion cells with a pair of sieve elements internal to them (Fig. 3.1A). A single phloem parenchyma cell abuts the sieve elements, but is not always present in the smallest veins (Haritatos *et al.*, 1996). Bundle sheath and companion cells always abut one another directly, have extensive contact, and their common walls are traversed by very large numbers of asymmetrically branched plasmodesmata (Turgeon *et al.*, 1975). This is the case in all plants that translocate RFOs in excess over sucrose (Turgeon, 2000). To distinguish such minor vein companion cells from the ordinary companion cells of sucrose-transporting plants, they are called intermediary cells (Turgeon *et al.*, 1993). The most likely pathway of sugar transport is very direct: mesophyll—bundle sheath—intermediary cell—sieve element. Ordinary companion cells are present in the larger minor veins of RFO-transporting species, raising the possibility of a second loading mechanism in the same leaf. This concept of mixed loading is discussed later.

Parenchyma cells make up a larger fraction of the minor vein phloem in tobacco (*Nicotiana tabacum*; Fig. 3.1B). Here, three companion cells alternate with phloem parenchyma cells in a ring, enclosing two sieve elements. On the basis of these contacts, sugar could pass to SE/CCCs directly from the bundle sheath, or via the phloem parenchyma cells.

In pea plants (*Pisum sativum*; Fig 3.1C), the minor veins often have a layered appearance (Wimmers and Turgeon, 1991). Phloem parenchyma cells alternate with SE/CCCs, the sieve elements toward the inside, as usual. In peas, the companion cells are specialized as transfer cells (i.e., they have wall ingrowths that greatly increase the surface area of the plasma membrane and therefore the capacity for trans-membrane solute flux) (Wimmers and Turgeon, 1991). These are type A transfer cells, as described by Gunning and Pate (1969). Type A transfer cells are companion cells that have ingrowths on all walls, but those on the wall facing the associated sieve elements are not as extensive.

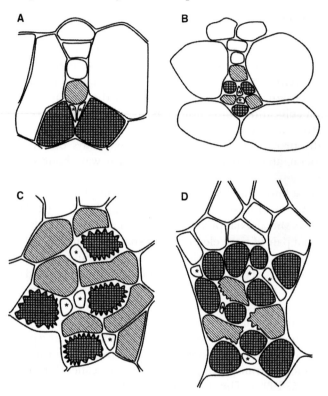

**Figure 3.1** Line tracings of minor veins illustrating the arrangement of cell types. The tracings are from electron micrographs: (A) *Cucurbita pepo* (Volk *et al.*, 1996); (B) *Nicotiana tabacum* (Ding *et al.*, 1995); (C) *Pisum sativum* (Wimmers and Turgeon, 1991); and *Arabidopsis thaliana* (Haritatos *et al.*, 2000b). The vein pictured in D is shown in Fig. 3.2. Diagonal hatching indicates phloem parenchyma cells, companion cells (and intermediary cells of *Cucurbita pepo*) are indicated in darker gray with crosshatching, and sieve elements are marked with asterisks. Unlabeled cells are of other types, including the xylem and bundle sheath. From Haritatos *et al. Planta* (2000) 211:105-111. Copyright Springer; figure 3 adapted with permission.

*Arabidopsis* minor veins are larger and more complex than in the preceding examples (Figs. 3.1D, 3.2). In transverse section, a minor vein typically has 5 sieve elements, approximately 10 companion cells, and approximately 7 phloem parenchyma cells (Haritatos *et al.*, 2000b). As in pea, there is a noticeable layering, with tiers of phloem parenchyma cells and SE/CCCs. In some of the images of phloem parenchyma cells in *Arabidopsis,* the walls closest to the sieve elements and companion cells have clustered ingrowths (Fig. 3.2; Haritatos *et al.*, 2000b). This orientation of wall ingrowths seems to be specific to phloem parenchyma cells in minor veins (Haritatos *et al.*, 2000b). Gunning and Pate (1969) called them type B transfer cells. They have not been studied extensively, but it is possible that the ingrowths facilitate sugar efflux into, rather than uptake from, the apoplast.

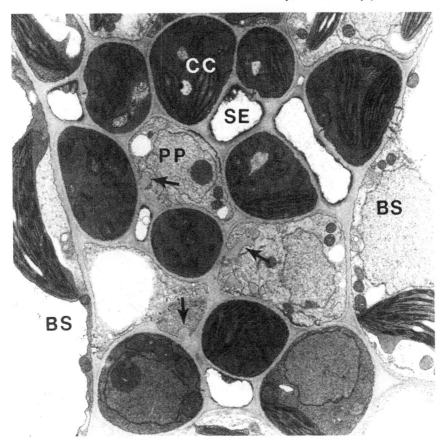

**Figure 3.2** Minor vein phloem of *Arabidopsis thaliana* in transverse section (from Haritatos *et al.*, 2000b). Arrows indicate transfer cell wall ingrowths in phloem parenchyma cells (PP). This vein is diagrammed in Fig. 3.1D. BS = bundle sheath cell; SE = sieve element; CC = companion cell.

Cell relationships in minor vein phloem have been reduced to a single schematic in Fig. 3.3. The interfaces and pathways thought to be especially critical in phloem loading are numbered and discussed next.

## Transport Between Mesophyll Cells

It is often assumed that photoassimilate in mesophyll cells reaches the minor veins symplastically, as indicted in Fig. 3.3 (step 1). In reality there

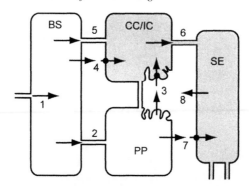

**Figure 3.3**  Schematic diagram of cell types in minor veins. Likely symplastic or apoplastic pathways of solute flux involved in phloem loading are shown as arrows. Connecting lines between cells indicate plasmodesmata. Solid lines indicate the plasma membrane. Convoluted lines in the companion cell (CC) and phloem parenchyma cell (PP) indicate elaboration of the plasma membrane caused by wall ingrowths. Circles on the lines are transmembrane sucrose/proton symporters. The solute content of the SE/CC complex is much higher than that of other cells, as represented by shading. Potential transport steps are numbered and referred to in the text. BS = bundle sheath; SE = sieve element; IC = intermediary cell.

is little direct evidence to support this assumption, aside from the fact that intercellular connections between mesophyll cells are relatively common, and the plasmodesmata are open to passage of microinjected dyes (Beebe and Russin, 1999). As long as there is a concentration gradient from mesophyll to veins in apoplastic and symplastic compartments, solute should diffuse along both routes. Two factors might favor the symplastic pathway: cytoplasmic streaming and pressure-driven transport through plasmodesmata. Although cytoplasmic streaming could conceivably facilitate transport by mixing the contents of adjacent cells, Pickard (2003) argues that this effect is minor for small hydrophilic particles such as sucrose. Pressure-driven transport would require that plasmodesmatal channels in the mesophyll of mature leaves are able to support convective transport and that there is a pressure gradient toward the veins. These are reasonable stipulations, but neither has been proven experimentally.

Since sucrose can pass into and out of cells (see later), a single sucrose molecule probably takes a complex route, spending time in both symplastic and apoplastic compartments. There is general agreement that, in $C_4$ plants, $C_3$ and $C_4$ compounds are exchanged between mesophyll cells and bundle sheath through the abundant plasmodesmata that connect them (Leegood, 1996).

## The Role of Phloem Parenchyma Cells

Once photoassimilate has reached the bundle sheath, it must enter the vein. Phloem parenchyma cells play a prominent role in the architecture of most minor veins and are apparently involved in the entry process. This is apparent in the staggered arrangement of companion cells and phloem parenchyma cells in tobacco minor veins, and in the tiered arrangement of cells in pea and *Arabidopsis*, resulting in extensive contact between the parenchyma cells and the SE/CCC.

To illustrate the amount of contact between different cell types in *Arabidopsis* minor veins, proportional interface lengths and plasmodesmatal numbers are shown in Fig. 3.4. Many plasmodesmata link the bundle

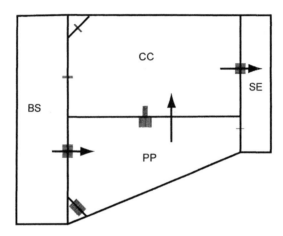

**Figure 3.4** Schematic diagram of the potential symplastic and apoplastic contact between the bundle sheath (BS) and the three cell types in the minor vein phloem of *Arabidopsis thaliana* (data from Haritatos *et al.*, 2000b). This figure is similar to a "plasmodesmogram" (van Bel and Oparka, 1995) with the additional feature of illustrating interface lengths across which apoplastic transfer may occur. The length of the line between different cell types is proportional to the total interface length between those cell types in a typical vein. For example, the interface between companion cells (CC) and phloem parenchyma (PP) cells is approximately 1.8 times that between companion cells and sieve elements (SE). Cell volumes are not proportional. Oblique lines indicate total interface length between cells of the same type: PP-PP on the bottom and CC-CC on the top. The most extensive cell-cell contact is at the PP-CC interface. Cell wall ingrowths in PP cells increase the proportional intercellular contact with both SEs and CCs (not shown). Relative frequency of plasmodesmata between the cell types (summed over all the cells in the vein) is indicated by the number of lines joining the cells. Plasmodesmata are most numerous between adjacent PP cells and least numerous between PP cells and SEs. Those between PP and CC are unequally branched, with more branches on the PP side. Based on the degree of intercellular contact, the immunolocalization of sucrose carriers on the plasma membranes of companion cells (see text), and the relative frequency of plasmodesmata, the most likely route for a sucrose molecule to follow in the course of phloem loading is apparently through the symplast into PP cells, via the apoplast into CCs, and through the symplast from CCs to SEs (*arrows*).

sheath and phloem parenchyma cells. Plasmodesmatal frequencies between one phloem parenchyma cell and another are especially high, a feature of this cell type noted in other plants, including barley (Evert *et al.*, 1996), potato (McCauley and Evert, 1989), *Moricandia* (Beebe and Evert, 1992), pea (Wimmers and Turgeon, 1991), and *Liriodendron* (Goggin *et al.*, 2001). From these considerations it appears that the phloem parenchyma acts as a conduit for symplastic transport into the vein (Fig. 3.3, step 2).

Dye-coupling studies corroborate the structural evidence for symplastic continuity through phloem parenchyma cells. For example, Lucifer Yellow microinjected into the mesophyll cells of corn leaves diffuses through the bundle sheath and into phloem parenchyma cells (Botha *et al.*, 2000). This pathway appears to be important in the normal distribution of photo-assimilate because in veins of the *sucrose export deficient 1* (*sxd1*) mutant, diffusion of dye stops at the bundle sheath-phloem parenchyma border (Botha *et al.*, 2000). In this mutant, anthocyanin and starch build up in the distal region of the lamina, indicating an excess of carbohydrate. The anthocyanin-containing tissue is unable to export sucrose. By electron microscopy it has been shown that the plasmodesmata linking bundle sheath and phloem parenchyma cells are defective (Russin *et al.*, 1996; Botha *et al.*, 2000). The *sxd1* mutation therefore provides clear genetic and ultrastructural evidence that, in corn, phloem parenchyma cells are particularly important in delivering sucrose to the sieve tubes.

## Entry into the SE/CCC via the Apoplast

As discussed previously, the energy-requiring step that loads solutes into the phloem, against a thermodynamic gradient, occurs at the SE/CCC. There is now compelling evidence, at least in certain species, for phloem loading from the apoplast. This conclusion, based initially on analysis of sucrose moving through the apoplast (Sovonick *et al.*, 1974), is now supported by several lines of evidence. For example, yeast invertase secreted into the apoplast of transgenic plants disrupts phloem loading (Von Schaewen *et al.*, 1990; Dickinson *et al.*, 1991). This is to be expected if sucrose destined for export enters the apoplast at some point and can only be pumped into the phloem from the apoplast by a carrier(s) that is specific for sucrose and not hexose molecules.

Where does the sucrose enter the apoplast? If, as discussed previously, the major transport route into veins is through the phloem parenchyma, then the most likely point of entry of sucrose into the apoplast is at the boundary between the phloem parenchyma and SE/CCCs (Fig. 3.3, steps 3 and 7); however, the boundary between bundle sheath cells and companion cells cannot be discounted (Fig. 3.3, step 4). As noted previously, many of the phloem parenchyma cells in *Arabidopsis* minor veins have transfer cell ingrowths (type B transfer cells) where they are in contact with sieve elements, which may facilitate efflux of sucrose to the apoplast before uptake by the phloem (Haritatos *et al.*, 2000b).

Efflux across the plasma membrane into the cell wall space might be carrier-mediated, but this carrier has not yet been identified. Since efflux could be rate limiting under certain circumstances and regulated, it is important that this step in the loading process be characterized more thoroughly.

In recent years, a family of sucrose/proton co-transporters (sucrose transporters) has been described (Lalonde *et al.*, 2003), some members of which are involved in phloem loading. In addition, polyol transporters have been identified in mature leaves of celery (Noiraud *et al.*, 2001a) and plantain (Ramsperger-Gleixner *et al.*, 2004), which supports the hypothesis that these compounds are also loaded into the phloem from the apoplast in these species.

To date, SUT4 is the likeliest sucrose transporter to play a prime role in phloem loading, while others may play a more prominent role in reclamation of leaked sucrose. These functions are not mutually exclusive. SUT4 has the highest capacity of the known carriers. The $K_m$ of SUT4 is ~12 mM, significantly above the average concentration of sucrose in the apoplast of leaves (~2 mM; Lohaus *et al.*, 2001). Although only speculation at this point, the high $K_m$ of SUT4 may suit it to loading sucrose that is present at higher than the average apoplastic concentration at a specific site in the apoplast, perhaps in the cell walls between phloem parenchyma cells and the SE/CCC. No matter how well suited SUT4 is to phloem loading, downregulation of *SUT1* in the Solanaceae, and its apparent ortholog *AtSUC2* in *Arabidopsis,* inhibits phloem transport, indicating an important role for this transporter (Kühn *et al.*, 1996; Bürkle *et al.*, 1998; Gottwald *et al.*, 2000).

None of the sucrose transporters are strictly localized to minor veins; they all have appreciable activity along the long-distance phloem pathway and in sink tissues. However, within the veins of the *Arabidopsis* leaf, AtSUT4 is localized in smaller, rather than larger, veins (Weise *et al.*, 2000).

The precise cellular localizations of the various transporters have not been fully established, but data indicate that there may be important differences between species. While several claims have been made concerning the role of various transporters in phloem loading, little of the focus to date has been on minor veins; most localization studies involve the long-distance path phloem. Some of these images come from leaves, but they are often from large veins, or the petiole, or the site is not indicated. Given the important differences in function between path and loading phloem, and distinct differences that can exist in expression of certain genes

(Haritatos *et al.*, 2000a) and the expression of different transporters (Ramsperger-Gleixner *et al.*, 2004) in different companion cells of the leaf, it is possible that expression patterns of sucrose transporters differ. More attention needs to be paid specifically to the minor venation, and in these studies the precise location of the interfaces should be thoroughly documented.

With these caveats in mind, it appears that sucrose transporters are present on the plasma membranes of sieve elements, not companion cells, in solanaceous species (Kühn *et al.*, 1997; Barker *et al.*, 2000; Weise *et al.*, 2000; Reinders *et al.*, 2002). It may be that the companion cells in minor veins of solanaceous plants are not directly involved in phloem loading. Instead, most or all of the sucrose may be delivered to the apoplast surrounding the sieve elements via phloem parenchyma cells (Fig. 3.3, pathway defined by steps 1, 2, and 7).

In contrast to the data on solanaceous species, *SUC2* has been localized to the plasma membrane of companion cells in *Arabidopsis* (Stadler and Sauer, 1996) and *Plantago major* (Stadler *et al.*, 1995). Phloem transport is severely inhibited in an *Arabidopsis* mutant deficient in this transporter (Gottwald *et al.*, 2000). This suggests loading defined by steps 1, 2, 3, and 4 in Fig. 3.3. The polyol transporters PmPLT1 and PmPLT2 localize to companion cells in *Plantago major* (Ramsperger-Gleixner *et al.*, 2004).

Structural data also support involvement of companion cells in phloem loading. The presence of type A transfer (companion) cells (Gunning and Pate, 1969) is a reasonable indication that these cells are directly involved in loading from the apoplast. When plants are grown under high light to increase the rate of photosynthesis and photoassimilate export, there is a proportional increase in the elaboration of wall ingrowths (Wimmers and Turgeon, 1991). *Plantago* minor vein companion cells are also specialized as transfer cells (Gunning and Pate, 1969), which is consistent with the immunolocalization of sucrose and polyol transporters to the companion cell. However, this type of transfer cell is not found in companion cells in the Solanaceae, or in *Arabidopsis*.

## Entry into the SE/CCC via the Symplast

So far we have considered only the apoplastic mode of entry into the SE/CCC. Is it possible for sucrose to enter the complex through plasmodesmata, either as a supplementary or alternate route? As a general rule, very few plasmodesmata link sieve elements to any cell type except companion cells, so symplastic loading directly into the sieve element seems unlikely. However, plasmodesmata link companion cells to both bundle sheath and phloem parenchyma cells in the minor veins of all species, and in some cases can be quite numerous.

Dye coupling studies have been conducted in several laboratories in an effort to demonstrate symplastic continuity between mesophyll cells and the SE/CCC (Erwee *et al.*, 1985; Madore *et al.*, 1986; Fisher, 1988; van Kesteren *et al.*, 1988; Turgeon and Hepler, 1989; Botha *et al.*, 2000). This technique is useful, but it must be kept in mind that it is extremely sensitive and difficult to quantify; thus positive results could overemphasize the importance of a symplastic route. Also, the experiments are complicated by the fact that minor vein companion cells are difficult to distinguish from phloem parenchyma cells in fresh tissue, and the sieve elements are extremely narrow and hard to see. Therefore, dye that has been recorded in minor veins after microinjection into the mesophyll may have been in the phloem parenchyma, not the SE/CCC. This was not a problem in a study of dye coupling in *Cucurbita pepo* (Turgeon and Hepler, 1989) because the simple structure of the minor veins allowed dye to be injected directly into intermediary cells. In this case the dyes moved into the bundle sheath and mesophyll. However, the existence of a symplastic connection from mesophyll to the SE/CCC, open to molecules the size of sucrose, has not been adequately proven in other species.

Nonetheless, plasmodesmata leading into minor vein companion cells are present in all species, and it is reasonable to conclude that they have some function since the potential cost of retaining them is high, given that viruses use them as avenues to invade the phloem and become systemic (Cheng *et al.*, 2000). Is it possible that they constitute a pathway for phloem loading? One way to answer this question is to apply the sulfhydryl-modifying compound *p*-chloromercuribenzenesulfonic acid (PCMBS), which inhibits sucrose transporters (Giaquinta, 1983), and to determine if there is any residual loading. PCMBS penetrates cells very slowly and thus exerts its effects in the apoplast. Many experiments of this type have been conducted, with two contrasting results. Depending on species, PCMBS either blocks phloem loading efficiently, or it has virtually no effect (Turgeon, 2000). As discussed later, phloem loading in all sucrose-translocating plants tested to date is sensitive to PCMBS. Phloem loading in RFO-transporting plants is insensitive to PCMBS.

In sucrose-translocating species, PCMBS is so effective that it seems unlikely that there is a substantial symplastic component to loading. A small amount of symplastic transfer into the translocation stream does appear to take place in *Ricinus communis* if apoplastic loading is inhibited with PCMBS (Orlich *et al.*, 1998). Whether this component of loading in *Ricinus* occurs in the absence of PCMBS, or is somehow stimulated by the compound, is not known. One possibility, pointed out by Orlich *et al.* (1998), is that the plasmodesmata leading into the companion cells are pressure regulated and that PCMBS, by inhibiting phloem loading, reduces the pressure inside companion cells to the point that the plasmodesmata open, thus allowing limited diffusion.

If the plasmodesmata in minor vein companion cells of sucrose-translocating plants are not used for phloem loading, what is their function? Why does the sucrose in the companion cells not leak out through these pores, thus dissipating the gradient established by apoplastic loading? Perhaps these plasmodesmata are involved in the traffic of solute molecules that do not enter the phloem via carrier systems. They might also traffic informational molecules and maintain electrical conductivity. Leakage may not be a substantial problem if the plasmodesmata are present in small numbers, or if they are regulated. They could be closed enough to keep sucrose in, yet remain sufficiently open to accommodate other functions. There is convincing structural evidence for occlusion of plasmodesmata that lead into the minor vein companion cells of *Moricandia* (Beebe and Evert, 1992).

Before dismissing the notion of symplastic loading in regard to species that translocate sucrose, as opposed to RFOs, we must consider the intriguing observation that some of these plants have high plasmodesmatal counts in the minor vein phloem. Gamalei (1989, 1991), who has categorized plants on the basis of minor-vein plasmodesmata, calls species with especially high numbers type 1, with open phloem, and they have long been considered putative symplastic loaders.

It is important to note that Gamalei's type 1 is composed of at least two subgroups. In one group, sucrose is the primary solute translocated, but in the other, RFOs dominate. Since the plasmodesmata in RFO-translocating plants are so specialized, and the evidence for a distinctive mechanism of phloem loading in these plants is so strong, we feel that it is best not to include them in Gamalei's system. The term *type 1* is used hereafter in this sense, excluding RFO plants.

Do type 1 plants load symplastically? If so, how? Since they typically transport sucrose, with smaller amounts of other oligosaccharides, or none at all, they do not load by a polymer trap. It has been suggested that loading occurs in these plants via a network of endoplasmic reticulum, joined in adjacent cells by the desmotubules of the connecting plasmodesmata (Gamalei *et al.*, 1994). However, no detailed mechanism or experimental data in support of this concept have been put forward.

Few experiments have been conducted on type 1 plants, but those that have do not support the hypothesis that they load via the symplast. Microautoradiography and exudation studies indicated that phloem loading in *Liriodendron tulipifera* (type 1) is sensitive to PCMBS (Goggin *et al.*, 2001).

Results with willow (*Salix babylonica*), another type 1 species, were more unusual in that no evidence for phloem loading was found (Turgeon and Medville, 1998). Sucrose apparently diffuses into the minor vein phloem. Nonetheless, it is at sufficiently high concentration that the resulting turgor pressure difference between source and sink drives the transport of phloem sap over long distances through the sieve tubes. Plasmolysis studies on *Populus* (Russin and Evert, 1985), in the same family as willow, are consistent with this interpretation. It should be noted that Münch (1930) did not include a loading step in his mass flow model. The concept of phloem loading arose in the 1930s and 1940s (Turgeon, 1996).

*Arabidopsis thaliana* is a type 1-2a species by Gamalei's definition (Haritatos *et al.*, 2000b), which means that it has fewer plasmodesmata than type 1 plants, but more than type-2 model species such as tobacco and *Vicia faba* that have very few minor vein plasmodesmata are known to load via the apoplast. Data from several laboratories are consistent with an apoplastic phloem loading mechanism in *Arabidopsis* (Truernit and Sauer, 1995; Gottwald *et al.*, 2000; Haritatos *et al.*, 2000b; Weise *et al.*, 2000).

In summary, there does not appear to be a correlation between plasmodesmatal frequencies in minor vein companion cells and the mechanism of phloem loading. Abundant plasmodesmata may be necessary to accommodate symplastic loading, but the presence of abundant plasmodesmata does not necessarily indicate that loading occurs through them.

## Symplastic Phloem Loading by the Polymer Trap

As noted previously, species that translocate RFOs have extremely numerous plasmodesmata between the bundle sheath and companion cells (intermediary cells) in minor veins (Fig. 3.3, step 5). The plasmodesmata are arranged in large fields, with more than 100 plasmodesmata in each (Volk *et al.*, 1996). According to Gamalei (1991), there are over *three orders of magnitude* more plasmodesmata at this interface in *Cucurbita* than in *Lactuca* and some other genera of apoplastic loaders. These are the intermediary cells referred to previously. Plasmodesmata between the bundle sheath and intermediary cells are highly branched, more so on the intermediary cell side. From these considerations, and the timing of plasmodesmata formation at the onset of the sink-source transition (Volk *et al.*, 1996), it appears that the numerous plasmodesmata at the bundle sheath-companion cell boundary are involved in phloem loading.

Symplastic phloem loading was first considered implausible because no mechanism of active transport of small molecules through plasmodesmata has been demonstrated. Therefore, on thermodynamic grounds, it would seem that symplastic loading involves diffusion against a concentration gradient, which is impossible. However, the observation that species with intermediary cells translocate raffinose family oligosaccharides (RFOs) (discussed in Turgeon, 1996) led to the development of a thermodynamically feasible model of symplastic loading termed the "polymer trap" (Turgeon and Gowan, 1990; Turgeon, 1991). According to this model, sucrose diffuses into intermediary cells down its concentration gradient and is converted to the larger oligosaccharides, raffinose and stachyose. Synthesis of these sugars reduces the concentration of sucrose to permit continued diffusion. A fundamental assumption of the model is that the branched plasmodesmata are slightly narrower than those at many other interfaces, with a size exclusion limit that allows diffusion of sucrose into the cell but does not permit diffusion of the larger sugars, raffinose and stachyose, in the opposite direction. Thus raffinose and stachyose accumulate to high concentrations, as measured either directly (Haritatos et al., 1996), or by plasmolysis (Turgeon and Hepler, 1989). Indeed, the numerous plasmodesmata between bundle sheath cells and intermediary cells are especially narrow on the intermediary cell side, to the point that the cytoplasmic annulus is not visible (Fisher, 1986; Turgeon et al., 1993).

Subsequent to the formulation of the polymer trap model, it was found that the RFO pathway in leaves of cucurbits is localized in intermediary cells, as the model predicts (Holthaus and Schmitz, 1991; Beebe and Turgeon, 1992). In some species, such as Ajuga, RFOs are also produced in the mesophyll, but the two pools are separate; export occurs only from the intermediary cell pool (Bachmann et al., 1994; Sprenger and Keller, 2000).

Knowledge that the RFO pathway resides in intermediary cells made it possible to test the second prediction of the model, that the plasmodesmata between intermediary cells and bundle sheath cells are too narrow to accommodate the passage of raffinose or stachyose. Minor veins were microdissected from melon leaves and the sugars analyzed (Haritatos et al., 1996). As predicted, both the trisaccharide and tetrasaccharide are high in concentration in intermediary cells, but almost undetectable in the surrounding mesophyll. In contrast, galactinol, the first intermediate in the synthesis of RFOs, is found in approximately the same concentration in both cell types. Galactinol is approximately the same size as sucrose and is synthesized in intermediary cells along with raffinose and stachyose. Since the enzymes of the RFO pathway are cytosolic, it seems reasonable to conclude that galactinol, raffinose, and stachyose are present in the cytosol. This being the case, it is clear that neither raffinose nor stachyose is able to pass from intermediary cells to the bundle sheath and then into the mesophyll through the cytoplasmic annuli of the connecting plasmodesmata (Fig. 3.3, step 5 in the direction opposite the arrow).

If galactinol is small enough to pass through the plasmodesmata of intermediary cells, it could theoretically be synthesized in the mesophyll and diffuse to the site of RFO synthesis in intermediary cells. This may be the case in certain plants such as Olea, Catalpa, and Ligustrum (Flora and Madore, 1993 and references therein).

It is also interesting that mannitol is translocated in *Olea* along with RFOs (Flora and Madore, 1993). This small compound (182 D) is synthesized in the mesophyll where it occurs at very high concentrations (8 times that of sucrose). The concentration in the phloem, on the other hand, is apparently lower than that of sucrose or RFOs (Flora and Madore, 1993). One interpretation of these data is that mannitol diffuses from the mesophyll to the sieve elements through the bundle sheath and intermediary cells in this species. Thus, the mode of entry of polyols into the phloem could occur by different mechanisms in different plants, by carrier-mediated uptake from the symplast in some, and by the symplastic route in others.

Fructans are transported in some plants. It has been suggested that fructan transport occurs by polymer trapping in *Agave* (Wang and Nobel, 1998).

## Mixed Loading

As noted previously, a consistent feature of species with intermediary cells is that they also have ordinary companion cells in the minor veins (i.e., companion cells without numerous plasmodesmata to surrounding cells). This raises the possibility that, in addition to polymer trapping, sucrose is pumped into the ordinary companion cells from the apoplast and therefore that translocated carbohydrate comes from two sources. This conclusion is further supported by the specialization of companion cells as transfer cells in *Acanthus* (van Bel *et al.*, 1992), and in *Nemesia* and *Rhodochiton* (Turgeon *et al.*, 1993). Since wall ingrowths in minor vein companion cells have been strongly implicated in apoplastic loading (Wimmers *et al.*, 1991), the appearance of these cells with intermediary cells, in the same vein, suggests that symplastic and apoplastic loading occur either simultaneously, or at different times, or under different environmental conditions.

Further evidence for mixed loading comes from sugar-uptake studies. In leaf discs of *Coleus* (Turgeon and Gowan, 1990) and basil (Flora and Madore, 1996) sucrose uptake is sensitive to PCMBS, which indicates the presence of a sucrose-proton co-transport system in the leaves of symplastic-loading plants. Recently, a sucrose transporter (AmSUT1) was cloned from *Alonsoa meridionalis* (Knop *et al.*, 2004). In minor veins, the transporter appears to be localized to ordinary companion cells rather than intermediary cells, suggesting that loading may take place by two mechanisms in the same vein. Again, in *Plantago*, sucrose and polyol appear to load from distinct subsets of cells in small veins (Ramsperger-Gleixner *et al.*, 2004).

Although these correlates are all consistent with apoplastic loading, they do not prove that this mechanism plays an appreciable role in species that load via the symplast. As far as ordinary companion cells and transfer cells are concerned, an analogous situation exists in grasses, which have two types of sieve elements in small veins: thin- and thick-walled. It has often been suggested that the thick-walled sieve elements function in retrieval, not in long-distance transport (see Chapter 6). This could explain the presence of ordinary companion cells, or transfer cells, in minor veins and it is certainly consistent with the presence of sucrose transporters. Since one role of transporters is to retrieve sucrose that leaks from the phloem into the apoplast, it is entirely reasonable for these proteins to be present in minor vein phloem, even if the pathway of photoassimilate loading is entirely symplastic (Grusak *et al.*, 1996).

One way to estimate the contribution of apoplastic loading in symplastic-loading species is to inhibit transport with PCMBS. This treatment should reduce the amount of apoplastically-loaded, but not symplastically loaded, sucrose in phloem exudate, and therefore increase the proportion of RFOs. In olive (Flora and Madore, 1993) and basil (Flora and Madore, 1996), PCMBS had no effect on the proportion of exuded sugars. In *Alonsoa*, PCMBS treatment reduced the absolute amount of transported sucrose by approximately 20% (Knop *et al.*, 2004). To date this is the best physiological evidence for mixed loading, although there are problems with the method used (EDTA–facilitated exudation) that limit confidence in quantitative results (Girousse *et al.*, 1991).

In summary, the evidence to date on mixed loading is equivocal. On the basis of structural data, it appears likely, but the contribution of the apoplastic component is apparently small, at least under the conditions in which experiments have been conducted, and thus it is difficult to demonstrate unequivocally.

## Solute Flux Between Companion Cells and Sieve Elements

Plasmodesmata-pore units (PPUs) are consistent features of the common walls between companion cells and sieve elements, including those in minor veins (Behnke and Sjolund, 1990). PPUs have branched plasmodesmata on the companion cell side and a larger pore on the sieve element side. Dye-coupling experiments indicate that they have very large size exclusion limits in path phloem (van Bel and Kempers, 1991; Oparka *et al.*, 1995; Kempers *et al.*, 1996; Kempers and van Bel, 1997), although this has not been confirmed in phloem of minor veins due to the technical limitations of working with small cells in a secluded location. Assuming a large size exclusion limit of PPUs in the SE/CCCs of minor veins, it seems reasonable to postulate that the dominant path for transfer to the sieve elements from the companion cells is through these pores.

Nonetheless, there is a considerable amount of cell wall contact between the two members of the complex (Figs. 3.2 and 3.4). Thus, it is possible that apoplastic transfer is responsible for at least some of the flux into the sieve element from companion cells.

As noted previously, localization of sucrose carriers indicates that sucrose bypasses companion cells and enters sieve elements directly from the apoplast in the Solanaceae (Reinders *et al.*, 2002 and references therein). However, the companion cells in tobacco veins have very high solute content, just as in other plants (Ding *et al.*, 1995). If sucrose is pumped into the sieve elements, and not into companion cells, it apparently diffuses into the latter through PPUs, increasing the turgor pressure to a level well above that of the mesophyll and phloem parenchyma (Fig. 3.3, reverse of step 6).

## Solute Exchange Between the Phloem and Flanking Tissues

If solutes from the metabolically rich companion cells are able to pass freely through PPUs to enter sieve elements, the composition of the translocation stream should be highly complex. Indeed, phloem sap is complex and contains many organic and inorganic substances at concentrations similar to those in the cytosolic fraction of surrounding cells (Lohaus *et al.*, 1995, and references therein). The implication of these findings is that a full range of metabolites within the companion cells is continually lost to the translocation stream and requires constant replacement. How is the metabolic integrity of companion cells maintained? In the path phloem, solute loss may be minimal if flux between the two cell types is near steady state. However, in minor veins, loss would be more acute, as flux is essentially unidirectional, relatively pure water entering the SE/CCC from the xylem and carrying away loaded photoassimilates and other solutes. Solute replacement from surrounding mesophyll might occur in species with prominent symplastic continuity, such as symplastic loaders and Gamalei's type 1 plants, but would be highly restricted in other species.

Is the transfer of solutes from companion cells to sieve elements unrestricted? Apparently not, since not all small solutes found in companion cells are present in the translocation stream. For example, very little galactinol is transported in species that translocate RFOs (Kandler, 1967; Flora and Madore, 1993; Turgeon *et al.*, 1993; Bachmann *et al.*, 1994). This is contrary to expectation since galactinol—an essential intermediate in the synthesis of RFOs—is present in the companion (intermediary) cells of minor veins in relatively high concentrations (Haritatos *et al.*, 1996), and appears to be an ideal transport compound in that it is a nonreducing disaccharide the approximate size of sucrose. Furthermore, galactinol appears to be largely in the cytosolic compartment, and should be able to pass freely through PPUs into the sieve elements and the translocation stream.

A study was recently undertaken that addressed the fate of small solutes in companion cells with respect to long-distance transport (Ayre *et al.,* 2003). Traceable solutes were synthesized specifically in the companion cells of minor veins by expressing the corresponding biosynthetic genes from the *CmGAS1* promoter in transgenic tobacco. The *CmGAS1* promoter naturally drives the expression of galactinol synthase in the intermediary cells of *Cucumis melo* and is specific to the minor vein companion cells in heterologous hosts such as tobacco and *Arabidopsis* (Haritatos, 2000a). The traceable solutes were galactinol and octopine; octopine is a small (246 D), ionized compound derived from arginine and pyruvate. In addition, the distribution of naturally occurring RFOs in *Coleus blumei* was addressed.

This study showed that small solutes in the cytoplasm of companion cells did indeed enter sieve elements and the translocation stream freely, but that these compounds were also distributed to surrounding tissues and the apoplast to varying extents. Galactinol was released from the phloem at a rate sufficient to rapidly clear from the translocation stream, and disperse throughout the leaf by water arriving in the xylem. The mode of release is not known, but may be via broad specificity channels, or simple leakage. Sucrose is also lost from the phloem at significant rates but is constantly retrieved from the apoplast by sucrose transporters (Thorpe and Minchin, 1996; see Chapter 10). Octopine was identified in the apoplast, but also demonstrated more efficient retention and transport in the phloem, probably because it is ionized. The larger RFOs, especially the tetrasaccharide stachyose, were well retained and efficiently transported, probably as a function of size.

For efficient translocation of small solutes, a mechanism of retrieval from the apoplast, as in the case of sucrose, or symplastic retention by charge or size, must be at play. Compounds that are lost from the phloem network, and not recovered by transporters, will be retained in the leaf and redistributed by the water entering the leaf in the adjacent xylem (see also Chapters 10 and 11).

## Conclusion

Phloem transport occurs by the bulk flow of water and dissolved nutrients from photosynthetic source tissues to heterotrophic sink tissues. Bulk flow results from the hydrostatic pressure difference in the phloem between source and sink tissues. This pressure difference is accentuated by phloem loading—the energized process of accumulating photoassimilate in the SE/CCC of minor veins. Phloem loading thereby contributes to the driving force of phloem transport and is a control point for nutrient distribution throughout the plant. Phloem loading is nearly ubiquitous among terrestrial plants and must therefore be highly advantageous. However, it is not essential, as demonstrated by the absence of loading in willow. Furthermore, the mechanism of phloem loading—symplastic, or apoplastic, or perhaps a mixture of the two—is not consistent among higher plants.

For phloem loading to be effective, the cell types accumulating photoassimilate must be symplastically isolated from the surrounding tissues, at least with regard to primary osmotica. Although the SE/CCC is generally accepted to be an isolated tissue, this is not obvious in the ultrastructural analysis of many species (i.e., Gamalei's type 1 plants). In the case of plants with intermediary cells, symplastic continuity exists, but the cells are apparently isolated nonetheless with respect to the primary osmotica (raffinose and stachyose) by narrowing of the plasmodesmatal channels.

The interface through which photoassimilate is loaded is poorly understood in most, if not all, species. In light of the considerations presented in this chapter, it is clear that analysis of selected "model" plants will not adequately illuminate the natural diversity in phloem loading mechanisms. To understand these processes more thoroughly, greater emphasis must be placed on the specific pathways sugars follow from mesophyll to sieve element, and along the translocation stream.

## Acknowledgments

We gratefully acknowledge financial support from the National Science Foundation and the NRI Competitive grants program of the USDA. We also thank Edwin Reidel and Ashlee McCaskill for critically reading the manuscript.

## References

Ayre, B. G., Keller, F. and Turgeon, R. (2003) Symplastic continuity between companion cells and the translocation stream: Long-distance transport is controlled by retention and retrieval mechanisms in the phloem. *Plant Physiol* **131:** 1518-1528.

Bachmann, M., Matile, P. and Keller, F. (1994) Metabolism of the raffinose family oligosaccharides in leaves of *Ajuga reptans* L: Cold acclimation, translocation, and sink to source transition: Discovery of chain elongation enzyme. *Plant Physiol* **105:** 1335-1345.

Barker, L., Kühn, C., Weise, A., Schulz, A., Gebhardt, C., Hirner, B., Hellmann, H., Schulze, W., Ward, J. M. and Frommer, W. B. (2000) SUT2, a putative sucrose sensor in sieve elements. *Plant Cell* **12**: 1153-1164.

Beebe, D. U. and Evert, R. F. (1992) Photoassimilate pathways and phloem loading in the leaf of *Moricandia arvensis* L. Dc. Brassicaceae. *Int J Plant Sci* **153**: 61-77.

Beebe, D. U. and Russin, W. A. (1999) Plasmodesmata in the phloem-loading pathway. In *Plasmodesmata: Structure, Function, Role in Cell Communication* (A. J. E. van Bel and W. J. P. van Kesteren, eds.) pp. 261-293. Springer-Verlag, Berlin, New York.

Beebe, D. U. and Turgeon, R. (1992) Localization of galactinol, raffinose, and stachyose synthesis in *Cucurbita pepo* leaves. *Planta* **188**: 354-361.

Behnke, H.-D. and Sjolund, R. D. (1990) *Sieve Elements: Comparative Structure, Induction and Development.* Springer-Verlag, Berlin, New York.

Botha, C. E. J., Cross, R. H. M., van Bel, A. J. E. and Peter, C. I. (2000) Phloem loading in the sucrose-export-defective (*SXD-1*) mutant maize is limited by callose deposition at plasmodesmata in bundle sheath-vascular parenchyma interface. *Protoplasma* **214**: 65-72.

Bürkle, L., Hibberd, J. M., Quick, W. P., Kühn, C., Hirner, B. and Frommer, W. B. (1998) The H+-sucrose cotransporter NtSUT1 is essential for sugar export from tobacco leaves. *Plant Physiol* **118**: 59-68.

Cheng, N.-H., Su, C.-L., Carter, S. A. and Nelson, R. S. (2000) Vascular invasion routes and systemic accumulation patterns of tobacco mosaic virus in *Nicotiana benthamiana*. *Plant J* **23**: 349-362.

Deeken, R., Geiger, D., Fromm, J., Koroleva, O., Ache, P., Langenfeld-Heyser, R., Sauer, N., May, S. T. and Hedrich, R. (2002) Loss of the AKT2/3 potassium channel affects sugar loading into the phloem of *Arabidopsis*. *Planta* **216**: 334-344.

Dickinson, C. D., Altabella, T. and Chrispeels, M. J. (1991) Slow-growth phenotype of transgenic tomato expressing apoplastic invertase. *Plant Physiol* **95**: 420-425.

Ding, X. S., Shintaku, M. H., Arnold, S. A. and Nelson, R. S. (1995) Accumulation of mild and severe strains of tobacco mosaic virus in minor veins of tobacco. *Mol Plant Microbe Interact* **8**: 32-40.

Erwee, M. G., Goodwin, P. B. and van Bel, A. J. E. (1985) Cell-cell communication in the leaves of *Commelina cyanea* and other plants. *Plant Cell Environ* **8**: 173-178.

Evert, R. F., Russin, W. A. and Botha, C. E. J. (1996) Distribution and frequency of plasmodesmata in relation to photoassimilate pathways and phloem loading in the barley leaf. *Planta* **198**: 572-579.

Fisher, D. G. (1986) Ultrastructure, plasmodesmatal frequency, and solute concentration in green areas of variegated *Coleus blumei* Benth. leaves. *Planta* **169**: 141-152.

Fisher, D. G. (1988) Movement of Lucifer yellow in leaves of *Coleus blumei* Benth. *Plant Cell Environ* **11**: 639-644.

Flora, L. L. and Madore, M. A. (1993) Stachyose and mannitol transport in olive (*Olea europaea* L.). *Planta* **189**: 484-490.

Flora, L. L. and Madore, M. A. (1996) Significance of minor-vein anatomy to carbohydrate transport. *Planta* **198**: 171-178.

Gamalei, Y. (1989) Structure and function of leaf minor veins in trees and herbs. *Trees* **3**: 96-110.

Gamalei, Y. (1991) Phloem loading and its development related to plant evolution from trees to herbs. *Trees* **5**: 50-64.

Gamalei, Y., van Bel, A. J. E., Pakhomova, M. V. and Sjutkina, A. V. (1994) Effects of temperature on the conformation of the endoplasmic reticulum and on starch accumulation in leaves with the symplastic minor-vein configuration. *Planta* **194**: 443-453.

Giaquinta, R. T. (1983) Phloem loading of sucrose. *Annu Rev Plant Physiol* **34**: 347-387.

Goggin, F. L., Medville, R. and Turgeon, R. (2001) Phloem loading in the tulip tree. Mechanisms and evolutionary implications. *Plant Physiol* **125**: 891-899.

Girousse, C, Bonnemain, J. L., Delrot, S. and Bournoville, R. (1991) Sugar and amino acid composition of phloem sap of *Medicago sativa*: A comparative study of two collecting methods. *Plant Physiol Biochem* **29**: 41-48.

Gottwald, J. R., Krysan, P. J., Young, J. C., Evert, R. F. and Sussman, M. R. (2000) Genetic evidence for the *in planta* role of phloem-specific plasma membrane sucrose transporters. *PNAS* **97**: 13979-13984.

Grusak, M. A., Beebe, D. U. and Turgeon, R. (1996) Phloem loading. In *Photoassimilate Distribution in Plants and Crops. Source-Sink Relationships* (E. Zamski, and A. A. Schaffer, eds.) pp. 209-227. Marcel Dekker, New York.

Gunning, B. E. S. and Pate, J. S. (1969). Transfer cells, plant cells with wall ingrowths, specialized in relation to short distance transport of solutes: Their occurrence, structure, and development. *Protoplasma* **68**: 107-133.

Haritatos, E., Ayre, B. G. and Turgeon, R. (2000a). Identification of phloem involved in assimilate loading in leaves by the activity of the galactinol synthase promoter. *Plant Physiol* **123**: 929-937.

Haritatos, E., Keller, F. and Turgeon, R. (1996) Raffinose oligosaccharide concentrations measured in individual cell and tissue types in *Cucumis melo* L. leaves: Implications for phloem loading. *Planta* **198**: 614-622.

Haritatos, E., Medville, R. and Turgeon, R. (2000b). Minor vein structure and sugar transport in *Arabidopsis thaliana*. *Planta* **211**: 105-111.

Holthaus, U. and Schmitz, K. (1991). Distribution and immunolocalization of stachyose synthase in *Cucumis melo* L. *Planta* **185**: 479-486.

Kandler, O. (1967) Biosynthesis of poly- and oligosaccharides during photosynthesis in green plants. In *Harvesting the Sun: Photosynthesis in Plant Life* (A. San Pietro, F. A Greer and T. J. Army, eds.) pp. 131-152. Academic Press, New York.

Kempers, R., Van, A. J. K., Achterberg, J. and van Bel, A. J. E. (1996) Exclusion limit of plasmodesmata between sieve element and companion cells in stem phloem of *Vicia faba* is at least 10 kDa. *J Exp Bot* **47**: 1298.

Kempers, R. and van Bel, A. J. E. (1997) Symplasmic connections between sieve element and companion cell in the stem phloem of *Vicia faba* L. have a molecular exclusion limit of at least 10 kDa. *Planta* **201**: 195-201.

Knop, C., Stadler, R., Sauer, R., and Lohaus, G. (2004). AmSUT1, a sucrose transporter in collection and transport phloem of the putative symplastic phloem loader *Alonsoa meridionalis*. *Plant Physiol* **134**: 204-214.

Komor, E. (2000) Source physiology and assimilate transport: The interaction of sucrose metabolism, starch storage and phloem export in source leaves and the effects on sugar status in the phloem. *Aust J Plant Physiol* **27**: 497-505.

Kühn, C., Franceschi, V. R., Schulz, A., Lemoine, R. and Frommer, W. B. (1997) Macromolecular trafficking indicated by localization and turnover of sucrose transporters in enucleate sieve elements. *Science* **275**: 1298-1300.

Kühn, C., Quick, W. P., Schulz, A., Riesmeier, J. W., Sonnewald, U. and Frommer, W. B. (1996) Companion cell-specific inhibition of the potato sucrose transporter SUT1. *Plant Cell Environ* **19**: 1115-1123.

Lansing, A. J. and Franceschi, V. R. (2000) The paraveinal mesophyll: A specialized path for intermediary transfer of assimilates in legume leaves. *Aust J Plant Physiol* **27**: 757-767.

Leegood, R. C. (1996) Primary photosynthate production: Physiology and metabolism. In *Photoassimilate Distribution in Plants and Crops. Source-Sink Relationships* (E. Zamski and A. A.Schaffer, eds.) pp. 21-41. Marcel Dekker, New York.

Lalonde, S., Tegeder, M., Throne-Holst, M., Frommer, W. R. and Patrick, J. W. (2003) Phloem loading and unloading of sugars and amino acids. *Plant Cell Environ* **26**: 37-56.

Lohaus, G., Pennewiss, K., Sattelmacher, B, Hussmann, M. and Muehling, K. H. (2001) Is the infiltration-centrifugation technique appropriate for the isolation of apoplastic fluid? A critical evaluation with different plant species. *Physiol Plant* **111**: 457-465.

Lohaus, G., Winter, H., Riens, B. and Heldt, H. W. (1995) Further studies of the phloem loading process in leaves of barley and spinach. The comparison of metabolite concentrations in the apoplastic compartment with those in the cytosolic compartment and in the sieve tubes. *Botanica Acta* **108**: 270-275.

Loomis, W. E. (1955). Resistance of plants to herbicides. In *Origin of Resistance to Toxic Agents* (M. G. Sevag, R. D. Reid and O. E. Reynolds, eds.) pp. 99-121. Academic Press, New York.

Madore, M. A., Oross, J. W. and Lucas, W. J. (1986) Symplastic transport in *Ipomoea tricolor* leaves. Demonstration of functional symplastic connections from the mesophyll to minor veins by a novel dye-tracer method. *Plant Physiol* **82**: 432-442.

McCauley, M. M. and Evert, R. F. (1989) Minor veins of the potato *Solanum tuberosum* L.: Leaf ultrastructure and plasmodesmatal frequency. *Botanical Gazette* **150**: 351-368.

Münch, E. (1930). *Die stoffbewegungen in der pflanze.* Gustav Fischer, Jena, Germany.

Noiraud, N., Maurousset, L. and Lemoine, R. (2001a). Identification of a mannitol transporter, AgMaT1, in celery phloem. *Plant Cell* **13**: 695-705.

Noiraud, N., Maurousset, L. and Lemoine, R. (2001b) Transport of polyols in higher plants. *Plant Physiol Biochem* **39**: 717-728.

Oparka, K. J., Prior, D. A. M. and Wright, K. M. (1995) Symplastic communication between primary and developing lateral roots of *Arabidopsis thaliana. J Exp Bot* **46**: 187-197.

Oparka, K. J. and van Bel, A. J. E. (1992) Pathways of phloem loading and unloading: A plea for uniform terminology. In *Carbon Partitioning Between Organisms* (C. J. Pollock, J. F. Farrar and A. J. Gordon, eds.) pp. 249-254. BIOS Scientific Publishers, Oxford.

Orlich, G., Hofbrueckl, M. and Schulz, A. (1998) A symplasmic flow of sucrose contributes to phloem loading in *Ricinus* cotyledons. *Planta* **206**: 108-116.

Pickard, W. F. (2003). The role of cytoplasmic streaming in symplastic transport. *Plant Cell Environ* **26**: 1-15.

Ramsperger-Gleixner, M., Geiger, D., Hedrich, R. and Sauer, N. (2004) Differential expression of sucrose transporter and polyol transporter genes during maturation of common plantain companion cells. *Plant Physiol* **134**: 147-160.

Reinders, A., Schulze, W., Kühn, C., Barker, L., Schulz, A., Ward, J. M. and Frommer, W. B. (2002) Protein-protein interactions between sucrose transporters of different affinities colocalized in the same enucleate sieve element. *Plant Cell* **14**: 1567-1577.

Russin, W. A. and Evert, R. F. (1985) Studies on the leaf of *Populus deltoides* (Salicaceae). Quantitative aspects and solute concentrations of the sieve-tube members. *Am J Bot* **72**: 487-500.

Russin, W. A., Evert, R. F., Vanderveer, P. J., Sharkey, T. D. and Briggs, S. P. (1996) Modification of a specific class of plasmodesmata and loss of sucrose export ability in the sucrose export defective maize mutant. *Plant Cell* **8**: 645-658.

Sovonick, S. A., Geiger, D. R. and Fellows, R. J. (1974) Evidence for active phloem loading in the minor veins of sugar beet. *Plant Physiol* **54**: 886-891.

Sprenger, N. and Keller, F. (2000) Allocation of raffinose family oligosaccharides to transport and storage pools in *Ajuga reptans*: The roles of two distinct galactinol synthases. *Plant J* **21**: 249-258.

Stadler, R., Brandner, J., Schulz, A., Gahrtz, M. and Sauer, N. (1995) Phloem loading by the PmSUC2 sucrose carrier from *Plantago major* occurs into companion cells. *Plant Cell* **7**: 1545-1554.

Stadler, R. and Sauer, N. (1996) The *Arabidopsis thaliana* AtSUC2 gene is specifically expressed in companion cells. *Botan Acta* **109**: 299-306.

Thompson, M. V. and Holbrook, N. M. (2003) Scaling phloem transport: Water potential equilibrium and osmoregulatory flow. *Plant Cell Environ* **26**: 1561-1577.

Thorpe, M. R. and Minchin, P. E. H. (1996) Mechanisms of long- and short-distance transport from sources to sinks. In *Photoassimilate Distribution in Plants and Crops. Source-Sink Relationships* (E. Zamski and A. A. Schaffer, eds.) pp. 261-282. Marcel Dekker, New York.

Truernit, E. and Sauer, N. (1995) The promoter of the *Arabidopsis thaliana SUC2* sucrose-H$^+$ symporter gene directs expression of β-glucuronidase to the phloem: Evidence for phloem loading and unloading by SUC2. *Planta* **196:** 564-570.

Turgeon, R. (1991) Symplastic phloem loading and the sink-source transition in leaves: A model. In *Recent Advances in Phloem Transport and Assimilate Compartmentation* (J.-L. Bonnemain, S. Delrot, J. Dainty and W. J. Lucas, eds.) pp. 18-22. Ouest Editions, Nantes, France.

Turgeon, R. (1996) Phloem loading and plasmodesmata. *Trends Plant Sci* **1:** 418-423.

Turgeon, R. (2000) Plasmodesmata and solute exchange in the phloem. *Aust J Plant Physiol* **27:** 521-529.

Turgeon, R., Beebe, D. U. and Gowan, E. (1993) The intermediary cell: Minor-vein anatomy and raffinose oligosaccharide synthesis in the Scrophulariaceae. *Planta* **191:** 446-456.

Turgeon, R. and Gowan, E. (1990) Phloem loading in *Coleus blumei* in the absence of carrier-mediated uptake of export sugar from the apoplast. *Plant Physiol* **94:** 1244-1249.

Turgeon, R. and Hepler, P. K. (1989) Symplastic continuity between mesophyll and companion cells in minor veins of mature *Cucurbita pepo* L. leaves. *Planta* **179:** 24-31.

Turgeon, R. and Medville, R. (1998) The absence of phloem loading in willow leaves. *PNAS* **95:** 12055-12060.

Turgeon, R., Webb, J. A. and Evert, R. F. (1975) Ultrastructure of minor veins in *Cucurbita pepo* leaves. *Protoplasma* **83:** 217-231.

van Bel, A. J. E., Gamalei, Y. V., Ammerlaan, A. and Bik, L. P. M. (1992) Dissimilar phloem loading in leaves with symplasmic or apoplasmic minor-vein configurations. *Planta* **186:** 518-525.

van Bel, A. J. E. and Kempers, R. (1991) Symplastic isolation of the sieve element-companion cell complex in the phloem of *Ricinus communis* and *Salix alba* stems. *Planta* **183:** 69-76.

van Bel, A. J. E. and Oparka, K. J. (1995) On the validity of plasmodesmograms. *Botan Acta* **108:** 174-182.

Vaughn, M. W., Harrington, G. N. and Bush, D. R. (2002) Sucrose-mediated transcriptional regulation of sucrose symporter activity in the phloem. *PNAS* **99:** 10876-10880.

Volk, G. M., Turgeon, R. and Beebe, D. U. (1996) Secondary plasmodesmata formation in the minor-vein phloem of *Cucumis melo* L. and *Cucurbita pepo* L. *Planta* **199:** 425-432.

Von Schaewen, A., Stitt, M., Schmidt, R., Sonnewald, U. and Willmitzer, L. (1990) Expression of a yeast-derived invertase in the cell wall of tobacco and *Arabidopsis* plants leads to accumulation of carbohydrate and inhibition of photosynthesis and strongly influences growth and phenotype of transgenic tobacco plants. *EMBO J* **9:** 3033-3044.

Wang, N. and Nobel, P. S. (1998) Phloem transport of fructans in the crassulacean acid metabolism species *Agave deserti. Plant Physiol* **116:** 709-714.

Weise, A., Barker, L., Kühn, C., Lalonde, S., Buschmann, H., Frommer, W. B. and Ward, J. M. (2000) A new subfamily of sucrose transporters, SUT4, with low affinity/high capacity localized in enucleate sieve elements of plants. *Plant Cell* **12:** 1345-1355.

Wimmers, L. E. and Turgeon, R. (1991) Transfer cells and solute uptake in minor veins of *Pisum sativum* leaves. *Planta* **186:** 2-12.

# 4

## Stomatal Control and Water Transport in the Xylem

*Peter Franks and Timothy J. Brodribb*

The movement of plants from aquatic to terrestrial habitats in the Ordovician placed new and conflicting demands on the structure and function of photosynthetic organs. Protection against desiccation led to the formation of an epidermis with low permeability to water, while the requirement to facilitate entry of gaseous $CO_2$ close to photosynthetic tissue favored an increase in epidermal porosity. This dilemma was solved early during the evolution of land plants with the advent of stomata, which, through their ability to regulate epidermal porosity, provided a means by which to optimize the tradeoff between carbon gain and water loss. However, without a reliable water supply to the sites of photosynthesis, stomata remain closed and cells starve.

Since their evolution, the stomata and xylem have been inextricably connected in the transpiration pathway. For a given environment and leaf photosynthetic biochemistry, the rates of carbon uptake and transpirational water loss are controlled by stomatal conductance. Thus, for the same environment, higher photosynthetic rates result in higher transpiration rates, and xylem investment represents a major part of the cost of maintaining a transpirational flux while keeping the photosynthetic tissue hydrated. This chapter examines coordination between xylem and stomata, focusing on how the evolution and function of these water-conducting tissues in plants has led to a linkage between living and ostensibly nonliving tissue.

## Origins of the Association Between Stomata and Xylem

### Evolution of Stomatal Function

Stomata, or at least stoma-like structures, are evident in living and fossilized representatives of some of the earliest known forms of land plants. These include the sporangia of some hornworts and mosses, as well as in fossils of the earliest known vascular plants, such as *Cooksonia* and *Zosterophyllum* from around 400 Myr ago (Edwards, 1993). Therefore, it seems that stomata have played a role since the very earliest attempts at land colonization by plants.

One of the earliest land plants, the thallose liverworts of the order Marchantiales, has anatomical features that resemble, in gross form, the leaves of higher plants. Proctor (1981) described them as the nearest analog to flowering-plant leaves among the bryophytes. Their special features include an epidermis covered in a waxy cuticle and a convoluted and highly porous internal layer of chlorenchyma cells. Most significant to the discussion here, however, are the pores in the epidermis through which $CO_2$ uptake and water loss occur (Fig 4.1A). These pores are formed by a ring of cells, rather than two kidney-shaped guard cells, and do not open and close with the dynamic range of movement exhibited by "true" stomata. For these reasons the pore-bearing Marchantialean liverworts are usually described together with all other liverworts as lacking stomata. However, given that these pores allow photosynthetic gas exchange between inner thallus and atmosphere, across what is an otherwise relatively impermeable cuticle, they are stomata in the very broadest sense. It is tempting to think that early, prearchetypal stomata could have resembled pores like these.

A major advantage for plants that photosynthesize in a gaseous rather than aqueous environment is that $CO_2$ diffusivity is about 10,000 times higher in air than in water. A problem for many of the earliest forms of land plants is that in a sense they never really left their aquatic environment. Those that retain an ectohydric structure (water conduction via surface capillary structures) are burdened with a relatively thick water film between atmosphere and sites of photosynthesis, creating a low diffusive conductance to photosynthetic gas exchange. Furthermore, these plants cannot regulate the rate of evaporative water loss, which is entirely dependent on atmospheric conditions and laminar boundary layer conductance. However, Marchantialean liverworts represent the beginning of a radically new water management paradigm for plants. Their development of a water-impermeable cuticle on the surface of the thallus necessitated two important changes: (1) pores through which photosynthetic gas exchange could take place, and (2) endohydry (water conduction via internal pathways). With this, humid air chambers can be maintained beneath the cuticle in which the absence of excessive extracellular water increases diffusive conductance to sites of photosynthesis. Green and Snelgar (1982) demonstrated how this structure improved photosynthetic productivity in thalli of pore-bearing versus non-pore-bearing liverworts. In principle, this scheme of endohydric water balance management has changed little over the course of terrestrial plant diversification: More elaborate vascular systems have placed chlorenchyma at ever greater distances from source water, and

A    B

C    D

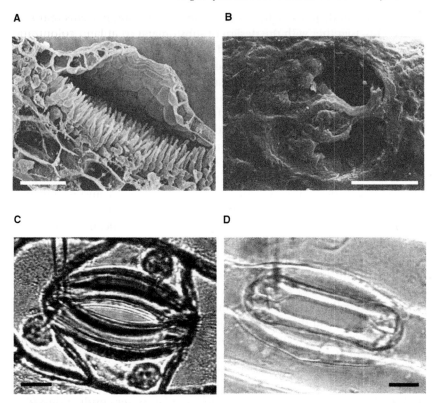

**Figure 4.1** Stomata have remained relatively unchanged in basic form (but possibly not physiologically) for several hundred million years. (A) Cross-section of the thallus of *Conocephalum*, a pore-bearing liverwort. The pores are not true stomata but serve a similar purpose (SEM courtesy of M.C.F. Proctor; scale bar = 100 μm). (B) SEM of two kidney-shaped guard cells forming a stoma in fossilized *Cooksonia pertoni* from around 400 Myr ago. This extinct leafless plant stood only a few centimeters tall and is regarded as an archetypal vascular plant (Edwards *et al.*, 1992; SEM courtesy of L. Axe; scale bar = 20 μm.). (C) Surface view of typical angiosperm stoma (*Tradescantia virginiana*) showing the extent of lateral displacement of inflated guard cells. Scale bar = 10 μm. (D) Surface view of typical graminoid stoma (*Triticum aestivum*), showing elongated guard cells. Scale bar = 10 μm.

better stomatal regulation has minimized water deficits in drier atmospheres, but cuticle, stomata, and endohydry are the key components underlying the success of plants on land.

Although the pores in liverwort thalli show some ability to reduce their aperture in response to unfavorable moisture status (Walker and Pennington, 1939; Proctor, 1981), their ability to restrict evaporative water loss is limited. It was the advent of the kidney-shaped (reniform) guard cell pair (Fig. 4.1B-D) that provided the mechanical means to open and close the stomatal pore and, potentially, to regulate gas exchange across the epidermis. This, together with the design of substomatal chambers for optimal $CO_2$ diffusion into the leaf (Pickard, 1982), is what sustains the high-photosynthetic-capacity homoiohydric tracheophytes that dominate the landscape today. However, the appearance of reniform guard cells is likely to have been only the beginning of a sequence of adaptations in the stomatal apparatus that transformed the gas exchange characteristics of plants during their colonization of land. The easiest of these to visualize are the mechanical characteristics of stomatal movement, which show a progression from very limited aperture range in moss sporophytes and pteridophytes, to a much larger range of movement in angiosperms (Ziegler, 1987). The selection pressures that drove this progression are not well understood, but they may be linked to accompanying evolutionary developments in plant water-conducting systems.

For the same stomatal pore depth, leaf conductance to water vapor is a function of both the mean stomatal pore width and the stomatal density (stomata per unit leaf area). Thus plants with similarly high leaf diffusive conductances may differ considerably in the dynamic range of movement of their guard cells (i.e., plants with limited range of guard cell movement may achieve high leaf diffusive conductance by having high stomatal density). Modifications to mechanical properties of the guard cells that enable a greater range of movement without increased energetic cost should yield a selective advantage for improved flexibility in the regulation of leaf gas exchange. Such flexibility would allow the plant to maintain desirable (safe) hydrodynamic conditions in the vascular system over a broader range of environmental conditions.

## Evolution of Xylem

Although xylem evolution has been covered comprehensively from a predominantly anatomical perspective by Bailey and Tupper (1918), a clearer understanding of these trends from the functional perspective is now emerging with more extensive physiological studies. The earliest endohydric plant water-conducting systems are likely to have incorporated hydroid-like cells similar to the hydroids that are polyphyletic among bryophytes (Ligrone et al., 2002). The important characteristic of these cells and all subsequent water-conducting cells is the process of cytoplasmic lysis, which greatly increases the volume of the apoplast, producing a hydraulic pathway with a specific conductivity many orders of magnitude higher than that of the symplasm (Raven, 1977). One consequence of drawing the transpiration stream through these nonliving tubes is compression of the walls of conducting cells (Carlquist, 1975) by tension in the water column. Hydroids lack lignin (a phenolic polymer resistant to compression) and as such are incapable of supplying water under significant tension, a serious limitation to the water-carrying capacity of hydroid-derived conducting tissue. The deposition of lignin in thickened bands on the walls of conducting cells was therefore an important innovation in xylem evolution, and was well established by the lower Devonian (Kenrick and Crane, 1991; Boyce *et al.*, 2002).

Examination of the preserved trachied walls from Devonian tracheophytes has shown that interconduit pits in these water-conducting cells were generally extremely small, in the range 50 to 200 nm and apparently simple (Kenrick and Crane, 1991; Edwards, 2003). Such perforations must have imposed a great resistance to water flow even without considering the added resistance of the primary wall (see Chapter 16). On the other hand, these pits are likely to have yielded high resistance to cavitation considering the pores themselves would resist several MPa of pressure even without the further support of the primary wall. The predominance of struts and projections into the lumena of these cells presumably resisted collapse under xylem tension. From these early tracheids, evolution has tended to favor increased efficiency of axial water flow, leading to progressively greater digestion of walls in the axial direction and membrane lysis in the end walls of conduits (Bailey and Tupper, 1918; Zimmermann, 1983; Carlquist and Schneider, 2002). Angiosperms possess the most efficient axial water conducting tissues, composed of vessels built from sequential tracheary elements in which both the end walls and cell membranes can be almost completely hydrolyzed, thus forming long tubes with pit membranes only between vessels. Vessels are not only longer, but can have far greater diameters than tracheids, and, because the hydraulic conductivity of tubes scales to the fourth power of the radius, can supply much greater flow rates than tracheid-based vascular systems at equivalent pressure gradients. Vessels thus require a smaller investment than tracheids for equivalent transport capacity. Evidence of this is seen in the generally higher leaf area: sapwood area (Huber value) found in angiosperm compared with conifer trees (Brodribb and Feild, 2000). Despite the apparent advantages of producing a low-resistance vascular network (Bond, 1989), few studies have illustrated these benefits in the context of competitive interactions between plants (Becker *et al.*, 1999).

Retention of the primary wall between neighboring conduits remains essential for the containment of embolisms in all plant vascular systems. The persistence of this high-resistance link (although in angiosperm wood, these high resistance connections are reduced; see Chapter 16) has driven pit evolution in a direction that maximizes hydraulic flow while protecting against pit membrane stretching or rupture. The most common example of this flow versus protection tradeoff is the bordered pit, which appears first in fossil lycophytes, and has since become widespread in xylem from ferns to the metaxylem of gymnosperms and angiosperms. The flared aperture of bordered pits yields a large area for water flow across the primary wall while supporting the membrane should neighboring conduits become embolized and the pit aspirated. Conifers have perfected this structure with the torus and margo assembly, in which hydrolysis of most of the pit membrane, except a central plug (the torus), enables maximum flow while providing protection against the spread of air embolisms. Another common example of pit evolution is the presence of vestures or papillae in the bordered pits of many angiosperms (Janzen, 1998). These projections into the pit chamber appear to support the pit membrane against distortion on aspiration, a process believed to result in decreased air-seeding vulnerability (Zweypfenning, 1978; Choat *et al.*, 2004). As such, vestured pits may circumvent the apparent tradeoff between pit-membrane permeability and vulnerability.

## Biophysical Properties of Stomata and Xylem

### Stomatal Mechanism

Stomata are essentially turgor-operated valves embedded in the epidermis for the purpose of regulating leaf (or plant) gas exchange. It is not known exactly how they do this, but much has been learned about several of the key biophysical and biochemical processes involved (Cowan, 1977; Farquhar *et al.*, 1980b; Assmann and Shimazaki, 1999; Schroeder *et al.*, 2001; Zeiger *et al.*, 2002). Using a complex system of sensors, stomatal guard cells transduce a suite of environmental and internal variables into the mechanical forces that bend them apart just far enough to allow gas exchange to proceed at, seemingly, a precisely controlled rate. There are several external variables or signals, the major ones being incident light, ambient $CO_2$ concentration, and humidity, that can influence guard cell turgor. Guard cells also respond to several compounds that occur naturally in the plant, including phytohormones like ABA (Raschke, 1975; Davies *et al.*, 2002). For leaves exposed to any combination of these variables, a complex interaction involving reception of the respective signals and their transduction determines guard cell solute content. This, in combination with the physical constraints imposed by guard cell wall elasticity, establishes in the steady state an equilibrium between guard cell osmotic pressure ($\Pi_g$), guard cell turgor pressure ($P_g$), and guard cell water potential ($\Psi_g$), such that $\Psi_g = P_g - \Pi_g$. For any given $\Psi_g$, changes in guard cell osmotic content, and hence $\Pi_g$, will lead to an equal change in $P_g$. Alternatively, for any given $\Pi_g$, a change in $\Psi_g$ will promote an equal change in $P_g$. In the regulation of stomatal aperture, $\Pi_g$ is *actively* adjusted using metabolic energy.

Transpirational fluxes and hydraulic conductances establish $\Psi_g$ somewhere below the baseline set by soil water potential, and $P_g$ follows passively. In many species stomatal aperture (and ultimately diffusive conductance) is determined by the balance of $P_g$ and epidermal turgor pressure, $P_e$, rather than $P_g$ alone. This is because the mechanical nature of the stomatal apparatus is such that an increase in $P_e$ will tend to push guard cells together, thereby counteracting stomatal opening. This closing effect by $P_e$ is referred to as a "mechanical advantage" because an equal increase in $P_g$ and $P_e$ will often cause a net reduction in stomatal aperture. Due to an incomplete understanding of the many physiological elements of stomatal function, and how they are integrated, mechanistic stomatal models are not well advanced. Several empirical models provide useful interpolations of stomatal behavior under a wide range of environmental conditions (Farquhar and Wong, 1984; Ball *et al.*, 1987; Tardieu and Davies, 1993; Leuning, 1995). However, further advances in the mechanistic modeling of stomatal function are essential to our understanding of plant function, as these models provide a framework for explaining differences between individuals, as well as the theoretical foundation on which predictive models should be based.

It is widely observed that in response to increasing transpiration demand (leaf-to-air water vapor pressure difference, $D$) stomata tend to close, sometimes almost completely, with the result that transpiration rate at higher $D$ is less than what it would be without stomatal closure (Darwin, 1898; Lange *et al.*, 1971; Hall *et al.*, 1976; Cowan, 1994). The benefit of this is that the transpiration-induced drop in leaf tissue water potential is less than what it would be without this response. Therefore, it is widely interpreted as a mechanism not only for avoiding dangerously low tissue water potentials, but for regulating leaf gas exchange in a way that tends toward optimizing carbon gain with respect to water loss (Cowan, 1977; Cowan and Farquhar, 1977). It has been difficult to identify a control mechanism that is physically capable of producing the wide range of response characteristics observed to date. By measuring stomatal conductance at different ambient water vapor pressures in air and helox (water diffuses 2.3 times more readily in helox), however, Mott and Parkhurst (1991) proved that it is a change in transpiration rate, not ambient vapor pressure per se, to which stomata respond. Therefore, any such control mechanism must incorporate the distribution of water potential throughout the leaf and be sensitive to changes that affect both liquid and vapor phases of the transpiration stream (e.g., $D$ and hydraulic conductances along the flow path from soil to roots to leaves). Although some species show sufficient stomatal sensitivity to $D$ to control transpiration rate at close to a constant level, the degree of transpiration control is highly varied across species. Figure 4.2 illustrates several typical response patterns, varying mainly by greater or lesser relative increases in

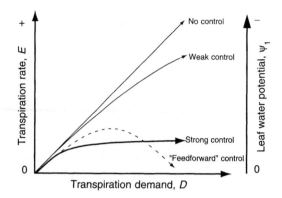

**Figure 4.2** Generalized relationship between transpiration demand ($D$), leaf transpiration rate, and leaf water potential when stomata exert control over transpiration rate. Most plants show a response that is somewhere between weak control and strong control, and sometimes the response pattern marked "feedforward" is observed. See text for a discussion of the possible mechanistic basis of these responses.

transpiration rate with increasing $D$. The response marked "feedforward" control, characterized by a region in which transpiration rate actually decreases with increasing $D$, was originally identified as such because it typified a system that could sense and respond to transpiration potential independently of bulk leaf water status (Cowan, 1977; Farquhar, 1978). A negative feedback system, in which increasing transpiration rate decreases guard cell turgor, could at best keep transpiration rate (and therefore bulk leaf water potential) constant with increasing $D$. It should be noted, however, that feedback and feedforward control of transpiration rate, as originally proposed, are not mutually exclusive. Feedback and feedforward control could be operating together in all of the control responses depicted in Fig. 4.2, but feedback alone could not produce the response whereby transpiration rate decreases with increasing $D$. Alternatively, if the gain of the feedback loop changes with transpiration rate (e.g., $\Pi_g$ is not held constant or hydraulic conductance changes due to xylem embolisms), then it will be possible for a negative feedback system to mimic feedforward. Though a feedforward signal has yet to be identified, Farquhar (1978) showed that it need only comprise a portion of the transpiration rate that is unaffected by stomatal aperture, such as transpiration from the surface of guard cells external to the stomatal pore.

Whether feedback or feedforward, or both, it is now becoming apparent that the mechanism of control of transpiration rate is more complex than the passive hydromechanical models originally proposed. In most cases involving a substantial stomatal closure in response to increasing $D$, the change to a new steady state stomatal conductance can take somewhere between 20 and 180 minutes, be dependent on time of day, and in many cases be only partly reversible in the short term (Grantz and Zeiger, 1986; Kappen and Haeger, 1991; Franks *et al.*, 1997). Hydraulic equilibrium of guard cells in epidermal strips exposed to osmotic shock is achieved in the time frame of a minute or two (Fischer, 1973), and so it is highly likely that in addition to passive hydromechanical processes, regulation of transpiration rate under changing $D$ involves active, metabolically driven adjustments to guard cell osmotic content, or possibly even changes in hydraulic conductances (Buckley *et al.*, 2003; Franks, 2004).

### Xylem Hydrodynamics

The conductivity of leaf, stem, or root xylem is not constant but rather a dynamic function of leaf water potential due to the effects of cavitation in the water column under tension. In the current year's xylem, the relationship between water potential and xylem embolism is almost a step function, with the transition between the nonembolized and highly embolized condition occurring over a narrow range of water potential (Melcher *et al.*, 2003). Hence there is considerable potential for positive feedback between decreasing water potential and increasing embolism, causing "runaway embolism" (Sperry and Pockman, 1993), a situation that would quickly render xylem nonfunctional if left unchecked. However, avoiding the risk of cavitation by producing cavitation-resistant xylem is a costly investment for plants. Xylem resistance to cavitation is determined by the wall and membrane properties of the conduits and varies widely among species. Increased cavitation resistance in species comes at the cost of increased carbon investment and reduced hydraulic conductivity (Hacke *et al.*, 2001). These costs are sufficiently high that despite the danger of runaway embolism leading to hydraulic isolation of leaves from roots, plants tend to operate close to the threshold for xylem cavitation.

The vulnerability of xylem appears to be a pivotal component in the coordination between xylem and stomatal dynamics. Our current state of knowledge is based on quantitative comparisons between declining stomatal and xylem conductivities in response to decreasing water potential. These comparisons have been made using "vulnerability curves," which indicate the percent decrease in xylem conductivity (or increase in acoustic emissions) due to embolism as decreasing water potentials are imposed (Sperry, 1986; Tyree and Sperry, 1988, 1989). Detailed examinations of xylem vulnerability in a selection of temperate and tropical trees have revealed correlation between the water potential found to induce xylem cavitation in stems (Sperry, 1986; Salleo *et al.*, 2000; Nardini *et al.*, 2001; Cochard *et al.*, 2002; Brodribb *et al.*, 2003) or leaves (Salleo *et al.*, 2001; Nardini *et al.*, 2001; Brodribb and Holbrook, 2003b), and a reduction in stomatal conductance. Stomatal closure has been found to correlate better with xylem vulnerability than any other physiological leaf parameter investigated, including photosynthetic response to water stress and even cell turgor relations (Brodribb *et al.*, 2003). The general consensus from these studies is that once leaf water potential falls to the threshold for xylem cavitation, stomatal closure is triggered to avoid significant embolism from occurring.

These observations have led to the formulation of a general model in which plant water use can be predicted by incorporating soil and plant hydraulics with xylem vulnerability, to calculate optimal water use (Sperry *et al.*, 1998, 2002). Nevertheless the mechanics of this association remain vague, with uncertainty in the location and proportion of xylem cavitation required to trigger a guard cell response. Variation in the relationship between xylem and stomatal conductances has also been observed, with some examples of complete stomatal closure and even drought-induced leaf shedding while stem xylem conductivity remained high (Brodribb and Holbrook, 2003a). These examples come from drought deciduous tropical forest, where the conservation of stem water appears to be a strategy for surviving extended periods without rain. In contrast to stems, the vulnerability of leaf veins to cavitation in these species remained tightly correlated with stomatal closure, suggesting that leaves rather than stems or roots represent the nexus between xylem and stomatal fluxes.

### A Changing Environment for Photosynthesis: Evolution of Xylem-Stomatal Coordination

Coordination between stomatal and xylem function was probably of little consequence for early land plants as they lived close to the ground where, due to the thick boundary layer, transpiration rates would have been low and largely determined by temperature. However, the evolutionary increase in plant size (particularly height) led to an increase in both the variability of the evaporative environment of photosynthetic organs and the cost of replacing transpired water (due to increased transport distances). This cost became compounded by a massive decline in atmospheric $CO_2$ in the late Devonian (Berner, 1998), adding pressure on plants to reduce metabolic expenditure on xylem to a minimum. This combination of factors may have led to the current situation in modern tracheophytes where hydraulic supply is minimized to such an extent that stomatal regulation of water loss is required to prevent xylem cavitation (Sperry *et al.*, 2002). The resultant correlation of xylem water supply capacity and stomatal water loss can be observed at scales from the daily regulation of gas exchange in response to evaporative demand, to the architecture of different tree species (Meinzer, 2002).

In extant tracheophytes, we see that at the species level, maximum stomatal conductance is coordinated with the xylem hydraulic capacity in individual branches (Sperry and Pockman, 1993) and whole trees (Whitehead *et al.*, 1984; Meinzer *et al.*, 1995; Nardini and Salleo, 2000). Plant communities growing under conditions of extremely stable water supply have demonstrated good correlation between photosynthetic and hydraulic capacity even among species selected from a large phylogenetic range (Brodribb and Feild, 2000; Fig. 4.3). Linkage at this scale appears largely due to genetic coordination between xylem and leaf synthesis and is evidenced by correlation between anatomical features of the stomata and xylem, such as pore length and conduit diameter among species (Aasamaa *et al.*, 2001). However, modification of these and other traits during leaf expansion may allow for considerable phenotypic tuning of maximum xylem and stomatal conductances.

Phenotypic tuning of the amount of leaf area per unit supporting xylem can be localized with branches or act at the whole plant level. At the branch scale it has been found that the area of expanding leaves is responsive to the water potential and evaporative demand during leaf expansion

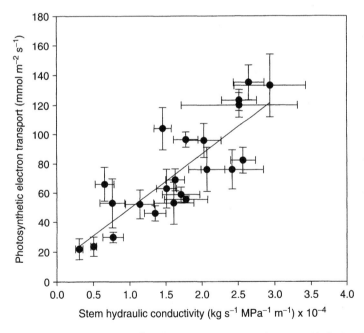

**Figure 4.3** Linear correlation ($r^2 = 0.70$) between mean leaf-area-specific hydraulic conductivity of stem xylem ($K_L$) and photosynthetic electron transport rate in 24 tree species from primary temperate to subtropical rainforest (adapted from Brodribb and Field, 2000). Conifers and vesselless angiosperms are included in the species sample.

(Zwieniecki *et al.*, 2004). The synthesis/release of ABA in response to water stress is well known, and the effects of ABA in reducing leaf expansion (Gollan *et al.*, 1992; Kramer and Boyer, 1995) and stomatal pore size (Franks and Farquhar, 2001) enable modification of the ratio of stomatal to hydraulic conductance. Franks and Farquhar, (2001) also found that in *Tradescantia virginiana*, elevated ABA during leaf development acted exclusively and permanently on stomatal geometry and mechanics to shift the photosynthetic operating point in the direction of improved water-use efficiency.

### Leaves: Hydrodynamics and Interfacing Xylem with Stomata

The close proximity of xylem and stomata in the leaf allows rapid communication of physical and chemical signals between these tissues. Moreover, the xylem in leaves appears to be highly resistive to hydraulic flow (Tyree *et al.*, 1999; Nardini, 2001; Sack *et al.*, 2002; Brodribb and Holbrook 2003a,b) and vulnerable to cavitation (Nardini *et al.*, 2001; Cochard *et al.*, 2002; Brodribb and Holbrook, 2003b), thus making this terminal part of the xylem transport pathway sensitive to changes in evaporation from the leaves.

Recent studies of the hydraulic properties of leaves have revealed that between 30% and 80% of whole plant hydraulic resistance of trees occurs in leaves (Becker *et al.*, 1999; Nardini, 2001; Sack *et al.*, 2002) over distances that typically represent less than 1% of the hydraulic path length of the plant xylem. This means that relative to other parts of the xylem pathway, leaf xylem exerts a disproportionately strong influence over leaf water potential and guard cell water potential. The additional drawdown in water potential due to tissue hydraulic resistance between xylem and sites of evaporation has not been extensively studied, but may also be significant when related to hydraulic path length. The potential for leaves to rapidly regulate their own water potential is most evident under nonstressed conditions of photosynthesis where stomatal closure can effectively collapse the water potential gradient in the leaf (Brodribb and Holbrook, 2003a,b). This latter observation is of particular importance when considering the hypothetical interaction of stomata and xylem through cavitation prevention. A relatively low hydraulic conductance in the leaf may indeed be adaptive in the sense that it will amplify the effect of changing evaporation on leaf water potential. This then gives stomata a greater sensitivity to atmospheric change while enabling them to rapidly and profoundly modify leaf (and leaf xylem) water potential by changing pore aperture.

Where it has been measured, the vulnerability of leaf xylem to cavitation by water stress is at least as high as other parts of the plant (Nardini *et al.*, 2001; Cochard *et al.*, 2002), and, when combined with a low intrinsic conductance to water flow, xylem cavitation in transpiring leaves has the potential to cause rapid leaf dehydration. Stomata appear responsive to this threat, closing at leaf water potentials that induce weak cavitation of

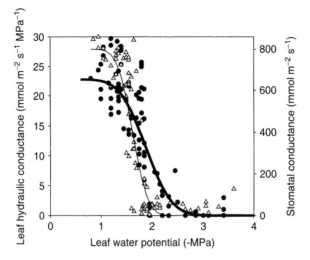

**Figure 4.4** Responses of stomatal conductance (triangles-thin line) and leaf hydraulic conductance (circles-bold line) to leaf water potential in the tropical legume *Glyricidia sepium.* In this species and in others investigated (replotted from data in Brodribb and Holbrook, 2003a), stomatal closure occurs during the initial phase of leaf vein cavitation illustrating the close association between xylem vulnerability and stomatal closure.

leaf xylem, apparently as a means of averting a precipitous decline in leaf hydraulic conductance (Fig. 4.4). These data suggest that the leaf xylem wields a disproportionately large influence over the dynamics of plant hydraulic conductance, and consequently over the coordination between liquid and vapor phase water transport in plants.

## Linking Hydraulics with Gas Exchange

### Why *A*, $g_w$, and Hydraulic Conductance Should Correlate

Photosynthetic capacity and stomatal conductance vary considerably across terrestrial plant communities (Schulze and Hall, 1982). What is more intriguing is that when steady state net $CO_2$ assimilation rate *A* and stomatal diffusive conductance $g_w$ are measured concurrently under similar conditions within a given species, they are positively and often linearly correlated, regardless of whether photosynthetic capacity is altered by age, growth irradiance, or nutrition (Björkman, 1973; Wong *et al.*, 1979, 1985). Field and Mooney (1986) observed a similar correlation across 21 species growing naturally. A corollary of this is that leaf intercellular $CO_2$ concentration $c_i$ tends to be rather conservative, typically between about 65% and 85% of ambient $CO_2$ concentration, when measured under saturating light in the absence of drought. In addition to this relationship, hydraulic conductance is found to correlate with $g_w$ (Meinzer and Grantz, 1990; Meinzer *et al.*, 1995; Saliendra *et al.*, 1995) and $A$ (Sober, 1997; Brodribb and Feild, 2000) within and across species. While a close association between $A$, $g_w$ and hydraulic conductance might seem intuitively to be so, it is not essential that they be linearly correlated. Indeed, under the influence of drought or increasing $D$, the plot of steady-state $g_w$ versus $A$ often shows a tendency for $g_w$ to decrease more rapidly than $A$ (Farquhar *et al.*, 1980a; Hall and Schulze, 1980; Meinzer, 1982). Furthermore, across 13 randomly selected $C_3$ species of varying growth form, Franks and Farquhar (1999) observed a positive correlation between $A$ and $c_i$, indicating a nonlinear relationship in which steady state $g_w$ increased at a greater rate than photosynthetic capacity across this group of species. Franks and Farquhar (1999) showed that this trend would arise if selection for higher photosynthetic capacity were accompanied by selection for higher chloroplast nitrogen-use-efficiency and higher hydraulic conductivity relative to leaf area. The implications for a positive linear or quasilinear relationship between $A$, $g_w$, and hydraulic conductance are that certain internal leaf physiological parameters are being controlled and coordinated. Although the mechanism of this interrelationship is not yet known, the following is a description of some of the functional implications.

***Photosynthetic Operating Point***    When hydraulic flow and transpiration rate are in steady state, stomatal conductance to water vapor, $g_w$, is related to hydraulic conductance and $CO_2$ assimilation rate as follows:

$$g_w = \frac{E}{D/P_a} = \frac{J_L}{D/P_a} = \frac{k_{L(s-1)}(\Psi_s - \Psi_1 - \rho gh)}{D/P_a} = 1.6 g_c = \frac{1.6A}{(c_a - c_i)} \tag{1}$$

where $E$ is transpiration rate (relative to leaf area), $J_L$ is hydraulic flow rate (relative to leaf area), $D$ is the difference in water vapor pressure between leaf interior and outside, $P_a$ is atmospheric pressure, $k_{L(s-1)}$ is the leaf-area-specific hydraulic conductance from soil to leaf, $\psi_s$ is soil water potential, $\Psi_1$ is leaf water potential, $\rho$ is the density of water, $g$ is acceleration due to gravity, $h$ is vertical height from soil to leaf, $g_c$ is stomatal conductance to $CO_2$, $A$ is $CO_2$ assimilation rate, $c_a$ is concentration of $CO_2$ in the air outside the leaf, and $c_i$ is $CO_2$ concentration in the air inside the leaf. The term $(\Psi_s - \Psi_1 - \rho gh)$ represents the hydraulic pressure difference between soil and leaf that is driving $J_L$. Hydraulic conductance is a function of both the hydraulic conductivity of the conducting medium (in this case relative to leaf area), $K_L$, and the hydraulic path length, (e.g., $k_{L(s-1)} = K_L/\Delta L_{s-1}$), where $\Delta L_{s\text{-}l}$ is the path length from soil to leaf. A full treatment of the linkage between stomatal and hydraulic conductances is given in Franks (2004), but the preceding summary provides a quantitative basis for the discussion here.

Under saturating light, ambient $CO_2$ concentration and any given humidity and temperature, leaf gas exchange will achieve a steady state or "operating point." This point is defined by the instantaneous $A$ and $c_i$, and reflects not only an equilibrium between $A$ and $g_c$, but between the many biochemical, biomechanical, and hydraulic feedback loops with which they are associated. The $C_3$ photosynthesis model of Farquhar *et al.*, 1980b and Farquhar and von Caemmerer (1982) defines the relationship between $A$ and $c_i$ as a function of both the capacity for the regeneration of the substrate for the enzyme ribulose bisphosphate ($RuP_2$) carboxylase-oxygenase (Rubisco; dominated by thylakoid electron transport capacity), and Rubisco activity itself. These two components are strongly governed by the relative investment of nitrogen in thylakoid and stromal proteins, respectively. Based on this model, the $A$ versus $c_i$ relationship for two different hypothetical combinations of nitrogen investment are shown in Fig. 4.5A (dark lines). It has been argued (Cowan, 1977; Cowan and Farquhar, 1977) that because there is often a large change in the ratio of the sensitivities $\partial E/\partial g_w$ and $\partial A/\partial g_w$ at the point of transition between Rubisco-limited and Rubisco-substrate-regeneration-limited $CO_2$ assimilation, the solution for optimizing carbon gain with respect to water loss is to operate near this transition point. This operating point is marked on the two $A$ versus $c_i$ curves in Fig. 4.5A as $A_{OP1}$ and $A_{OP2}$. The rate of $CO_2$ assimilation at each of these operating points is sustained by the stomatal conductance to $CO_2$, $g_c$, denoted respectively as $g_{c1}$ and $g_{c2}$ in Fig. 4.5A. Although it is not fully understood why, the $c_i$ that corresponds to these operating points is highly conservative. What this means is that when photosynthetic capacity is altered by a change in the investment in thylakoid and stromal proteins, stomatal conductance must change accordingly to achieve the optimum operating point: With greater nitrogen investment $g_c$ will increase, and with less nitrogen investment it will decrease. This will lead to a positive correlation between $A$ and $g_c$ (or $A$ and $g_w$). However, hydraulic constraints limit the magnitude of $g_c$ and hence photosynthetic capacity.

*Hydraulic Conductance*  Unlike stomatal conductance, which a plant can actively regulate according to environmental conditions, $k_{L(s-l)}$ is relatively static. It can increase with loss of leaf area or decrease with xylem embolisms, but these changes are potentially costly and difficult to reverse in the short term. It is therefore important that $k_{L(s-l)}$ be matched appropriately with leaf evaporative conditions from the outset. So, what is the appropriate $k_{L(s-l)}$? Figure 4.5B illustrates the linear relationship between $J_L$ and $(\psi_s - \psi_l - \rho gh)$ for two different values of $k_{L(s-l)}$ (abbreviated as $k_1$ and

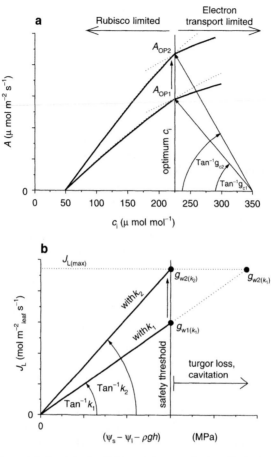

**Figure 4.5** A model showing the linkage between photosynthetic gas exchange and hydraulic fluxes, and an explanation for the observed correlation between $A$, $g_w$ and $k_{L(s-1)}$. In (a), conservation of $c_i$ requires $g_w$ to increase with $A$. In (b), an increase in $g_w$ requires an increase in $k_{L(s-1)}$ to keep $(\Psi_s-\Psi_1-\rho gh)$ below a safety threshold. Note that this model is for conditions of constant $\Psi_s$ and $h$, so that increases in $(\Psi_s-\Psi_1-\rho gh)$ relate entirely to decreasing $\Psi_1$. Further explanation is provided in the text.

$k_2$; dark lines). For the purpose of this exercise, we assume that $g_{w1(k1)}$ is the stomatal conductance that corresponds to $A_{OP1}$ in Fig. 4.5A, and that with this stomatal conductance $E$, and hence $(\Psi_s-\Psi_1-\rho gh)$, is such that nowhere in the leaf or along the stem xylem is water potential more negative than what is deemed to be the safety threshold. In this case, $\Psi_s$ and h are constant, so that changes in $(\Psi_s-\Psi_1-\rho gh)$ relate entirely to changes in $\Psi_1$. The safety threshold may be, for example, the point of turgor loss or the onset of cavitation. With the large range in observed cell osmotic pressures and embolism susceptibility, it is difficult to say with any certainty what this threshold might be. Within an individual plant, $k_{L(s-l)}$ and $g_w$ are tightly coupled: Changing $k_{L(s-l)}$ by notching or inducing xylem embolisms brings about a rapid readjustment of $g_w$ (Sperry *et al.*, 1993; Hubbard *et al.*, 2001). However, if the observed positive correlation between $A$, $g_w$ and $k_{L(s-l)}$ is general, it suggests that $(\psi_s - \psi_l - \rho g h)$ is somewhat conserved. This being the case, any increase in steady state $g_w$ associated with an increase in photosynthetic capacity would have to be accompanied by an increase in $k_{L(s-l)}$. This is illustrated in Fig. 4.5B where maintenance of $(\psi_s - \psi_l - \rho g h)$ on attainment of higher photosynthetic capacity requires a shift from $k_1$ to $k_2$, and operation at $g_{w2(k2)}$ rather than the more dangerous $g_{w2(k1)}$. The provision of higher hydraulic conductance with the move to $k_2$ requires additional resource investment in wood structure. However, with the evolution of tracheary elements that provide higher hydraulic conductivity per unit carbon invested, this cost would have been offset. That those plants with highest photosynthetic capacity tend to be vessel-bearing angiosperms suggests that this hydraulic trait has been selected for in the quest for high photosynthetic capacity.

## Summary

Internal water conduction and stomata evolved approximately at the same time, providing a water supply to photosynthetic organs distant from their water source. Consequently, the hydraulic transport and photosynthetic capacity of tracheophytes remain closely linked. Increasing photosynthetic capacity necessitated a parallel enhancement of the conductivity of xylem tissue during the course of evolution of tracheophytes, although this may have come at the cost of exposing conduits to an increased risk of cavitation. As a result, modern tracheophytes often operate close to the water potential that induces xylem cavitation, so stomatal regulation of minimum leaf water potential is critical to protect the xylem from cavitation-induced dysfunction.

Linkage between stomata and xylem by a simple hydropassive process is insufficient to explain all stomatal responses to changing transpiration demand. Complex interactions between epidermal and guard cells are still poorly understood, but active ion exchange is almost certainly involved when stomata close in response to increasing $D$. Together, xylem and stomata constrain the water potential gradient in the vascular system within a safe range.

# References

Aasamaa, K., Sõber, A. and Rahi, M. (2001) Leaf anatomical characteristics associated with shoot hydraulic conductance, stomatal conductance and stomatal sensitivity to changes of leaf water status in temperate deciduous trees. *Aust J Plant Physiol* **28:** 765-774.

Assmann, S. M. and Shimazaki, K. (1999) The multisensory guard cell. Stomatal responses to blue light and abscisic acid. *Plant Physiol* **119:** 809-815.

Bailey, I. W. and Tupper, W. N. (1918) Size variation in tracheary cells: I. A comparison between the secondary xylems of vascular cryptograms, gymnosperms and angiosperms. *Am Acad Arts Sci Proc* **54:** 149-204.

Ball, J. T., Woodrow, I. E. and Berry, J. A. (1987) A model predicting stomatal conductance and its contribution to the control of photosynthesis under different environmental conditions. *Prog Photosynth Res* **4:** 221-224.

Becker, P., Tyree, M. T. and Tsuda, M. (1999) Hydraulic conductances of angiosperms versus conifers: Similar transport sufficiency at the whole-plant level. *Tree Physiol* **19:** 445-452.

Berner, R. A. (1998) The carbon cycle and $CO_2$ over phanerozoic time: The role of land plants. *Philosophical Transactions of the Royal Society of London - Series B: Biol Sci* **353:** 75-81.

Björkman, O. (1973) Comparative studies in photosynthesis in higher plants. In *Current Topics on Photobiology and Photochemistry* (A Giese, ed.) Vol. VIII, pp. 1-63. Academic Press, New York.

Bond, W. J. (1989) The tortoise and the hare: Ecology of angiosperm dominance and gymnosperm persistence. *Biol J Linnean Soc* **36:** 227-249.

Boyce, C. K., Cody, G. D., Feser, M., Jacobsen, C., Knoll, A. H. and Wirick, S. (2002) Organic chemical differentiation within fossil plant cell walls detected with X-ray spectromicroscopy. *Geology* **30:** 1039-1042.

Brodribb, T. J. and Feild, T. S. (2000) Stem hydraulic supply is linked to leaf photosynthetic capacity: Evidence from New Caledonian and Tasmanian rainforests. *Plant Cell Environ* **23:** 1381-1388.

Brodribb, T. J. and Holbrook, N. M. (2003a) Changes in leaf hydraulic conductance during leaf shedding in seasonally dry tropical forest. *New Phytol* **158:** 295-303.

Brodribb, T. J. and Holbrook, N. M. (2003b) Stomatal closure during leaf dehydration, correlation with other leaf physiological traits. *Plant Physiol* **132:** 2166-2173.

Brodribb, T. J., Holbrook, N. M., Edwards, E. J. and Gutierrez, M. V. (2003) Relation between stomatal closure, leaf turgor and xylem vulnerability in eight tropical dry forest trees. *Plant Cell Environ* **26:** 443-450.

Buckley, T. N., Mott, K. A. and Farquhar, G. D. (2003) A hydromechanical and biochemical model of stomatal conductance. *Plant Cell Environ* **26:** 1767-1785.

Carlquist, S. (1975) *Ecological Strategies of Xylem Evolution.* University of California Press, Berkeley.

Carlquist, S. and Schneider, E. L. (2002) The tracheid-vessel element transition in angiosperms involves multiple independent features: Cladistic consequences. *Am J Botany* **89:** 185-195.

Choat, B., Jansen, S., Zwieniecki, M. A., Smets, E. and Holbrook, N. M. (2004) Changes in pit membrane porosity due to deflection and stretching: The role of vestured pits. *J Exp Botany* **55:** 1569-1575.

Cochard, H., Coll, L., Le Roux, X. and Ameglio, T. (2002) Unraveling the effects of plant hydraulics on stomatal closure during water stress in walnut. *Plant Physiol* **128:** 282-290.

Cowan, I. R. (1977) Stomatal behaviour and environment. *Adv Botan Res* **4:** 117-228.

Cowan, I. R. (1994). As to the mode of action of the guard cells in dry air. In *Ecophysiology of Photosynthesis* (M.M. Caldwell, ed.) pp 205-229. Springer-Verlag, Berlin.

Cowan, I. R. and Farquhar, G. D. (1977) Stomatal function in relation to leaf metabolism and environment. In *Integration of Activity in the Higher Plant* (D.H. Jennings, ed.) pp 471-505. Cambridge University Press, Cambridge.

Darwin, F. (1898) Observations on stomata. *Philosophical Transactions of the Royal Society of London Series B* **190:** 531-621.

Davies, W. J., Wilkinson, S. and Loveys, B. (2002) Stomatal control by chemical signalling and the exploitation of this mechanism to increase water use efficiency in agriculture. *New Phytologist* **153:** 449-460.

Edwards, D. (1993) Cells and tissues in the vegetative sporophytes of early land plants. *New Phytologist* **125:** 225-247.

Edwards, D. (2003). Xylem in early tracheophytes. *Plant Cell and Environment* 26, 57-72.

Farquhar, G.D. (1978) Feedforward responses of stomata to humidity. *Aust J Plant Physiol* **5:** 787-800.

Farquhar, G. D., Schulze, E. D. and Kuppers, M. (1980a). Responses to humidity by stomata of *Nicotiana glauca L.* and *Corylus avellana L.* are consistent with the optimization of carbon dioxide uptake with respect to water loss. *Aust J Plant Physiol* **7:** 315-327.

Farquhar, G. D. and von Caemmerer, S. (1982) Modelling of photosynthetic response to environmental conditions. In *Physiological Plant Ecology* II, Volume 12B, *Encyclopedia of Plant Physiology* (H. Ziegler, ed.) pp. 550-587. Springer-Verlag, Berlin.

Farquhar, G. D., von Caemmerer, S. and Berry, J. A. (1980b). A biochemical model of photosynthetic $CO_2$ assimilation in leaves of $C_3$ plants. *Planta* **14:** 78-90.

Farquhar, G. D. and Wong, S. C. (1984). An empirical model of stomatal conductance. *Aust J Plant Physiol* **11:** 191-210.

Field, C. and Mooney, H. A. (1986). The photosynthesis-nitrogen relationship in wild plants. In *On the Economy of Plant Form and Function* (T. J. Givnish, ed.) pp. 25-55. Cambridge University Press, Cambridge.

Fischer, R. A. (1973). The relationship of stomatal aperture and guard cell turgor pressure in *Vicia faba. J Exp Botany* **24:** 387-399.

Franks, P. J. (2004). Stomatal control and hydraulic conductance, with special reference to tall trees. *Tree Physiol* **24:** 865-878.

Franks, P. J., Cowan, I. R., and Farquhar, G. D. (1997). The apparent feedfoward response of stomata to air vapour pressure deficit: Information revealed by different experimental procedures with two rainforest trees. *Plant Cell Environ* **20:** 142-154.

Franks, P. J. and Farquhar, G. D. (1999). A relationship between humidity response, growth form and photosynthetic operating point in C-3 plants. *Plant Cell Environ* **22:** 1337-1349.

Franks, P. J. and Farquhar, G. D. (2001) The effect of exogenous abscisic acid on stomatal development, stomatal mechanics, and leaf gas exchange in *Tradescantia virginiana. Plant Physiol* **125:** 935-942.

Gollan, T., Schurr, U. and Schulze, E.-D. (1992) Stomatal response to drying soil in relation to changes in the xylem sap composition of *Helianthus annuus*. I. The concentration of cations, anions, amino acids and pH of the xylem sap. *Plant Cell Environ* **15:** 551-559.

Grantz, D. A. and Zeiger, E. (1986) Stomatal responses to light and leaf-air water vapour pressure differences show similar kinetics in sugarcane and soybean. *Plant Physiol* **81:** 865-868.

Green, T. G. A. and Snelgar, W. P. (1982) A comparison of photosynthesis in two thalloid liverworts. *Oecologia* **54:** 275-280.

Hacke, U., Sperry, J. S., Pockman, W. T., Davis, S. D. and McCulloh, K. A. (2001) Trends in wood density and structure are linked to prevention of xylem embolism by negative pressure. *Oecologia* **126:** 457-461.

Hall, A. E. and Schulze, E.-D. (1980) Stomatal response to environment and possible interrelations between stomatal effects on transpiration and $CO_2$ assimilation. *Plant Cell Environ* **3:** 467-474.

Hall, A. E., Schulze, E.-D. and Lange, O. L. (1976) Current perspectives of steady state stomatal response to environment. In *Water and Plant Life. Ecological Studies 19* (E.-D. Schulze, ed.) pp. 169-185. Springer, Berlin.

Hubbard, R. M., Ryan, M. G., Stiller, V. and Sperry, J. S. (2001) Stomatal conductance and photosynthesis vary linearly with plant hydraulic conductance in ponderosa pine. *Plant Cell Environ* **24:** 113-121.

Janzen, S. (1998) Vestures in woody plants: A review. *IAWA Bull* **19:** 347-382.

Kappen, L. and Haeger, S. (1991) Stomatal response of *Tradescantia albiflora* to changing air humidity in light and in darkness. *J Exp Botany* **42:** 979-986.

Kenrick, P. and Crane, P. R. (1991) Water conducting cells in early fossil land plants. *Botanical Gazette* **152:** 335-356.

Kramer, P. J. and Boyer, J. S. (1995). *Water Relations of Plants and Soils.* Academic Press, San Diego.

Lange, O. L., Lösch, R., Schulze, E.-D., and Kappen, L. (1971) Responses of stomata to changes in humidity. *Planta* **100:** 76-86.

Leuning, R. (1995) A critical appraisal of a combined stomatal-photosynthesis model. *Plant Cell Environ* **18:** 339-355.

Ligrone, R., Vaughn, K. C., Renzaglia, K. S., Knox, P. J. and Duchett, J. G. (2002) Diversity in the distribution of polysaccharide and glycoprotein epitopes in the cell walls of bryophytes: New evidence for the multiple evolution of water conducting cells. *New Phytologist* **156:** 491-508.

Meinzer, F. C. (1982). The effect of vapor pressure on stomatal control of gas exchange in Douglas fir (*Pseudotsuga menziesii*) saplings. *Oecologia* **54:** 236-242.

Meinzer, F. C. (2002) Co-ordination of vapour and liquid phase water transport properties in plants. *Plant Cell Environ* **25:** 265-274.

Meinzer, F. C., Goldstein, G., Jackson, P., Holbrook, N. M., Gutierrez, M. V. and Cavelier, J. (1995) Environmental and physiological regulation of transpiration in tropical forest gap species: The influence of boundary layer and hydraulic properties. *Oecologia* **101:** 514-522.

Meinzer, F. C. and Grantz, D. A. (1990). Stomatal and hydraulic conductance in growing sugarcane: Stomatal adjustment to water transport capacity. *Plant Cell Environ* **13:** 383-388.

Melcher, P. J., Zweiniecki, M. A. and Holbrook, N. M. (2003) Vulnerability of xylem vessels to cavitation in sugar maple. Scaling from individual vessels to whole branches. *Plant Physiol* **131:** 1775-1780.

Mott, K. A. and Parkhurst, D. F. (1991) Stomatal responses to humidity in air and helox. *Plant Cell Environ* **14:** 509-515.

Nardini, A. (2001) Are sclerophylls and malacophylls hydraulically different? *Biol Plantarum* **44:** 239-245.

Nardini, A. and Salleo, S. (2000) Limitation of stomatal conductance by hydraulic traits: Sensing or preventing xylem cavitation? *Trees-Structure Function* **15:** 14-24.

Nardini, A., Tyree, M. T. and Salleo, S. (2001) Xylem cavitation in the leaf of *Prunus laurocerasus* and its impact on leaf hydraulics. *Plant Physiol* **125:** 1700-1709.

Pickard, W. F. (1982) Why is the substomatal chamber as large as it is. *Plant Physiol* **69:** 971-974.

Proctor, M. C. F. (1981) Diffusion resistances in bryophytes. In *Plants and Their Atmospheric Environment* (P. G. Jarvis, ed.) pp. 219-229. Blackwell Scientific Publications, London.

Raschke, K. (1975). Simultaneous requirement of carbon dioxide and abscisic for somatal closing in *Xanthium strumarium* L. *Planta* **125:** 243-259

Raven, J. A. (1977) Evolution of vascular land plants in relation to supracellular transport processes. *Adv Botanical Res* **5:** 153-219.

Sack, L., Melcher, P. J., Zwieniecki, M. A. and Holbrook, N. M. (2002) The hydraulic conductance of the angiosperm leaf lamina: A comparison of three measurement methods. *J Exp Bot* **53:** 2177-2184.

Saliendra, N. Z., Sperry, J. S. and Comstock, JP. (1995) Influence of leaf water status on stomatal response to hydraulic conductance, atmospheric drought, and soil drought in *Betula occidentalis. Planta* **196:** 357-366.

Salleo, S., Lo Gullo, M. A., Raimondo, F. and Nardini, A. (2001) Vulnerability to cavitation of leaf minor veins: Any impact on leaf gas exchange? *Plant Cell Environ* **24:** 851-859.

Salleo, S., Nardini, A., Pitt, F. and Lo Gullo, M.A. (2000) Xylem cavitation and hydraulic control of stomatal conductance in Laurel (*Laurus nobilis* L.). *Plant Cell Environ* **23:** 71-79.

Schroeder, J. I., Allen, G. J., Hugouvieux, V., Kwak, J. M. and Waner, D. (2001) Guard cell signal transduction. *Annu Rev Plant Physiol Plant Mol Biol* **52:** 627-658.

Schulze, E.-D., and Hall, A. E. (1982) Stomatal responses, water loss and $CO_2$ assimilation rates of plants in contrasting environments. In *Physiological Plant Ecology* II, Volume 12B, *Encyclopedia of Plant Physiology* (H. Ziegler, ed.) pp 181-230. Springer-Verlag, Berlin.

Sober, A. (1997) Hydraulic conductance, stomatal conductance, and maximal photosynthetic rate in bean leaves. *Photosynthetica* **34:** 599-603.

Sperry, J. S. (1986) Relationship of xylem embolism to xylem pressure potential, stomatal closure, and shoot morphology in the palm *Rhapis excelsa*. *Plant Physiol* **80:** 110-116.

Sperry, J. S., Alder, N. N. and Eastlack, S. E. (1993) The effect of reduced hydraulic conductance on stomatal conductance and xylem cavitation. *J Exp Botany* **44:** 1075-1082.

Sperry, J. S., Hacke, U. G., Oren, R. and Comstock, J. P. (2002) Water deficits and hydraulic limits to leaf water supply. *Plant Cell Environ* **25:** 251-263.

Sperry, J. S. and Pockman, W.T. (1993) Limitation of transpiration by hydraulic conductance and xylem cavitation in Betula occidentalis. *Plant Cell Environ* **16:** 279-287.

Tardieu, F. and Davies, W. J. (1993) Integration of hydraulic and chemical signalling in the control of stomatal conductance and water status of droughted plants. *Plant Cell Environ* **16:** 341-349.

Tyree, M. T., Sobrado, M. A., Stratton, L. C. and Becker, P. (1999) Diversity of hydraulic conductances in leaves of temperate and tropical species: Possible causes and consequences. *J Trop Forest Sci* **11:** 47-60.

Tyree, M. T. and Sperry, J. S. (1988) Do woody plants operate near the point of catastrophic xylem dysfunction caused by dynamic water stress? *Plant Physiol* **88:** 574-580.

Tyree, M. T. and Sperry, J. S. (1989). Vulnerability of xylem to cavitation and embolism. *Annu Rev Plant Physiol Plant Mol Biol* **40:** 19-38.

Walker, R. and Pennington, W. (1939) The movements of the air pores of *Preissia quadrata* (Scop.). *New Phytologist* **38,** 62-68.

Whitehead, D., Jarvis, P. G. and Waring, R. H. (1984) Stomatal conductance, transpiration, and resistance to water uptake in a *Pinus sylvestris* spacing experiment. *Can J Forest Res* **14:** 692-700.

Wong, S. C., Cowan, I. R. and Farquhar, G. D. (1979) Stomatal conductance correlates with photosynthetic capacity. *Nature* **282:** 424-426.

Wong, S.-C., Cowan, I. R. and Farquhar, G. D. (1985). Leaf conductance in relation to rate of $CO_2$ assimilation. *Plant Physiol* **78:** 830-834.

Zeiger, E., Talbott, L. D., Frechilla, S., Srivastava, A. and Zhu, J. X. (2002) The guard cell chloroplast: A perspective for the twenty-first century. *New Phytologist* **153:** 415-424.

Ziegler, H. (1987) The evolution of stomata. In *Stomatal Function* (E. Zeiger, G. D. Farquhar, G. D. Cowan and I. R. Cowan, eds.) pp. 29-57. Stanford University Press, Stanford.

Zimmermann, M. H. (1983). *Xylem Structure and the Ascent of Sap.* Springer-Verlag, Berlin.

Zweypfenning, R. C. V. J. (1978) A hypothesis on the function of vestured pits. *IAWA Bull* **1:** 13-15.

Zwieniecki, M. A., Boyce, C. K. and Holbrook, N. M. (2004) Hydraulic limitations imposed by crown placement determine final size and shape of *Quercus rubra* L. leaves. *Plant Cell Environ* **27:** 357-365.

# Part II

## Transport Attributes of Leaves, Roots, and Fruits

# 5

# Leaf Hydraulics and Its Implications in Plant Structure and Function

*Lawren Sack and Melvin T. Tyree*

Each year more than 40 trillion tons of water move through plant leaves, about 10% of the water leaving the planet's surface, two thirds of that from land (Hordon, 1998; USGS, 2004). This component of water's passage through the hydrologic cycle, the microhydrological process inside the leaf, contains many outstanding mysteries. Water flow through leaves has important implications for understanding whole-plant hydraulics and plant growth, as well as leaf structure, function, and ecology. Major questions relating to leaf hydraulics were reviewed in the 1980s (e.g., Boyer, 1985; Davies, 1986). Since the late 1990s, research has increased, including new perspectives from physics, anatomy, modeling, and ecology. In this chapter we describe the importance of the leaf in determining whole-plant hydraulic conductance, which can strongly influence gas exchange and growth. Next, we examine the basis of leaf hydraulic conductance in the pathways of water movement in the leaf. In the following sections, we describe how leaf hydraulic conductance is linked with other aspects of leaf structure/function, including venation design, leaf shape, water storage, and structural features important in plant carbon economy and drought tolerance. We survey recent work on the dynamics of leaf hydraulic conductance across plant growth conditions, as well as diurnally, and across leaf ages. Finally, we propose topics for future research.

## Leaf Hydraulic Conductance in the Whole-Plant System

The leaf hydraulic conductance ($K_{leaf}$) is the ratio of flow rate ($F_{leaf}$)—through the petiole to the sites of evaporation (where the liquid phase ends and the vapor phase begins)—to the driving force for flow, which is the water potential difference across the leaf ($\Delta\Psi_{leaf}$). The hydraulic resistance is the inverse of conductance, defined as $\Delta\Psi_{leaf}/F_{leaf}$. Either definition models bulk liquid flow through a whole leaf as though through a piece of tubing of fixed length, where the hydraulic conductance indicates the diameter of the equivalent tubing. $K_{leaf}$ is typically normalized by leaf area (i.e., $F_{leaf}/\Delta\Psi_{leaf}$ is further divided by lamina area; units of kg [or mmol of water] $s^{-1}$ $MPa^{-1}$ $m^{-2}$), based on the assumption that the number of parallel pathways for water flux will scale linearly with leaf area for a given type of leaf. This idea has been supported for several species (Meinzer and Grantz, 1990; Martre *et al.*, 2001; Sack *et al.*, 2004). $K_{leaf}$ is apparently a substantial bottleneck in the whole-plant hydraulic pathway, and a potentially major determinant of gas exchange. The leaf lamina hydraulic conductance [$K_{lamina} = (K_{leaf}^{-1} - $ petiole hydraulic conductance$^{-1})^{-1}$] varies at least 30-fold across species according to published measurements (Fig. 5.1A), indicating potentially strong ecological importance.

How does $K_{leaf}$, as a component of the whole-plant hydraulic system, impact gas exchange? The resistance of the stomata in the path of vapor diffusion out of the leaf is typically greater than two orders of magnitude higher than the hydraulic resistance to bulk flow through the whole plant, including the leaf, and so the hydraulic properties of the liquid path do not *directly* determine transpiration rates. However, while the transpiration rate is dictated by the stomatal aperture, at a given soil water supply and transpiration rate, the leaf water potential is determined by the plant hydraulic conductance ($K_{plant}$; Cowan, 1972; Tyree and Zimmermann, 2002) and stomata close at low leaf water potentials to prevent leaf desiccation. Consequently, stomatal conductance is frequently suboptimal at typical midday leaf-water potentials because of hydraulic constraints on gas exchange (Tyree, 2003a). $K_{plant}$ often correlates with maximum stomatal conductance within and across species (e.g., Küppers, 1984; Meinzer and Grantz, 1990; Nardini and Salleo, 2000).

Published data show that $K_{leaf}$ scales with $K_{plant}$ (Fig. 5.1A). This relationship indicates that on average the leaf lamina constitutes $\approx \frac{1}{4}$ of whole-plant resistance (Fig. 5.1A; Sack *et al.*, 2003b). Further, $K_{leaf}$ itself is correlated with both stomatal conductance and maximum photosynthetic rate per unit area across species for temperate deciduous trees (Aasamaa *et al.*, 2001), apparently via a tight correlation with total stomatal pore area, a major determinant of stomatal conductance (Fig. 5.1B; Sack *et al.*, 2003b). The coordination of $K_{leaf}$ with $K_{plant}$ and with gas exchange indicates the importance of $K_{leaf}$ in defining plant function and driving differences in species function.

As described previously, the measurement of $K_{leaf}$ models the leaf as though it were a piece of tubing. However, water flows through a leaf in a multitude of pathways. Past the petiole insertion point, water moves

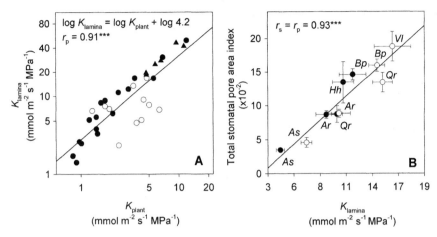

**Figure 5.1** (A) The scaling of leaf lamina and whole plant hydraulic conductance, for 34 species; $K_{lamina} = 4.2 \times K_{plant}$. [$K_{lamina} = (K_{leaf}^{-1} - $ petiole hydraulic conductance$^{-1})^{-1}$]. Filled triangles represent herbs, filled circles woody seedlings and saplings, and open circles mature trees and shrubs (data of Becker *et al.*, 1999; Nardini and Tyree, 1999; Nardini *et al.*, 2000; Nardini and Salleo, 2000; Tsuda and Tyree, 2000). (B) Coordination of total stomatal pore area index and $K_{lamina}$. Total stomatal pore area index = stomatal guard cell length$^2$ × stomatal density. Open/filled symbols represent sun/shade leaves. Species: *Ar, Acer rubrum; As, Acer saccharum; Bp, Betula papyrifera; Hh, Hedera helix; Qr, Quercus rubra; Vl, Vitis labrusca*. Error bars = 1 SE; ***: $P < 0.001$. From Sack *et al.* (2003b). Copyright Blackwell Publishing; adapted with permission.

through leaky xylem conduits within a typically reticulate venation wrapped in membrane-bound cells, and water leaving the veins flows potentially through multiple cell walls and perhaps membranes throughout the lamina before evaporating at many loci. Thus, $K_{leaf}$ is an aggregate measure for a complex hydrologic system. To understand the basis of measuring $K_{leaf}$, it is necessary to understand the flow pathways through the leaf. There are several methods in common use for quantifying $K_{leaf}$ (Table 5.1), and in applying these methods there remain outstanding issues. The most important unknown is the degree to which the different methods measure the conductance of the same flow pathways through the leaf, and how similar these flow pathways are to those of natural transpiration.

## How Does Water Flow from the Petiole to the Sites of Evaporation?

Water enters the petiole through branch xylem bundles, which reorganize in a complex junction at the leaf insertion (Esau, 1965) and enter the

**Table 5.1**  $K_{leaf}$ measurement methods for leaves transpiring *in vivo*, and for excised leaves

| Method | | | Sample references |
|---|---|---|---|
| (a) *In vivo* methods | *Measurement of flow rate* | *Measurement of driving force* | |
| | Transpiration; measured as pot water loss rate | $\Delta\Psi_{leaf}$ measured as difference in pressure bomb water potential between transpiring leaf and bagged nontranspiring leaf[b] | Tsuda and Tyree, 2000; Brodribb and Holbrook, 2003a |
| | *or* as porometer or gas exchange system measured 'transpiration'[a] | *or* as difference in water potential between a leaf sampled for psychrometry while transpiring, and a leaf measured for vein water potential[c] using *in situ* psychrometry[d] | Matzner and Comstock, 2001 |
| | *or* as sapflow proximally to the leaf | | |
| | | *or* by measuring within leaf gradients in water potential using psychrometry for vein water potential[c] and direct measurement of epidermal cell turgor and osmotic potential | Shack et al and Brinckmann, 1985; Shack *et al.*, 1987 |
| (b) Excised leaf methods | | | |
| i. Evaporative flux methods | $F_{leaf}$ is estimated for excised leaves transpiring under natural or artificial conditions | $\Delta\Psi_{leaf}$ as for (a) above | Sack *et al.*, 2002 and references therein |
| ii. High pressure flow meter | Water driven through the leaf under positive pressure[e], and flow rate and pressure driving force measured simultaneously proximally to the leaf | | Tyree *et al.*, 1993; Yang and Tyree, 1994; Tyree *et al.*, 1999; Sack *et al.*, 2002 |
| iii. Vacuum pump method | Water is pulled out of the leaf under several levels of partial vacuum[e], and $K_{leaf}$ is estimated as slope of $F_{leaf}$ vs levels of vacuum | | Kolb *et al.*, 1996; Nardini *et al.*, 2001; Sack *et al.*, 2002 |

**Table 5.1** *—cont'd*

| Method | | Sample references |
|---|---|---|
| iv. Rehydra-<br> tion kinetics<br> method | $K_{leaf}$ estimated analytically from rehydration<br> kinetics of leaf water potential for partially<br> desiccated leaves imbibing water | Tyree *et al.*, 1975;<br> Tyree and Cheung,<br> 1977; Brodribb and<br> Holbrook, 2003b |

For all methods but (b) iv, (the rehydration kinetics method), $K_{leaf}$ determined as measured flow rate $(F_{leaf})$/driving force. There are theoretical grounds to speculate that the different methods might produce slightly different values of $K_{leaf}$. Tests for methods (b)i, (b)ii, and (b)iii have shown that when applied under specific conditions, similar values are produced (Tsuda and Tyree, 2000; Nardini *et al.*, 2001; Sack *et al.*, 2002).

[a]The potential overestimation of transpiration using gas exchange systems (due to removal of boundary layer resistance, and potential misestimation of leaf temperature; Tyree and Wilmot, 1990), need to be considered in the study design when using this method.

[b]The non-transpiring leaf is thus assumed to have equilibrated with the branch water potential, at the leaf insertion.

[c]In methods using vein water potential and an estimate of transpiring leaf water potential, the conductance estimated is, of course, that between vein and evaporation sites, rather than that of the whole-leaf.

[d]We note that the water potential measured for 'transpiring' leaves taken for the pressure bomb, or sampled for psychrometry, are for leaves equilibrated and non-transpiring at the time of water potential determination. Hence the total water potential difference between 'transpiring' and non-transpiring leaves may not give the total driving force between base of the petiole and the 'average loci' of evaporation.

[e]Whether a pressure gradient is equivalent to $\Delta\Psi_{leaf}$ to describe the driving force of water movement depends on the precise pathway of water movement in leaves—i.e., whether osmosis is a driving force to be considered—and should really be dealt with in terms of irreversible theromodynamics. The general assumption is that during steady-state flow through the leaf, osmosis is a negligible contribution to the driving force (but see Tyree *et al.*, 1999; Tyree, 2003b).

major veins. From this point onward, the exact water flow paths are uncertain. In most angiosperm leaves, water flows through several orders of major veins before entering the minor vein network (i.e., the network of small veins throughout the lamina embedded in the mesophyll, which in cross-section typically contain one or two xylem conduits). The reticulation of the venation means that water can exit major veins of any order into the minor veins, or, potentially, water might exit the major veins directly into the surrounding tissue. However, the density of minor veins accounts for 93% to 96% of the total vein density in temperate tree species (e.g., Armacost, 1944; Plymale and Wylie, 1944; Wylie, 1951; Dengler and Mackay, 1975; Russin and Evert, 1984). Thus, the bulk of water flow to the mesophyll and epidermis is probably supplied principally through the minor vein network.

The minor veins are typically wrapped in a bundle sheath of parenchymatous cells along most of their length (Esau, 1965). Classical anatomy and histology on a range of species suggest that a large part of the water leaving the minor veins necessarily passes through the bundle sheath cells; the perpendicular cell walls may be suberized and might thus constitute a barrier analogous to the root Casparian strip, which forces water to move from apoplast to symplast (Van Fleet, 1950; Lersten, 1997). For a range of species, dye experiments have also suggested that the bulk of transpired water tends to exit the vein xylem into the bundle sheath cells—in leaves transpiring a solution of sulphorodamine G, an apoplastic dye, crystals form in the minor veins, indicating that water is diffusing out across the membrane (Canny, 1990a, 1990b). Other evidence comes from the temperature response of measured $K_{leaf}$. When water moves through leaves or leafy shoots, the conductance increases or declines as temperature is respectively increased or decreased, and the slope of the response is stronger than that expected to arise simply from changes in the viscosity of water (Tyree *et al.*, 1973; Boyer, 1974; Tyree and Cheung, 1977; Fredeen and Sage, 1999; Cochard *et al.*, 2000; Matzner and Comstock, 2001; Sack *et al.*, 2004). This extra-viscosity response suggests at least some water passes through membranes; indeed, such temperature sensitivity established the fact of water moving through membranes in root water transport (Kramer and Boyer, 1995). It is unclear whether *all* water leaves the xylem through membranes, and more research is needed to establish the details of this crucial stage.

What happens to water once out of the xylem and in the bundle sheath? According to early anatomical studies, water movement through the mesophyll would mostly occur between spongy mesophyll cells, which are in contact to a far greater degree than are palisade cells (Wylie, 1946). Water might move through the mesophyll to the sites of evaporation apoplastically (i.e., never crossing a membrane) or pass cell-to-cell, whether symplastically (i.e., via plasmodesmata, though water must pass one membrane to enter the symplasm and another membrane to exit to the evaporative surface), or transcellularly (i.e., crossing cell walls and membranes twice for every cell in the path). Theoretical considerations and early experiments on leaves of trees and herbs suggested that water movement from veins to evaporation sites was primarily via the apoplast, in cell wall nanochannels (Weatherley, 1963; Boyer, 1977), but other studies on one of the same species, sunflower, reported that symplastic or cell-to-cell movement cannot be excluded (Cruiziat *et al.*, 1980; Tyree *et al.*, 1981). Furthermore, in addition to water flow across the mesophyll, in many leaves an important potential pathway for water movement to the epidermis is through bundle sheath extensions, cells that bridge the epidermis and minor veins and that are commonly found in temperate tree species (Armacost, 1944; Wylie, 1952; McClendon, 1992). Evidence for the importance of a direct flow path to the epidermis includes the fact that in several species, areas of the epidermis remain hydrated even when having little vertical contact with the underlying mesophyll (LaRue, 1931; Warrit *et al.*, 1980). Additionally, sudden hydraulic pressure changes can be transferred quickly from the petiole to epidermis in *Tradescantia* leaves, which possess bundle sheath extensions (Sheriff and Meidner, 1974).

Finally, there is the question of where in the leaf water evaporates. If water evaporates preferentially from a certain group of cells, then pathways to those cells will be most important during transpiration. Some have presented arguments for water evaporating principally from the cells adjacent to the stomata—from the epidermal cells around the stomata, and/or from the layer of mesophyll (usually spongy) directly above the stomata, and/or from the guard cells themselves (Byott and Sheriff, 1976; Tyree and Yianoulis, 1980). In the archetypal dicotyledonous leaf, with only abaxial stomata and layers of spongy mesophyll above, this would mean that the cells most active photosynthetically, the palisade cells, which run beneath the adaxial surface, would not lie in the primary pathways of transpiration, and would be kept more turgid than the spongy mesophyll in normal transpiration, and thus buffered to a degree from potential water limitation. On the other hand, several have proposed an opposite scenario: that water tends to evaporate throughout the lamina, making the diffusion pathway of water vapor similar to that of $CO_2$ (though opposite in direction). Indeed the computation of intercellular $CO_2$ concentration using typical photosynthesis systems relies on this assumption (Field *et al.*, 2000). Further circumstantial evidence for this alternative scenario is the sheer amount of exposed cell wall throughout the leaf relative to that simply around the stomata (Davies, 1986), and the fact that, in at least several species, there is suberization inside the leaf around the stomata and adjacent mesophyll, which would reduce evaporation (Pallardy and Kozlowski, 1979; Nonami and Schulze, 1989). In studies using a cell pressure probe on transpiring leaves of *Tradescantia*, the water potential of the bulk mesophyll was reported to be lower than that in the epidermis, which might indicate that there is greater evaporation from the bulk mesophyll (Nonami and Schulze, 1989). However, the same finding might arise even if most water evaporated from cells close to the epidermis, if water is supplied to these cells with relatively high conductance (e.g., via the bundle-sheath extensions in this species; Sheriff and Meidner, 1974). The question of where water principally evaporates within the leaf is still open.

Such a linear description of flow paths, while important for mechanistic reduction, of course oversimplifies the reality. Water flux occurs simultaneously through many pathways, with local flow rates and pressure drops determined by resistances of each component in the context of the whole system. For full elucidation, the system could be modeled as a complex electronic circuit, looped in multiple places, but at present the component resistances are largely unknown. For instance, we do not know the resistance to flow through minor vein conduits, through vein xylem pit membranes, bundle sheath membranes, mesophyll cell membranes, plasmodesmata, or cell wall microchannels. Further, because of uncertainty about where water evaporates, we cannot precisely locate the driving forces. However, even when all the resistances and driving forces are determined, a model will not be completely satisfactory in itself. Confirmation will be delayed until technology allows pressure probing of cells throughout the leaf during transpiration, or, most ideally, the direct visualization of the pressures within cells throughout the leaf.

Water flow pathways through the leaf have important bearing on the measurement and interpretation of $K_{leaf}$. Several methods for determining $K_{leaf}$ (e.g., the evaporative flux method; Table 5.1) rely on estimating the overall driving force using the pressure bomb water potential for the transpiring leaf. It is unclear, however, how well the pressure bomb can estimate the overall driving force. If water evaporates preferentially from a population of cells near the stomata, a possibility described previously, then the pressure bomb water potential, as a volume-averaged water potential for all the leaf's cells, will underestimate the driving force, and thus overestimate $K_{leaf}$, though it is uncertain by how much. For instance, let us suppose that water evaporates principally from spongy mesophyll and epidermides, and negligibly from the palisade cells, located further from the stomata. During transpiration, the water potential of palisade cells would be nearly equilibrated with that of the adjacent minor venation, and the water potential in the spongy mesophyll and epidermides would be lower by some amount. The pressure bomb water potential will underestimate the driving force, because it integrates the water potential of the palisade, which is not in the transpiration path. The pressure bomb would estimate the driving force *well* only if the difference in water potential between palisade cells and sites of evaporation is not substantial. Such would be the case if the hydraulic conductance from the minor venation to the sites of evaporation is very high relative to the conductance of the xylem and bundle sheath; in this case, the water potential would not differ substantially among leaf cells.

Notably, the evaporative flux method produces $K_{leaf}$ values similar to those measured using other methods that do not rely on the pressure bomb to determine driving force (Table 5.1; Sack *et al.*, 2002), the high pressure flowmeter (HPFM) and the vacuum chamber method (Sack *et al.*, 2002). These methods also have been hypothesized to overestimate $K_{leaf}$, for a different reason; these methods establish new flow paths through the leaf; once out of the veins, water may short-circuit transpirational paths, flowing through lacunae of the mesophyll airspaces. It is unlikely, however, that these three methods would all overestimate $K_{leaf}$ to the same extent. The similarity among methods suggests that the flow paths are similar or that the differences in the flow paths are in components that contain minimal resistance (Sack *et al.*, 2002, 2004). Such would be the case if most of the hydraulic resistance in the leaf (i.e., $1/K_{leaf}$) were in the venation system, rather than in the paths of water flow distal to the veins.

## Coordination of $K_{leaf}$, Venation System Design, and Leaf Shape

How much does whole-leaf hydraulic function depend on properties of the venation? Leaves vary tremendously in venation architecture, from the gridlike system of grasses, to the radial venation of *Ginkgo*, to the radiator-like system of trees of the tropical genus *Calophyllum* (possessing midrib with close-set parallel secondaries that branch off at right angles), to the diverse dendritic architectures common in dicotylendonous trees. If the leaf venation is an important component of the leaf hydraulic resistance (i.e., an important bottleneck in $K_{leaf}$) then variation in venation architecture might significantly impact leaf water use. The importance of the venation in water transport would be coordinated with its role in nutrient supply, sugar distribution (see Chapter 6), as well as in support and defense against mechanical damage and herbivory (e.g., Niklas, 1999; Choong *et al.*, 1992). Additionally, as found for tropical tree species, major veins protruding from the lamina can play a significant role in breaking up the boundary layer of still air around the leaf, facilitating gas exchange (Grace *et al.*, 1980). Through selection, these roles together may be coordinated and co-optimized with respect to construction costs (Rosen, 1967; Givnish, 1987), though such optimization is contingent on the genes and developmental sequences available during evolution, and constrained by potential trade-offs arising among structures and functions.

The importance of the leaf venation to the hydraulics of the whole-leaf depends on how much of the total resistance to transpirational water flow is in the vein xylem, and how much is outside the xylem. Recent work on venation pressure distribution, using the xylem pressure probe, indicated a large resistance within the leaf vasculature in *Laurus nobilis* (Zwieniecki *et al.*, 2002). Vein cutting experiments also have indicated a large proportion of the leaf's resistance in the vasculature, 64% in *Acer saccharum* and 74% in *Quercus rubra* (Fig. 5.2; see also Color Plate section; Sack *et al.*, 2004). In these experiments, leaves were submitted to $K_{leaf}$ determination by the HPFM, and subsequently veins were cut, beginning with the minor veins, and then the higher orders of major veins, such that water leaked out of the cuts, shorting out downstream resistance. A major location of resistance is between the major and minor veins, or in the minor veins themselves. The lower order major veins typically contain little of the leaf resistance (Yang and Tyree, 1994; Tyree *et al.*, 2001; Sack *et al.*, 2004). The proportion of the leaf resistance that is vascular depends on the xylem conduit number and dimensions, and on the

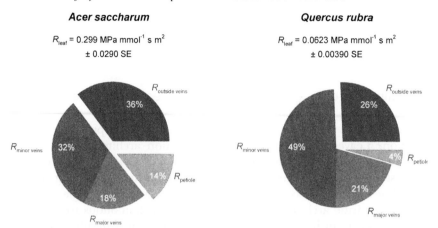

**Acer saccharum**

$R_{leaf}$ = 0.299 MPa mmol$^{-1}$ s m$^2$
± 0.0290 SE

**Quercus rubra**

$R_{leaf}$ = 0.0623 MPa mmol$^{-1}$ s m$^2$
± 0.00390 SE

**Figure 5.2**    The distribution of hydraulic resistance (=1/hydraulic conductance) in the leaves of sugar maple (*Acer saccharum*) and red oak (*Quercus rubra*). (See also Color Plate section.) From Sack *et al.* (2004). Copyright American Society of Plant Biologists; adapted with permission.

extravascular pathway (i.e., the permeability of the membranes, the plasmodesmata, or the nanochannels formed by the cell wall microfibrils), as well as the transfer areas of the cell-to-cell pathways. The possible coordination of $K_{leaf}$ with membrane properties such as aquaporin populations is a current field for pioneering research (Martre *et al.*, 2002; Siefritz *et al.*, 2002).

The leaf lamina is supplied by the petiole. This single-point supply to the lamina raises the problem of supplying water equitably, so that gas exchange rates and cell water deficits are distributed uniformly (i.e., bringing the water potential of each square centimeter of the leaf within a narrow range). Data from porometer studies (Sack *et al.*, 2003a) and thermal photography (Jones, 1999) on temperate tree species and grapevine suggest that for an evenly lit leaf under well-watered conditions, stomatal conductance is relatively even, from petiole insertion to leaf tip (though transpiration can be locally highly patchy, especially from stomate to stomate throughout the leaf (Terashima, 1992; Mott and Buckley, 1998). If transpiration is indeed relatively homogeneous across the lamina from leaf base to tip (at say square centimeter resolution), for water potential to show a similarly narrow spatial variation from leaf base to tip, water must be distributed relatively equitably. Several models of leaf hydraulic design proposed in the literature might account for equitable distribution, based on the relative hydraulic conductances of vein orders, and the conductance of the vasculature relative to the pathways from veins to evaporation sites (Tyree and Cheung, 1977; Canny, 1993; Nardini *et al.*, 2001; Roth-Nebelsick *et al.*, 2001; Zwieniecki *et al.*, 2002; Sack *et al.*, 2004).

Another key feature of the leaf venation is its damage tolerance. The lower order major veins are essential in water supply. Such a conclusion was at odds with the classical demonstration that leaves of many temperate woody species survive the cutting of major veins, including primaries, in apparently perfect health (Plymale and Wylie, 1944). More recent work has shown that the apparent health in *Quercus rubra* leaves with severed midribs belies a strong loss of function; $K_{leaf}$, transpiration, and photosynthetic quantum yield were reduced dramatically (Fig. 5.3; Sack *et al.*, 2003a). However, leaves with different venation architectures might be able to sustain different amounts of major vein severing while maintaining function at different amounts. Redundancy in the vasculature—including the multiple conduits in parallel within veins, multiple veins of a given order, and the reticulation of the minor vein network—may provide alternative pathways for water flow, buffering to some extent the effects of vascular damage, as well as of drought-induced embolism (see discussion below).

Venation system design is tantalizingly linked with leaf shape, in *Arabidopsis* mutants and in leaf development (Dengler and Kang, 2001). The linkage of venation and shape may have a hydraulic basis (Thoday 1931; Sack *et al.*, 2002; Zwieniecki *et al.*, 2002). Leaves with higher outline complexity (i.e., lobing) have their mesophyll regions at the leaf edge closer to low-order "supply" veins (e.g., primaries and secondaries). By contrast, in more entire leaves, the relatively larger areas of mesophyll far from the lower-order supply veins are supplied via a greater length of low-conductance higher-order veins, and these areas may thus contribute to a lower overall $K_{leaf}$ than for lobed leaves. Such a linkage between $K_{leaf}$ and lobing may be termed structural—arising from common anatomy. Additionally, $K_{leaf}$ and lobing might be functionally linked (i.e., without sharing a common structural basis), if exposed conditions lead to selection of a high $K_{leaf}$ to maintain water supply for high transpiration rates, in parallel

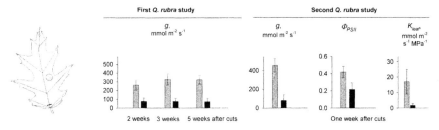

**Figure 5.3** Two studies of the effects of severing the midrib on *in vivo* stomatal conductance, photosynthesis, and $K_{leaf}$ in *Quercus rubra*. Stomatal conductance ($g$) and the quantum yield of photosystem II ($\Phi_{PSII}$) were measured in the indicated area; black line represents the cut. Black bars represent treated leaves and gray bars paired control leaves (from Sack *et al.*, 2003a). Copyright Botanical Society of America; adapted with permission.

with selection for lobing. Lobing may be an advantage in exposed conditions, because the less well-supplied mesophyll regions in entire leaves are prone to desiccation under high evaporative demand or limited water supply (Thoday, 1931; Zwieniecki *et al.*, 2002, 2004). Further, lobed leaves have a thinner still-air boundary layer, which minimizes lamina overheating (Vogel, 1968, 1970). A few recent studies support a linkage of $K_{leaf}$ and leaf outline complexity. *Quercus rubra* sun leaves have greater outline complexity than shade leaves, as well as greater $K_{leaf}$ (Sack *et al.*, 2003b; Zwieniecki *et al.*, 2004). Across species of *Quercus*, leaves with deeper lobes had higher $K_{leaf}$ (Sisó *et al.*, 2001). In a study of sun and shade leaves of six temperate deciduous species, leaf perimeter/area (an index of outline complexity) correlated with $K_{leaf}$, though one species, *Vitis labrusca*, was a strong outlier (Sack *et al.*, 2003b). One hypothesis is that leaf shape and vascular architecture may be structurally coordinated within species, and across species that have similar vascular design (i.e., similar conduit numbers and diameters in each vein order). Across sets of species that differ strongly in vascular design, however, the leaf shape and venation architecture might well be structurally independent, though functionally coordinated for their advantages in exposed conditions.

## Coordination of $K_{leaf}$ and Leaf Water Storage

Leaf water storage occurs in the vacuoles of mesophyll or epidermal cells, and/or in specialized thin-walled water storage cells, which in some species are achlorophyllous, and/or in apoplastic mucilage (Roberts, 1979; Nobel and Jordan, 1983; Morse, 1990; Sack *et al.*, 2003b). The leaf-area specific capacitance—the water released per change in leaf water potential, per leaf area—is the product of tissue-specific capacitance, leaf dry mass per area, and leaf water content per dry mass. These characters vary strongly, though smoothly, across species. For instance, in a set of semidesert species ranging from nonsucculents, to semileaf succulents, to full leaf succulents, leaf water content per dry mass varied smoothly by nearly 20-fold along a continuum (Von Willert *et al.*, 1990).

Commonly, water storage in the leaf has been associated with desiccation avoidance (or desiccation delay)—i.e., for sustaining cuticular transpiration when the stomata have closed (Levitt, 1980; Lamont and Lamont, 2000). Could stored water play a role in normal transpiration? Certainly stored leaf water in most cases cannot supply typical mid-day rates of transpiration under high illumination if water were not supplied from the soil. Leaf water storage often accounts for less than 1% of daily transpiration needs in tropical trees to up to 16% in *Thuja* trees (Tyree *et al.*, 1991; Machado and Tyree, 1994). On the other hand, stored water may play an important role in sustaining a relatively stable leaf water potential as transpiration rate and root water supply fluctuate. In principle, the lower the water storage, the greater impacts these fluctuations will have in driving transient changes in leaf water potential, desiccating leaf tissue, and causing stomatal closure. Leaves with high $K_{leaf}$ and high maximum transpiration rates would benefit most from stored water as a buffer of fluctuations in water potential. Consistent with this principle, in leaves of temperate woody species, leaf area-specific capacitance was correlated with $K_{leaf}$ (Fig. 5.4; Sack et al., 2003b). A large capacitance might be especially useful in semidesert leaves, which would benefit by maintaining stably high stomatal conductance, when moisture is available. A linkage between high $K_{leaf}$ and water storage might contribute to the finding for temperate tree species that at a given transpiration rate excised leaves with high $K_{leaf}$ close their stomata relatively slowly (Aasamaa et al., 2001). The two potential roles of leaf capacitance, drought survival and buffering of water potential fluctuations, will need to be teased apart—perhaps most profitably in a phylogenetic context, especially in relation to the frequent evolution of succulence.

## Coordination of $K_{leaf}$ with Other Aspects of Leaf Structure, Carbon Economy, and Drought Tolerance

How is $K_{leaf}$ integrated in the complex of characters involved in leaf structure and function? It is clear that within and across biomes leaves vary

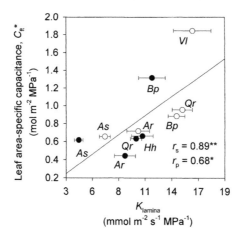

**Figure 5.4**   Coordination of leaf area-specific capacitance with $K_{leaf}$ in northern temperate tree and climber species. Symbols as in Fig. 5.1 (B) $^*P < 0.05$; $^{**}0.01 > P > 0.001$ (from Sack et al., 2003b). Copyright Blackwell Publishing; adapted with permission.

tremendously in venation, shape, and water storage, as discussed previously, and also in area, thickness, gas exchange, and desiccation tolerance. The underlying basis for variation in leaf traits related to carbon economy has been extensively investigated, and several of these traits are correlated across diverse leaves. One example is the correlation of high nitrogen per unit mass, high light-saturated photosynthetic rate per unit mass, low leaf mass per area (LMA), high leaf water content, and short leaf lifespan (e.g., Field and Mooney, 1986; Reich *et al.*, 1999; Wright and Westoby, 2002).

There appears to be similar coordination of leaf characters associated with water flux through the leaf, which also bear on carbon economy (Grubb, 1984; Sack *et al.*, 2003b). Indeed, $K_{leaf}$ may constrain gas exchange per area (see earlier discussion), and thus play a direct role in carbon economy. Some work suggests that $K_{leaf}$ may be strongly coordinated with leaf thickness (Sack *et al.*, 2003b; Fig. 5.5). Thicker leaves tend to have higher mesophyll surface area per leaf area for $CO_2$ fixation, and thus, higher rates of drawdown of intercellular $CO_2$. Thus, thickness is expected to be positively coordinated with stomatal pore area, which in turn should be coordinated with $K_{leaf}$, so that water supply matches demand. Stomatal pore area would turn in constrain maximum photosynthetic rate per unit area. Photosynthetic rate per unit leaf mass, probably the more appropriate measure of gain in leaf-level carbon economy, depends additionally on leaf mass per area, which seems to be independent of $K_{leaf}$ (Tyree *et al.*, 1999; Salleo and Nardini, 2000; Sack *et al.*, 2003b), despite the relationship of $K_{leaf}$ to leaf thickness, a component of LMA ($K_{leaf}$ is apparently independent of leaf density, the other component of LMA). Further, $K_{leaf}$ appears to be unrelated to leaf lifespan (Tyree *et al.*, 1999; Nardini, 2001; but see Sobrado, 1998). Thus, $K_{leaf}$ represents a potential constraint on carbon economy independent of other inter-related characters that influence carbon economy (LMA, nitrogen concentration, leaf lifespan). The overall importance of $K_{leaf}$ and associated traits in defining species' ecology, and potential crop performance requires further investigation.

Is $K_{leaf}$ important in determining drought tolerance? A high $K_{leaf}$ in itself may contribute a measure of drought tolerance (i.e., because a lower drop in water potential occurs across the leaf at a given transpiration rate), and, as seen previously, $K_{leaf}$ is positively coordinated in temperate woody species with capacitance leaf area-specific, an index of water storage. For leaves of temperate woody species growing in moist soil, however, $K_{leaf}$ is unrelated to other leaf features that contribute to the ability to function at low water potential—a low turgor loss point, high modulus of elasticity, and low cuticular transpiration rate (Sack *et al.*, 2003b). The independence of $K_{leaf}$ and drought tolerance is consistent with the fact that both drought tolerant and intolerant species can have high maximum stomatal conductances and transpiration rates when moisture supply is high (Maximov,

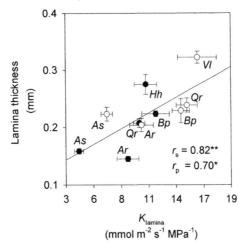

**Figure 5.5**    Coordination of $K_{leaf}$ and leaf thickness. Symbols as in Fig. 5.1(B) and Fig. 5.4 (from Sack *et al.*, 2003b). Copyright Blackwell Publishing; adapted with permission.

1931; Wright *et al.*, 2001). More work is needed to elucidate how leaf hydraulics, structure, and function may be coordinated, co-selected, and co-optimized across the diversity of leaf types.

## Variability of $K_{leaf}$ Across Environments, Diurnally, and with Leaf Age

The coordination of $K_{leaf}$ and other leaf characters has typically been studied for plants grown under high-irradiance, in moist soil. However, $K_{leaf}$ is a highly variable trait, even among leaves of the same species, or, indeed, of the same plant. First, $K_{leaf}$ is environmentally plastic, higher for leaves of given species grown in higher resource conditions: sun leaves of temperate deciduous trees have been found to have $K_{leaf}$ values up to 67% higher than shade leaves—consistent with the higher degree of exposure and associated water flux (Sack *et al.*, 2003b). Similarly, leaves of temperate deciduous tree seedlings grown in higher water or nutrient supplies may tend to have higher $K_{leaf}$ (Aasamaa *et al.*, 2001), as do leaves of *Quercus ilex* seedlings associated with ectomycorrhizae, relative to those of nonsymbiotic plants (Nardini *et al.*, 2000). Additionally, $K_{leaf}$ declines with leaf age, moving down the stem from young to old leaves in sunflowers, and seasonally in tree species (Salleo *et al.*, 2002; Brodribb and Holbrook, 2003a; Sack, L., unpublished data). More work is needed, but the age-related decline of

$K_{leaf}$ might result from the accumulation of embolism in leaf veins, from the formation of tyloses (Salleo et al., 2002), or from changes in the permeabilities of membranes and cell walls in the leaf (Van Fleet, 1950).

Evidence is accumulating that $K_{leaf}$ is also reduced during diurnal water stress and during drought. One study reports diurnal changes in petiole hydraulic conductivity in two of three temperate tree species tested (Zwieniecki et al., 2000). Acoustic and dye studies show that embolisms in minor veins arise during desiccation (Kikuta et al., 1997; Salleo et al., 2001, 2003; Lo Gullo et al., 2003). An ideal method for measuring $K_{leaf}$ in desiccated leaves is elusive. Methods that measure $K_{leaf}$ on desiccated leaves by driving flow through the leaf at higher pressure than ambient may redissolve embolisms rather quickly. However, using such methods, considerable declines in $K_{leaf}$ due to desiccation have been reported (Nardini et al., 2001, 2003; Salleo et al., 2001; Brodribb and Holbrook, 2003b; Lo Gullo et al., 2003; Trifilo et al., 2003a and b). The redundancy of conduits within veins and the reticulation of the minor veins may buffer $K_{leaf}$ from the effects of cavitation-induced vein embolism to some degree, but at high levels of embolism, the decline of leaf water potential associated with decline in $K_{leaf}$ may drive stomatal closure (Salleo et al., 2001; Nardini et al., 2003; Trifilo et al., 2003b; Brodribb and Holbrook, 2003b). In excised sunflower leaves allowed to rehydrate, cavitated conduits in the leaf veins will apparently refill (Trifilo et al., 2003a). The question of whether, in vivo, leaves may refill embolized conduits (i.e., redissolve embolism either when under high tension, or when tension is low) is currently entirely open.

Because the leaf represents a substantial component of the whole-plant resistance, reduction of $K_{leaf}$ would have significant consequences at the whole-plant level (Meinzer, 2002). Declines of $K_{leaf}$ may trigger stomatal closure and may play an important role in driving the well-recognized declines of gas exchange late in the day, late in the season, and in aging leaves (Jones, 1992; Kitajima et al., 1997).

## Summary of Directions for Future Research

The field of leaf hydraulics is still in discovery phase, and new work is needed to remove the many unknowns. More research is needed in examining the methods for measuring $K_{leaf}$, as well as in each of the areas described previously. Our understanding of pathways of water flow through the leaf remains rudimentary, and new approaches are needed, based either on probes or on visualization. The possible coordination of $K_{leaf}$ with venation architecture and/or with particular aspects of membrane physiology (e.g., aquaporins and plasmodesmata) is an exciting field for future research. Also, the coordination of $K_{leaf}$ and other leaf features merits further study, as do the dynamics of $K_{leaf}$ in changing environmental conditions, at time scales ranging from minutes, to hours, to seasons.

Finally, it will be important to examine the possibility of coordination of $K_{leaf}$ with other organs and systems in the plant, within and across species. Do plants with high $K_{leaf}$ have high root conductance? Is $K_{leaf}$ coordinated with particular root morphologies and anatomies? Research is needed on the coordination of nutrient and water supply within the leaf, and, of course, the integration of the water transport system with phloem (see Chapters 10 and 11). Because leaf hydraulics is important on the global scale and intricate at the smallest scale, its clarification will illuminate many topics in leaf physiology, structure, evolution, and ecology.

## Acknowledgments

Lawren Sack was supported by The Arnold Arboretum of Harvard University (Putnam Fellowship), and the Smithsonian Tropical Research Institute (Short Term Fellowship), and Mel Tyree by the USDA Forest Service and the Andrew Mellon Foundation. We thank our project collaborators for generous insights and discussion, and Kevin Boyce, Matthew Gilbert, Missy Holbrook, and Ken Shack et al for comments on the manuscript.

## References

Aasamaa, K., Sober, A. and Rahi, M. (2001) Leaf anatomical characteristics associated with shoot hydraulic conductance, stomatal conductance and stomatal sensitivity to changes of leaf water status in temperate deciduous trees. *Aust J Plant Physiol* **28**: 765-774.

Armacost, R. R. (1944) The structure and function of the border parenchyma and vein-ribs of certain dicotyledon leaves. *Proc Iowa Acad Sci* **51**: 157-169.

Becker, P., Tyree, M. T. and Tsuda, M. (1999) Hydraulic conductances of angiosperms versus conifers: Similar transport sufficiency at the whole-plant level. *Tree Physiol* **19**: 445-452.

Boyer, J. S. (1974) Water transport in plants: Mechanism of apparent changes in resistance during absorption. *Planta* **117**: 187-207.

Boyer, J. S. (1977) Regulation of water movement in whole plants. *Symp Soc Exp Biol* **31**: 455-470.

Boyer, J. S. (1985) Water transport. *Annu Rev Plant Physiol* **36**: 473-516.

Brodribb, T. J. and Holbrook, N. M. (2003a) Changes in leaf hydraulic conductance during leaf shedding in seasonally dry tropical forest. *New Phytologist* **158**: 295-303.

Brodribb, T. J. and Holbrook, N. M. (2003b) Stomatal closure during leaf dehydration, correlation with other leaf physiological traits. *Plant Physiol* **132**: 2166-2173.

Byott, G. S. and Sheriff, D. W. (1976) Water movement into and through *Tradescantia virginiana* (L.) leaves. 2. Liquid flow pathways and evaporative sites. *J Exp Bot* **27**: 634-639.

Canny, M. J. (1990a) Fine veins of dicotyledon leaves as sites for enrichment of solutes of the xylem sap. *New Phytologist* **115:** 511-516.

Canny, M. J. (1990b) What becomes of the transpiration stream? *New Phytologist* **114:** 341-368.

Canny, M. J. (1993) The transpiration stream in the leaf apoplast: Water and solutes. *Philos Transactions R Soc London Series B-Biological Sciences* **341:** 87-100.

Choong, M. F., Lucas, P. W., Ong, J. S. Y., Pereira, B., Tan, H. T. W. and Turner, I. M. (1992) Leaf fracture toughness and sclerophylly: Their correlations and ecological implications. *New Phytologist* **121:** 597-610.

Cochard, H., Martin, R., Gross, P. and Bogeat-Triboulot, M. B. (2000) Temperature effects on hydraulic conductance and water relations of *Quercus robur* L. *J Exp Bot* **51:** 1255-1259.

Cowan, I. R. (1972) Electrical analog of evaporation from, and flow of water in plants. *Planta* **106:** 221-226.

Cruiziat, P., Tyree, M. T., Bodet, C. and Lo Gullo, M. A. (1980) Kinetics of rehydration of detached sunflower leaves following substantial water loss. *New Phytologist* **84:** 293-306.

Davies, W. J. (1986) Transpiration and the water balance of plants. In *Plant Physiology* (F. C. Steward, ed.) pp. 49-154. Academic Press, Orlando, FL.

Dengler, N. and Kang, J. (2001) Vascular patterning and leaf shape. *Curr Opin Plant Biol* **4:** 50-56.

Dengler, N. G. and Mackay, L. B. (1975) Leaf anatomy of beech, *Fagus grandifolia. Can J Bot* **53:** 2202-2211.

Esau, K. (1965) *Plant Anatomy,* 2nd Ed. John Wiley, New York.

Field, C. B., Ball, J. T. and Berry, J. A. (2000) Photosynthesis: Principles and field techniques. In *Plant Physiological Ecology: Field Methods and Instrumentation* (R. W. Pearcy, J. R. Ehleringer, H. A. Mooney and P. W. Rundel, eds.) pp. 209-253. Kluwer, Dordrecht, the Netherlands.

Field, C. B. and Mooney, H. A. (1986) The photosynthesis: Nitrogen relationship in wild plants. In *On the Economy of Plant Form and Function* (T. J. Givnish, ed.) pp. 25-55. Cambridge University Press, Cambridge.

Fredeen, A. L. and Sage, R. F. (1999) Temperature and humidity effects on branchlet gas-exchange in white spruce: An explanation for the increase in transpiration with branchlet temperature. *Trees-Structure and Function* **14:** 161-168.

Givnish, T. J. (1987) Comparative studies of leaf form: Assessing the relative roles of selective pressures and phylogenetic constraints. *New Phytologist* **106:** 131-160.

Grace, J., Fasehun, F. E. and Dixon, M. (1980) Boundary layer conductance of the leaves of some tropical timber trees. *Plant Cell Environ* **3:** 443-450.

Grubb, P. J. (1984) Some growth points in investigative plant ecology. In *Trends in Ecological Research for the 1980s* (J. H. Cooley and F. B. Golley, eds.) pp. 51-74. Plenum, New York.

Hordon, R. M. (1998) Hydrological Cycle. In *Encyclopedia of Hydrology and Water Resources* (R. W. Herschy and R. W. Fairbridge, eds.) pp. 367-371. Kluwer, Dordrecht, the Netherlands.

Jones, H. G. (1992) *Plants and Microclimate,* 2nd Ed. Cambridge University Press, Cambridge.

Jones, H. G. (1999) Use of thermography for quantitative studies of spatial and temporal variation of stomatal conductance over leaf surfaces. *Plant Cell Environ* **22:** 1043-1055.

Kikuta, S. B., LoGullo, M. A., Nardini, A., Richter, H. and Salleo, S. (1997) Ultrasound acoustic emissions from dehydrating leaves of deciduous and evergreen trees. *Plant Cell Environ* **20:** 1381-1390.

Kitajima, K., Mulkey, S. S. and Wright, S. J. (1997) Decline of photosynthetic capacity with leaf age in relation to leaf longevities for five tropical canopy tree species. *Am J Bot* **84:** 702-708.

Kolb, K. J., Sperry, J. S. and Lamont, B. B. (1996) A method for measuring xylem hydraulic conductance and embolism in entire root and shoot systems. *J Exp Bot* **47:** 1805-1810.

Kramer, P. J. and Boyer, J. S. (1995) *Water Relations of Plants and Soils.* Academic Press, San Diego.

Küppers, M. (1984) Carbon relations and competition between woody species in a central European hedgerow. 2. Stomatal responses, water-use, and hydraulic conductivity in the root-leaf pathway. *Oecologia* **64:** 344-354.

Lamont, B. B. and Lamont, H. C. (2000) Utilizable water in leaves of 8 arid species as derived from pressure-volume curves and chlorophyll fluorescence. *Physiologia Plantarum* **110**: 64-71.

LaRue, C. D. (1931) The water supply of the epidermis of leaves. *Michigan Acad Sci Arts Letters* **12**: 131-139.

Lersten, N. R. (1997) Occurrence of endodermis with a casparian strip in stem and leaf. *Botan Rev* **63**: 265-272.

Levitt, J. (1980) *Responses of Plants to Environmental Stresses, Vol. II: Water, Radiation, Salt and Other Stresses.* Academic Press, New York.

Lo Gullo, M. A., Nardini, A., Trifilo, P. and Salleo, S. (2003) Changes in leaf hydraulics and stomatal conductance following drought stress and irrigation in *Ceratonia siliqua* (carob tree). *Physiol Plantarum* **117**: 186-194.

Machado, J. L. and Tyree, M. T. (1994) Patterns of hydraulic architecture and water relations of two tropical canopy trees with contrasting leaf phenologies: *Ochroma pyramidale* and *Pseudobombax septenatum. Tree Physiol* **14**: 219-240.

Martre, P., Cochard, H. and Durand, J. L. (2001). Hydraulic architecture and water flow in growing grass tillers (*Festuca arundinacea* Schreb.). *Plant, Cell Environ* **24**: 65-76.

Martre, P., Morillon, R., Barrieu, F., North, G. B., Nobel, P. S. and Chrispeels, M. J. (2002) Plasma membrane aquaporins play a significant role during recovery from water deficit. *Plant Physiol* **130**: 2101-2110.

Matzner, S. and Comstock, J. (2001) The temperature dependence of shoot hydraulic resistance: Implications for stomatal behaviour and hydraulic limitation. *Plant Cell Environ* **24**: 1299-1307.

Maximov, N. A. (1931) The physiological significance of the xeromorphic structure of plants. *J Ecol* **19**: 279-282.

McClendon, J. H. (1992) Photographic survey of the occurrence of bundle sheath extensions in deciduous dicots. *Plant Physiol* **99**: 1677-1679.

Meinzer, F. C. (2002) Co-ordination of vapour and liquid phase water transport properties in plants. *Plant Cell Environ* **25**: 265-274.

Meinzer, F. C. and Grantz, D. A. (1990) Stomatal and hydraulic conductance in growing sugarcane: Stomatal adjustment to water transport capacity. *Plant Cell Environ* **13**: 383-388.

Morse, S. R. (1990) Water balance in *Hemizonia luzulifolia*: The role of extracellular polysaccharides. *Plant Cell Environ* **13**: 39-48.

Mott, K. A. and Buckley, T. N. (1998) Stomatal heterogeneity. *J Exp Bot* **49**: 407-417.

Nardini, A. (2001) Are sclerophylls and malacophylls hydraulically different? *Biol Plantarum* **44**: 239-245.

Nardini, A. and Salleo, S. (2000) Limitation of stomatal conductance by hydraulic traits: Sensing or preventing xylem cavitation? *Trees-Structure and Function* **15**: 14-24.

Nardini, A., Salleo, S. and Raimondo, F. (2003) Changes in leaf hydraulic conductance correlate with leaf vein embolism in *Cercis siliquastrum* L. *Trees-Structure and Function* **17**: 529-534.

Nardini, A., Salleo, S., Tyree, M. T. and Vertovec, M. (2000) Influence of the ectomycorrhizas formed by *Tuber melanosporum* Vitt. on hydraulic conductance and water relations of *Quercus ilex* L. seedlings. *Ann Forest Sci* **57**: 305-312.

Nardini, A. and Tyree, M. T. (1999) Root and shoot hydraulic conductance of seven *Quercus* species. *Ann Forest Science* **56**: 371-377.

Nardini, A., Tyree, M. T. and Salleo, S. (2001) Xylem cavitation in the leaf of *Prunus laurocerasus* and its impact on leaf hydraulics. *Plant Physiol* **125**: 1700-1709.

Niklas, K. J. (1999) A mechanical perspective on foliage leaf form and function. *New Phytologist* **143**: 19-31.

Nobel, P. S. and Jordan, P. W. (1983) Transpiration stream of desert species: Resistances and capacitances for a $C_3$, a $C_4$, and a CAM plant. *J Exp Bot* **34**: 1379-1391.

Nonami, H. and Schulze, E. D. (1989) Cell water potential, osmotic potential, and turgor in the epidermis and mesophyll of transpiring leaves: Combined measurements with the cell pressure probe and nanoliter osmometer. *Planta* **177**: 35-46.

Pallardy, S. G. and Kozlowski, T. T. (1979) Stomatal response of *Populus* clones to light intensity and vapor pressure deficit. *Plant Physiol* **64:** 112-114.

Plymale, E. L. and Wylie, R. B. (1944) The major veins of mesomorphic leaves. *Am J Bot* **31:** 99-106.

Reich, P. B., Ellsworth, D. S., Walters, M. B., Vose, J. M., Gresham, C., Volin, J. C. and Bowman, W. D. (1999) Generality of leaf trait relationships: A test across six biomes. *Ecology* **80:** 1955-1969.

Roberts, S. W. (1979) Properties of internal water exchange in leaves of *Ilex opaca* Ait. and *Cornus florida* L. *J Exp Bot* **30:** 955-963.

Rosen, R. (1967) *Optimality Principles in Biology*. Plenum, London.

Roth-Nebelsick, A., Uhl, D., Mosbrugger, V. and Kerp, H. (2001) Evolution and function of leaf venation architecture: A review. *Ann Bot* **87:** 553-566.

Russin, W. A. and Evert, R. F. (1984) Studies on the leaf of *Populus deltoides* (Salicaceae): Morphology and anatomy. *Am J Bot* **71:** 1398-1415.

Sack, L., Cowan, P. D. and Holbrook, N. M. (2003a) The major veins of mesomorphic leaves revisited: Tests for conductive overload in *Acer saccharum* (Aceraceae) and *Quercus rubra* (Fagaceae). *Am J Bot* **90:** 32-39.

Sack, L., Cowan, P. D., Jaikumar, N. and Holbrook, N. M. (2003b) The 'hydrology' of leaves: Co-ordination of structure and function in temperate woody species. *Plant Cell Environ* **26:** 1343-1356.

Sack, L., Melcher, P. J., Zwieniecki, M. A. and Holbrook, N. M. (2002) The hydraulic conductance of the angiosperm leaf lamina: A comparison of three measurement methods. *J Exp Bot* **53:** 2177-2184.

Sack, L., Streeter, C. M. and Holbrook, N. M. (2004) Hydraulic analysis of water flow through leaves of sugar maple and red oak. *Plant Physiol* **134:** 1824-1833.

Salleo, S., Lo Gullo, M. A., Raimondo, F. and Nardini, A. (2001) Vulnerability to cavitation of leaf minor veins: Any impact on leaf gas exchange? *Plant Cell Environ* **24:** 851-859.

Salleo, S. and Nardini, A. (2000) Sclerophylly: Evolutionary advantage or mere epiphenomenon? *Plant Biosystems* **134:** 247-259.

Salleo, S., Nardini, A., Lo Gullo, M. A. and Ghirardelli, L. A. (2002) Changes in stem and leaf hydraulics preceding leaf shedding in *Castanea sativa* L. *Biol Plantarum* **45:** 227-234.

Salleo, S., Raimondo, F., Trifilo, P. and Nardini, A. (2003) Axial-to-radial water permeability of leaf major veins: A possible determinant of the impact of vein embolism in leaf hydraulics? *Plant Cell Environ* **26:** 1749-1758.

Shack et al, K. A. (1987) Direct measurement of turgor and osmotic potential in individual epidermal cells: Independent confirmation of leaf water potential as determined by *in situ* psychrometry. *Plant Physiol* **83:** 719-722.

Shack et al, K. A. and Brinckmann, E. (1985) In situ measurement of epidermal cell turgor, leaf water potential, and gas exchange in *Tradescantia virginiana* L. *Plant Physiol* **78:** 66-70.

Sheriff, D. W. and Meidner, H. (1974) Water pathways in leaves of *Hedera helix* L. and *Tradescantia virginiana* L. *J Exp Bot* **25:** 1147-1156.

Siefritz, F., Tyree, M. T., Lovisolo, C., Schubert, A. and Kaldenhoff, R. (2002) PIP1 plasma membrane aquaporins in tobacco: From cellular effects to function in plants. *Plant Cell* **14:** 869-876.

Sisó, S., Camarero, J. J. and Gil-Pelegrín, E. (2001) Relationship between hydraulic resistance and leaf morphology in broadleaf *Quercus* species: A new interpretation of leaf lobation. *Trees-Structure and Function* **15:** 341-345.

Sobrado, M. A. (1998) Hydraulic conductance and water potential differences inside leaves of tropical evergreen and deciduous species. *Biol Plantarum* **40:** 633-637.

Terashima, I. (1992) Anatomy of non-uniform leaf photosynthesis. *Photosynthesis Res* **31:** 195-212.

Thoday, D. (1931) The significance of reduction in the size of leaves. *J Ecol* **19:** 297-303.

Trifilo, P., Gasco, A., Raimondo, F., Nardini, A. and Salleo, S. (2003a) Kinetics of recovery of leaf hydraulic conductance and vein functionality from cavitation-induced embolism in sunflower. *J Exp Bot* **54:** 2323-2330.

Trifilo, P., Nardini, A., Lo Gullo, M. A. and Salleo, S. (2003b) Vein cavitation and stomatal behaviour of sunflower (*Helianthus annuus*) leaves under water limitation. *Physiol Plantarum* **119:** 409-417.

Tsuda, M. and Tyree, M. T. (2000) Plant hydraulic conductance measured by the high pressure flow meter in crop plants. *J Exp Bot* **51:** 823-828.

Tyree, M. T. (2003a) Hydraulic limits on tree performance: Transpiration, carbon gain and growth of trees. *Trees—Structure and Function* **17:** 95-100.

Tyree, M. T. (2003b) Hydraulic properties of roots. In *Root Ecology Ecological Studies, Vol.168* (E. J. W. Visser and H. de Kroon, eds.) pp. 125-150. Springer-Verlag, Berlin.

Tyree, M. T., Benis, M. and Dainty, J. (1973) Water relations of hemlock (*Tsuga canadensis*). 3. Temperature dependence of water exchange in a pressure bomb. *Can J Bot* **51:** 1537-1543.

Tyree, M. T., Caldwell, C. and Dainty, J. (1975) Water relations of hemlock (*Tsuga canadensis*). 5. Localization of resistances to bulk water flow. *Can J Bot* **53:** 1078-1084.

Tyree, M. T. and Cheung, Y. N. S. (1977) Resistance to water flow in *Fagus grandifolia* leaves. *Can J Bot* **55:** 2591-2599.

Tyree, M. T., Cruiziat, P., Benis, M., Logullo, M. A. and Salleo, S. (1981) The kinetics of rehydration of detached sunflower leaves from different initial water deficits. *Plant Cell Environ* **4:** 309-317.

Tyree, M. T., Nardini, A. and Salleo, S. (2001) Hydraulic architecture of whole plants and single leaves. In *L'arbre 2000 the Tree* (M. Labrecque, ed.) pp. 215-221. Isabelle Quentin Publisher, Montreal.

Tyree, M. T., Sinclair, B., Lu, P. and Granier, A. (1993) Whole shoot hydraulic resistance in *Quercus* species measured with a new high-pressure flowmeter. *Annales Des Sciences Forestieres* **50:** 417-423.

Tyree, M. T., Snyderman, D. A., Wilmot, T. R. and Machado, J. L. (1991) Water relations and hydraulic architecture of a tropical tree (*Schefflera morototoni*): Data, models, and a comparison with two temperate species (*Acer saccharum* and *Thuja occidentalis*). *Plant Physiol* **96:** 1105-1113.

Tyree, M. T., Sobrado, M. A., Stratton, L. J. and Becker, P. (1999) Diversity of hydraulic conductance in leaves of temperate and tropical species: Possible causes and consequences. *J Trop Forest Sci* **11:** 47-60.

Tyree, M. T. and Wilmot, T. R. (1990) Errors in the calculation of evaporation and leaf conductance in steady-state porometry: The importance of accurate measurement of leaf temperature. *Can J Forest Res* **20:** 1031-1035.

Tyree, M. T. and Yianoulis, P. (1980) The site of water evaporation from sub-stomatal cavities, liquid path resistances and hydroactive stomatal closure. *Ann Bot* **46:** 175-193.

Tyree, M. T. and Zimmermann, M. H. (2002) *Xylem Structure and the Ascent of Sap.* Springer, Berlin.

U.S. Geological Survey (2004) The water cycle. http://ga.water.usgs.gov/edu/watercycle.html.

Van Fleet, D. S. (1950) The cell forms, and their common substance reactions, in the parenchyma-vascular boundary. *Bull Torrey Botanical Club* **77:** 340-353.

Vogel, S. (1968) Sun leaves and shade leaves differences in convective heat dissipation. *Ecology* **49:** 1203-1204.

Vogel, S. (1970) Convective cooling at low airspeeds and shapes of broad leaves. *J Exp Bot* **21:** 91-101.

Von Willert, D. J., Eller, B. M., Werger, M. J. A. and Brinckmann, E. (1990) Desert succulents and their life strategies. *Vegetatio* **90:** 133-143.

Warrit, B., Landsberg, J. J. and Thorpe, M. R. (1980) Responses of apple leaf stomata to environmental factors. *Plant Cell Environ* **3:** 13-22.

Weatherley, P. E. (1963) The pathway of water movement across the root cortex and leaf mesophyll of transpiring plants. In *The Water Relations of Plants* (A. J. Rutter and F. H. Whitehead, eds.) pp. 85-100. John Wiley, New York.

Wright, I. J., Reich, P. B. and Westoby, M. (2001) Strategy shifts in leaf physiology, structure and nutrient content between species of high- and low-rainfall and high- and low-nutrient habitats. *Functional Ecol* **15:** 423-434.

Wright, I. J. and Westoby, M. (2002) Leaves at low versus high rainfall: Coordination of structure, lifespan and physiology. *New Phytologist* **155:** 403-416.

Wylie, R. B. (1946) Relations between tissue organization and vascularization in leaves of certain tropical and subtropical dicotyledons. *Am J Bot* **33:** 721-726.

Wylie, R. B. (1951) Principles of foliar organization shown by sun-shade leaves from ten species of deciduous dicotyledonous trees. *Am J Bot* **38:** 355-361.

Wylie, R. B. (1952) The bundle sheath extension in leaves of dicotyledons. *Am J Bot* **39:** 645-651.

Yang, S. D. and Tyree, M. T. (1994). Hydraulic architecture of *Acer saccharum* and *A. rubrum:* Comparison of branches to whole trees and the contribution of leaves to hydraulic resistance. *J Exp Bot* **45:** 179-186.

Zwieniecki, M. A., Boyce, C. K. and Holbrook, N. M. (2004) Hydraulic limitations imposed by crown placement determine final size and shape of *Quercus rubra* L. leaves. *Plant Cell Environ* **27:** 357-365.

Zwieniecki, M. A., Hutyra, L., Thompson, M. V. and Holbrook, N. M. (2000) Dynamic changes in petiole specific conductivity in red maple (*Acer rubrum* L.), tulip tree (*Liriodendron tulipifera* L.) and northern fox grape (*Vitis labrusca* L.). *Plant Cell Environ* **23:** 407-414.

Zwieniecki, M. A., Melcher, P. J., Boyce, C. K., Sack, L. and Holbrook, N. M. (2002) Hydraulic architecture of leaf venation in *Laurus nobilis* L. *Plant Cell Environ* **25:** 1445-1450.

# 6

## Interaction of Phloem and Xylem During Phloem Loading: Functional Symplasmic Roles for Thin- and Thick-Walled Sieve Tubes in Monocotyledons

*C. E. J. Botha*

The phloem in monocotyledons differs from that found in leaves of dicotyledons and eudicotyledons in that two distinct sieve tube types are recognized in intermediate and small vascular bundles. These are the early metaphloem sieve elements, which are characterized by their association with companion cells, and a second type, called the late-formed sieve elements, which characteristically have thickened walls and lack companion cell associations. The existence of two distinct sieve tube types raises questions of their structural and functional roles in phloem loading. Companion cells are known to play a central role in the phloem of dicotyledonous plants. Indeed, Oparka and Turgeon (1999) have likened the companion cells to the phloem's traffic control centers. Thus, the absence of companion cells in thick-walled sieve tubes raises the question of whether, if traffic control is absent, thick-walled sieve tubes in monocots are necessarily nonfunctional. Alternatively, is the control function simply taken over by the adjacent large vascular parenchyma cells, which could possibly function as pseudo companion-cells?

This chapter focuses on structure-function relationships that influence phloem loading in monocotyledonous leaves. These major issues are considered: (1) What are the structural implications associated with the coexistence of two distinct sieve tube types? (2) Do companion cells function in the same manner as those associated with sieve elements in dicotyledonous plants or are the companion cell's functions taken over by parenchymatous elements associated with the thick-walled sieve elements? (3) Is the vascular parenchyma implicated in solute retrieval from the xylem and does retrieved solute end up in both thin- and thick-walled sieve tubes? The use of fluorescent dyes to visualize symplasmic pathways for solute movement in monocotyledonous leaves is reviewed with emphasis on understanding the functional roles of the two types of sieve tubes, as well as the existence of a solute retrieval pathway between the xylem and the phloem.

## Structural Considerations of the Loading Pathway

Our understanding of vasculature within the leaf blades of monocotyledons, with the exception of the grasses, remains poor (see Botha *et al.*, 1982 and references cited; Botha, 1992; Cartwright *et al.*, 1977; Dannenhoffer *et al.*, 1990; Evert, 1986; Evert *et al.*, 1978, 1996a, 1996b; Evert and Russin, 1993; Farrar *et al.*, 1992; Haupt *et al.*, 2001; Kuo and O'Brien, 1974; Lush, 1976; Matsiliza and Botha, 2002; Russell and Evert, 1985). Most monocotyledonous leaf blades contain but three longitudinal vein sizes or classes, excluding the marginal vein network, the midrib vein system and the cross-veins. Longitudinal vein structure within a typical grass leaf can be defined according to vein size, presence or absence of supporting mechanical tissues, and arrangement of the vascular tissues within the veins themselves (Botha *et al.*, 1982). The large veins contain large metaxylem vessels and a protoxylem lacuna and are always associated with hypodermal fibrous sclerenchymatous girders, which interrupt the chlorenchymatous bundle sheath and form a support structure connected to the adaxial and abaxial epidermis. In contrast, intermediate veins lack large metaxylem vessels and are usually supported by hypodermal sclerenchyma associated with one surface only. Small veins are usually entirely surrounded by a chlorenchymatous bundle sheath, and are thus structurally analogous to the minor veins in dicotyledonous leaves. Like intermediate veins, the xylem in mature small veins is not associated with a protoxylem lacuna (see Botha *et al.*, 1982; Evert *et al.*, 1996b for further explanation).

The phloem in monocotyledons is unique in that two types of sieve tubes exist in mature vascular bundles. The first fits the classical description of phloem sieve elements (SE) being associated with companion cells (Evert *et al.*, 1977, 1985, 1996a, 1996b, and references cited; Botha and Cross, 1997 and references cited; Haupt *et al.*, 2001). These sieve tubes have thin walls. The second type, and the last cells to differentiate in the leaf blade bundles, is usually closely associated with the metaxylem vessels. These late-formed sieve elements have thick walls and lack recognizable companion cells (even though this is incorrectly reported in sink leaf tissue in Haupt *et al.*, 2001). In all but two instances (Cartwright *et al.*, 1977; Kuo and O'Brien, 1974), they are reportedly not lignified. Thick-walled sieve tubes were first reported in wheat (*Triticum aestivum*) by Kuo and O'Brien (1974) and later found to occur in all Gramineae (see Walsh, 1974; Miyake and Maeda, 1976; Cartwright *et al.*, 1977; Evert, *et al.*, 1977, 1978, 1996a; Evert, 1986 and references cited; Botha, 1992; Botha and van Bel, 1992; Evert and Russin 1993; Botha and Cross, 1997; Haupt *et al.*, 2001 and literature cited) as well as other monocotyledons, such as the Commelinaceae (van Bel *et al.*, 1988). Structural relationships within the vascular bundles are illustrated in the section of an intermediate vascular bundle of *Eragrostis curvula,* a common veld grass in southern Africa (Fig. 6.1; see also Color Plate section). The bundle contains two late-formed thick-walled sieve tubes (cells shown with solid dots) that are in close spatial proximity to the metaxylem. A column of

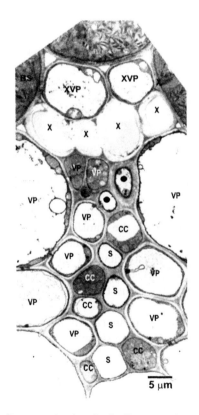

**Figure 6.1** Intermediate vascular bundle in *Eragrostis plana,* a $C_4$ grass, showing the phloem, which is composed of two thick-walled (*solid dots*) and four functional thin-walled sieve tubes. Thick-walled sieve tubes are close to xylem. The thin-walled sieve tubes (S) are associated with companion cells (CC) as well as vascular parenchyma (VP) cells. The thick-walled sieve tubes (*solid circles*) are associated with vascular parenchyma only. Note the very large lateral vascular parenchyma cells situated on either side of the central phloem core in this bundle. (See also Color Plate section.)

four thin-walled metaphloem sieve tubes occurs centrally, surrounded by contiguous companion cells and phloem parenchyma cells. Note also the vascular parenchyma above the tracheary elements that are in contact with bundle sheath cells, as well as the large vascular parenchyma cells below the xylem elements that have wall contact with the thick-walled sieve tubes.

The proportion of thin- to thick-walled SEs changes with vein order. In the small longitudinal veins, the ratio of thin- to thick-walled sieve tubes is much lower (between 1 and 3 thin-walled:1 thick-walled) than is commonly observed in the large longitudinal veins (up to 5 to 10 thin-walled:1 to 2 thick-walled; Colbert and Evert, 1982; Russell and Evert, 1985; Dannenhoffer *et al.*, 1990). The smallest veins usually contain one thin-walled and one thick-walled sieve tube, but Dannenhoffer *et al.* (1990) report that the smallest of the small veins in barley (*Hordeum vulgare*) may contain only one thick-walled sieve tube and no thin-walled sieve tubes.

Xylem and phloem elements within all three vein size classes are associated with vascular parenchyma cells. These cells are often quite large and are frequently dominant features of the vascular bundles (Fig. 6.1). Xylem vessels are generally closely coupled to parenchymatic elements via pit membranes (Fig. 6.2; see also Color Plate section), suggesting that this interface may have an important physiological role in solute exchange. Figure 6.3 (see also Color Plate section) shows the highly fenestrated, swollen wall area between the vascular parenchyma cell and the tracheary element (boxed area in Fig. 6.2) in more detail. This structure was first reported by Evert *et al.* 1978 (see their Fig. 2). The hydrolyzed, loosely-fibrillar wall appears to be highly porous in the pit region, which could provide a relatively easy passage for the transfer of water and solutes from the tracheary element to the vascular parenchyma cell and vice versa. This route may also be available for retrieval of solutes from the xylem.

Even though numerous transverse (also termed lateral or cross) veins interconnect the longitudinal veins, there is good evidence from a number of sources suggesting that these lateral veins are not involved in assimilate uptake, but instead function primarily in lateral transfer of assimilate. [14]C-labeled photosynthate has been shown to accumulate in the small longitudinal veins of *Panicum*, with the label being recorded in the transverse veins some time after the main pulse of [14]C had passed out of the leaves (Lush, 1976). A subsequent paper by Fritz *et al.* (1983) showed conclusively that the thin-walled sieve tubes of the small and intermediate longitudinal veins in maize (*Zea mays*) were the channels into which [14]C-labeled photosynthate was loaded. However, it is important to note that lack of evidence for involvement in active phloem loading of the lateral veins does not preclude a role in temporary storage or in redirecting assimilate streams. It is also possible that these lateral veins may serve to coordinate the function of the different longitudinal vein orders through an as yet unknown mechanism (Lush, 1976). At the very least, cross-veins function

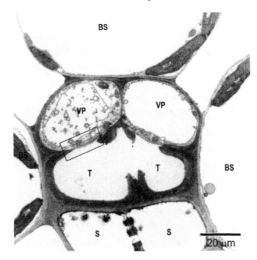

**Figure 6.2**  Ultrastructure of a small vein in a wheat leaf, which contains two vascular parenchyma cells (VP) associated with two tracheary elements (T), below which are two sieve tubes (S). Note that no companion or vascular parenchyma cells are visible in this micrograph. (See also Color Plate section.)

as transfer conduits between the small, intermediate, and large veins in the leaf blade, which we know to be involved in uptake and transfer, transport, and export of assimilate, respectively.

## Role of Thin- and Thick-Walled Sieve Tubes

Phloem loading has been described as apoplasmic, symplasmic, or mixed-mode apoplasmic-symplasmic (see review by van Bel, 1993), with one basis

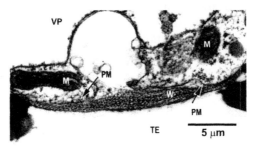

**Figure 6.3**  Details of the highly fenestrated wall between vascular parenchyma cells and tracheary elements in wheat (boxed region in Fig. 6.2). The plasmamembrane (PM) is evident at the boundary between the pit and the cytoplasm of the parenchyma cell. Parenchyma cell contains several mitochondria (M). Note the highly porous wall in the pit region, which could allow the transfer of water and solutes from the vascular parenchyma cell to the tracheary element and vice versa. (See also Color Plate section.)

for classification being the presence or absence of plasmodesmata along the loading pathway. If plasmodesmata are present along the loading pathway, then the loading process is assumed to be, at least in part, symplasmic. Alternatively, if plasmodesmata are absent at one or more interfaces, the loading process is considered to be apoplasmic. A cross-section of a mature intermediate vascular bundle from barley (Fig. 6.4; see also Color Plate section) illustrates that the thick-walled sieve tubes are poorly connected to associated vascular parenchyma cells (0.24% plasmodesma per micrometer common cell wall), yet the parenchymatic elements within the bundle are well connected to each other (29.8% plasmodesma per micrometer cell wall) and to the mestome sheath (33.2% plasmodesma per micrometer cell wall). Similarly, the thin-walled sieve tubes are poorly connected with their companion cells (0.2% plasmodesma per micrometer cell wall), which are in turn poorly connected to adjacent vascular parenchyma cells (0.5% plasmodesma per micrometer cell wall between the companion cells and surrounding vascular parenchyma). These data suggest a substantial degree of symplasmic isolation of both thin- and thick-walled sieve tubes consistent with a predominantly apoplastic loading mechanism. Symplasmic isolation of thick-walled sieve elements from the surrounding vascular parenchyma has also been inferred from electrophysiological studies using barley (Farrar et al., 1992; Botha and Cross, 1997). Low plasmodesmatal connectivity between parenchymatous elements and sieve tubes appears to be the general pattern in monocotyledons (Table 6.1). Zea mays stands out as a marked exception, with thick-walled sieve tubes that are well connected to adjacent vascular parenchyma, while a similar frequency of plasmodesma exists at the interface between the thin-walled sieve tubes and their companion cells (Table 6.1). However, even in Zea, thin-walled sieve tubes and their companion cells have few symplasmic connections with adjacent vascular parenchyma.

Grasses are known to host a range of systemic viruses, many of which move in the phloem (see Oparka et al., 1999 and references cited). Because viral transmission requires functional plasmodesmata along the transport route (Roberts et al., 2003 and references cited), the phloem in monocotyledons cannot be as symplasmically isolated as is suggested by plasmodesmal frequency studies. Clearly, some intercellular connections via plasmodesmata must exist in mature leaves to accommodate virus transmission to the phloem.

A number of studies indicate that phloem loading in monocotyledons involves only the thin-walled sieve tubes. Cartwright et al. (1977) and later Fritz et al. (1983) used microautoradiography to suggest that phloem loading is limited to thin-walled sieve tubes. This conclusion is supported by plasmolytic studies (Evert et al., 1978), which demonstrated that osmotic potentials are much higher in thin-walled than in thick-walled SE in Zea. Matsiliza

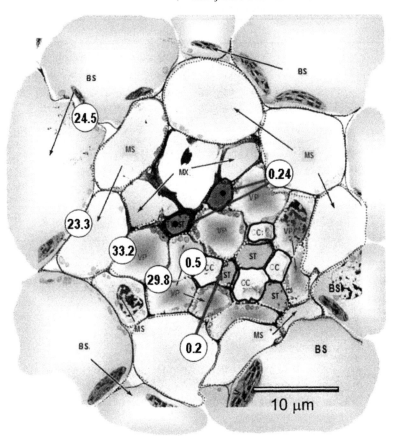

**Figure 6.4**  Plasmodesmal frequency calculated on a per micrometer common cell wall interface basis for a mature intermediate leaf vascular bundle from barley (*Hordeum vulgare* L. cv. Dyan). Note the close spatial association between the thick-walled sieve tubes (*solid dots*) and the metaxylem vessels (MX). Thick-walled sieve tubes are separated from the earlier thin-walled metaphloem, which comprises parenchymatic elements (VP), sieve tubes (ST), and associated companion cells (CC). MS = mestome sheath cells; BS = bundle sheath cells. (See also Color Plate section.)

and Botha (2002) and Botha and Matsiliza (2004) provided the first direct evidence that the aphid, *Sitobion yakini*, has a distinct preference for thin-walled sieve tubes in the small longitudinal veins of barley. Given that the aphids' stylets often passed through xylem conduits, as well as thick-walled sieve tubes, along the way to their preferred feeding site, it is unlikely that wall composition or thickness influenced their preference for thin-walled sieve tubes. Together, these studies indicate that, in both barley and in wheat, thick-walled sieve tubes are not involved in normal pathways of phloem loading, raising the question of what function, if any, these cells might provide.

**Table 6.1**  Summary of plasmodesmal frequencies between bundle sheath/mestome sheath cells

| Species | Cell-Cell Interface | | | |
|---|---|---|---|---|
| | **Bundle Sheath/Mestome Sheath** | | | |
| | VP | CC | ST | TWST |
| *Zea mays*[1] | **** | ⊕ | ⊗ | ∅ |
| *Themeda triandra*[2] | **** | ∅ | ∅ | – |
| *Hordeum vulgare* cv Morex[3] | **** | ⊗ | ⊕ | ⊕ |
| *Hordeum vulgare* cv Dyan[4] | **** | ⊕ | ∅ | – |
| *Eragrostis plana*[5] | **** | ∅ | ∅ | – |
| *Panicum maximum*[5] | **** | ⊗ | ∅ | – |
| *Bromus unioloides*[5] | **** | ∅ | ∅ | – |
| | **Vascular Parenchyma** | | | |
| | VP | CC | ST | TWST |
| *Zea mays*[1] | *** | ⊕ | ∅ | *** |
| *Themeda triandra*[2] | | ⊕ | ⊗ | ⊕ |
| *Hordeum vulgare* cv Morex[3] | **** | ⊕ | ∅ | ⊗ |
| *Hordeum vulgare* cv Dyan[4] | **** | ⊕ | ⊗ | ⊗ |
| *Eragrostis plana*[5] | ** | * | ⊕ | ⊗ |
| *Panicum maximum*[5] | ** | ⊕ | ⊗ | ⊕ |
| *Bromus unioloides*[5] | **** | * | ⊕ | ⊗ |
| | **Companion Cells** | | | |
| | CC | ST | TWST | |
| *Zea mays*[1] | ⊕ | *** | | |
| *Themeda triandra*[2] | ⊗ | ⊗ | ∅ | |
| *Hordeum vulgare* cv Morex[3] | ⊗ | ** | ⊗ | |
| *Hordeum vulgare* cv Dyan[4] | ⊕ | ⊗ | ∅ | |
| *Eragrostis plana*[5] | ⊕ | ⊕ | ∅ | |
| *Panicum maximum*[5] | ∅ | ⊕ | ∅ | |
| *Bromus unioloides*[5] | ⊗ | ⊕ | ∅ | |

VP = vascular parenchyma; CC = companion cells; ST = thin-walled sieve tubes; TWST = thick-walled sieve tubes in source leaves

[a] = Bundle sheath or vascular parenchyma, concomitant with vascular parenchyma

**** = plentiful/high

*** = abundant

** = low

⊕ = scarce

⊗ = rare

∅ = absent

– = not structurally possible

1 = Evert *et al.*, 1977, aggregated data for small and intermediate vascular bundles; 2 = Botha and Evert, 1988; 3 = Evert *et al.*, 1996a; 4 = Botha and Cross, 1997; 5 = Botha, 1992.

# Experimental Evidence for Apoplast/Symplast Transfer Between Xylem and Phloem

Resolution of issues surrounding plasmodesmal continuity, symplasmic loading and unloading pathways, the role of cross-veins in phloem loading, and the functionality of thick-walled and thin-walled sieve tubes with contiguous parenchymatic elements requires visualization of movement within the symplast. The fluorescent probe 5,6-carboxyfluorescein (5,6-CF) has proved to be extremely useful in previous studies of phloem physiology (see Grignon *et al.*, 1989). It does not fluoresce when applied in the diacetate form and is readily taken up by plant cells, usually across walls and membranes of damaged cells. Once contained within physiologically intact systems, the molecule is cleaved, releasing free 5,6-CF, which fluoresces. 5,6-CF is polar and therefore reasonably membrane-impermeant, although it is known to accumulate in some cell vacuoles (see Wright *et al.*, 1996). Appearance in contiguous cells is thought to require plasmodesmal trafficking. Grignon *et al.* (1989) reported that 5,6-CF remained confined to the phloem for up to 4 days, demonstrating both the stability and tissue-level specificity of this molecule. A word of caution may be necessary, however, as the $pK_a$ of 5,6-CF (estimated at 6.3-6.4 at pH 7) suggests that a small amount (less than 0.1%, Wright *et al.*, 1996) could possibly remain undissociated in the phloem, which generally has a pH of approximately 8 (Delétage-Grandon *et al.*, 2001). In longer term experiments, even this small amount could make a significant contribution to the apparent movement of 5,6-CF (Wright and Oparka, 1996). In addition, because 5,6-CF is soluble in water, it will diffuse away from cut or damaged cells, thereby potentially affecting interpretation of its localization.

Application of 5,6-carboxyfluorescence diacetate (5,6-CFDA) to a mature barley leaf demonstrates that the fluorescence appears initially in both the cytoplasm and the vacuole, suggesting that some diacetate was transported across the tonoplast (Fig. 6.5; see also Color Plate section). However, several cells distant from the point of application, fluorescence is localized to the cytoplasm, with little evidence of vacuolar sequestration of 5,6-CFDA. Five centimeters from the point of application, fluorescence is associated only with the vascular tissue. No evidence of outward leakage from the bundle sheath is evident, indicating that local unloading does not occur via the bundle sheath after uptake into the phloem (Fig. 6.6; see also Color Plate section). The appearance of the dye in the phloem of cross-veins supports their involvement in assimilate transport in mature leaves (Figs. 6.7 and 6.8; see also Color Plate section).

Experiments carried out by Heyser *et al.* (1978) confirmed that sucrose is loaded into the companion cell-sieve tube complexes from the apoplast of the vascular bundles in *Z. mays* leaf strips. Subsequently, Fritz *et al.* (1983) showed that $^{14}$C-sucrose was retrieved from the xylem in strips of *Z.*

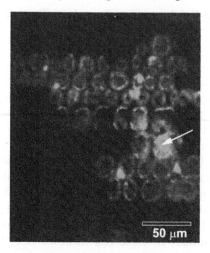

**Figure 6.5**   Surface view of a barley leaf 10 minutes after gentle scraping and application of 5,6-CF diacetate to the abraded portion of the leaf (*arrow*). Note that the fluorescence has spread to adjacent cells, and that those further away from the point of application show 5,6-carboxyfluorescine within the cytosol and less evidently associated with the vacuole. (See also Color Plate section.)

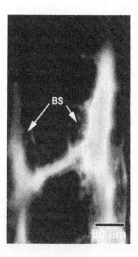

**Figure 6.6**   Epifluorescence micrograph showing uptake of 5,6-CF and symplasmic transport into mesophyll cells in *Hordeum vulgare* leaf 10 minutes after application to a lightly-abraded region of the leaf, unlabelled arrow points to abraded cell to which 5,6-carboxyfluorescein diacetate was applied. Note that 5,6-CF has spread to a large number of cells during this short period, but that the bundle sheath cells (BS) are essentially devoid of 5,6-CF. Thin strands of fluorescence are cytoplasm-related; no evidence of vacuolar uptake is evident in this or other similarly-treated tissue. (See also Color Plate section.)

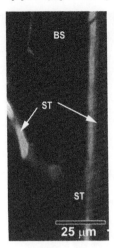

**Figure 6.7**   Confocal image showing localization of 5,6-CF within the single sieve tube in a cross-vein (*left*) and a sieve tube in a small intermediate longitudinal leaf blade vascular bundle (*right*) in wheat. (See also Color Plate section.)

*mays* leaf. [14]C-sucrose was introduced to the xylem via the cut ends of mature leaves, where it was transported in the transpiration stream. The authors demonstrated that vascular parenchyma cells abutting vessels retrieved [14]C-sucrose from the xylem and transferred it to the thick-walled sieve tubes, presumably symplasmically, given the abundant plasmodesma between vascular parenchyma cells and the thick-walled sieve tubes in this species (Table 6.1). With time, the thin-walled sieve tubes also became labeled. Based on their work, Fritz *et al.* (1983) suggested that the [14]C-sucrose in the thin-walled sieve tubes must have been loaded from the apoplast, as the thin-walled sieve tubes and their associated companion cells have been shown to be virtually isolated symplasmically from the thick-walled sieve tubes. Because there are so few plasmodesmata at the vascular parenchyma-companion cell interface, Fritz *et al.* (1983) concluded that the thin-walled sieve tubes can accumulate sucrose directly from the apoplast, without the involvement of companion cells. The large volume of solute introduced into the apoplast in these experiments is not normally present in the xylem, making it difficult to say with any certainty the extent to which this pathway operates in an intact leaf. Nevertheless, these observations demonstrate that solute can move from the xylem into the surrounding vascular parenchyma cells, and from there to the phloem.

The experiments of Heyser *et al.* (1978) raise intriguing questions regarding the existence of a retrieval pathway from the xylem to the vascular parenchyma. To explore the potential role of an apoplast:symplast transfer

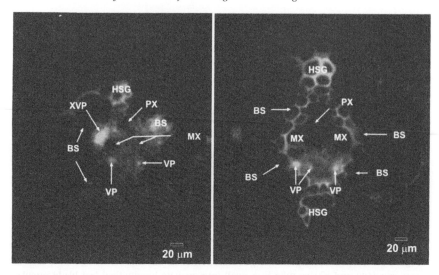

**Figures 6.8 and 6.9** Distribution of 5,6-CF in small and intermediate respectively, vascular bundles in *Zea mays* leaf blade tissue, showing recovery of 5,6-CF from the xylem after 120 (Fig. 6.8, *left*) and 180 (Fig. 6.9, *right*) minutes uptake of 5,6-CFDA from the cut end of a severed source leaf of *Zea mays*. Sections cut in silicone oil and viewed with a narrow-band filter set specific for 5,6-CF without interference from autofluorescence associated with lignin. Fluorescence is intense in the parenchyma located on the xylem side of the bundle (XVP), as well as in bundle sheath (BS) and vascular parenchyma cells, but is absent from the xylem, suggesting that dissociation of 5,6-CFDA occurred within the xylem parenchyma. One vascular parenchyma and one thin-walled sieve element contain intense label. After 180 minutes (Fig. 6.9, *right*) the fluorescence intensity is greatest in the phloem, with vascular parenchyma and thin-walled sieve elements showing intense label. Hypodermal sclerenchyma girder cells (HSG) contain label as well. (See also Color Plate section.)

process linking xylem and phloem in monocotyledonous leaves, mature leaves of *Z. mays* were severed near their base and immediately placed in a solution of 5,6-CFDA (1.78 or 3.56 mM) and allowed to transpire freely. After various time intervals, the leaves were removed and examined with an epifluorescence microscope. Regions of interest (~6 cm from the cut basal end of the leaf) were sectioned at 15 μm into silicone oil (which prevents diffusion of the fluorochrome) and examined using an appropriate Chroma (# 41028 yellow GFP 10C/Topaz) filter set, which limited lignin autofluorescence. After 120 minutes, fluorescence was visible in the parenchyma cells on the xylem side of the vascular bundle and also in some of the bundle sheath cells (Fig. 6.8). After 180 minutes, the fluorescence appears mostly in the parenchymatous elements of the phloem (Fig. 6.9; see also Color Plate section). It is not possible to identify sieve tubes in these thick sections; nevertheless, it is likely that some of the cells that are fluorescing in the phloem may be sieve tubes, as sieve tubes containing 5,6-CF have frequently been seen with the use of confocal microscopy (Fig. 6.7).

Given the progression from the xylem through the parenchymatous cells abutting the tracheary elements, and the relatively slow movement of the fluorophore, it is highly unlikely that any significant level of undissociated 5,6-CFDA persists beyond the parenchymatous elements abutting the xylem elements. The most likely pathway for 5,6-CFDA movement on exiting the xylem vessel via pit membranes (Fig. 6.3) is across the plasmamembrane of adjacent vascular parenchyma cells, from which point the dye could be carried passively via the symplast to other, more distant cells. The uptake experiments described here demonstrate that 5,6-CF appears first in the xylem parenchyma; subsequently, some is transferred radially via the bundle sheath cells, before fluorescence finally appears in the phloem tissue in parenchymatous elements, but also probably within sieve elements as well (Fig. 6.9). Functional plasmodesma must be present, or this retrieval pattern could not occur. Retrieval of solutes from the xylem via the hydrolyzed pit membranes is not restricted to sucrose as reported by Heyser *et al.* (1978), but can be accessed by other substances as well. Opalka *et al.* (2003) confirmed that the pit membranes between xylem vessels and vascular parenchyma cells offer a pathway through which rice yellow mottle virus can enter the symplast of concomitant vascular parenchyma cells in *Oryza sativa* leaves.

## Concluding Remarks

Structurally as well as functionally, phloem loading in monocotyledons is complicated by the presence of thin-walled sieve tubes associated with companion cells, and thick-walled sieve tubes more spatially allied with the xylem.

The xenobiotic probe 5,6-CFDA is taken up from the mesophyll and, once dissociated to its fluorescent (5,6-CF) form, ends up in functional sieve tubes, some distance (>6 cm in leaves; and also in sieve tubes in roots, after 24 hours). The slow rate of uptake (measured in hours) is strong evidence for the presence of a symplasmically loaded, plasmodesmata-requiring pathway, which must function in tandem with the more important and efficient apoplasmic phloem loading system. Conventional fluorescence and confocal microscopy leaves no doubt that cross-veins are involved in the transfer of the phloem-mobile 5,6-CF between longitudinal veins, or that it is the sieve tubes that ultimately transport the fluorochrome.

Experimental evidence presented here suggests that vascular parenchyma cells are involved in the recovery of solute from the xylem. Application of the probe 5,6-carboxyfluorescein into the transpiration stream shows that the undissociated tracer exits the xylem via hydrolyzed pit membranes, then enters living parenchyma cells and moves toward the phloem within the symplast. The extent to which this xylem:phloem retrieval process operates in intact leaves is not known. However, as all grasses studied to date exhibit the same or similar xylem to parenchyma offloading patterns, it is highly probable that this route exists broadly across this family. Nevertheless, it is important to stress that the xylem to phloem pathway described here for the first time is unlikely to be a major transport route, as assimilates and other solutes are unlikely to be present in the xylem in such high concentrations as in these experiments.

The role of the thick-walled sieve tubes remains uncertain. Higher resolution techniques are required, including perhaps a "probe cocktail" with a selective symplasmic state, before we can ascribe a role for the thick-walled sieve tubes and, more specifically, to symplasmically-mediated transport between the vascular parenchyma and the thick-walled sieve tubes. For the moment, the thick-walled sieve tubes of monocotyledons remain an enigma. Perhaps the thick-walled sieve tubes have lost functionality. Scanning electron micrograph images of fossil leaf fragments of a $C_4$ species some 7 to 5 million years old, (Thomasson et al., 1986), which are similar to modern-day chloridoid grasses, suggests that thick-walled sieve tubes have occurred within the phloem for a long time. Much remains unanswered and further work involving real-time confocal microscopic studies of the movement of xenobiotic fluorochromes is needed. Such studies may provide further clarification of the phloem loading pathway and intercellular transport and symplasmic connectivity within the vascular bundles of monocot leaves. Finally, the slow loading and uptake rate observed under experimental conditions leads to speculation that it may represent a relic of some inefficient ancestral symplasmic loading system, which, as we have seen, is still exploited by viruses today.

## Acknowledgments

Dr. Michael Knoblauch and Professor A. J. E. van Bel of the Botanical Institute Giessen University are gratefully acknowledged for assistance and for making the laser scanning confocal microscope available with which Fig. 6.7 was obtained.

## References

Botha, C. E. J. (1992) Plasmodesmatal distribution structure and frequency in relation to assimilation in $C_3$ and $C_4$ grasses in southern Africa. *Planta (Heidelb)* **187:** 348-358.

Botha, C. E. J. and Cross, R.H.M. (1997) Plasmodesmatal frequency in relation to short-distance transport and phloem loading in leaves of barley (*Hordeum vulgare*). Phloem is not loaded directly from the symplast. *Physiol Plant* **99**(3): 355-362.

Botha, C. E. J. and Evert, R.F. (1988) Plasmodesmatal destribution and frequency in vascular bundles and continguous tissues of the leaf of *Themeda triandra*. *Planta (Heidelb)* **173**: 433-441.

Botha, C. E. J., Evert, R. F., Cross, R. H. M. and Marshall, D. (1982) Comparative anatomy of mature *Themeda triandra* Forsk. Leaf blades: a correlated light and electron microscope study. *J. S. Afr. Bot* **48**: 311-328.

Botha, C. E. J. and Matsiliza, B. (2004) Reduction in transport in wheat (*Triticum aestivum*) is caused by sustained phloem feeding by the Russian wheat aphid (*Diuraphis noxia*). *South Af J Bot.* **70**(2): 249-254.

Botha, C. E. J. and van Bel, A.J.E. (1992) Quantification of symplastic continuity as visualised by plasmodesmograms: Diagnostic value for phloem-loading pathways. *Planta* **187**: 359-366.

Cartwright, S. C., Lush, W. M. and Canny, M. J. (1977) A comparison of translocation of labelled assimilate by normal and lignified sieve elements in wheat leaves. *Planta* **134**: 207-208.

Colbert, J. T. and Evert, R. F. (1982) Leaf vasculature in sugarcane (*Saccharum officinarum* L.). *Planta* **156**: 136-151.

Dannenhoffer, J. M., Ebert, W. and Evert, R. F. (1990) Leaf vasculature in barley, *Hordeum vulgare* (Poaceae). *Am J Bot* **77**: 636-652.

Delétage-Grandon, C., Chollet, J.-F., Faucher, M., Rocher, F., Komor, E. and Bonnemain, J.-L. (2001) Carrier-mediated uptake and phloem system of a 350-dalton chlorinated xenobiotic with an α-amino acid function1. *Plant Physiol* **125**(4): 1620-1632.

Evert, R. F. (1986) Phloem loading in maize. In *Regulation of Carbon and Nitrogen Reduction and Utilization in Maize* (J. C. Shannon, D. P. Knievel and C. D. Bayer, eds.) pp. 67-81. American Society of Plant Physiologists, Rockville, MD.

Evert, R. F., Botha, C. E. J. and Mierzwa, R. J. (1985) Free space marker studies on the leaf of *Zea mays* L. *Protoplasma* **126**: 62-73.

Evert, R. F., Eschrich, W. and Heyser, W. (1977) Distribution and structure of the plasmodesmata in mesophyll and bundles sheath cells of *Zea mays* L. *Planta* **136**: 77-89.

Evert, R. F., Eschrich, W. and Heyser, W. (1978) Leaf structure in relation to solute transport and phloem loading in *Zea mays* L. *Planta* **138**: 279-294.

Evert, R. F. and Russin, W. A. (1993) Structurally, phloem unloading in the maize leaf cannot be symplastic. *Am J Bot* **80**: 1310-1317.

Evert, R. F., Russin, W. A. and Bosabalidis, A. M. (1996b) Anatomical and ultrastructural changes associated with sink-to-source transition in developing maize leaves. *Int J Plant Sci* **157**: 247-261.

Evert, R. F., Russin, W. A. and Botha, C. E. J. (1996a) Distribution and frequency of plasmodesmata in relation to photoassimilate pathways and phloem loading in the barley leaf. *Planta* **198**: 572-579.

Farrar, J., van der Schoot, C., Drent, P. and van Bel, A. J. E. (1992) Symplastic transport of Lucifer Yellow in mature leaf blades of barley: Potential mesophyll-to-sieve-tube transfer. *New Phytol* **120**: 191-196.

Fritz, E., Evert, R.F. and Heyser, W. (1983) Microautoradiographic studies of phloem loading and transport in the leaf of *Zea mays* L. *Planta* **159**: 193-206.

Grignon, N., Touraine, B. and Durand, M. (1989) 6(5) carboxyfluorescein as a tracer of phloem sap translocation. *Am J Bot* **76**: 871-877.

Haupt, S., Duncan, G. H., Holzberg, S. and Oparka, K. J. (2001) Evidence for symplastic phloem unloading in sink leaves of barley. *Plant Physiol* **125**: 209-218.

Heyser, W., Evert, R. F., Fritz, E. and Eschrich, W. (1978) Sucrose in the free space of translocating maize leaf bundles. *Plant Physiol* **62**: 491-494.

Kuo, J. and O'Brien, T. P. (1974) Lignified sieve elements in the wheat leaf. *Planta* **117**: 349-353.

Lush, W. M. (1976). Leaf structure and translocation of dry matter in a $C_3$ and a $C_4$ grass. *Planta* **130:** 235-244.

Matsiliza, B. and Botha, C. E. J. (2002) Aphid (*Sitobion yakini*) investigation suggests thin-walled sieve tubes in barley (*Hordeum vulgare*) to be more functional than thick-walled sieve tubes. *Physiol Plant* **115:** 137-143.

Miyake, H. and Maeda, E. (1976) The fine structure of plastids in various tissues in the leaf blade of rice. *Ann Bot* **40:** 1131-1138.

Opalka, N., Brugidou, C., Bonneau, C., Nicole, M., Beachy, R. N., Yeager, M. and Fauquet, C. (2003) Movement of rice yellow mottle virus between xylem cells through pit membranes. *Proc Natl Acad Sci USA* **95:** 3323-3328.

Oparka, K. and Turgeon, R. (1999) Sieve elements and companion cells-traffic control centers of the phloem. *Plant Cell* **11:** 739-750.

Roberts, I. M., Wang, D., Thomas, C. L. and Maule, A. J. (2003) Pea seed-borne mosaic virus seed transmission exploits novel symplastic pathways to infect the pea embryo and is, in part, dependent upon chance. *Protoplasma* **222:** 31-43.

Russell, S. H. and Evert, R. F. (1985) Leaf vasculature in *Zea mays* L. *Planta* **164:** 448-458.

Thomasson, J. R., Nelson, M. E. and Zakrzewski, R. J. (1986) A fossil grass (Gramineae: Choridoideae) from the Miocene with Kranz anatomy. *Science* **233:** 876-878.

van Bel, A. J. E. (1993) Strategies of phloem loading. *Ann Rev Plant Physiol Plant Mol Biol* **44:** 253-281.

van Bel, A. J. E., Van Kesteren, W. J. P. and Papenhuijzen, C. (1988) Ultrastructural indications for coexistence of symplastic and apoplastic phloem loading in *Commelina benghalensis* leaves. Differences in ontogenic development, spatial arrangement and symplastic connections of the two sieve tubes in the minor vein. *Planta* **176:** 159-172.

Walsh, M. A. (1974) Late-formed metaphloem sieve-elements in *Zea mays* L. *Planta* **121:** 17-25.

Wright, K. M., Horobin, R. W. and Oparka, K. J. (1996) Phloem mobility of fluorescent xenobiotics in *Arabidopsis* in relation to their physicochemical properties. *J Exp Bot* **47:** 1779-1787.

Wright, K. M. and Oparka, K. J. (1996) The fluorescent probe HPTS as a phloem-mobile symplastic tracer: An evaluation using laser scanning microscopy. *J Exp Bot* **47(296):** 439-445.

# 7

## Water Flow in Roots: Structural and Regulatory Features

*Gretchen B. North and Carol A. Peterson*

When considering water flow through plants, roots are often treated as black boxes for simplicity (Fiscus, 1975; Dalton *et al.*, 1975). Axial flows within roots (i.e., the longitudinal movement through the xylem) can be modeled and measured relatively easily, and rates of flow for mature, well-watered roots are usually within 20% to 50% of the rates predicted by the Hagen-Poiseuille equation (Huang and Nobel, 1994). In herbaceous plants in moist soil, rates of axial flow are usually high enough that the primary limitation for water transport occurs in the radial pathway (i.e., from the ambient solution into the root xylem; Steudle and Peterson, 1998). The details of radial water flow are less well understood than those of axial flow, however, because roots are highly complex, and the proportions of water moving in the various available pathways may be altered depending on the driving forces. The anatomy of roots changes along their length as various structures develop, mature, and, perhaps, die. Their structure can change in response to different environmental conditions (reviewed by Enstone *et al.*, 2003). In addition, there is wide variation in structure amongst species. Despite the complexities of roots and the challenges involved in determining how water moves through them, recent progress has been made on several fronts.

In this chapter, we first examine the structural components of the radial pathway for water flow in roots. Then we discuss possible regulators of water movement through both radial and axial pathways, specifically, aquaporins and embolism, respectively, and how these change during development and water stress. Finally, we consider how water stress might also affect the proportional limitations imposed by the two pathways on overall water transport by roots.

## Structural Components of the Radial Pathway

### Water Flow Through Parenchyma Tissue

We begin by considering the pathways of water flow through parenchyma tissue, as this is typically the predominant cell type in young roots. Three pathways can be considered: apoplastic, symplastic, and transcellular. The apoplast of a group of living parenchyma cells is the continuum of cell walls. The intermicrofibrillar spaces in the wall through which water and ions can move occupy 20% to 50% of the total wall volume (Lüttge and Higinbotham, 1979; Peterson, 1987). The symplast is the continuum of cytoplasms interconnected by plasmodesmata. When a molecule (like water) passes through membranes relatively easily, it moves predominantly by the transcellular path that includes the cell walls, cytoplasms, and vacuoles of the cells. The symplastic and transcellular pathways have proved difficult to separate experimentally and have been lumped together as the cell-to-cell pathway (Steudle and Peterson, 1998). The relative areas of the wall, cytoplasm, and vacuole change as water passes through a cell. This is most easily visualized by considering successive longitudinal sections through a cell. To take a specific example, in a typical cortical parenchyma cell (Fig. 7.1) of onion root, the cell diameter is 41 μm, length 400 μm, wall thickness 0.34 μm, and cytoplasm thickness 0.68 μm (Fengshan Ma, personal communication). The first longitudinal section consists entirely of wall material except for the small plasmodesmata running through it (Fig. 7.1A,Ba). The area occupied by the apoplast is 98.2% of the total, and the remaining 1.8% is occupied by the plasmodesmata of the symplast (Fengshan Ma, personal communication). The next section is composed of parietal cytoplasm bordered by a thin wall (Fig. 7.1A,Bb). Using the previous cellular dimensions, the area occupied by the apoplast has been reduced to 1.7% of the total, and that of the symplast increased to 98.3%. The third (median) section through the cell mainly consists of vacuole with thin layers of cytoplasm and cell wall (Fig. 7.1A,Bc). Here, the area of the vacuole is 95.0%, that of the cytoplasm is 3.3%, and that of the wall is 1.7%. Continuing through the cell, the predominance of the cytoplasm (Fig. 7.1A,Bd) and cell wall (Fig. 7.1A,Be) appear again. Note that the areas actually available for transport in the wall (i.e., the intermicrofibrillar spaces) are 50% or less than those of the total wall areas given above.

The symplastic pathway is believed to play a minor role in water movement across the root owing to the small dimensions of the plasmodesmata; thus, the transcellular path is estimated to contribute up to 95% to the cell-to-cell path (Tobias Henzler and Ernst Steudle, personal communication). Water movement through membranes, as distinct from ion movement, is facilitated by aquaporins (see later), making the transcellular path a major one.

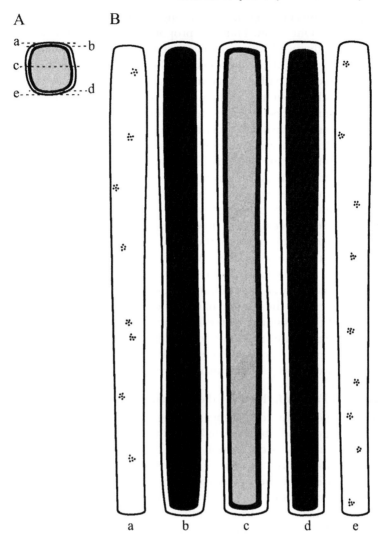

**Figure 7.1** Diagrams of a cortical cell in cross section (A), and longitudinal sections (B) through the cell as indicated by lower-case letters. Walls are white, cytoplasm is black, and vacuole is gray. Cell outlines are drawn to scale, but not the three compartments within them. Ba. Section through the wall. This region is totally apoplast except for small plasmodesmata (indicated by black dots). Bb. Section through the cytoplasm. The area is mainly symplast, and this is surrounded by a small area of apoplast (wall). Bc. Section through the mid-region of the cell to include the vacuole at its widest. Most of the area is vacuole; areas of apoplast and symplast are small by comparison. Bd as in Bb. Be as in Ba.

The proportions of water moving through these pathways are governed by an analog of Ohm's law. Thus, the proportions depend on (1) the resistance of the pathway, (2) the area that it presents to the movement of water (e.g., Fig. 7.1), and (3) the path length. Clearly, throughout most of the cell, the transcellular path commands by far the largest area for water flow compared to the other two pathways (Fig. 7.1). However, this advantage may be counterbalanced by differences in resistances among the paths.

Water movement through roots can be characterized as hydraulic or osmotic, the latter referring to water movement through membranes as a result of osmotic forces. Both forces operate in the acquisition of water, which then moves through the root system and up through the plant as described by the cohesion-tension theory (Steudle, 2001). During times of high transpiration, and consequently rapid water intake through the radial pathways, the resistance to radial flow in the root is relatively low compared to the resistance that occurs when the flow is slower, at least in some species (see Steudle and Peterson, 1998). According to the composite transport model, the low root resistance is due to a greater use of the apoplastic path. Conversely, during lower rates of transport, osmotic influences are proportionally greater; thus, the fraction of water flowing through the membranes increases and the resistance of the root increases (Steudle and Peterson, 1998). An increase in transpiration can lead to a bulk flow of water in the walls of some species (Aloni and Peterson, 1998). The consequences of this movement have not as yet been included in models of water flow in roots.

### Anatomical Factors Influencing the Resistances of the Radial Pathways

The structure of many cells changes during root development, and some of these changes would be expected to influence the hydraulic conductivity of the three paths as outlined in Table 7.1. Testing these hypotheses is difficult because at present there are no techniques to measure the separate conductivities of the apoplastic, symplastic, and transcellular paths directly. Hence, relative contributions of the apoplastic and cell-to-cell paths must be determined from measurements of the hydraulic conductivities of roots and individual cells (Steudle and Jeschke, 1983; Steudle and Brinkmann, 1989; Azaizeh *et al.*, 1992). In addition, it has proved challenging to design an experiment to assess the effect of a specific developmental change on the radial hydraulic conductivity of the root. Typically, a comparison is made between conductivities of roots with and without a certain feature, but since there can be many additional variables involved, the results are difficult to interpret.

### *Casparian Bands*

**A. In the Endodermis**    Casparian bands are modifications of the primary walls that are laid down within the hydraulically isolated root tip before the

**Table 7.1**    The predicted effects of aquaporins and selected anatomical features on the movement of water through various paths in roots

| Anatomical Feature | Pathway Influenced | Direction of Influence on Hydraulic Conductivity | Rationale |
|---|---|---|---|
| Casparian bands | Apoplastic | Decrease | Occlusion of intermicrofibrillar spaces by lignin and suberin |
| Suberin lamellae (a) preserving plasmodesmata | Transcellular | Decrease | Hydrophobic material blocks movement from walls to plasmalemma |
| (b) severing plasmodesmata | Symplastic | Decrease | Symplastic connections between cells are broken |
| | Transcellular | Decrease | Hydrophobic material blocks movement from walls to lumen |
| Thickened walls | Apoplastic | Increase | Area is increased |
| | Symplastic | Decrease | Lengths of some plasmodesmata (i.e., those not in primary pit fields) are increased |
| Lignified walls | Apoplastic | Decrease | Occlusion of some or all intermicrofibrillar spaces |
| | Transcellular | Decrease | Occlusion of some or all intermicrofibrillar spaces |
| Aquaporins | Transcellular | Increase | Membrane conductivity is increased |
| Aerenchyma | Apoplastic | Decrease | Many cells in the radial path are no longer present |
| | Symplastic | Decrease | Many cells in the radial path are no longer present |
| | Transcellular | Decrease | Many cells in the radial path are no longer present |
| Cell death | Apoplastic | Increase | Area formerly occupied by the protoplast has become part of this compartment |
| | Symplastic | Decrease | Compartment no longer exists |
| | Transcellular | Decrease | Compartments no longer exist |

tracheary elements of the xylem become functional (Peterson and Lefcourt, 1990). In their unmodified state, primary walls are 50% water, some of which is present in channels ranging from 1 to 10 nm in diameter (Frey-Wyssling, 1969). The water in these channels is mobile, as opposed to the tightly bound water in smaller channels (Frey-Wyssling, 1969). Casparian bands consist of a wall region located in the anticlinal (radial and transverse) walls of the endodermis and exodermis (Fig. 7.2A,B), where lignin and suberin have been deposited in the intermicrofibrillar spaces. According to Schreiber and his associates, who first isolated and analyzed Casparian bands chemically, the walls are 6.5% lignin and 0.5% suberin in the monocot *Monstera deliciosa*, 5.2% lignin and 1.4% suberin in the monocot *Clivia miniata*, and 2.7% lignin and 2.5% suberin in the dicot *Pisum sativum*. There is no detectable aromatic suberin in the last species (Zeier and Schreiber, 1977, 1998; Zeier *et al.*, 1999a). There is good evidence that Casparian bands are linked from one cell to the next (Pierson and Dumbroff, 1969) and to the plasmalemma of the protoplast (Bonnett, 1968), so that the entire apoplastic path into the root stele (that houses the xylem) is affected by this structurally minute wall modification. For more information on endodermal (and exodermal) development, see Ma and Peterson (2003).

Since most of the apoplastic movement of water may occur in the inter-microfibrillar spaces, occlusion of these with any material is expected to diminish the efficiency of the apoplast as a transport pathway, and this should be evident in the endodermis (Table 7.1). However, several measurements of total root radial hydraulic conductivity have produced values larger than expected if the apoplastic path were efficiently sealed at the endodermis (e.g., Melchior and Steudle, 1993). Does this mean that the apoplastic path is closed to ions but open, at least to some extent, for water and uncharged solutes? The well-documented lack of permeability of this layer to ions, with hydrated radii varying between 0.378 nm for sulfate (Peterson, 1987) and 0.447 nm for iron (de Rufz de Lavison, 1910), indicates that all the intermicrofibrillar spaces should have been modified. It is possible that channels remain in these spaces that are small enough to allow water (diameter 0.28 nm; Stein, 1986) to pass but will not allow ions to do so. Further, the chemical composition of the modification (i.e., more lignin than suberin) means that the walls may be less hydrophobic than previously believed. Another possible explanation for the relatively high permeability of the endodermis to water is that the apoplastic path for water is blocked but is compensated for by unusually high hydraulic conductivities of the endodermal membranes. This could be achieved by a larger number of aquaporins in the plasmalemmae and tonoplasts of the endodermis. Supporting evidence for this idea is provided by a study of aquaporin mRNA synthesis in maize roots. More of this RNA is expressed in the cytoplasms of endodermal and xylem parenchyma cells than in central cortical cells (Barrieu *et al.*, 1998). It is reasonable to assume that this leads to the production of an increased number of aquaporins in the membranes of the endodermal cells (see later). Mechanically wounding a small area of the endodermis in maize does not alter the root's hydraulic conductivity, indicating that the conductivity of this layer is already quite high

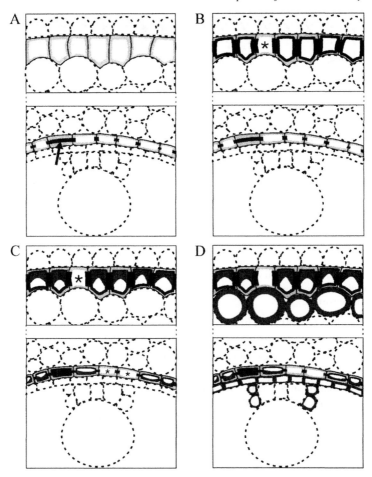

**Figure 7.2**   Diagrams of roots in cross-section (not to scale) to illustrate various anatomical features. Endodermis and exodermis are drawn with solid lines; outlines of other cells are indicated with dotted lines. The entire width of the cortex is not shown except in F. Primary walls light gray; Casparian bands and suberin lamellae black; lignified walls medium gray. (A) Section through a young root zone in which the endodermis has developed a Casparian band in the anticlinal walls of its cells. In one cell (*arrow*) of the endodermis, a transverse wall is shown; the Casparian band extends tangentially through this wall. (B) Section through an older part of the root in which all cells of the exodermis have developed Casparian bands in their anticlinal walls and most cells have also developed suberin lamellae. The cell shown without a lamella (∗) is a passage cell. (C) The endodermis has developed suberin lamellae in most cells (extending across the face of the wall in the cell showing the transverse wall). Two endodermal cells (∗) without lamellae are passage cells. Cells of the exodermis and endodermis, except for the passage cells, have also developed lignified, tertiary walls. (D) Cortical cells adjacent to the exodermis have developed thick, lignified walls. In the stele, the pericycle cells and xylem parenchyma also have thick, lignified walls.

*Continued*

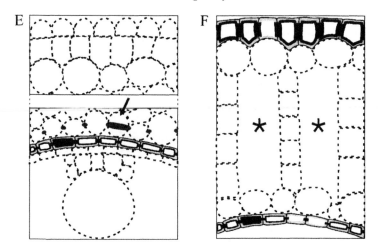

**Figure 7.2** *Cont'd* (E) Cortical cells external to the endodermis have developed phi thickenings on their anticlinal walls. The transverse wall of one cell (*arrow*) is illustrated; the phi thickenings extend through this wall. (F) Aerenchyma with large, air-filled lacunae (∗) is present in the cortex.

(Peterson *et al.*, 1993). The direct measurement needed to settle this debate (i.e., the hydraulic conductivity of isolated Casparian bands to water) has not been made because of technical difficulties.

**B. In the Exodermis**  Many angiosperm species develop a second set of Casparian bands in the exodermis (Fig. 7.2B). These resemble endodermal Casparian bands in their responses to histochemical tests and impermeability to fluorescent dyes and ions (de Rufz de Lavison, 1910; Peterson *et al.*, 1982; Peterson, 1987; Cholewa and Peterson, 2004). They occupy a greater wall expanse than in the endodermis. For example, in onion roots, 9.5 µm in the exodermal anticlinal wall is filled with a Casparian band compared with only 1.2 µm in the endodermis (Fengshan Ma, personal communication). It has not been possible to characterize the Casparian bands of the endodermis chemically because they develop along with the suberin lamellae in walls of the same cells. Further, when exodermal roots are subjected to enzymic digestion (with which the endodermal Casparian bands have been isolated), the exodermal walls remain firmly attached to those of the epidermis (Peterson *et al.*, 1978). The isolated product thus consists of exodermal Casparian bands, suberin lamellae, and epidermal walls. The same issues raised in connection with water transport through the apoplast of the endodermis apply to the exodermis. Further details on water (and solute) transport through the exodermis can be found in Hose *et al.* (2001).

*Suberin Lamellae*  Cells of the endodermis of many species, especially monocots, develop suberin lamellae some time after depositing their

Casparian bands; in the exodermis, these wall modifications develop nearly concurrently with the Casparian bands (Fig. 7.2B,C). Suberin lamellae are thin layers of material usually deposited on all internal faces of a primary wall. The lamellae stain with lipophilic dyes and can be visualized with a light microscope (see Brundrett *et al.*, 1991). With the transmission electron microscope, they appear as highly contrasting, alternating electron-dense and electron-lucent layers when the specimen has been fixed (in this case also stained) with osmium (Scott and Peterson, 1979). Frequently the wall modifications of the endodermis and exodermis continue, so that a thick cellulosic wall is laid down, and this wall is usually further lignified. In many cases this wall is thicker on the inner tangential side in the endodermis, and on the outer tangential side in the exodermis (Fig. 7.2C). Some endodermal cells lack both lamellae and thick walls and are known as passage cells. These are usually situated near the xylem poles. In some species (e.g., barley; Robards *et al.*, 1973), all endodermal cells eventually develop suberin lamellae and thickened walls. Passage cells also occur in some types of exodermis (see review by Ma *et al.*, 2003).

A comparison of the chemical composition of isolated walls with Casparian bands versus isolated walls with both Casparian bands and suberin lamellae indicates that the lamellae have more pure suberin than the bands (Zeier *et al.*, 1999b). Suberin is a complex polymer consisting of aliphatic and aromatic domains (see review by Bernards, 2002). Its putative effect on water movement would be related to the proportions of the two materials (the aliphatic regions being more hydrophobic than the aromatic) and the spatial orientation of the two domains. It is usually assumed that an addition of suberin lamellae to a wall reduces its permeability to water. Because of their position in the endodermal and exodermal cells, suberin lamellae could be expected to reduce the hydraulic conductivity of the transcellular path but not the apoplastic path (Fig 7.2B,C; Table 7.1). Their effect on the symplastic path depends on whether or not the plasmalemmae remain intact, and the enclosed cells alive (Table 7.1). In all endodermal layers studied to date, the plasmodesmata are not severed and the cells do remain alive. (This is why they are subsequently able to lay down a thick, tertiary wall.) The situation in the exodermis is more variable. In onion, which has a dimorphic exodermis, suberin lamella form first in the long cells; they sever the plasmodesmata and the enclosed cells die (Ma and Peterson, 2000). Thus, the symplastic path is changed in this layer. The path is reduced to only the short (passage) cells of the exodermis, which occupy about 13% of the exodermal surface (Ma and Peterson, 2001). On the other hand, in lateral roots in maize, which has a uniform exodermis, the plasmodesmata remain intact, as in the endodermis (Wang *et al.*, 1995). It is safe to assume that exodermal cells that lay down a thickened cellulosic (tertiary) wall are alive and have intact plasmodesmata linking them to their surrounding cells. Some plants develop several exodermal layers or other layers associated with them, *Typha* being a good example. The Casparian band begins in the radial walls of the outer layer of the cortex and expands through the walls to form an H-shaped structure. Then suberin lamellae and tertiary walls are laid down as well (Seago *et al.*, 1999).

Studies aimed at correlating hydraulic conductivity with suberin lamella development in the endodermis or exodermis have yielded a great variety of results. According to Talcisnik *et al.* (1999), roots in exodermal species lose water more slowly than those without an exodermis. Sanderson (1983) correlated endodermal anatomy with water uptake in barley roots using a minipotometer system. He found that a reduction of water uptake was correlated with development of the tertiary walls, not with the development of the suberin lamellae. Similarly, Barrowclough *et al.* (2000) observed no reduction in the hydraulic conductivity of onion roots due to the development of suberin lamellae in either the endodermis or exodermis. North and Nobel (1996) determined the hydraulic conductivities of specific root layers in *Opuntia* by comparing the hydraulic conductances of root segments before and after dissecting off the layer in question. They found that the hydraulic conductivity of an endodermis with suberin lamellae in 50% or less of its cells was the same as that for eight ranks of suberized phellem cells. To explain the surprising permeability of the phellem, they suggested that moved through primary walls, the middle lamella, pit connections, and an increased apoplastic pathway associated with lateral root development. Zimmermann *et al.* (2000) promoted exodermal maturation by growing plants in aeroponic culture. By comparing the hydraulic conductivity of these roots with those grown in hydroponics, they found that the conductivity of the exodermal part of the root was reduced by a factor of four (Zimmermann and Steudle, 1998; see also discussion in Miyamoto *et al.*, 2001). This reduction correlated with an increased amount of aliphatic suberin components in the aeroponically grown roots. On the other hand, Clarkson *et al.* (1987) noted that maize roots were more permeable to water in regions with exodermal suberin lamellae than without. A much less permeable exodermis (as measured from isolated sleeves of tissue) was generated from the root bases that had been suspended in moist air above the hydroponic solution. This led Clarkson *et al.* (1987) to propose that exposure to air or perhaps oxygen may have changed the chemical nature of the suberin. Another possibility is that the epidermal cells had died during exposure to air and the suberin lamellae of the exodermis had sealed over the wall openings originally occupied by plasmodesmata. Ranathunge *et al.* (2003) made a direct measurement of the hydraulic conductivity of the outer part of the rice root using an ingenious method of perfusing the underlying aerenchyma and forcing water to move through the outer cell layers. The four cell layers through which the water flowed included an immature exodermis in the younger zone tested and a mature exodermis (with Casparian bands and suberin lamellae) in the older zone tested. Despite this anatomical difference, the hydraulic conductivities of the two zones were not significantly different. From the diversity of results obtained, sometimes even within the same species, it appears that several factors, both developmental and environmental, determine the hydraulic conductivity of suberized root cells. To sort it out, a comprehensive study in which all the variables are either controlled or measured is needed.

***Thickening and Lignification of Walls*** The addition of secondary wall material to the inner faces of the primary walls may occur in some cells of the endodermis and exodermis as mentioned previously (Fig. 7.2C), the pericycle, stelar parenchyma, and layers in the central cortex (Fig. 7.2D). A special case is that of phi thickenings, which occupy only parts of the central cortical cell walls and are typically situated in the walls like Casparian bands (Fig. 7.2E). These thickenings are known to occur in members of several gymnosperm taxa (Cycadales, Ginkgoales, Taxales, Cupressaceae, and Araucariaceae) where they appear to be a family characteristic (Gerrath *et al.*, 2002). Their distribution among angiosperms is less well studied, but Brundrett *et al.* (1990) documented their presence in some species (yellow birch, white birch, and black cherry), and they are also known in apple and geranium (see Peterson *et al.*, 1981). They can be induced to form in the epidermis of maize roots (Degenhart and Gimmler, 2000).

In roots, thickening of the walls is usually associated with lignification, so the expected increase (due to increased area of the pathway) in conductivity of the apoplastic path could be counterbalanced by the occlusion of the intermicrofibrillar channels by lignin (Table 7.1). Some of the cells remain alive but others die, so involvement of the symplast as a transport pathway is variable. In the case of phi thickenings, those of apple and geranium proved permeable to an apoplastic, fluorescent tracer dye, so one would not expect them to pose a major barrier to water movement through the apoplast (Peterson *et al.*, 1981).

***Features Influencing Connection to the Soil*** Water moves through the plant in a continuum that begins in the soil. Therefore, contact between the root and soil solution is critical for water flow into and through the root. The most important (and least ambiguous) effect of root hairs on water uptake may be that they help maintain such contact (Hofer, 1996). Roots can shrink in response to diurnal transpirational demands (Faiz and Weatherley, 1982) or during longer periods of water stress (Nobel and Cui, 1992), opening an air gap between roots and soil that can limit water transport. In the case of desert succulents, the limitation posed by such a gap is greatest in moderately dry soils, before a reduction in soil hydraulic conductivity occurs due to more extreme drying. The development of an air gap can be delayed or prevented by the presence of a rhizosheath, which consists of soil particles bound together by root hairs, mucilage, and sloughed root cells, and occurs in a number of xerophytic species such as desert succulents (North and Nobel, 1992) and dune grasses (Wullstein *et al.*, 1979), as well as in more mesophytic species such as maize (McCully and Canny, 1988). The root exodermis and associated sclerified cortical cells could be structural features important in this context. Because of their mechanical strength, these layers could reduce cortical shrinkage after death of the epidermis and most or all of its cortical cells, thereby helping maintain root-soil contact. This idea begs to be tested experimentally. Measuring water transfer between soil and roots *in situ* remains difficult, because otherwise valuable techniques such as the high-pressure flow meter (Tyree *et al.*, 1995) tend to eliminate rhizosphere resistance.

*Aerenchyma*    Roots of aquatic and wetland plants typically have very large spaces or lacunae in their central cortex (Fig. 7.2F). In some species, exposure to low oxygen levels will induce the formation of this tissue. The result is a change in the architecture of the radial pathway. Removal of much of the cortex may leave radial files of cells, "spokes" of collapsed cells, or cells arranged in other patterns (Justin and Armstrong, 1987). The expected result of such changes would be a reduction in the hydraulic conductivity of the roots (Table 7.1). Miyamoto *et al.* (2001) measured the hydraulic conductivity of rice roots and found it to be relatively low, comparable to that measured for other species when an osmotic difference rather than a hydrostatic pressure difference was the driving force. However, there are other species without aerenchyma (e.g., scarlet runner bean and barley) that have as low or lower hydraulic conductivities as measured by hydrostatic pressure differences (Miyamoto *et al.*, 2001). It is technically difficult to measure the hydraulic conductivity of the aerenchyma, but Ranathunge *et al.* (2003) have estimated its conductivity in rice by considering both water vapor diffusion across the spaces and water moving through the cells (in radial alignment in the aerenchyma of these roots). The value obtained, $8.9 \times 10^{-8}$ m s$^{-1}$ MPa$^{-1}$, is twice as large as that of the overall transport of the root and about equal to the calculated hydraulic conductivity of the endodermis and cells internal to it. These results indicate that the aerenchyma system may not contribute an especially large resistance to the radial flow of water.

*Cell Death*    There is good evidence that the death of cells increases their hydraulic conductivity. Technically, when a cell dies, the apoplastic pathway is enlarged to include the whole cell. Destruction of membranes associated with cell death makes the areas occupied by the former protoplasts highly conductive. In fact, the conductivity of this zone should now be higher than that of the apoplast, as large water channels can be envisaged (Table 7.1). Killing part of a root with hot water increased the hydraulic conductivity of that zone by eightfold compared to a normal, living root (Peterson and Steudle, 1993). North and Nobel (1996) estimated that the conductivity of a living cortex of *Opuntia* roots was $2 \times 10^{-7}$ m s$^{-1}$ MPa$^{-1}$, whereas the conductivity of a dead cortex was nearly three times higher. When the living cells of the outer part of the rice root were heat-killed, the conductivity of the assembly of cells was doubled (Ranathunge *et al.*, 2003).

## Regulation of Radial Hydraulic Conductivity by Aquaporins

With respect to water transport, roots can be modeled as leaky cables, with root hydraulic conductivity ($L_p$; m s$^{-1}$ MPa$^{-1}$) having both radial ($L_R$: same units as $L_p$) and axial components ($K_h$, m$^4$ s$^{-1}$ MPa$^{-1}$; Landsberg and Fowkes, 1978; Hsiao and Xu, 2000). Unlike ions, water moves through roots passively, driven by differences in water potential, yet roots should not be seen as strictly passive conduits. At the level of the root system, root proliferation and abscission allow plants to respond to changes in the spatial and temporal availability of soil moisture. However, root turnover can be considered a relatively costly, long-term response to environmental heterogeneity. A wealth of recent evidence points to aquaporins as a less expensive, quicker means of regulating radial water flow, at least in metabolically active root regions (North *et al.*, 2004). Aquaporins are a family of membrane-spanning proteins present in both the plasmalemma and tonoplast of plant cells. Their presence and probable functions in roots have been reviewed recently and comprehensively (Javot and Maurel, 2002; Tyerman *et al.*, 2002), and a number of studies confirm both the role of aquaporins in radial water transport and the predominance of the transcellular pathway (Barrowclough *et al.*, 2000; Quigley *et al.*, 2001; Siefritz *et al.*, 2002). Here, we focus on variation in aquaporins due to changes in water availability. In addition, we consider stress-induced, reversible reductions in aquaporin activity as critical in regulating root hydraulic conductivity.

### Effects of Mercurial Compounds on Root Radial Hydraulic Conductivity

The role of aquaporins in water transport at the whole-root level has been investigated for several species through the use of mercurial agents, such as HgCl$_2$, which presumably bind with cysteine residues in aquaporins in the plasmalemma (plasmalemma integral proteins, or PIPs) or tonoplast (tonoplast integral proteins, or TIPs; Quigley *et al.*, 2001), thereby inhibiting water flow through the channels. As with most poisoning techniques, the effects of mercurials may be less specific than desired; however, the reopening of water channels by a scavenging agent such as mercaptoethanol improves the credibility of the technique. In addition, roots of several species respond in a consistent manner when hydraulic conductivity is measured after exposure to a mercurial compound. For distal (young) root segments of the desert succulent *Agave deserti* under wet conditions (Fig. 7.3, solid line), $HgCl_2$ reduces $L_P$ by about 60%, and 2-mercaptoethanol restores $L_P$ to 80% to 90% of its premercurial value (North and Nobel, 2000). Such a reduction in $L_P$ is near the middle of values reported by Javot and Maurel (2002), which range from 32% for the cactus *Opuntia acanthocarpa* (Martre *et al.*, 2001), to 47% for aspen (*Populus tremuloides*; Wan and Zwiazek, 1999), and up to 90% for barley (*Hordeum vulgare*; Tazawa *et al.*, 1997).

The sensitivity of $L_P$ to mercurial agents can vary, depending on root development, time of day, and various environmental stresses. With respect to development, $HgCl_2$ does not reduce $L_P$ in a 3-mm long zone at the tips of roots of *Zea mays*, either because water flow is primarily through plasmodesmatal connections that are more abundant here than in more proximal regions (Hukin *et al.*, 2002; see Chapter 8), or because cell walls in the apical region are relatively impermeable to ions (Lüttge and Weigl, 1962; Enstone and Peterson, 1992). For roots of onion, $HgCl_2$ reduces $L_P$

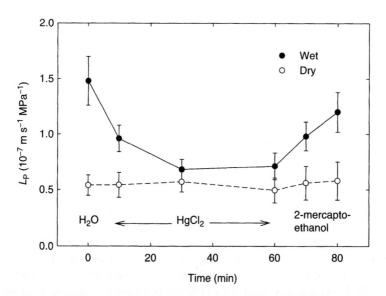

**Figure 7.3**  Root hydraulic conductivity ($L_P$) for young (distal) root segments of *Agave deserti* under wet conditions (*filled circles*) and after soil drying (*open circles*). Root segment was initially placed in water, transferred at 0 minutes to 50 µM $HgCl_2$, and then transferred at 60 minutes to 10 mM 2-mercaptoethanol for sequential measurements of $L_P$. Data are means ± SE for four plants. (From North and Nobel, 2000.)

in successive zones from the tip to 120 mm proximal, with greatest inhibition occurring in the 70 to 120 mm zone, where suberin lamellae are present in both the endodermis and the exodermis (Barrowclough *et al.*, 2000). Exposing the $HgCl_2$-treated roots to $H_2S$ gas revealed that the mercurial agent (which formed a visible precipitate) did not penetrate past the exodermis, suggesting that inhibition is disproportionately due to water channels in epidermal cells and short, living cells in the exodermis. Roots of the desert succulents *Agave deserti* (North and Nobel, 2000; North *et al.*, 2004) and *Opuntia acanthocarpa* (Martre *et al.*, 2001) show a slightly different pattern, in that $HgCl_2$ reduces $L_P$ for distal root segments (0 to 70 mm back from the tip) under wet conditions, but not for more proximal midroot segments (Fig. 7.4). It should be noted that the midroot segments of these desert succulents are farther from the root tip and considerably older than the most proximal root zone of onion examined by Barrowclough *et al.* (2000). Possible explanations for the lack of sensitivity in the older root regions of *A. deserti* and *O. acanthocarpa* are that aquaporins are absent or nonfunctional in older tissues, or that the greater amount of suberization and lignification at midroot block the ingress of mercurial agents.

The role of aquaporins in regulating root hydraulic conductivity may be most critical during exposure to and recovery from environmental stresses. As shown in Fig. 7.4, root $L_P$ for *O. acanthocarpa* after a period of soil drying is no longer inhibited by $HgCl_2$. One possible explanation is that water stress has already closed the channels by inactivating the proteins or inhibiting their synthesis; thus, $HgCl_2$ has no further inhibitory effect. For these roots, aquaporin inhibition could account for about 50% of the decrease in $L_P$ due to drying, with the remainder perhaps due to increased resistance in the periderm and reduction in axial conductivity due to embolism (Martre *et. al.*, 2001). Other environmental stresses that reduce $L_P$ of roots or root cells, such as hypoxia (Zhang and Tyerman, 1999), nutrient deprivation (Clarkson *et al.*, 2000), and salinity (Carvajal *et al.*, 2000) also eliminate further reduction in $L_P$ by $HgCl_2$. The coupling between stress-induced decreases in $L_P$ and aquaporin inhibition could be caused by reductions in metabolism, leading to lower rates of phosphorylation of the aquaporin proteins, as suggested by work with spinach leaves (Johansson *et al.*, 1998). Work on *Chara* internodes suggests that high concentrations of nonpermeating osmotica can close aquaporins, perhaps by causing tension-induced deformation of the channel within the protein (Ye *et al.*, 2004). Although a number of mechanisms may be involved, aquaporin closure or inactivation allows a reduction in root hydraulic conductivity during conditions such as soil drying or high salinity that might otherwise lead to reverse water flow. However, water stress actually increases mercury sensitivity for two cultivars of rice, despite decreases in

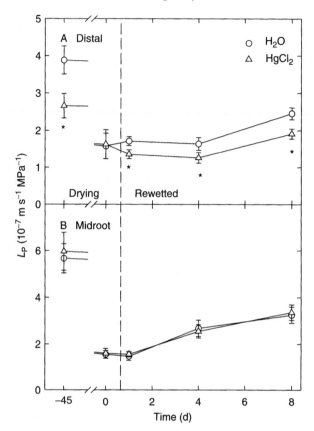

**Figure 7.4**    $L_p$ for distal (A) and midroot (B) segments of roots of *Opuntia acanthocarpa* in wet soil and during soil drying and rewetting, measured first in water then after transferal to 50 μM $HgCl_2$. Data are means ± SE for four to six plants. The asterisks indicate a significant difference due to $HgCl_2$ treatment (paired *t*-test, $P < 0.05$). (From Martre *et al.*, 2001.)

root $L_p$ (Lu and Neumann, 1999), and patterns of gene expression for both PIPs and TIPs with respect to stressful conditions are decidedly mixed (Tyerman *et al.*, 2002).

Resuming water uptake after drying is as critical for plants as preventing water loss, and aquaporins are implicated in the few cases where responses to rewetting have been examined. Although rewetting does not fully restore $L_p$ to its predrying value for distal roots of *O. acanthocarpa*, renewed aquaporin sensitivity to $HgCl_2$ is apparent within 1 day (Fig. 7.4). A more complete recovery in $L_p$ occurs for young roots of *A. deserti* after 14 days of soil drying as opposed to 45 days for *O. acanthocarpa*, and aquaporin sensitivity is greater as well (North *et al.*, 2004). For both *O. acanthocarpa* and

*A. deserti*, renewed aquaporin sensitivity after 1 day of rewetting occurs in the absence of new root growth. Whether the recovery in $L_p$ is due to the reopening of existing water channels, perhaps by renewed metabolism and phosphorylation, or to the synthesis of new aquaporins remains to be determined. Rapid aquaporin turnover is indicated in some studies, for example, in an investigation of proteins in the plasma membrane and tonoplast of radish (Suga *et al.*, 2001).

### Evidence from Antisense Plants for Aquaporin Involvement in Root Hydraulic Conductivity

The engineering of plants with antisense constructs allows the effects of aquaporins to be pinpointed without some of the problems associated with the use of mercurial and other toxic compounds. As more PIP and TIP genes are sequenced for more species, the use of antisense or gene-knockout plants should help resolve many of the remaining questions about aquaporin regulation of water transport. However, the large number of aquaporins identified so far (e.g., 38 aquaporin-like sequences in the *Arabidopsis* genome; Quigley *et al.*, 2001) and the diversity and uncertainty of their subcellular and tissue locations mean that recent findings based on engineered plants are still more suggestive than definitive. One consistent result from such studies is that knocking out aquaporins reduces the water permeability of root cell protoplasts (Kaldenhoff *et al.*, 1998; Martre *et al.*, 2002; Siefritz *et al.*, 2002). At the organ level, specific root hydraulic conductivity (expressed on a root area basis) for tobacco plants with an antisense construct for a plasmalemma aquaporin is only 42% of that for control plants (Siefriz *et al.*, 2002). Unlike control plants, these antisense plants also wilt when roots are exposed to an osmotic shock, leading the authors to conclude that the presences of aquaporins reduces water stress. In double antisense plants of *Arabidopsis*, in which two PIPs are not expressed, root hydraulic conductivity on a root mass basis is 32% of that for control plants, although a nearly threefold increase in root mass relative to leaf mass in the antisense plants results in whole plant water uptake similar to that of control plants under wet conditions (Martre *et al.*, 2002). However, double antisense plants have lower leaf turgor pressure and lower transpiration than do control plants during 8 days of soil drying, and transpiration, whole plant hydraulic conductivity, and new leaf production are lower after rewetting for antisense plants than for control plants.

At first glance, the conclusions drawn from antisense tobacco plants and *Arabidopsis* appear somewhat contradictory to results from $HgCl_2$ experiments that demonstrate reduced aquaporin sensitivity during drying and other environmental stresses. If, as experiments with antisense plants suggest, aquaporins help diminish or delay the onset of water stress, then their inhibition (or engineered absence) should be detrimental for plants during drying. One explanation that might reconcile such results is that aquaporins are sensitive to a precise level of water stress, with closure of water channels occurring only when a critical low water potential is reached. Before this point, aquaporins in both the plasmalemma and the tonoplast may be essential in regulating intracellular and intercellular water movement. After this point, aquaporin closure may help prevent further water loss.

### Location of Aquaporins Within Roots

Aquaporin location within root regions and tissues has been investigated by examining a number of phenomena, including gene expression, immunolocalization, and water uptake. Given the number of aquaporin proteins and the diversity of their subcellular locations, it is not surprising that they occur in most cell types, where they may have quite different functions. For example, several studies show a high concentration of aquaporins near the root tip, yet these may be involved more in cellular elongation than in root hydraulic conductivity (Hukin *et al.*, 2002). In several species, notably high concentrations of aquaporins occur in the endodermis and in vascular parenchyma ( Javot and Maurel, 2002), cells that play prominent roles in root water transport. The epidermis and the short cells of the exodermis in onion are also rich in aquaporins (Barrowclough *et al.*, 2000), as is also the case for the dimorphic exodermis of *A. deserti* (North *et al.*, 2004). Recent work with $HgCl_2$ suggests that aquaporins can be expected to occur in any metabolically active root cell and that they are less likely to be present in older, heavily suberized and lignified root regions. The possible role (or even presence) of aquaporins in older woody roots with extensive parenchyma in the periderm, rays, and vascular tissues remains to be investigated.

## Regulation of Root Axial Hydraulic Conductivity

As discussed in this chapter and in much of the literature on root hydraulic conductivity, the radial pathway is the primary site of limitation and regulation of water transport in most cases. That said, there are three common and important exceptions to this generalization: Root $L_p$ can be limited by axial hydraulic conductivity, $K_h$ due to (1) immaturity of xylem conduits, (2) embolism, and (3) increased path length resulting from deep rooting. Immature late metaxylem can persist along main roots of *Z. mays* up to 30 cm from the root tip (St. Aubin *et al.*, 1986), but the so-called zone of hydraulic isolation (where $K_h$ limits water uptake) is likely to be less than 20 mm long (Frensch and Steudle, 1989). For woody species in particular, it is difficult to determine how much of the root system might be hydrauli-cally isolated or at least limited in axial water transport by the immaturity of xylem conduits. The number, turnover rate, and developmental states of fine roots remain unknown and challenging to investigate for most species.

Like aquaporins, embolism can be considered a form of hydraulic regu-lation because it represents a graded and frequently reversible set of responses to environmental heterogeneity. Unlike the case for above-ground plant parts, a mechanism whereby embolism in roots can be reversed is easy to envision and likely to occur; emboli can be dissolved by root pressure, which can develop when transpiration is absent or reduced (Steudle, 2001). Although investigation of embolism in roots has lagged behind that of shoots, several studies show that roots tend to embolize more readily than shoots in the laboratory and often, though not always, in the field (Sperry *et al.*, 2002). Not surprisingly, fine roots near the surface of drying soil can be the most vulnerable of plant organs to embolism (Sperry and Hacke, 2002). Hacke *et al.* (2000) suggest that fine roots may be considered as "hydraulic fuses," protecting the hydraulic continuum within a plant by localizing failure to a repairable or replaceable set of peripheral organs. Although desert succulents may be unique in the extent of hydraulic buffering provided by the shoot, the pattern of $K_h$ and percent embolism in organs from fine roots to the stem of *A. deserti* suggest not only that fine roots might act as fuses, but also that root junctions might act as circuit breakers (Table 7.2). For this monocotyledonous species lacking the ability to make secondary xylem, repair of embolized conduits must occur by refilling. Fine lateral roots of *A. deserti*, which are almost com-pletely embolized during prolonged drying, tend to be abscised (North *et al.*, 1993), thereby acting as fuses. After fine lateral roots, the most highly embolized zones during soil drying are the junctions between lateral root and main roots and between roots and the stem (Table 7.2). These junc-tions behave more as circuit breakers than as fuses in that they may iso-late regions of hydraulic failure and prevent possible reverse flow from plant regions with higher water potentials (e.g., main roots and stem); in addition, they don't need to be replaced. When soil water is resupplied, $K_h$ for main roots of *A. deserti* and of other desert succulents readily returns to predrying values, indicating that embolism has been reversed (North and Nobel, 1991, 1992). For cold desert shrubs and other woody species that experience prolonged drought, embolism reversal may not occur, and native embolism in the roots can exceed 50% (Sperry and Hacke, 2002).

Given that embolism in roots can be extensive, when might it prove lim-iting to overall root hydraulic conductivity? The answer, unfortunately, depends on several variables that are nonuniform for a single root, let alone for a root system in heterogeneous soil. Two equations based on

**Table 7.2**  Axial hydraulic conductivity for root and stem
regions of *Agave deserti* after 21 days in drying soil, measured before
($K_h^{\text{initial}}$) and after pressurization ($K_h^{\text{maximal}}$) to remove air emboli

| Organ or zone | $K_h^{\text{initial}}$ (m$^4$ s$^{-1}$ MPa$^{-1}$) | $K_h^{\text{maximal}}$ (m$^4$ s$^{-1}$ MPa$^{-1}$) | PLC (%) |
|---|---|---|---|
| Lateral root | $0.83 \times 10^{-12}$ | $46.5 \times 10^{-12}$ | 98 |
| Root/root jct. | $3.08 \times 10^{-12}$ | $12.32 \times 10^{-12}$ | 75 |
| Young main root | $2.63 \times 10^{-12}$ | $4.07 \times 10^{-12}$ | 35 |
| Older main root | $7.47 \times 10^{-9}$ | $9.21 \times 10^{-9}$ | 19 |
| Root/stem jct. | $1.77 \times 10^{-12}$ | $4.80 \times 10^{-12}$ | 63 |
| Stem cylinder | $2.83 \times 10^{-12}$ | $3.5 \times 10^{-12}$ | 19 |

Percent loss of conductivity (PLC) due to embolism was calculated as $[(K_h^{\text{maximal}} - K_h^{\text{initial}})/K_h^{\text{maximal}}] \times 100$.
From North and Nobel, 1991; Ewers *et al.* 1992; and North *et al.* 1992.

leaky cable theory and modified from Landsberg and Fowkes (1978) can
be used to generate at least a preliminary answer regarding roots for which
$K_h$, $L_p$, and root dimensions (radius, $r_{\text{root}}$, and length, $l$) are known. Root
radial conductivity ($L_R$) can be calculated by iteration, with $L_R$ initially set
equal to $L_p$ (North and Nobel, 1992):

$$L_R = L_p\, \alpha l\, /\, \tanh\,(\alpha l) \tag{1}$$

where $\alpha = (2\pi\, r_{\text{root}}\, L_R\, /\, K_h)^{1/2}$
Values for $L_R$ and $K_h$ can be used to determine the effective root length for
water uptake ($l_{\text{eff}}$), which can then be compared to actual root length $l$:

$$l_{\text{eff}} = l \tanh\,(\alpha(r_{\text{root}})l)\, /\, \alpha(r_{\text{root}})l \tag{2}$$

When $l_{\text{eff}}$ is a great deal smaller than $l$ (as can occur when $K_h$ is small relative to $L_R$), then water uptake is confined to the upper part of the root. A
young root of *A. deserti* under wet conditions can have a diameter of 3.5
mm ($r_{\text{root}} = 1.75$), $L_p$ of $1.5 \times 10^{-7}$ m s$^{-1}$ MPa$^{-1}$ (a value typical for roots of
many species), and $K_h$ of $1.0 \times 10^{-11}$ m$^4$ s$^{-1}$ MPa$^{-1}$. For such a root,
$l_{\text{eff}}$ increases with increasing $l$ up to about 9 cm (when $l_{\text{eff}} = 5.5$ cm). When
$l = 10$ cm, $l_{\text{eff}}$ decreases to 5.4 cm, suggesting that any further increase in
new root length would be wasted. To justify such expense (and to allow
more efficient water absorption), $K_h$ must increase, for example, to about
$5.0 \times 10^{-11}$ m$^4$ s$^{-1}$ MPa$^{-1}$ for water uptake to occur along about 90% of the
root length. For a 10-cm long lateral root of *A. deserti* with a diameter of 1.5
mm, $L_p$ of $2.0 \times 10^{-7}$ m s$^{-1}$ MPa$^{-1}$, and $K_h$ of $8.3 \times 10^{-13}$ m$^4$ s$^{-1}$ MPa$^{-1}$ (the
value with embolism in Table 7.2), $l_{\text{eff}}$ is only 0.9 cm. Plugging in the
embolism-free value for $K_h$ ($4.7 \times 10^{-11}$ m$^4$ s$^{-1}$ MPa$^{-1}$) increases $l_{\text{eff}}$ to 9.3 cm,

or nearly the actual root length. Thus, embolism appears to limit overall root hydraulic conductivity in this case.

## Conclusions and Directions for Future Research

Of all the structural root features that have been correlated with radial hydraulic conductivity, results from aquaporins and cell death are the most reproducible and clearly explained in terms of their effect on the pathway of water movement. It is somewhat surprising that deposition of hydrophobic materials in the cells' walls of the endodermis and exodermis does not always have a strong effect on water uptake by the root. This finding suggests two possibilities. First, these structures really do not inhibit water movement. Second, these structures do inhibit water movement, but this is compensated for by an increased water flow resulting from lowered resistance in another pathway. If we eliminate the symplastic pathway as a minor one for water, this means that the compensation must occur in the transcellular path. Further, since membranes are the major resistance in this path, the modification should be lowering their resistances to the passage of water. The prime candidates for this purpose are water channels (aquaporins). Perhaps because of the involvement of these proteins, the root is able to produce suberized structures whereby it regulates ion movement, protects against pathogens, and retains oxygen in the aerenchyma of aquatic plant roots (see Enstone *et al.*, 2003) without overly compromising the water-absorbing capability of the system.

With respect to axial water transport in roots, embolism may be regarded as a regulator in that it can both localize system failure and often be reversed to restore system function. In some cases, particularly for woody species with extensive root systems, embolism can also shift the primary limitation on root water transport from the radial to the axial pathway.

Future research into root water transport might profitably occur at several levels. As genes for aquaporins become known for more species, the use of antisense and knockout strains can help identify the role of these proteins in intercellular water movement, particularly during water stress and subsequent recovery. At the cellular and tissue level, the resistances conferred by Casparian bands and suberin lamellae still need to be examined, preferably in experiments that allow discrimination between apoplastic and transcellular water flow. At the whole root level, the likely changes in hydraulic conductivity resulting from developmental state and soil moisture need to be investigated for more species, particularly shrubs and trees. Finally, more complete mapping of root systems in the field is needed so that predictions of water uptake can be based on the proportional contributions of roots that can differ greatly along their lengths and in response to water availability.

## Acknowledgments

We thank Prof. Ernst Steudle for critically reading the manuscript and making helpful contributions, Ms. Meghan Hayter for drawing Figs. 7.1 and 7.2, Drs. Pierre Martre and Park S. Nobel for fruitful collaborations, and Drs. Uwe Hacke and J. S. Pritchard for valuable editorial suggestions. Some of the work described in this chapter was supported by the National Science Foundation (grant no. IBN-9975163, to P. S. Nobel).

## References

Aloni, R. and Peterson, C. A. (1998) Indirect evidence for bulk water flow in root cortical cell walls of three dicotyledonous species. *Planta* **207**: 1-7.

Azaizeh, H., Gunse, B. and Steudle, E. (1992) Effects of NaCl and $CaCl_2$ on water transport across root cells of maize (*Zea mays* L.) seedlings. *Plant Physiol* **99**: 886-894.

Barrieu, F., Chaumont, F. and Chrispeels, M. J. (1998) High expression of the tonoplast aquaporin *ZmTIP1* in epidermal and conducting tissues of maize. *Plant Physiol* **117**: 1153-1163.

Barrowclough, D. E., Peterson, C. A. and Steudle, E. (2000) Radial hydraulic conductivity along developing onion roots. *J Exp Bot* **51**: 547-557.

Bernards, M. A. (2002) Demystifying suberin. *Can J Bot* **80**: 227-240.

Bonnett, H. T., Jr. (1968) The root endodermis: Fine structure and function. *J Cell Biol* **37**: 199-205.

Brundrett, M., Murase, G. and Kendrick, B. (1990) Comparative anatomy of roots and mycorrhizae of common Ontario trees. *Can J Bot* **68**: 551-578.

Brundrett, M. C., Kendrick, B. and Peterson, C. A. (1991) Efficient lipid staining in plant material with Sudan red 7B or Fluorol yellow 088 in polyethylene glycol-glycerol. *Biotechnic Histochem.* **66**: 111-116.

Carvajal, M., Cerda, A. and Martinez, V. (2000) Does calcium ameliorate the negative effect of NaCl on melon root water transport by regulating aquaporin activity? *New Phytol* **145**: 439-447.

Cholewa, E. and Peterson, C. A. (2004) Evidence for symplastic involvement in the radial movement of calcium in onion roots. *Plant Physiol* **134**: 1793-1802.

Clarkson, D. T., Carvajal, M., Henzler, T., Waterhouse, R. N., Smyth, A. J., Cooke, D. T. and Steudle, E. (2000) Root hydraulic conductance: Diurnal aquaporin expression and the effects of nutrient stress. *J Exp Bot* **51**: 61-70.

Clarkson, D. T., Robards, A. W., Stephens, J. E. and Stark, M. (1987) Suberin lamellae in the hypodermis of maize (*Zea mays*) roots; development and factors affecting the permeability of hypodermal layers. *Plant Cell Environ* **10**: 83-93.

Dalton, F. N., Raats, P. A. C. and Gardner, W. R. (1975) Simultaneous uptake of water and solutes by plant roots. *Agron J* **67**: 334-339.

Degenhart, B. and Gimmler, H. (2000) Cell wall adaptations to multiple environmental stresses in maize roots. *J Exp Bot* **51**: 595-603.

de Rufz de Lavison, J. (1910) Du mode de pénétration de quelques sels dans la plante vivante. Rôle de l'endoderms. *Rev Gén Bot* **22**: 225-241.

Enstone, D., Ma, F. and Peterson, C. A. (2003) Root endodermis and exodermis: Structure, function, and responses to the environment. *J Plant Growth Reg* **21**: 335-351.

Enstone, D. E. and Peterson, C. A. (1992) The apoplastic permeability of root apices. *Can J Bot* **70**: 1502-1512.

Ewers, F. W., North, G. B. and Nobel, P. S. (1992) Root-stem junctions of a desert monocotyledon and a dicotyledon: Hydraulic consequences under wet conditions and during drought. *New Phytol* **121**: 377-385.

Faiz, S. M. A. and Weatherley, P. E. (1982) Root contraction in transpiring plants. *New Phytol* **92**: 333-343.

Fiscus, E. L. (1975) The interaction between osmotic- and pressure-induced water flow in plant roots. *Plant Physiol* **55**: 917-922.

Frensch, J. and Steudle, E. (1989) Axial and radial hydraulic resistance to roots of maize (*Zea mays* L.). *Plant Physiol* **91**: 719-726.

Frey-Wyssling, A. (1969) The structure and biogenesis of native cellulose. *Fortschr Chem Org Naturst* **26**: 1-30.

Gerrath, J. M., Covington, L., Doubt, J. and Larson, D. W. (2002) Occurrence of phi thickenings is correlated with gymnosperm systematics. *Can J Bot* **80**: 852-860.

Hacke, U. G., Sperry, J. S., Ewers, B. E., Ellsworth, D. S., Schäfer K. V. R. and Oren, R. (2000) Influence of soil porosity on water use in *Pinus taeda*. *Oecologia* **124**: 495-505.

Hofer, R.-M. (1996) Root hairs. In *Plant Roots: The Hidden Half*, 2nd Ed. (Y. Waisel, E. Amram and U. Kafkafi, eds.) pp. 111-126. Marcel Dekker, Inc. New York.

Hose, E., Clarkson, D.T., Steudle, E., Schreiber, L. and Hartung, W. (2001) The exodermis: A variable apoplastic barrier. *J Exp Bot* **52**: 2245-2264.

Hsiao, T. C. and Xu, L.-K. (2000) Sensitivity of growth of roots versus leaves to water stress: Biophysical analysis and relation to water transport. *J Exp Bot* **51**: 1595-1616.

Huang, B. and Nobel, P. S. (1994) Root hydraulic conductivity and its components, with emphasis on desert succulents. *Agron J* **86**: 767-774.

Hukin, D., Doering-Saad, C., Thomas, C. R. and Pritchard, J. (2002) Sensitivity of cell hydraulic conductivity to mercury is coincident with symplasmic isolation and expression of plasmalemma aquaporin genes in growing maize roots. *Planta* **215**: 1047-1056.

Javot, H. and Maurel, C. (2002) The role of aquaporins in root water uptake. *Ann Bot* **90**: 301-313.

Johansson, I., Karlsson, M., Shukla, V. K., Chrispeels, M. J., Larsson, C. and Kjellbom, P. (1998) Water transport activity of the plasma membrane aquaporin PM28A is regulated by phosphorylation. *Plant Cell* **10**: 451-459.

Justin, S. H. F. W. and Armstrong, W. (1987) The anatomical characteristics of roots and plant response to soil flooding. *New Phytol* **106**: 465-495.

Kaldenhoff, R., Grote, K., Zhu, J. J. and Zimmermann, U. (1998) Significance of plasmalemma aquaporins for water-transport in *Arabidopsis thaliana*. *Plant J* **14**: 121-128.

Landsberg, J. J. and Fowkes, N. D. (1978) Water movement through plant roots. *Ann Bot* **42**: 493-508.

Lu, Z. and Neumann, P. M. (1999) Water stress inhibits hydraulic conductance and leaf growth in rice seedlings but not the transport of water via mercury-sensitive water channels in the root. *Plant Physiol* **120**: 143-151.

Lüttge, U. and Higinbotham, N. (1979) *Transport in Plants*. Springer-Verlag, Berlin.

Lüttge, U. and Weigl, J. (1962) Mikroautoradiographische Untersuchungen der Aufnahme und des Transportes von $^{35}SO_4^{--}$ und $^{45}Ca^{++}$ in Keimwurzeln von *Zea mays* und *Pisum sativum* L. *Planta* **58**: 113-126.

Ma, F. and Peterson, C.A. (2000) Plasmodesmata in onion (*Allium cepa* L.) roots: A study enabled by improved fixation and embedding techniques. *Protoplasma* **211**: 103-115.

Ma, F. and Peterson, C. A. (2001) Frequencies of plasmodesmata in *Allium cepa* L. roots: Implications for solute transport pathways. *J Exp Bot* **79:** 577-590.

Ma, F. and Peterson, C. A. (2003) Recent insights into the development, structure and chemistry of the endodermis and exodermis. *Can J Bot* **81:** 405-421.

Martre, P., Morillon, R., Barrieu, F., North, G. B., Nobel, P. S. and Chrispeels, M. J. (2002) Plasma membrane aquaporins play a significant role during recovery from water deficit. *Plant Physiol* **130:** 2101-2110.

Martre, P., North, G. B. and Nobel, P. S. (2001) Hydraulic conductance and mercury-sensitive water transport for roots of *Opuntia acanthocarpa* in relation to soil drying and rewetting. *Plant Physiol* **126:** 352-362.

McCully, M. E. and Canny, M. J. (1988) Pathways and processes of water and nutrient movement in roots. *Plant Soil* **111:** 159-170.

Melchior, W. and Steudle, E. (1993) Water transport in onion (*Allium cepa* L.) roots. Changes of axial and radial hydraulic conductivities during root development. *Plant Physiol* **101:** 1305-1315.

Miyamoto, N., Steudle, E., Hirasawa, T. and Lafitte, R. (2001) Hydraulic conductivity of rice roots. *J Exp Bot* **52:** 1835-1846.

Nobel, P. S. and Cui, M. (1992) Hydraulic conductances of the soil, the root-soil air gap, and the root: Changes for desert succulents in drying soil. *J Exp Bot* **43:** 319-326.

North, G. B., Ewers, F. E. and Nobel, P. S. (1992) Main root-lateral root junctions of two desert succulents: Changes in axial and radial components of hydraulic conductivity during drying. *Am J Bot* **79:** 1039-1050.

North, G. B., Huang, B., and Nobel, P. S. (1993) Changes in structure and hydraulic conductivity for root junctions of desert succulents as soil water status varies. *Bot Acta* **106:** 126-135.

North, G. B., Martre P., and Nobel, P. S. (2004) Aquaporins account for variations in hydraulic conductance for metabolically active root regions of *Agave deserti* in wet, dry, and rewetted soil. *Plant Cell Environ* **27:** 219-228.

North, G. B. and Nobel, P. S. (1991) Changes in hydraulic conductivity and anatomy caused by drying and rewetting roots of *Agave deserti* (*Agavaceae*). *Am J Bot* **78:** 906-915.

North, G. B. and Nobel, P. S. (1992) Drought-induced changes in hydraulic conductivity and structure in roots of *Ferocactus acanthodes* and *Opuntia ficus-indica*. *New Phytol* **120:** 9-19.

North, G. B. and Nobel, P. S. (1996) Radial hydraulic conductivity of individual root tissues of *Opuntia ficus-indica* (L.) Miller as soil moisture varies. *Ann Bot* **77:** 133-142.

North, G. B. and Nobel, P. S. (2000) Heterogeneity in water availability alters cellular development and hydraulic conductivity along roots of a desert succulent. *Ann Bot* **85:** 247-255.

Peterson, C. A. (1987) The exodermal Casparian band of onion roots blocks the apoplastic movement of sulphate ions. *J Exp Bot* **32:** 2068-2081.

Peterson, C. A., Emanuel, M. E., and Weerdenburg, C.A. (1981) The permeability of phi thickenings in apple (*Pyrus malus*) and geranium (*Pelargonium hortorum*) roots to an apoplastic fluorescent dye tracer. *Can J Bot* **59:** 1107-1110.

Peterson, C. A., Emanuel, M. E. and Wilson, C. (1982) Identification of a Casparian band in the hypodermis of onion and corn roots. *Can J Bot* **60:** 1529-1535.

Peterson, C. A. and Lefcourt, B. E. M. (1990) Development of endodermal Casparian bands and xylem in lateral roots of broad bean. *Can J Bot* **68:** 2729-2735.

Peterson, C. A., Murrmann, M. and Steudle, E. (1993) Location of the major barriers to water and ion movement in young roots of *Zea mays* L. *Planta* **190:** 127-136.

Peterson, C. A., Peterson, R. L., and Robards, A. E. (1978) A correlated histochemical and ultrastructural study of the epidermis and hypodermis of onion roots. *Protoplasma* **96:** 1-21.

Peterson, C. A. and Steudle, E. (1993) Lateral hydraulic conductivity of early metaxylem vessels in *Zea mays* L. roots. *Planta* **189:** 288-297.

Pierson, D. R. and Dumbroff, E. B. (1969) Demonstration of a complete Casparian strip in *Avena* and *Ipomoea* by a fluorescent staining technique. *Can J Bot* **47:** 1869-1871.

Quigley, F., Rosenberg, J. M., Shachar-Hill, Y. and Bohnert, H. J. (2001) From genome to function: The *Arabidopsis* aquaporins. *Genome Biol* **3**: 1-17.

Ranathunge, K., Steudle, E. and Lafitte, R. (2003) Control of water uptake by rice (*Oryza sativa* L.): role of the outer part of the root. *Planta* **217**: 193-205.

Robards, A. W., Jackson, S. M., Clarkson, D. T. and Sanderson, J. (1973) The structure of barley roots in relation to the transport of ions into the stele. *Protoplasma* **77**: 291-311.

St. Aubin, G., Canny, M. J. and McCully, M. E. (1986) Living vessel elements in the late metaxylem of sheathed maize roots. *Ann Bot* **58**: 577-588.

Sanderson, J. (1983) Water uptake by different regions of the barley root. Pathways of radial flow in relation to development of the endodermis. *J Exp Bot* **34**: 240-253.

Scott, M. G. and Peterson, R. L. (1979) The root endodermis in *Ranunculus acris*. I. Structure and ontogeny. *Can J Bot* **57**: 1040-1062.

Seago, J. L., Peterson, C. A., Enstone, D. E. and Scholey, C. A. (1999) Development of the endodermis and hypodermis in *Typha glauca* Godr. and *Typha angustifolia* L. roots. *Can J Bot* **77**: 122-134.

Siefritz, F., Tyree, M. T., Lovisolo, C., Schubert, A. and Kaldenhoff, R. (2002) PIP1 plasma membrane aquaporins in tobacco: From cellular effects to function in plants. *Plant Cell* **14**: 869-876.

Sperry, J. S. and Hacke, U. G. (2002) Desert shrub water relations with respect to soil characteristics and plant functional type. *Funct Ecol* **16**: 367-378.

Sperry, J. S., Stiller, V. and Hacke, U. G. (2002) Soil water uptake and water transport through root systems. In *Plant Roots: The Hidden Half*, 3rd Ed. (Y. Waisel, E. Amram and U. Kafkafi, eds.) pp. 663-681. Marcel Dekker, Inc. New York.

Stein, W. D. (1986) *Transport and Diffusion Across Cell Membranes*. Academic Press, Orlando.

Steudle, E. (2001) The cohesion-tension mechanism and the acquisition of water by plant roots. *Annu Rev Plant Physiol Plant Mol Biol* **52**: 847-875.

Steudle, E. and Brinkmann, E. (1989) The osmotic model of the root: Water and solute relations of roots of *Phaseolus coccineus*. *Bot Acta* **15**: 85-95.

Steudle, E. and Jeschke, W. D. (1983) Water transport in barley roots. *Planta* **158**: 237-248.

Steudle, E. and Peterson, C. A. (1998) How does water get through roots? *J Exp Bot* **49**: 775-788.

Suga, S., Imagawa, S. and Maeshima, M. (2001) Specificity of the accumulation of mRNAs and proteins of the plasma membrane and tonoplast aquaporins in radish organs. *Planta* **212**: 294-304.

Taleisnik, E., Peyrano, G., Córdoba, A. and Arias, C. (1999) Water retention capacity in roots segments differing in the degree of exodermis development. *Ann Bot* **83**: 19-27.

Tazawa, M., Ohkuma, E., Shibasaka, M. and Nakashima, S. (1997) Mercurial-sensitive water transport in barley roots. *J Plant Res* **110**: 435-442.

Tyerman, S. D., Niemietz, C. M. and Bramley, H. (2002) Plant aquaporins: Multifunctional water and solute channels with expanding roles. *Plant Cell Environ* **25**: 173-194.

Tyree, M. T., Patiño, S., Bennink, J. and Alexander, J. (1995) Dynamic measurements of root hydraulic conductance using a high-pressure flowmeter in the laboratory and field. *J Exp Bot* **46**: 83-94.

Wan, X. and Zwiazek, J. J. (1999) Mercuric chloride effects on root water transport in aspen seedlings. *Plant Physiol* **121**: 939-946.

Wang, X. L., McCully, M. E. and Canny, M. J. (1995) Branch roots of *Zea*. V. Structural features that may influence water and nutrient transport. *Bot Acta* **108**: 209-219.

Wullstein, L. H., Bruening, M. L. and Bollen, W. B. (1979) Nitrogen fixation associated with sand grain root sheaths (rhizosheaths) of certain xeric grasses. *Physiol Plant* **46**: 1-4.

Ye, Q., Wiera, B. and Steudle, E. (2004) A cohesion/tension mechanism explains the gating of water channels (aquaporins) in *Chara* internodes by high concentration. *J Exp Bot* **55**: 449-461.

Zeier, J., Goll, A., Yokoyama, M., Karahara, I. and Schreiber, L. (1999a) Structure and chemical composition of endodermal and rhizodermal/hypodermal walls of several species. *Plant Cell Environ* **22:** 271-279.

Zeier, J., Ruel, K., Ryser, U. and Schreiber, L. (1999b) Chemical analysis and immunolocalization of lignin and suberin in endodermal and hypodermal/rhizodermal cell walls of developing maize (*Zea mays* L.) primary roots. *Planta* **209:** 1-12.

Zeier, J. and Schreiber, L. (1997) Chemical composition of hypodermal and endodermal cell walls and xylem vessels isolated from *Clivia miniata*. *Plant Physiol* **113:** 1223-1231.

Zeier, J. and Schreiber, L. (1998) Comparative investigation of primary and tertiary endodermal cell walls isolated from the roots of five monocotyledonous species: Chemical composition in relation to fine structure. *Planta* **206:** 349-361.

Zhang, W. H. and Tyerman, S. D. (1999) Inhibition of water channels by $HgCl_2$ in intact wheat root cells. *Plant Physiol* **120:** 849-857.

Zimmermann, H. M, Hartmann, K., Schreiber, L. and Steudle, E. (2000) Chemical composition of apoplastic transport barriers in relation to radial hydraulic conductivity of corn roots (*Zea mays* L.). *Planta* **210:** 302-311.

Zimmermann, H. M. and Steudle, E. (1998) Apoplastic transport across young maize roots: Effect of the exodermis. *Planta* **206:** 7-19.

# 8

## Roots as an Integrated Part of the Translocation Pathway

*Jeremy Pritchard, Brian Ford-Lloyd, and H. John Newbury*

The growing root tip is at the end of the translocation pathway and is highly dependent on the import of solutes for its continued growth and other metabolism. The delivery of organic solutes into growing root cells has been considered to occur by two completely different mechanisms: uptake from the apoplast across membranes or symplastic delivery through plasmodesmatally connected domains. The pathways have very different hardware and fundamentally different driving forces. Apoplastic uptake of solutes requires membrane transporters or channels, whereas in symplastically connected cells solutes and water can enter together without crossing a membrane.

Both existing and emerging data indicate that, far from being alternatives, the two routes for solute delivery to root tips may operate together in the same root tip. This pattern can be compared with the situation in developing fruits where a temporal separation of symplastic and apoplastic loading routes can occur.

The goals of this chapter are to consider the relationship between root extension and solute deposition, to review the evidence for the existence of symplastic and apoplastic pathways within the growing zone, and to assess the consequences of these two pathways for solute transport to growing sinks. Because qualitative changes in the pathway for solute delivery within the root-growing zone predict associated differences in the presence of specific transporters for both solutes and water, some speculations are offered as to the location of specific transporters within the root tip. To help link studies of solute delivery and growth in roots with studies on other tissues, parallels are drawn between the transport processes occurring in source, sink, and pathway.

## Root Growth and Solute Deposition

### Roots as a Sink for Solutes

Roots provide the exploratory mechanism that allows plants to overcome their sedentary nature and thus are central to plant function and success in any environment. Root extension occurs as cells immediately behind the root tip increase in length (Pritchard, 1994). Cell extension, in turn, is driven by turgor pressure; a reduction in turgor pressure causes a reduction in root extension (Pritchard *et al.*, 1991). Turgor is generated osmotically by the accumulation of solutes in the protoplast, which results in water uptake due to the water potential difference between the apoplast and symplast.

Extension growth is triggered when biochemical events in the cell wall loosen the wall and lower turgor (Cosgrove, 1986; Boyer, 2001). The reduction in turgor generates a more negative cell water potential in the cell. The subsequent influx of water results in an increase in volume, or in other words, growth. If the source of water is solely apoplastic, then the entry of water into the cell will cause a dilution of the cell contents. Such a reduction in cell solute concentration will reduce both the osmotic pressure and turgor pressure. Thus, sustained extension growth requires continual and balanced uptake of water and solutes. Countering dilution by growth is not the only sink for solutes in the root tip. Resources are additionally required to construct new dry matter (biomass deposition), to fuel respiration, and to replace materials lost due to leakage and sloughing off of root material (see Chapter 13). Thus, the root tip is a considerable and potentially complex sink for solutes.

### Growth Varies Along the Root Axis

Extension growth is a significant component of the sink for solutes along the root tip. As the relative elemental growth (REG) rate increases, cell length begins to increase as the rate of cell division declines. Cell expansion rate reaches a maximum, then declines in the region of decreasing REG rate (Fig. 8.1). The apical region contains the root meristem where cell volume is dominated by cytoplasm. During the initial phase of cell expansion in the accelerating zone, cells become highly vacuolate.

The length of the growing zone varies with both species and environment. For example, in *Arabidopsis* the length of the growing zone is 2 to 3 mm (Freixes *et al.*, 2002). Under optimal conditions the length of the growing zone in maize is about 10 mm long and, as occurs in most species, the distribution of the regions of increasing and decreasing REG are approximately symmetrical about the point of maximum growth. Recently, some workers have analyzed the distribution of growth along the root axis at greater resolution than has been possible previously (van der Weele *et al.*, 2003). They indicate that the smooth bell-shaped curves indicated in the

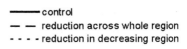

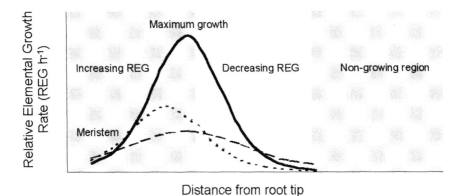

**Distance from root tip**

**Figure 8.1** Distribution of relative elemental growth rate (REG) along a root tip. Relevant functional regions are identified. Changes in the distribution of REG observed following experimentally induced changes in whole root extension are also illustrated. A reduction predominantly in the decreasing REG region is more commonly observed than a reduction in extension in both increasing and decreasing REG regions. A reduction predominantly in the region of increasing REG is never observed.

older literature may be artifacts and that the transition from increasing through decreasing REG to growth cessation may be far more abrupt than previously thought. The work of Sharp and co-workers (Sharp *et al.*, 1988) on maize indicates that during drought the growing zone shortens as the rate slows in the region of decreasing REG. Importantly, in many species, local growth is often maintained at control rates in the region of increasing REG while environmental perturbation can reduce growth in the region of decreasing REG. Thus, the dilution of solutes by extension growth varies along the root.

### Environmental Effects on Solute Accumulation in the Root Tip

Turgor pressure is generally constant within the root tip. However, the solutes that generate this pressure can vary both longitudinally and radially. The main solutes accumulated in osmotically significant amounts along the root tip are potassium salts, hexoses (glucose and fructose), and amino acids. In contrast to the high concentrations found in the sieve element, sucrose accumulation is low along the root apex, although a slightly higher concentration can occur in the apical, meristematic region. Hexose concentration is low in the region of increasing REG, but increases markedly in basal regions and reaches a constant value through the region of decreasing REG (Sharp *et al.*, 1990; Walter *et al.*, 2003). Radial gradients in turgor pressure are not commonly seen across root tips (but see Croser *et al.*, 2000), although the composition of solutes contributing to this constant osmotic pressure may vary (Pritchard *et al.*, 1996).

While the distribution of solutes within root tips provides important clues about their route of delivery, concentration does not provide information on the rate of solute import. Import of solutes into root tips can be followed directly using short-lived radiotracer such as 11C. Analysis of import into maize root tips indicated that over 29% of the carbon recently fixed in photosynthesis that entered the root was delivered to the apical 15 mm (Fig. 8.2), with the majority of this being localized to the growing region of the root tip (apical 10 mm).

The considerable bias in solute delivery toward the apical region of increasing REG in Fig. 8.2 is in marked contrast with the net deposition of hexose calculated using the continuity equation. Such calculations demonstrate an increase and subsequent decrease away from the root tip that parallels the distribution of local growth (Silk *et al.*, 1986; Sharp *et al.*, 1990), with the 0 to 5 and 5 to 10 mm regions of the root tip having the same rate of net solute import. The difference between directly observed solute deposition measured using 11C and that calculated using the continuity equa-

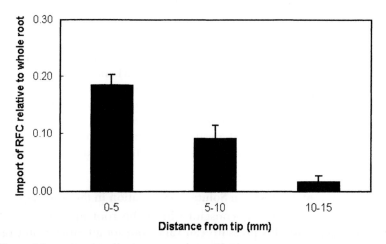

**Figure 8.2**  Import of recently fixed carbon (RFC) to the apical region of maize roots measured using the method of Minchin and Thorpe (1996). Ten-day-old plants were supplied with $^{11}$C labeled $CO_2$. Import of RFC into the indicated regions was followed using appropriately placed scintillation detectors. 0-5 mm is the accelerating region, 5-10 is the decelerating region of the growing zone, and 10-15 is the newly mature nongrowing cells. Data are expressed as the proportion of tracer in each region relative to that entering the whole root. Each bar represents the mean of 5 determinations ± SD.

tion presumably reflects higher metabolism of organic solutes in the apical regions of the root and possibly a greater efflux of organic solutes toward the root apex (see Chapter 13).

Changes in the pattern of solute deposition can occur rapidly after a perturbation in growth conditions. For example, switching maize roots from pH 7 to pH 3.4 rapidly increased root extension rate, with the increase in growth being greatest in the apical (increasing REG) region of the root tip (Winch and Pritchard, 1999). Despite a fourfold increase in extension, no decrease in turgor pressure or osmotic pressure could be detected. These data indicate a tight relationship between solute import and extension growth and suggest an extremely responsive system (or systems) of solute delivery. Differences in the spatial distribution of response(s) are consistent with the operation of different mechanisms in apical regions of the growing zone.

The pattern of solute distribution is affected by environmental conditions. For example, total solute accumulation, measured as osmotic pressure, increases during drought (e.g. Sharp *et al.*, 1990; Pritchard *et al.*, 1996), although the accumulation of both potassium and sucrose remains unaffected (Sharp *et al.*, 1990; Pritchard *et al.*, 1996; Muller *et al.*, 1998). Net deposition rate of hexose increased in the apical region of the maize root growing zones after exposure to 1.6 MPa drought for 48 hours, but interestingly there was no increase in potassium deposition in the same region (Sharp *et al.*, 1990). Thus drought specifically affects hexose accumulation (Sharp *et al.*, 1990; Pritchard *et al.*, 1996). Other authors have also noted a qualitatively different behavior between sucrose and hexose. In maize roots grown under low light, there was no change in the levels of either sucrose or hexoses along the growing zone. However, at higher light intensities, hexose concentration increased away from the tip and sucrose levels exhibited a small but significant increase at the start of the region of increasing REG (Muller *et al.*, 1998). The qualitatively different behavior of organic and inorganic solutes along the growing region underlies the tight homeostasis of the water relations in the growing zone and suggests a complex and integrated regulation of solute deposition. This regulation must reside at least in part in the pathways of solute and water delivery.

### Root Extension Growth Varies with Carbon Availability

Over a short time scale (seconds to minutes) extension growth is the consequence of a dynamic relationship between turgor pressure and cell wall properties (Frensch and Hsiao, 1994; Pritchard, 1994). However, because solutes are required to generate turgor, over longer time periods (hours to days) root growth rates often correlate with carbon availability. In naturally growing roots, the only source of carbon is from the sieve elements. Thus, manipulation of the source and quantity of carbon can alter the root extension rate. In maize seedlings, seed removal reduced the elongation of the primary roots, interpreted as the result of eliminating the primary carbon source for growth (Muller *et al.*, 1998). In the same study, slightly older plants not dependent on seed reserves showed elevated root extension at higher light intensities. Similarly, reducing leaf area by pruning reduced the extension rate of wheat roots and their sugar content within 1 hour (Bingham *et al.*, 1996). In some species, elevated $CO_2$ can result in increased root biomass. For example, in a range of native British grass land species, elevated $CO_2$ led to an increase in both root growth rates (extension) and the number of lateral roots (Ferris and Taylor, 1994).

*Arabidopsis* is an increasingly utilized experimental system to study root growth. Using an agar growth substrate, root extension can be easily manipulated by varying both the type and source of carbon, as well as the light intensity (Freixes *et al.*, 2002). In these studies, sucrose is usually included in the agar, effectively short circuiting acropetal delivery through the sieve element. However, this provides a convenient way to manipulate plant and root carbon status and to explore the relationship between carbon supply and root extension growth. At low light intensities, *Arabidopsis* roots with no sucrose in the agar medium grew slower than those supplied with 2% sucrose (Freixes *et al.*, 2002). This relationship was eliminated at higher light intensities, presumably because the leaves could now supply an excess of carbon. Thus, over longer time scales (hours to days), there is a link between root carbon levels and extension growth. It is sensible to assume that a mechanistic link between these two phenomena will be mediated by the phloem.

## Roots Have Symplastic and Apoplastic Domains of Unloading

### A Chimeric Model of Unloading

A cell in the root apex has two potential sources of solutes: uptake from the apoplast across the plasma membrane, and alternatively or additionally, solutes and water can enter cells directly from the sieve element if a symplastic pathway exists. A range of approaches has demonstrated that growing cells at the root tip are symplastically connected to each other and ultimately to sieve elements. Photoassimilate accumulates in root tips despite the presence of *p*-chloro-mercuribenzenesulfonic acid thought to block sucrose uptake from the apoplasm (Dick and Après, 1975; Giaquinta *et al.*, 1983). Oparka and co-workers (e.g., Oparka *et al.*, 1994) have more directly demonstrated such symplastic continuity by observing transport of phloem localized tracers to the root tip. The use of [11]C to follow the movement of recently fixed carbon also provides evidence supporting symplastic transport into root tips. For example, the movement of $^{11}$C into root tips of barley was unaffected by metabolic inhibitors (Farrar and Minchin, 1991) consistent with the hypothesis that no carriers are involved.

However, there is evidence of an apoplastic route within the growing region. When examining the delivery of fluorescent symplastic tracer to the tips of *Arabidopsis* roots, older cells lose symplastic continuity with the sieve element (Duckett *et al.*, 1994). In this study, the transition from a symplastic domain to apoplastically isolated cells was well defined and occurred at around 1.2 mm from the root tip. A similar movement of symplastic tracer was shown in *Arabidopsis* plants in which GFP was produced in companion cells when the GFP coding region was controlled by a SUC2 promoter (Imlau *et al.*, 1999). GFP fluorescence moved systemically to the root tip and the zone of unloading extended to approximately 300 μm from the root tip.

In these GFP studies, it is not clear whether label in the basal regions is present as a result of transport through open symplastic connections, or whether it is present in newly isolated cells as they are displaced away from the tip by extension in the apex. Residual tracer would be present in newly isolated cells but would be progressively diluted by apoplastic water uptake during expansion and in the case of GFP by metabolism, predicting a more diffuse boundary. During the experiment of Duckett *et al.* (1994), the size of the symplastic zone revealed by the fluorescent tracer increased from about 600 μm at 43 minutes from loading to 1200 μm at 163 minutes, though the boundary remained well defined. In similar studies of symplastic unloading of *Arabidopsis* roots, the maximum length of the symplastic zone was approximately 400 μm 42 minutes after photobleaching of incoming fluorescent dye (Oparka *et al.*, 1994). Thus the fluorescence data indicate that the symplastic zone extends to a maximum of 1.2 mm from the root tip. However, the *Arabidopsis* growing zone extends to over 2 mm from the *Arabidopsis* root tip (Freixes *et al.*, 2002). Therefore, if data from these different studies are comparable, regions of the extending zone are symplastically isolated from the sieve elements. The suggestion that the transition between different growth zones of the root may be more abrupt than previously thought (van der Weele *et al.*, 2003) makes it imperative to take care when attempting to co-localize these zones with the different regions of unloading.

Data from maize roots are also consistent with the hypothesis that apoplastic uptake is the only unloading route in some regions of the growing zone. Symplastic isolation began approximately 3 mm from the root tip, whereas maximal growth was at 5 mm and the region of decreasing REG extended a further 5 to 10 mm from the root tip (Hukin *et al.*, 2002). Even after prolonged periods of tracer loading, no fluorescence was observed in the region of decreasing REG. Thus the region of decreasing

REG is potentially isolated from the sieve element despite considerable solute deposition in this region (Sharp *et al.*, 1990). It has been suggested that the size of the symplastic domain may change under different solute availability regimens (Patrick and Offler, 1996), a process that would be facilitated by the documented gating behavior of plasmodesmata (Baluska *et al.*, 2001; Schulz, 1995).

### Driving Forces for Solute Unloading Are Qualitatively Different in Apoplastic and Symplastic Regions

It has been calculated that diffusion alone cannot explain solute import within the root tip (Bret_Harte and Silk, 1994). In symplastically connected regions the driving force for solute import is the pressure difference between the sieve element and the receiving cell (see Chapter 2; Patrick, 1997). Differences in pressures have been demonstrated indirectly as osmotic pressures (Warmbrodt, 1987; Pritchard, 1996). Direct measurements of turgor pressure in sieve elements have confirmed the existence of a pressure difference between sieve element and growing cells in barley roots. A reduction in solute import to the growing zone roots induced by low $K^+$ correlated with a lower sieve element turgor pressure (Gould *et al.*, 2004).

In apoplastic regions water enters cells moving down water potential gradients created by the accumulation of solutes, and in growing cells by cell wall loosening. Solutes are taken up, often actively through $H^+$ co-transporters energized by $H^+$ ATPase, or move down electrochemical gradients through channels. Thus, in contrast to the symplastic route, both the pathway and driving force for water and solute movement are uncoupled owing to the different mechanisms for water and solute movement across semipermeable membranes.

The decrease in osmotic pressure necessary to maintain a turgor difference in the translocation pathway between sieve elements and root cells is achieved by a combination of metabolism (biomass deposition and respiration), dilution by growth, and leakage (see Chapter 13). The size of the differential between sieve element and companion cell osmotic pressure is modified by the hydrolysis of sucrose in the sink, presumably by invertases, doubling its osmotic contribution. Both the longitudinal and subcellular location of this hydrolysis, in either apoplast or symplast, will have a marked effect on sink cell turgor. Import through the symplastic pathway would be affected by any change in the sink cell turgor.

The existence of different pathways and driving forces along root tips predict differences in solute delivery to root tips after environmental perturbation. In particular, changes in turgor pressure will have a direct effect in symplastic regions but a less direct, qualitatively different effect in regions of apoplastic unloading. It is a common observation that turgor pressure recovers more quickly in apical regions of the growing zone (Pritchard *et al.*, 1991; Frensch and Hsiao, 1994) consistent with a direct supply of solute to this region. Similarly, the increase in hexose deposition in apical regions of the maize root noted by Sharp *et al.* (1990) occurred concomitant with an inferred increase in the pressure difference between sieve element and expanding cells, as cell turgor was decreased by some 0.3 MPa (Spollen and Sharp, 1991). No such increase in hexose deposition occurred in the region of decreasing REG despite a similar turgor change. In the apical region of the same roots, there was no increase in the deposition of $K^+$ following drought. The uncoupling of hexose and $K^+$ import is not consistent with direct flow of sieve element sap into these apical cells, although the possibility of differential mobility of hexose and $K^+$ following a similar route of initial unloading cannot be ruled out. The simplest explanation for the observation is that the two solutes are removed from the apoplast by different transporter/channel systems.

Manipulation of turgor in the different regions of the root tip provides further evidence of the different behavior of solute import in symplastic and apoplastic regions. Immersion in 200 mM mannitol decreased the turgor of the cortical cells in the maize root tip by about 0.2 to 0.3 MPa (Fig. 8.3A).

Import of recently fixed carbon increased immediately after turgor reduction along the apical 15 mm (Fig. 8.3B). However, the increase was greater in the region 0 to 5 mm from the root tip that encompasses the symplastic regions of unloading. The response was sustained despite a gradual increase in turgor (Fig. 8.3A), perhaps reflecting a decrease in the resistance of the symplastic pathway. In drought-stressed pea roots, an increase in solute import was accompanied by an increase in the diameter of the plasmodesmatal pores (Schulz, 1995). In growing apoplastic regions and in the nongrowing region at 10 to 15 mm from the root tip, an initial rise in import of recently fixed carbon was followed by a decline in import that correlated with the rise in turgor as the roots accumulated solutes.

Qualitative differences between the behavior of solute import in symplastic and apoplastic regions were also seen when, after 140 minutes of exposure to 200 mM mannitol, the root was returned to 0 mM mannitol. A transient elevation in turgor pressure (caused by a drought-induced accumulation of solutes) was accompanied by a rapid decrease in solute import in the symplastic region at 0 to 5 mm from the root tip. Thus the pattern of solute import in symplastically connected cells at 0 to 5 mm is as predicted from the change in pressure difference between sieve elements and growing cells. The smaller increase in import seen in both the growing and nongrowing apoplastic regions is presumably due to increased rates of uptake across the membrane, which is not as rapid or as marked as the bulk flow in symplastic regions. The marked increase in import in the 5 to 10 mm region

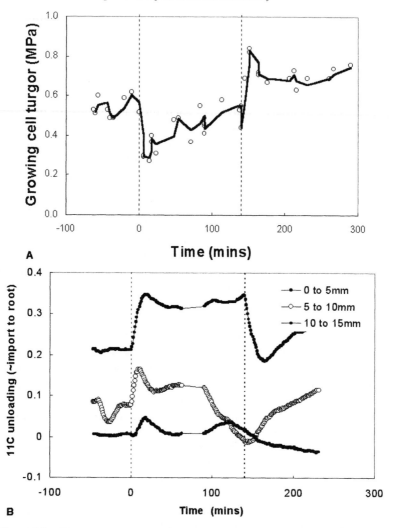

**Figure 8.3** Change in turgor pressure, measured with the pressure probe (3A) and import of recently fixed carbon (RFC) measured as import of [11]C (3B), of cortical cells along the growing region of maize roots. Roots were grown in hydroponics, and external water potential was reduced by addition of 200 mM mannitol (water potential of − 0.48 MPa), to the growing root medium at the time indicated by the dashed vertical lines. Each point is a single determination. For turgor pressure, no significant differences were seen in the response along the growing region so data are combined into a single plot. For [11]C measurements data shown are from a single root.

after elevation of turgor could again be due to increased uptake across membranes, although it is not easy to see a clear adaptive reason for an increase in uptake at higher turgor pressure. It is interesting to note a similar but delayed increase in solute uptake in the symplastic region.

## Distribution of Cell Water Potential and Solutes as Evidence for Unloading

For an apoplastic step to be involved in the delivery of solutes to root tips, they must be unloaded so that they are present in the cell wall. An indirect method of determining apoplastic solute concentrations is by measuring cell water potentials in nontranspiring plants. Nonzero cell water potentials can be due to transpiration tension, non unity reflection coefficients, growth-induced differences in water potential, or solutes in the cell's wall. In growing root tips, it was concluded that the presence of solutes in the cell wall dominates root cell water potential (Pritchard *et al.,* 1996). Such nonzero water potentials have been noted in roots of a number of species (see references in Pritchard *et al.,* 1996). In maize roots water potential was a constant −0.2 MPa along the region encompassing both symplastic and apoplastic unloading (Pritchard *et al.,* 1996). A higher (less negative) water potential would be predicted in regions of symplastic unloading, but these are not seen. Pea roots had a similar water potential, around −0.2 MPa along the growing zone; this was doubled to over −0.4 MPa after restriction of root extension by growing them in compacted medium (Croser *et al.,* 2000). These observations are consistent with increased solute leakage (or active unloading) into the apoplast following impedance, but similar to maize, there was no difference in changes in water potential of apical (symplastic) and basal (apoplastic) regions of the growing zone. In pea roots growing in compacted medium, turgor was elevated, which may be a cause of the increased solute leakage. The constant water potential along the root axis indicated similar levels of apoplastic solutes in both regions, consistent with an exclusively apoplastic unloading step in basal regions, which extends into the root apex. In the apical region apoplastic unloading would operate in parallel with symplastic unloading.

Heterogeneity of solute distribution within the growing region may also provide evidence of the delivery route. Outer cortical cells of pea roots had lower turgor than the inner cells (i.e., lower solute accumulation assuming water potential was unchanged) at 2 mm but not 40 mm, consistent with a source of solutes from the sieve elements in the center of the root (Croser *et al.,* 2000). Indeed, if the resistance of the symplastic unloading pathway is significantly high, a measurable radial turgor gradient would be expected to allow bulk flow to the outer cells. However, rapid changes in solute import can be observed with no detectable drop in cell turgor (e.g., Winch and Pritchard, 1999), indicating either that the resistance of the pathways is very low or that solute delivery occurs by a method other than pressure driven bulk flow. In mature nongrowing regions of maize roots, hexose concentration was higher in deeper cells, whereas a reciprocal distribution of potassium was observed (Pritchard *et al.,* 1996). Despite the observations of Croser (2000), most reports do not find any radial gradients within root-growing zones (Pritchard, 1994).

It is worth mentioning that symplastic delivery of solutes will, by definition, load incoming solutes directly into the cytosol; however, apoplastic loading requires uptake across the plasmalemma. In both cases tonoplast transport is required because, in most of the growing zone (excluding the meristem), the vacuole makes up the majority of the cell volume and therefore dominates the water relations of the cell. In addition subcellular solute distribution must be considered when comparing data obtained by different methods. For example, whole tissue analysis will include apoplastic, cytoplasmic, and vacuolar components, whereas single cell sampling will be dominated by the vacuole (Tomos *et al.*, 1994).

## Does Transporter Distribution Fit a Chimeric Distribution of Unloading?

The previous discussion divides the root apex into three functional regions: symplastic growing, apoplastic growing, and apoplastic non-growing regions. The physiological analysis predicts different transporters and enzymes of associated metabolism in different regions of the root. For example, in apoplastic regions the relevant plasmalemma transporter function must be present. Because sieve elements contain high concentrations of sucrose but root tips do not, an apoplastic step will require the presence of enzymes such as invertase to process the incoming sucrose. A wide range of organic, inorganic, and water channels will also be involved.

### Predictions of Transporter Gene Expression Within the Symplastic Growing Region

The chimeric model predicts that neither sucrose transporters nor cell wall invertases will be required in large numbers in the symplastic growing region. However, there is little sucrose in the growing cells compared to the large amount in the sieve elements. Single cell analysis has indicated the absence of vacuolar sucrose along the root apex (Pritchard *et al.*, 1996). Whole tissue analysis indicates elevated levels toward the root tip (Muller *et al.*, 1998; Walter *et al.*, 2003). However the concentrations are on the order of 30 mM, already at least an order of magnitude below that being imported in the sieve element. These observations are consistent with the hydrolysis of sucrose in the cytosol followed by uptake of hexoses across the tonoplast into the vacuole. Therefore, alkaline invertases and tonoplast localized hexose transporters are predicted in this region (Fig. 8.4 transport process #2). Alternatively, hydrolysis of sucrose may occur in the vacuole, in which case the presence of vacuolar (acid) invertases would be predicted.

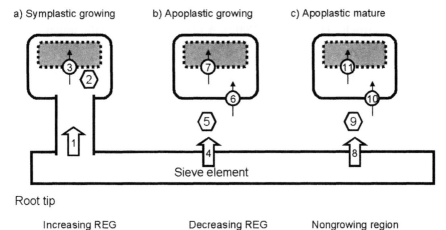

a) Symplastic growing    b) Apoplastic growing    c) Apoplastic mature

Root tip

Increasing REG          Decreasing REG          Nongrowing region

**Figure 8.4**   A model for solute unloading in the growing and nongrowing regions of the root tip based on current physiological data. Numbers and arrows refer to the different transport processes that occur in the different functional regions of the root. For further explanation and numbering, see text.

In maize roots there was an increase in activity of the vacuolar, but not the cell wall invertases following a water stress induced increase in hexose accumulation. Gene expression and protein levels increased in vacuoles of cortical cells but not in the stele; information on longitudinal distribution of gene expression was not presented (Kim *et al.*, 2000). An invertase characterized in carrot is also a potential candidate for localization in this region, as repression of activity in carrot showed altered solute partitioning to the root (Tang *et al.*, 1999). There are up to 11 Open Reading Frame (ORFs) in *Arabidopsis* that encode invertases (Henrissat *et al.*, 2001), and these include a number of strong candidates for root localization including At1g62660.

The analysis of the *Arabidopsis* genome suggests many possible candidates for hexose transporters active in the root tip (Maathius *et al.*, 2003; TAIR web site, http://www.arabidopsis.org/info/genefamily/genefamily.html). *Arabidopsis* genes include the hexose transporters AtSTP4 and AtSTP2. These have been hypothesized to have a role in unloading solutes in the root tip and have significant expression in the root tip of *Arabidopsis* (Williams *et al.*, 2000). Other hexose transporters, such as STP3, have no reported expression in roots but high expression in green tissues. Based on current knowledge, genes such as STP3 would be predicted to be absent from the symplastic region of the root tip.

Our analysis has focused on apoplastic and symplastic pathways operating in series; however, there is no reason why both should not operate in the symplastic regions, blurring the predictions made here. Notwithstanding this possibility, a study of maize roots demonstrates that the symplastic and apoplastic regions of the root tip can differ qualitatively in the presence of specific channels (Hukin *et al.*, 2002). A plasmalemma-specific aquaporin was located in the apoplastic but not the symplastic region of the root, whereas a tonoplast aquaporin, facilitating vacuolar uptake, was expressed equally in both regions. It was hypothesized that in symplastically connected regions, plasmalemma transporters are not required, as water (and the dissolved solutes it contains) bypasses the plasmalemma.

### Apoplastic Growing Region

The chimeric model predicts a greater reliance on membrane transporters in apoplastic than in symplasmic regions of unloading. Unloading of sucrose from the symplast via antiporters has been predicted (Williams *et al.*, 2000), although so far none have been identified. It may be that the processes of phloem-xylem transfer that occur during the transport pathway (see Chapter 10) are the same as those that occur in root tip sinks. If so, their identification in apical regions will facilitate a greater understanding of their function in root tips. It has also been suggested that alternatively, or perhaps additionally, sucrose influx carriers such as SUC1 could operate unenergized, facilitating efflux of sucrose down the concentration gradient (Williams *et al.*, 2000). However, it is not clear how transporters unloading from sieve element to apoplast and $H^+$ symporters loading hexose into the cells from the same apoplast could both operate when they require qualitatively different driving forces.

The lack of firm experimental evidence for active unloading of sucrose leads many authors to imply, or directly suggest, passive efflux of sucrose from sieve element to the apoplast (Patrick, 1997, Fig. 8.4, transport process #4). In the source and pathway portion of the translocation system, this may be termed leakage. By whatever route the sucrose reaches the apoplast, it has been proposed that unloaded sucrose is cleaved by cell wall invertases (Fig. 8.4, transport process #5 ). A carrot wall invertase altered partitioning to the root when downregulated (Tang *et al.*, 1999). Additionally, a cell wall invertase was found in a cDNA library made from *Ricinus* phloem sap, its sieve element specificity providing another potential candidate for a sink invertase (Doering Saad *et al.*, unpublished). Cell wall invertases have also been shown to be expressed in the maize root (Roitsch, 1999) and in *Arabidopsis* roots (Tymowska-Lalanne and Kries, 1998), although localization was not done at high spatial resolution.

Subsequent to sucrose hydrolysis, glucose and fructose are predicted to be taken up into cells across the plasmalemma via hexose transporters (Fig. 8.4, transport process #6). Some transporters may also be present in the tonoplast to facilitate uptake into the vacuole (Fig. 8.4, transport process #7 ); candidates are the *Arabidopsis* transporter AtSTP4 and AtSTP2 (Truernit *et al.*, 1999). STP1 is another strong candidate, being expressed in the root tip of *Medicago truncatula* (Harrison, 1996). In *Arabidopsis*, STP1 was found mainly in the seedling root and a knock-out mutant grew and developed normally (Sherson *et al.*, 2000).

### Nongrowing Region

In mature, nongrowing regions, the demand for solutes is lower than in either the symplastic or apoplastic growing regions (e.g., Fig. 8.2). Here the same qualitative distribution of transporters as in the apoplastic growing region is predicted but at a lower level of expression. In this region the processes of loading and unloading may well be the same as those that occur in the transport pathway (see Chapter 10), and leakage from sieve elements may supply the necessary unloading (Fig. 8.4, transport process #8, 1). The SUC2 symporter is central to loading of the sieve element companion cell complex, and so is not predicted in the symplastic or apoplastic unloading regions of the root. The sucrose/$H^+$ symporter SUC1 may fulfill a sucrose scavenging role (Barker *et al.*, 2000) and thus might occur in the transport region with levels of expression decreasing closer to the root apex. If sucrose hydrolysis occurs in the apoplast, these sucrose symporters will have a less important role in apoplastic uptake. Thus, the possibility remains that both transporters have a role in unloading in the root. If sucrose is hydrolyzed in the apoplast low levels of expression of cell wall, but not neutral, invertases are predicted in this region (Fig. 8.4, transport process #9) along with hexose transporters such as STP1, 2 and 4 (Fig. 8.4, transport process #10 & 11).

### Which Transporter Genes Are Involved in Unloading?
### A Transcriptomics Approach

The preceding analysis focuses on information available in the literature, which is at best fragmented so that interpretation can be subjective. However, molecular techniques such as transcriptomics are providing new approaches, and there is an increasing amount of information available in public databases. Clearly, expression in roots is essential if the gene plays a role in the chimeric model of unloading. Transciptomics data, such as Affymetrix datasets, are available on the NASC web site (http:// affymetrix.arabidopsis.info/narrays/experimentbrowse.pl). One example is the data set of Smirnof and Jones who extracted mRNA from mature primary *Arabidopsis* root tissue (corresponding to our nongrowing region) and analyzed it using an Affymetrix Arabidopsis Gene Chip. Filtering their data set with the *Arabidopsis* transporter genes documented by Maathuis *et al.* (2003) indicates that the highest expression signal in the root is obtained from a vacuolar invertase (At1g62660). A putative xylose $H^+$ symporter (At1g20840) has the second highest expression. The sucrose $H^+$ symporters SUC1 (At1g71880), SUC3 (At2g02860), and the hexose transporter STP4 (At3g19930) have significant expression in this region of the root. The data also indicate a number of these sugar transporters that have low expression, including an invertase (At3g52600) and a number of putative sucrose transporters (e.g., At5g06170). This simple analysis therefore provides candidates for further localization in this region of the root and eliminates others. Unfortunately, there is no examination of gene expression in the more apical regions of the root, so that a more detailed developmental analysis cannot be undertaken with this data set.

The analysis described here merely allows identification of presence or absence of transcripts of particular genes in the roots and does not allow examination of differential expression at the resolution required for our analysis. However, another study used transcriptomics with Affymetrix chips to examine differential gene expression along the *Arabidopsis* root tip at high spatial resolution (Birnbaum *et al.*, 2003). Indeed two of the three zones they examined corresponded well to our symplastic and apoplastic zones.

We have obtained the relevant Affymetrix data files corresponding to the symplastic and apoplastic zones of the root from the Benfey laboratory web site (Birnbaum *et al.* 2003; The *Arabidopsis* Gene Expression Database, http://www.arexdb.org/index.jsp). These data were processed using Genespring (version 6 – Silicon Genetics). Expression levels were normalized per chip (to 50th percentile) and per gene (to median). A twofold level of expression between the two zones' (symplastic and apoplastic zones of unloading) filter was applied, as well as a filter on flags (present or marginal). This resulted in a list of 2826 genes that passed the filters. When expression in the two zones was compared using a one-way ANOVA (not assuming variances equal) with multiple testing correction (Benjamini and Hochberg false discovery rate) and the Genespring cross-gene error model applied, 2826 genes passed the restriction with a $P$-value cut-off of 0.05. This list of genes was then filtered using the transporter list of Maathius *et al.* (2003). This analysis produced a subset of 143 transporter related genes that showed significant and differential expression between the two regions. A selection of these is shown in Table 8.1A.

A greater number of genes showed elevated expression in the apoplastic zone compared to the symplastic region. The greatest increase in expression was shown by At3g45710, annotated as a peptide transporter. A number of specific ABC transporters (e.g., At1g53270), ATPases (At3g60330), and potassium channels (At3g02850) were also upregulated in this region. However there was not a qualitative difference between the zones as a different set of transporters, but from the same functional categories, were upregulated in the symplastic compared to the apoplastic region (Table 8.1B, e.g., ABC At3g59140, ATPase, At3g42640, potassium channel At5g62410). Similarly, while there was differential expression of a number of the STP (hexose) transporters in the two regions, some had increased expression in the symplastic regions (e.g., STP30, At2g48020, STP25,

**Table 8.1** Selected genes that are differentially expressed between the zones of symplastic and apoplastic unloading along *arabidopsis* roots

| Gene Name | AMT Definition | Functional Category | Significance P-value | Ratio Expression Apoplastic/ Symplastic Zone |
|---|---|---|---|---|
| AT3g45710 | PTR46 | Peptide transporter | 0.012 | 119 |
| At1g53270 | ATWBC10 | ABC | 0.032 | 41 |
| At1g77380 | ATAAP3 | Aminoacid transporter | 0.019 | 40 |
| At3g02850 | SKOR | K channel | 0.036 | 26 |
| At1g64170 | CHX16 | Cation-H antiporter | 0.028 | 20 |
| At1g71960 | ATWBC26 | ABC | 0.043 | 19 |
| AT5g27350 | STP 23 | Sugar transporter | 0.028 | 13 |
| At4g17340 | TIP2,2 | Aquaporin | 0.033 | 12 |
| At5g47450 | TIP2,3 | Aquaporin | 0.020 | 12 |
| At1g18880 | PTR23 | Peptide transporter | 0.023 | 11 |
| At3g16340 | ATPDR1 | ABC | 0.018 | 10 |
| At4g23700 | CHX17 | Cation-H antiporter | 0.035 | 9 |
| At3g47740 | ATATH2 | ABC | 0.035 | 8 |
| AT5g09220 | ATAAP2 | Aminoacid transporter | 0.019 | 8 |
| AT3g45700 | PTR47 | Peptide transporter | 0.023 | 8 |
| At3g60330 | AHA7 | P-type pump | 0.031 | 7 |
| At4g23400 | PIP1,5 | Aquaporin | 0.026 | 6 |
| AT5g62680 | PTR40 | Peptide transporter | 0.040 | 6 |
| At5g57350 | AHA3 | P-type pump | 0.023 | 5 |
| At2g26650 | AKT1 | K channel | 0.036 | 4 |
| At3g61430 | PIP1,1 | Aquaporin | 0.022 | 4 |
| At5g60660 | PIP2,4 | Aquaporin | 0.022 | 4 |
| At3g43190 | | Sucrose Synthase | 0.039 | 3 |
| At1g71880 | SUC1 | Sugar transporter | 0.048 | 3 |
| AT3g20460 | STP 20 | Sugar transporter | 0.039 | 3 |
| At2g36830 | TIP1,1 | Aquaporin | 0.041 | 3 |
| At2g37180 | PIP2,3 | Aquaporin | 0.028 | 3 |
| At3g53420 | PIP2,1 | Aquaporin | 0.024 | 3 |
| At3g16240 | TIP2,1 | Aquaporin | 0.033 | 3 |
| At1g22710 | SUC2 | Sugar transporter | 0.025 | 3 |
| At2g34020 | | 19 TMS putative | 0.033 | 5 |

*Continued*

**Table 8.1**   Selected genes that are differentially expressed between the zones of symplastic and apoplastic unloading along *arabidopsis* roots—*Cont'd*

| Gene Name | AMT Definition | Functional Category | Significance P-value | Ratio Expression Apoplastic/ Symplastic Zone |
|---|---|---|---|---|
| At1g79360 | | Sugar transporter | 0.023 | 5 |
| At3g59140 | ATMRP14 | ABC | 0.045 | 4 |
| At2g48020 | STP 30 | Sugar transporter | 0.033 | 3 |
| At1g07290 | SUGAR6 | Sugar transporter | 0.018 | 3 |
| AT5g18840 | STP 25 | Sugar transporter | 0.046 | 3 |
| At2g28070 | ATWBC3 | ABC | 0.038 | 2 |
| At2g41560 | ACA4 | P-type pump | 0.038 | 2 |
| At5g37500 | GORK | K channel | 0.030 | 2 |
| At5g62410 | ATSMC4 | ABC | 0.025 | 2 |
| At3g42640 | AHA8 | P-type pump | 0.021 | 2 |
| At4g34860 | | Invertase | 0.046 | 2 |
| At1g64550 | ATGCN3 | ABC | 0.028 | 2 |

Transporter related genes identified by Maathius *et al.* (2003) were analyzed within the dataset of Birnbaum *et al.* (2003). Raw data from three replicate Affymetrics chips of gene expression from the regions corresponding to our symplastic and apoplastic regions of unloading were downloaded from the AREX web site (Birnbaum *et al.* [2003]; The *Arabidopsis* Gene Expression Database, http:// www.arexdb.org/index.jsp). These data were analyzed using Genespring to identify genes that were significantly and differentially expressed between the two regions and 2826 genes were identified. A subset of 143 genes were in common with the transporter-related genes described by Maathuis *et al.* (2003). A number of these genes are presented in Table 8.1, selected on their functional significance to water and solute unloading in the root tip. Genes that showed greater expression in the apoplastic region compared to the symplastic region are found in Table 8.1A and those with greater expression in symplastic compared to apoplastic zones are presented in Table 8.1B.

At5g18840), whereas others increased expression in the apoplastic region (STP23, At5g27350, STP20, At3g20460). Thus, contrary to the predictions of the model, transcripts of hexose transporters are present in both functional regions.

In contrast, the sucrose transporter genes SUC1 (At1g71880) and SUC2 (At1g22710) showed significantly elevated expression in the apoplastic region, consistent with the predictions of the model. Similarly, two members of the cation/proton exchanger family (CHX16 and CHX17) are upregulated in the apoplastic region as would be expected due to the need for membrane transport of solutes, including cations, in this region. No members of the CHX family were upregulated in the symplastic region, suggesting that cation import to cells was dominated by direct inflow/import from the sieve element. Similarly, all aquaporins had elevated expression in the apoplastic region; none showed higher expression in the symplastic regions. These included both plasmalemma (PIP2;1 At3g53420) and tonoplast (TIP;1:1, At2g36830). This expression pattern is consistent with the need for greater transport of water across membranes in apoplastic regions and supports the observation that PIP2 expression was increased in the apoplastic region of maize roots (Hukin *et al.*, 2002). Thus members of three very different transporter gene families (cations; CHX, sucrose; SUC and water; aquaporins) showed elevated expression in the apoplastic, and not the symplastic region, consistent with the requirement for greater membrane transport functions.

A neutral, cytoplasmic, invertase (At4g34860) had elevated expression in the symplastic region consistent with a role in processing the sucrose entering the cells through plasmodesmata. However, none of the acid (cell wall) invertases had any difference in expression between the two regions. Thus they may be required in both zones, perhaps supporting the suggestion that apoplastic unloading occurs in parallel with symplastic unloading.

Interestingly, another enzyme involved in metabolism of sucrose, a member of the sucrose synthase family (At3g43190), had elevated expression in the apoplastic region. The expression patterns of the sucrose-processing genes must be treated with caution as the roots were grown in the presence of 4.5% sucrose, effectively short circuiting a solely symplastic route of delivery. A different pattern of expression might be predicted in the absence of external sucrose and could provide a further means of testing the model.

This brief analysis highlights the utility of a transcriptomics approach. A more systematic analysis of these data sets is outside the scope of this chapter but may reveal the role of more transporters in unloading in root tips. While molecular advances will clearly aid testing of the model, some care in interpretation must be taken. For example, absence of significant gene expression detected by transcriptomics does not mean that the gene is not expressed in the target tissue. Immunolocalization confirmed the presence of a $Na^+/H^+$ transporter in the root endodermal membrane (Hall, Newbury, and Pritchard, unpublished), but significant expression of this gene cannot be detected in a number of transcriptomics experiments (e.g., Maser *et al.*, 2001). If such a lack of sensitivity occurs frequently, analysis and interpretation will be distorted; thus care must be taken when choosing candidate genes and information assimilated from a wide range of available sources. In addition, a solely transcriptomics approach is not the full solution to testing the model, as the boundaries of the symplastic and apoplastic regions are probably altered by environmental conditions (Patrick and Offler, 1996). Further analysis localizing gene expression under different environmental conditions using immunolocalization and reverse transcription polymerase chain reaction and examining the phenotype of relevant gene knockouts is required.

## Conclusion

The growing root tip is an important sink for solutes, although the route and thus the driving forces responsible for the delivery of water and solutes to growing cells are not well established. Currently available data are consistent with a chimeric pattern of unloading, symplastic near the root apex switching to apoplastic in older regions of the growing region. Such a pattern of delivery can provide a mechanistic explanation for the common observation that the region of increasing REG is more responsive to environmental perturbation, including changes in turgor pressure, than more basal regions. The model for pathways of water and solute movement within root tips predicts a heterogeneous distribution of transporters and proteins of associated metabolism that mediates the interface between the sieve element and a terminal sink cell. The transport process that occurs in root tips may well be the same as occurs in source and pathways, but the relative importance of different processes may differ. Identification of the specific transporters and their localization will clarify this.

It is not clear whether the presence of two different unloading pathways, in series or possibly in parallel, in the root tip delivers a selective advantage to the plant. It is unreasonable to assume that the chimeric pattern of unloading is a simple consequence of developmental or historical evolutionary constraints, as the genetic variation for both pathways is clearly present in plants. Phloem loading via the symplastic pathway is considered to be ancestral to the apoplastic pathway (Van Bel and Gamalei, 1992). The apoplastic route of loading in source leaves allows a greater control of phloem composition and more flexibility under nonideal environments. The same argument could be advanced for apoplastic unloading in roots. Indeed the region of apoplastic unloading is those areas where growth and solutes levels are sacrificed under conditions of greater environmental stress (e. g., Pritchard et al., 1991). The corollary is that the apical regions, where turgor homeostasis is much greater, are those receiving the direct supply of solute through the symplast. The lower resistance of this route may facilitate sufficient supply to the regions of meristematic activity and initial expansion essential for maintenance of both root form and extension growth. In addition symplastic connection may facilitate exudation of carbon that may confer a selective advantage by making available nutrients in the rhizosphere (see Chapter 13).

Associated with the transition from symplastic to apoplastic unloading, there must be a major change in the portfolio of gene expression encoding transporters and invertases to allow uptake from the apoplast. Thus the movement of cells between the different regions of the growing root provides a model to investigate the developmental regulation of gene expression (Birnbaum et al., 2003). In addition, the switch from symplastic to apoplastic transport mirrors that occurring in fruit (Lu *et al.*, 2000) and may therefore be informed by data from these systems. Advances in molecular techniques are beginning to allow testing of the model. Examining the responses of specific knockout mutants will reveal important information about the transport processes that underpin root extension.

## Acknowledgments

I acknowledge the award of a Leverhulme study abroad Fellowship and a Royal Society Travel Grant. Thanks are due to Nick Gould, Peter Minchin, and Mike Thorpe for many useful discussions.

## References

*Arabidopsis* Gene Expression Database (http://www.arexdb.org/index.jsp).

Baluska, F., Cvrckova, F., Kendrick-Jones, J. and Volkmann, D. (2001) Sink plasmodesmata as gateways for phloem unloading. Myosin VIII and calreticulin as molecular determinants of sink strength? *Plant Physiol* **126**: 39-46.

Barker, L., Kuhn, C., Weise, A., Schulz, A., Gebhardt, C., Hirner, B., Hellmann, H., Schulze, W., Ward, J. M. and Frommer, W. B. (2000) SUT2 a putative sucrose sensor in sieve elements. *Plant Cell* **12**: 1153-1164.

Bingham, I. J., Panico, A. and Stevenson, E. A. (1996) Extension rate and respiratory activity in the growth zone of wheat roots: Time course for adjustments after defoliation. *Physiol Plantarum* **98**: 201-209.

Birnbaum, K., Shasha, D. E., Wang, J. Y., Jung, J. W., Lambert, G. M., Galbraith, D. W. and Benfey, P. N. (2003) A gene expression map of the *Arabidopsis* root. *Science* **302**: 1956-1960.

Boyer, J. S. (2001) Growth induced water potentials originate from wall yielding during growth. *J Exp Bot* **52**: 1483-1488.

Bret_Harte, M. S. and Silk, W. K. (1994) Non vascular, symplastic diffusion of sucrose cannot satisfy the carbon demands of growth in the primary root tip of *Zea Mays. Plant Physiol* **105**: 19-33.

Cosgrove, D. J. (1986) Biophysical control of plant cell growth. *Annu Rev Plant Physiol* **37**: 377-405.

Croser, C., Bengough, A. G. and Pritchard, J. (2000) The effect of mechanical impedance on root growth in pea (*Pisum sativum*). II Cell expansion and wall rheology during recovery. *Physiol Plantarum* **109**: 150-159.

Dick, P. S. and Après, T. (1975) The pathway of sugar transport in roots of *Pisum sativum. J Exp Bot* **26**: 306-314.

Duckett, C. M., Oparka, K. J., Prior, D. A. M., Dolan, L. and Roberts, K. (1994) Dye coupling in the root epidermis of *Arabidopsis* is progressively reduced during development. *Development* **120**: 3247-3255.

Farrar, J. F. and Minchin, P. E. H. (1991) Carbon partitioning in split root systems of barley-relation to metabolism. *J Exp Bot* **42**: 1261-1269.

Ferris, R. and Taylor, G. (1994) Increased root growth in elevated $CO_2$: A biophysical analysis of root elongation. *J Exp Bot* **45**: 1603-1612.

Freixes, S., Thibaud, M. C., Tardieu, F. and Muller, B. (2002) Root elongation and branching is related to local hexose concentration in *Arabidopsis thaliana* seedlings. *Plant Cell Environ* **25:** 1357-1366.

Frensch, J. and Hsiao, T. C. (1994) Transient responses of cell turgor and growth of maize roots as affected by changes in water potential. *Plant Physiol* **104:** 247-254.

Giaquinata, R. T., Lin, W., Sadler, N. L. and Franceschi, V. R. (1983) Pathway of phloem unloading of sucrose in corn roots. *Plant Physiol* **72:** 362-367.

Gould, N., Thorpe, M. R., Minchin, P. E. H., Pritchard, J. and White, P. J. (2004) Solute is imported to elongating root cells of barley as a pressure driven flow of solution. *Funct Plant Biol* **31:** 391-397.

Harrison, M. J. (1996) A sugar transporter from *Medicago truncatula:* Altered expression pattern in roots during vesicular arbuscular (VA) mycorrhizal associations. *Plant J* **9:** 491-503.

Henrissat, B., Coutinho, P. M. and Davies, D. J. (2001) A census of carbohydrate active enzymes in the genome of *Arabidopsis thaliana. Plant Mol Biol* **47:** 55-72.

Hukin, D., Doering-Saad, C., Thomas, C. R. and Pritchard, J. (2002) Sensitivity of cell hydraulic conductivity to mercury is coincident with symplastic isolation and expression of plasmalemma aquaporin genes in growing maize roots. *Planta* **215:** 1047-1056.

Imlau, A., Truernit, E. and Sauer, N. (1999) Cell to cell and long distance trafficking of the green fluorescent protein in the phloem and symplastic unloading of the protein into sink tissues. *Plant Cell* **11:** 309-322.

Kim, J. Y., Mahe, A., Brangeon, J. and Prioul, J. L. (2000) A maize vacuolar invertase, IVR2, is induced by water stress. Organ/tissue specificity and diurnal modulation of expression. *Plant Physiol* **124:** 71-84.

Lu, Y. M., Zhang, D. P. and Yan, H. Y. (2000) Ultrastructure of phloem and its surrounding parenchyma cells in the developing apple fruit. *Acta Botan Sinica* **42:** 32-42.

Maathuis, F. J. M., Filatov, V., Herzyk, P., Krijger, G. C., Axelsen, K. B., Chen, S. X., Green, B. J., Li, Y., Madagan, K. L., Sanchez-Fernandez, R., Forde, B. G., Palmgren, M. G., Rea, P. A., Williams, L. E., Sanders, D. and Amtmann, A. (2003) Transcriptome analysis of root transporters reveals participation of multiple gene families in the response to cation stress. *Plant J* **35:** 675-692.

Maser, P., Thomine, S., Schroeder, J. I., Ward, J. M., Hirschi, K., Sze, H., Talke, I. N., Amtmann, A., Maathuis, F. J. M., Sanders, D., Harper, J. F., Tchieu, J., Gribskov, M., Persans, M. W., Salt, D. E., Kim, S. A. and Guerinot, M. L. (2001) Phylogenetic relationships within cation transporter families of *Arabidopsis. Plant Physiol* **126:** 1646-1667.

Minchin, P. E. H. and Thorpe, M. R. (1996) A method for monitoring g-radiation from an extended source with uniform sensitivity. *Appl Radiat Isotopes* **47:** 693-696.

Muller, B., Stosser, M. and Tardieu, F. (1998) Spatial distributions of tissue expansion and cell division rates are related to irradiance and to sugar content in the growing zone of maize roots. *Plant Cell Environ* **21:** 149-158.

Oparka, K. J., Duckett, C. M., Prior, D. A. M. and Fisher, D. B. (1994) Real-time imaging of phloem unloading in the root tip of *Arabidopsis. Plant J* **6:** 756-766.

Patrick, J. W. (1997) Phloem unloading: Sieve element unloading and post sieve element transport. *Annu Rev Plant Phys* **48:** 191-222.

Patrick, J. W. and Offler, C. E. (1996) Post sieve element transport of photosassimilates in sink regions. *J Exp Bot* **47:** 1165-1177.

Pritchard, J. (1994) The control of cell expansion in roots. Tansley review 68. *New Phytol* **127:** 3-26.

Pritchard, J. (1996) Aphid stylectomy reveals an osmotic step between sieve tube and cortical cells in barley roots. *J Exp Bot* **47:** 1519-1524.

Pritchard, J., Fricke, W., and Tomos, A. D. (1996) Turgor regulation during extension growth and osmotic stress of maize roots. An example of single cell mapping. *Plant Soil* **187:** 11-21.

Pritchard, J., Wyn Jones, R. G. and Tomos, A. D. (1991) Turgor growth and rheological gradients of wheat roots following osmotic stress. *J Exp Bot* **42:** 1043-1049.

Roitsch, T. (1999) Source sink regulation by sugar adds stress. *Curr Opin Plant Biol* **2**: 198-206.

Schulz, A. (1995) Plasmodesmatal widening accompanies the short term increase in symplastic unloading in pea root tips under osmotic stress. *Protoplasma* **188**: 22-37.

Sharp, R. E., Hsiao, T. C. and Silk, W. K. (1990) Growth of the primary root at low water potentials. II. Role of growth and deposition of hexose and potassium in osmotic adjustment. *Plant Physiol* **93**: 1337-1346.

Sharp, R. E., Silk, W. K. and Hsiao, T. (1988) Growth of the maize primary root at low water potentials 1 spatial distribution of expansive growth. *Plant Physiol* **87**: 50-57.

Sherson, S. M., Hemmann, G., Wallace, G., Forbes, S., Germain, V., Stadler, R., Bechtold, N., Sauer, N. and Smith, S. M. (2000) Monosaccharide/proton symporter AtSTP1 plays a major role in uptake and response of *Arabidopsis* seeds and seedlings to sugars. *Plant J* **24**: 849-857.

Silk, W. K., Hsiao, T. C., Diedenhofen, U. and Matson, C. (1986) Spatial distributions of potassium, solutes, and their deposition rates in the growth zone of the primary corn root. *Plant Physiol* **82**: 853-858.

Spollen, W. G. and Sharp, R. E. (1991) Spatial distribution of turgor and root growth at low water potentials. *Plant Physiol* **96**: 438-443.

TAIR web site (http://www.arabidopsis.org/info.genefamily/genefamily.html)

Tang, G. Q., Luscher, M. and Sturm, A. (1999) Antisense repression of vacular and cell wall invertase in transgenic carrot alters early plant development and sucrose partitioning. *Plant Cell* **11**: 177-189.

Tomos, A. D., Hinde, P. S. S., Richardson, P. B. R., Pritchard, J. and Fricke, W. (1994) Microsampling and measurements of solutes in single cells. In *Practical Approach Series* (N. Harris and K. J. Oparka, eds.) pp. 297-314. IRL Press, Oxford.

Truernit, E., Stadler, R., Baier, K. and Sauer, N. (1999) A male gametophyte-specific monosaccharide transporter in *Arabidopsis*. *Plant J* **17**: 191-201.

Tymowska-Lalanne, Z. and Kreis, M. (1998) Expression of the *Arabidopsis thaliana* invertase gene family. *Planta* **207**: 259-265.

Van Bel, A. J. E. and Gamalei, Y. V. (1992) Ecophysiology of phloem loading in source leaves. *Plant Cell Environ* **15**: 265-270.

van der Weele, C. M., Jiang, H. S., Palaniappan, K. K., Ivanov, V. B., Palaniappan, K. and Baskin, T. I. (2003) A new algorithm for computational image analysis of deformable motion at high spatial and temporal resolution applied to root growth. Roughly uniform elongation in the meristem and also, after an abrupt acceleration, in the elongation zone. *Plant Physiol* **132**: 1138-1148.

Walter, A., Feil, R. and Schurr, U. (2003) Expansion dynamics, metabolite composition and substance transfer of the primary root growth zone of *Zea mays* L. grown in different external nutrient availabilities. *Plant Cell Environ* **26**: 1451-1466.

Warmbrodt, R. D. (1987) Solute concentrations in the phloem and apex of the root of *Zea Mays*. *Am J Bot* **74**: 394-402.

Williams, L. E., Lemoine, R. and Sauer, N. (2000) Sugar transporters in higher plants: A diversity of roles and complex regulation. *TIPS* **5**: 283-290.

Winch, S. K. and Pritchard, J. (1999) Acid-induced wall loosening is confined to the accelerating region of the root growing zone. *J Exp Bot* **50**: 1481-1487.

# 9

# Growth and Water Transport in Fleshy Fruit

*Mark A. Matthews and Ken A. Shackel*

Fruit volumetric growth is primarily the result of water accumulation, and hence maintenance of fruit growth requires coordination between long-distance water and solute transport through the vascular tissue, and short-distance water and solute uptake at the level of individual cells. One hypothesized coordinating principle is that for fruit growth to occur, there must be a favorable difference in total water potential between the fruit and the rest of the plant (e.g., Grange and Andrews, 1994). Essentially all of the previous studies of fruit water relations, including our own, have been based on whole fruit measurements of total water potential, and many of these are consistent with this principle. The current understanding of fruit water relations, however, particularly at the cellular level, is quite rudimentary, and it may be premature to generalize about the water relations of the fruit as a whole before we have a clear understanding of the water relations of the component cells. In this chapter we give a brief overview of the water relations of fleshy fruits as they relate to the process of fruit growth, particularly in grapes, and present experimental measurements of fruit water balance, whole fruit osmotic potential, and intact fruit cell turgor in the grape berry. Our turgor measurements indicate that fruit cell turgor is substantially depressed by the presence of solutes in the apoplast. The consequences that apoplastic solutes may have on water exchange between the fruit and the plant are discussed, as well as questions and directions for future research.

## Fleshy Fruit Growth, Expansion, and Contraction

Fleshy fruit arise from a wide range of tissues (see Coombe, 1976), and in some respects the interaction between xylem and phloem transport during fruit growth is similar to that experienced by most other growing plant structures. Diurnal variations in size (expansion and contraction), and hence volume, of fleshy fruit are common (Elfving and Kaufmann, 1972; Johnson *et al.*, 1992; Tromp, 1984), superimposed on an overall size increase over time. The diurnal variations are usually attributed to changes in hydration, presumably due to changes in fruit cell turgor; under these conditions it is difficult, or perhaps impossible, to unequivocally separate growth (by definition the irreversible change in volume) from these reversible changes in volume. We assume that growth of the fruit reflects growth of the component cells (in the case of the grape berry these are mostly mesocarp cells), that growth occurs when nighttime expansion exceeds daytime contraction (in some cases daytime contraction may be absent), and that both processes arise from water flows into and out of the fruit. Hence, one approach to describing fruit growth is to quantify these flows as a "water balance" (net growth = flow in – flow out). We will further assume that the exchange of water between the fruit and the parent plant occurs primarily, if not exclusively, through the phloem and xylem tissues in the fruit pedicel. Inflow must come from the parent plant, but outflow can occur either via fruit transpiration or, in some cases (e.g., cowpea, Pate, 1988; apple, Lang, 1990) via water "backflow" through the xylem, although the prevalence in fleshy fruits of this kind of water flow, sometimes referred to as "water recycling," has not been established. This is an important phenomenon since a water potential gradient generated both by transpiration (Lee *et al.*, 1989; Lee, 1990) and by an osmotic gradient (Lang and Düring, 1991) has been implicated in the control of fruit growth and water economy.

## Multiphasic Growth Habit

Growth of many fleshy fruit, including *Prunus* sp. and *Vitis* sp., exhibit a double-sigmoid pattern in which there are two periods of rapid growth. The growth phases are separated by a period of variable length during which little or no expansive growth occurs. The growth habit of grape berries has long been divided arbitrarily into three stages (e.g., Connors, 1919) (Fig. 9.1A). Stage I extends from anthesis to the slowed growth of stage II, and stage III extends from the onset of the second growth phase until maturity. During stage I there is some cell division, but by stage III, all fruit growth is due to the expansion of existing cells (Coombe, 1976; Ojeda *et al.*, 1999). The duration of stage II is under both genetic and environmental control (Coombe, 1976), and in grape, much attention is given to the relatively abrupt transition from stage II to stage III, referred to as veraison. In addition to the rapid changes in growth rate at veraison (Fig. 9.1A), around this time the fruit ripening process begins, identified by the appearance of red color in red and black grape varieties, fruit softening, and a decline in fruit solute potential, with the decline in solute potential reflecting the accumulation of substantial concentrations of sugars (Fig. 9.1B). Hence the relatively abrupt transition from little or no growth to rapid growth is associated with an apparently abrupt increase in the rate of transport of sugars into the berry. The increase in sugar transport is substantial because both volume and solute concentration are increasing. Coombe (1992) has observed the resumption of growth to follow by a few days a measurable increase in hexose concentrations in berries at veraison.

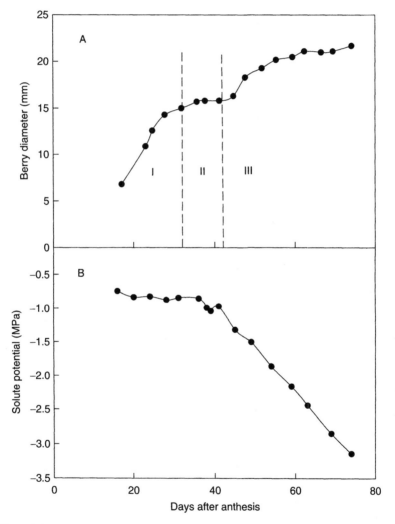

**Figure 9.1** Developmental pattern of (A) fruit diameter and (B) fruit solute potential for grape (*Vitis vinifera* L. cv Cardinal). The arbitrary designation of growth stages (I, II, and III) is described in the text (from Matthews *et al.* 1987).

It is axiomatic that expansive cell growth requires the presence of cell turgor, and in many, although not in all instances, a direct relation between turgor and the rate of growth has been found (e.g., Serpe and Matthews, 2000). It is also generally believed, however, that at any given turgor, the rate of growth is determined by the rate of weakening of the structure of the cell wall (e.g., Cosgrove, 2000). Hence the resumption of berry growth at veraison could, in theory, be associated with an increase in berry cell turgor, possibly as a result of solute accumulation in the absence of a change to cell wall strength, or to a weakening of the cell wall in the absence of a turgor change, or to a combination of both processes. It should be noted here that each of these processes would have different consequences for whole fruit water potential, and they may also have different consequences for the transport of water from the plant to the fruit, depending on which water transport pathway (phloem or xylem) is predominant. For instance, if turgor increases because of an increase in solute accumulation by mesocarp cells, then this may occur without any change in fruit total water potential, and hence no change in the water potential gradient between the fruit and the parent plant. In fact, if the sugar concentration in the mesocarp cells is higher than that of the phloem, then there may be an increase in fruit water potential owing to the presence of excess phloem-derived water, and water may flow from the fruit to the plant through the xylem, as occurs in other systems (Pate, 1988; Lang, 1990).

The steady state value of fruit cell turgor and total water potential during growth would depend on the hydraulic conductivity of the vascular pathway between the fruit and the parent plant. If there is excess phloem-derived water and the pathway for water return to the parent plant (presumably the xylem) had a low conductivity, then both cell turgor and cell total water potentials would substantially increase. If growth resumption were due to a weakening of the cell wall, however, then an initial decline would be expected in either cell turgor or cell total water potential (or both). The steady state value reached by these potentials would also depend on the hydraulic conductivity of the vascular pathway connecting the fruit and the parent plant, but, in this case, a low conductance xylem pathway would be associated with a substantial decrease in cell turgor and/or total water potential. These two alternative scenarios (turgor increase versus wall loosening) to explain increased fruit growth at veraison illustrate the importance of considering both short- and long-distance transport processes and their possible interaction.

### Diurnal Expansion and Contraction

Many fleshy fruit exhibit a diurnal pattern of expansion and contraction even when well watered (Shimomura, 1967; Coombe and Bishop, 1980; Nakano, 1989; Johnson *et al.*, 1992). Figure 9.2 shows the diurnal behavior of grape berry diameter around veraison. Virtually all berry growth occurs during the night; day periods are characterized by either contraction or absence of expansion. Initially (i.e., during stage II) daily contraction and nightly expansion are approximately equal, and little net growth is observed. The transition to stage III is characterized by a dramatic reduction in the daily contractions over 2 to 3 days. The contractions decreased from approximately 0.125 mm/day on days −5 and −4 to less than 0.025 mm/day on days 0 and 1 (Fig. 9.2). Nightly expansion also increased approximately 50% compared to the pre-veraison expansion. That this transition occurred before color development indicates that the resumption of growth and altered diurnal water relations are early events in the transition from stage II to stage III. What causes expansion to increase and daily contraction to be reduced at this stage?

Daily contraction of fleshy fruit is generally well correlated with decreasing leaf water potential, as was the case in citrus (Elfving and Kaufmann, 1972; Maotani *et al.*, 1977) and apple (Tromp, 1984). This is also the case in pre-veraison grape, where the daily contraction increased from about 4% to 10% as midday leaf water potential declined from about −1.0 to −1.5 MPa (Fig. 9.3). When shoot transpiration was restricted before veraison, a marked reduction of diurnal contraction also occurred (Greenspan *et al.*, 1994). Hence, it is generally assumed that fruit contraction simply indicates that a conductive apoplastic pathway (i.e., xylem) is connecting the fruit to

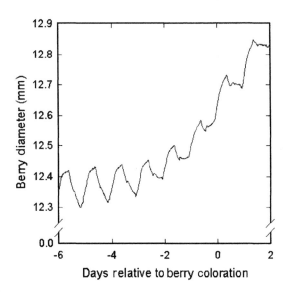

**Figure 9.2** Diurnal pattern of fruit diameter in grape (*Vitis vinifera* L. cv. Cabernet Sauvignon) at the transition of stage II to stage III (veraison).

the rest of the plant, and that when soil drying or plant transpiration causes a decline in overall plant water potential, a decrease in berry cell total water potential (and hence berry cell turgor pressure) is a direct result. Assuming that soil drying has minimal effect on berry transpiration, however, the increased diurnal contraction of berries during pre-veraison soil water deficits (Fig. 9.3) could be due either to backflow out of the fruit to the parent plant through the xylem, or to a reduced net vascular inflow to the fruit. In contrast to pre-veraison fruit, however, for the same range of midday leaf water potential, the daily contraction in fruit diameter of the post-veraison berry was about 1% and insensitive to changes in midday leaf water potential (Fig. 9.3). The relative insensitivity of post-veraison berries may indicate an improved water transport to post-veraison berries or a hydraulic isolation that prevents the berry from experiencing the diurnal water deficits occurring in the stem.

## Leaf, Stem, and Fruit Water Potential

Fruits and leaves can be considered as competing sinks for water, and in some cases it has been assumed that the difference between fruit and leaf water potential determines the directionality of water exchange between

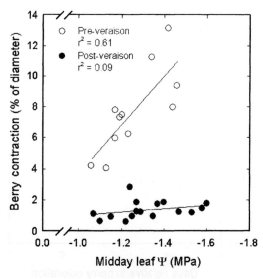

**Figure 9.3** Amplitude of diurnal berry contraction at various midday Ψ leaf for pre- and post-veraison Cabernet Sauvignon berries. Pre-veraison correlation is significant for $P < 0.01$; post-veraison correlation is not significant.

the fruit and the parent plant. For instance, backflow of vascular water from fruit to transpiring leaves has been suggested from diurnal measurements in which $\Psi_{leaf} < \Psi_{fruit}$ during midday for various crops (Klepper, 1968; Tyvergyak and Richardson, 1979; Syvertsen and Albrigo, 1980; Yamamoto, 1983), including grape (van Zyl, 1987). However, since fruits and leaves only share the stem portion of the water transport pathway, it is the fruit-to-stem water potential difference that is appropriate for this analysis. In grapes, cluster water potential ($\Psi_{peduncle}$) was lower than stem water potential at most times in the diurnal cycle and under wet and dry soil conditions (Greenspan *et al.*, 1996), indicating that a favorable gradient for water flow existed, at least from the stem to the cluster. Using *in situ* psychrometers, Guichard *et al.* (1999) showed a close and straightforward relation between the fruit-to-stem water potential gradient and the rate of water flow through the fruit pedicel in tomato over a diurnal course, with the lowest flow through the pedicel at midday, when fruit and stem water potentials were essentially equivalent. Tomatoes are unusual, however, in that they show little or no midday contraction (Johnson *et al.*, 1992) compared to pre-veraison grapes (Fig. 9.2) and other fruits (e.g., apple, Tromp, 1984). Consistent with the lack in midday contraction in tomato, however, Guichard *et al.* (1999) also found little diurnal change in fruit total water potential. Hence the pre-veraison diurnal oscillations in fruit size that are observed in grape may indicate that diurnal changes do occur in fruit total water potential and in fruit cell turgor (but see the section Fruit Turgor and Apoplastic Solutes) during this period. These pre-veraison diurnal oscillations in fruit size are also consistent with those expected on the basis of the diurnal pattern in stem and cluster total water potential.

The transition to little or no midday fruit contraction at veraison suggests that an uncoupling of fruit water potential from cluster or stem water potential occurs at this time, but the basis for this change is not clear. A similar change has been observed in cotton fruit (bolls, Anderson and Kerr, 1943), but in this case the rapid growth period with no midday contraction occurs first, and is followed by a period of no growth and marked midday contraction. The absence of midday contraction in growing organs may be difficult to interpret, however, as it may indicate either that the turgor and possibly the total water potential of the organ is stable over the diurnal course, or that organ growth is simply adjusting to the relatively gradual diurnal changes in cell turgor (e.g., Shackel *et al.*, 1987). A diurnally stable fruit water potential would result if fruit and plant water potential were uncoupled, and such uncoupling has been hypothesized in a number of systems, mostly based on the assumption that close coupling in the first place is due to xylem water transport. In grape, the apparent change in coupling at veraison has been studied using dye uptake methods and attributed to an irreversible disruption of xylem transport by many authors (Delrot *et al.*, 2001), although there is some evidence that the xylem remains functional in post-veraison berries (Rogiers et al., 2001). Hence, it is not yet clear whether the apparent change in coupling *in situ* is due to the xylem becoming irreversibly disrupted, or indirectly due to another mechanism, such as a reduction in the driving force that is responsible for xylem transport (hydrostatic gradient) between the fruit and the stem.

## Vascular Flows in Developing Fruit

Net flows into and out of fruit have been estimated using various mass balance approaches. From measurements of fruit transpiration and mass, and assumptions about phloem and xylem composition, Pate and co-workers estimated that water was supplied by xylem and phloem, approximately 60:40 (v/v), to white lupin fruit (Pate *et al.*, 1977), and that the phloem supply was greater than 100% of daily fruit water gain in cowpea (Peoples *et al.*, 1985; Atkins *et al.*, 1985). From measurements of fruit mass, transpiration, and Ca influx, Ho *et al.* (1987) estimated that the water supply to tomato fruit was predominately via the phloem, increasing from approximately 85% to approximately 95% of the total water uptake during fruit development. Lang and Thorpe (1989) introduced a novel approach to analysis of the fruit water budget in which the xylemic, phloemic, and transpirational flows can be separated. Diurnal measurements of intact (all 3 flows), detached (transpiration only), and phloem girdled (transpiration and xylem flow) fruit allow all three flows to be obtained by subtraction. This approach assumes independence of xylem and phloem transport, and a sensitivity analysis indicated that the assumption leads to little error except at very low flow rates (Fishman *et al.*, 2001). In apple, which typically exhibits daytime contraction, Lang (1990) found that the xylem contribution to overall fruit water balance progressively decreased as fruit developed.

Using the approach of Lang and Thorpe (1989), Matthews and colleagues evaluated the diurnal water budget of the grape berry before and after veraison (Greenspan *et al.*, 1994) and in comparison with leaf and stem water status during water deficits (Greenspan *et al.*, 1996). Before veraison and under well-watered conditions, the bulk of vascular water flow from parent plant to the berry occurs via the xylem, and the phloem contributes less than 10% of total inflow (Fig. 9.4). After veraison, phloem becomes the primary source of berry water, contributing more than 80% of total inflow under well-watered conditions. In apple, tomato, and grape, particularly late in fruit development, most authors have attributed the relatively small xylem contribution to the development of a high xylem hydraulic resistance within the fruit, often as a result of growth-induced xylem stretching or breakage (Lang and Ryan, 1994; Malone and Andrews, 2001; Düring *et al.*, 1987; Findlay *et al.*, 1987). In many of these studies (1) staining of lignin in the berry revealed evidence of stretching or gaps in the xylem appearing at or after veraison; and (2) dye uptake from cut pedicels and transport into the berry was diminished after veraison. Disrupted xylem within the berry would presumably contribute to a reduced conductance to the parent plant, but the direct implication of these observations within berries on water exchange between berry and parent plant are not clear. For instance, in the study of Düring *et al.* (1987), dye movement declined gradually over 2 weeks, whereas based on the diurnal patterns of expansion and contraction (Fig. 9.2), the transition from xylemic to phloemic transport is apparently rapid, occurring over 2 days. Xylem flow is also reduced, but not eliminated, at veraison, and the xylem appears to contribute to the recovery of berry volume during the night, as indicated by the positive nighttime xylem flow in the irrigated, post-veraison vine (Greenspan *et al.*, 1994). Hence, a purely anatomical explanation (i.e., xylem disruption) for the change from xylemic to phloemic transport at veraison may be oversimplified.

In addition to vascular flow estimates using intact/girdled/excised fruit, the diurnal behavior of berry size when restricting transpiration of leaves or fruit, and when imposing water deficits, all are consistent with an abrupt

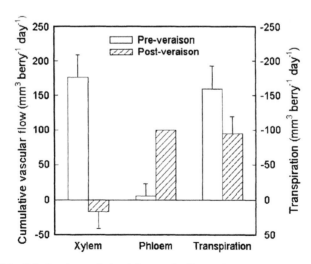

**Figure 9.4** Calculated cumulative daily vascular flow and transpiration for pre- and post-veraison berries of irrigated vines. Water flow was estimated from a linear combination of cumulative diurnal volume changes in intact, excised, and girdled berries (as in Fig. 9.6). Error bars indicate 1 SE, n = 3 (from Greenspan *et al.*, 1996).

shift from xylem-to-phloem-sourced water at veraison. It is clear, however, that at this time there is also an increased rate of sucrose translocation to the fruit (Brown and Coombe, 1985; Coombe *et al.*, 1987), and this must be associated with either an increased transport of the solvent water through the phloem or a greatly increased concentration of sucrose in the phloem sap. The estimated inflow rate through the phloem increased approximately 10-fold after veraison (Greenspan *et al.*, 1994). Although hexose accumulation during ripening constitutes osmotic adjustment, it is unlikely that the observed change in water flow patterns is primarily due to increasing sugar concentrations because berry osmotic potentials decrease only approximately 0.067 MPa.d$^{-1}$ (Fig. 9.1B; Matthews *et al.*, 1987). Furthermore, the water budget data indicate a decrease in xylem inflow after veraison, whereas an effective increased osmotic gradient should increase flow. Consequently, the post-veraison insensitivity of the berry-to-plant water deficits (Fig. 9.3) may be a manifestation of a strong phloem component at veraison, rather than a loss of xylem conductance. Using the same methods in prune (*Prunus domestica*), xylem flows to fruits were clearly dominant during stage II (mid-June), and phloem flows were becoming dominant at the start of stage III (early July) (Shackel, unpublished). However, later in stage III (mid-July) both flows were comparable again, implying the presence of functional xylem. These data suggest that the patterns exhibited during fruit development, in the relative importance of xylem or phloem flows, may be reversible, perhaps depending more on source and/or sink behavior than on a physical loss in xylem continuity/conductance.

## Fruit Turgor and Apoplastic Solutes

One important aspect of water relations in fleshy fruits is that substantial levels of solutes, primarily sugars, are accumulated toward the end of fruit development. This is particularly true in grapes, with many varieties reaching solute potentials as low as −3 MPa (Fig. 9.1B). With such low solute potentials, it is reasonable to hypothesize that there may be a mechanism that prevents the development of excessive fruit cell turgor, particularly under environmental conditions that would favor the occurrence of high total water potentials in the plant, such as wet soil and low evaporative demands. Indirect estimates of turgor in Cardinal grapes indicated that turgor was relatively constant (0.2 to 0.4 MPa), because the marked decline in solute potential during development was matched by a parallel decline in fruit total water potential (Matthews *et al.*, 1987). Direct measurements of turgor with the cell pressure probe in Cabernet Sauvignon grapes indicated that turgor was within this range (0.3 MPa) at 50 to 60 days after anthesis, but that a marked decline in turgor occurred after this time, with values stabilizing at about 0.03 MPa by 70 days after anthesis (Fig. 9.5). For these berries, the decline in turgor with fruit development was apparently uniform for cells from 100 μm to almost 1 mm below the fruit surface (Fig. 9.6). If fruit cell turgor is low as a result of low fruit total water potential, then it should be possible to hydrate fruit artificially and increase cell turgor. Excised grape berries will hydrate through the fruit pedicel, and after a 24-hour hydration period there was a clear increase in fruit cell turgor (Table 9.1). This indicates that one of the reasons for low cell turgor was a low total water potential (water deficit) in the fruit as a whole, consistent with the low solute potential data shown in Fig. 9.1B. However, the low cell turgor that occurred late in fruit development was also apparent after 24 hours of hydration (Table 9.1), indicating either that full hydration requires substantially longer than 24 hours, or that some other mechanism is responsible for low cell turgor in fruit cells.

One mechanism that has been proposed to explain lower than expected turgor in tomato fruit pericarp cells, despite tissue hydration, is the presence of apoplastic solutes (Shackel *et al.*, 1991). Evidence for the presence of apoplastic solutes has also been found in growing stems of pea, soybean, and cucumber (Cosgrove and Cleland, 1983) and in stems of sugarcane (Welbaum and Meinzer, 1990), which, like fruit tissue, accumulate substantial concentrations of sugars as a normal part of development.

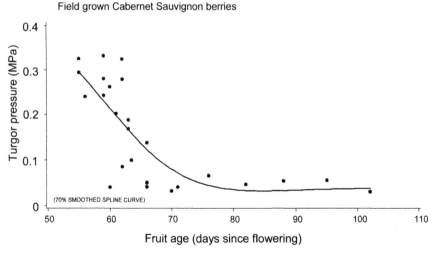

**Figure 9.5** Mesocarp cell turgor at various days after flowering for Cabernet Sauvignon berries.

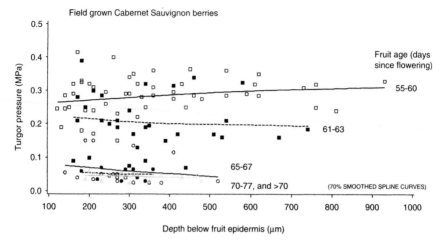

**Figure 9.6** Mesocarp cell turgor at various depths beneath surface of Cabernet Sauvignon berries. Curves are for data collected at different days after flowering.

The presence of solutes in the apoplast of a sink tissue, such as a fruit, are broadly consistent with models of phloem transport that involve an apoplastic sugar unloading step from the phloem, and are also consistent with the Münch pressure-flow hypothesis in that phloem turgor will be reduced, and hence phloem transport increased, by the presence of apoplastic solutes; however, the consequences of apoplastic solutes for water transport in the xylem are less clear. It may be that there must be a favorable stem-to-fruit total water potential gradient (fruit lower than stem) for growth to occur. However, even though apoplastic solutes will

**Table 9.1** Increase in berry weight after a 24-hour hydration period with the fruit pedicel immersed in tap water and the turgor of berry cells before and after hydration

| Berry Age (Days from Anthesis) | 24-Hour Increase in Berry FW (mg) | Cell Turgor Pressure (MPa) | |
|---|---|---|---|
| | | Prehydration | Posthydration |
| 30 | 56 | 0.31 | 0.44 |
| 45 | 64 | 0.25 | 0.42 |
| 60 | 56 | 0.23 | 0.34 |
| 75 | 32 | 0.09 | 0.19 |
| 90 | 26 | 0.02 | 0.07 |

Hydration was performed in a humidified chamber under laboratory conditions (20-25° C).

reduce fruit total water potential and also fruit cell turgor, they should not contribute to a matric (tension) gradient in the xylem, and hence should not directly influence water transport through the xylem. Unfortunately, when fruit total water potential is measured psychrometrically (Matthews *et al.*, 1987), it is impossible to separate apoplastic solute effects from apoplastic matric effects.

A key unresolved issue is the spatial scale at which apoplastic solutes may be regulated in plant reproductive tissues such as fruits and seeds. Bradford (1994) proposed the concept of a localized semipermeable apoplast barrier near the embryo, so that unloaded solutes would be confined to the apoplast in the vicinity of the embryo, rather than being free to flow back toward the leaves in the xylem. In the case of seeds, which convert most of the imported soluble carbohydrate into starch or other insoluble compounds, most evidence indicates that excess water arrives via phloem import, and that this excess water is recirculated back to the stem via the xylem (Pate *et al.*, 1985). In the case of fruits, however, particularly grapes, most of the imported carbohydrate remains soluble and is at high concentrations throughout essentially all of the fruit flesh. Hence, a similar mechanism would require either for the majority of the fruit apoplast to be semipermeable or for the barrier to be between the plant and the fruit as a whole, perhaps associated with the fruit pedicel. In theory, the presence of a semipermeable apoplast would solve the problem of xylem tension causing backflow of solutes from the fruit to the plant, but under conditions of minimal xylem tension (wet soil, low evaporative demand) such a barrier should also cause an osmotic pressure to develop within the fruit apoplast, presumably causing either apoplastic swelling or a phenomenon such as guttation.

It may be possible that the entire fruit xylem and apoplastic space do experience a normal diurnal range in tension, but that a low and consistent turgor is maintained by an exchange of solutes between the symplast and the apoplast at the cellular level. Because the volume of the apoplast is very small relative to the symplast, a minimal export or import of a suitably mobile solute (potassium?) across the plasma membrane would be required to cause a substantial change in apoplastic solute potential for any given cell. Under conditions of increasing water potential (declining apoplastic tension), solute export would prevent cell turgor from increasing, and even though the exported solutes would begin diffusing into the vascular system, the loss of solute from the fruit as a whole would be limited by the relatively long distance and limited cross section of the vascular pathway, compared to the size of the fruit. As water potential decreased (apoplastic tensions increase), either during the daytime or as a result of limited soil moisture, solute import from the apoplast would maintain turgor. If fruit cell turgor were maintained in this way, then fruit volume would also be maintained, and this should limit the quantity of water (and solute) that may be withdrawn from the fruit via the xylem as xylem tensions increase. Hence, a localized process of apoplastic solute regulation, perhaps based simply on cell turgor homeostasis, could account for a number of important aspects of post-veraison fruit water relations in grape: (1) low cell turgor and fruit tissue water potential that are largely independent of changes in plant (and cluster) water potential (uncoupling hypothesis), (2) inability to substantially increase fruit cell turgor by artificial hydration, and (3) a substantial reduction in the flows of water through the xylem. This conceptual model predicts that the sum of the tension (hydrostatic potential) and the osmotic potential in the apoplast will be maintained at a relatively constant value, either with soil drying or with normal diurnal changes in plant transpiration, by compensating changes in both components. It should be possible to test at least one of these components if direct measurements of xylem tension (Wei *et al.*, 2001) can be accomplished in the tracheids of grape berries.

## Conclusions

Fleshy fruit growth and development is of great practical importance, as well as representing a challenge for our understanding of long and short distance water and solute transport. Based on many lines of evidence, the water transport that is required for fruit growth changes from a xylem to a phloem-dominated pathway over the course of fruit development, and in grape this change in the relative importance of the two pathways appears to occur rather abruptly. This change is associated with an apparent uncoupling of fruit water status from plant water status, and may play a role in the well-known ability of grapes to continue to accumulate substantial concentrations of solutes (sugars) under limited soil water conditions, but the basis for the change in pathway dominance and for water status coupling is unclear. Direct measurements have demonstrated that fruit cell turgor is relatively low in grape, and that it declines progressively during development, despite a progressive increase in whole fruit, and presumably fruit cell, solute concentration. Because cell turgor also remains low after a 24-hour hydration, we propose that turgor is being depressed by the presence of apoplastic solutes. We further suggest that a regulated partitioning of these solutes between the symplast and apoplast, for the purposes of turgor homeostasis, may account for the change in vascular pathway dominance and the apparent uncoupling of fruit from plant water status.

# References

Anderson, D. B. and Kerr, T. (1943) A note on the growth behavior of cotton bolls. *Plant Physiol* **18**: 261.

Atkins, C. A., Pate, J. S. and Peoples, M. B. (1985) Water relations of cowpea fruits during development. In *Fundamental and Ecological Aspects of Nitrogen Metabolism in Higher Plants* (H. Lambers, J. J. Neeteson and I. Stulen, eds.) pp 235-238. Nijhoff, The Hague.

Bradford, K. J. (1994) Water stress and the water relations of seed development: A critical review. *Crop Sci* **34**: 1-11.

Brown, S. C. and Coombe, B. G. (1985) Solute accumulation by grape pericarp cells. III. Sugar changes *in vivo* and the effects of shading. *Biochemie und Physiolgie der Pflanzen* **179**: 157-171.

Connors, C. H. (1919) Growth of fruits of peach. *N J Agr Expt Sta Annu Rpt* **40**: 82-88.

Coombe, B. G. (1976) The development of fleshy fruits. *Annu Rev Plant Physiol* **27**: 507-528.

Coombe, B. G. (1992) Research on development and ripening of the grape berry. *Am J Enol Viticulture* **43**: 101-110.

Coombe, B. G. and Bishop, G. R. (1980) Development of the grape berry. II. Changes in diameter and deformability during veraison. *Aust J Agric Res* **31**: 499-509.

Coombe, B. G., Bovio, M. and Schneider, A. (1987) Solute accumulation by grape pericarp cells. V. Relationship to berry size and the effects of defoliation. *J Exp Bot* **38**: 1789-1798.

Cosgrove, D. J. (2000) Loosening of plant cell walls by expansins. *Nature* **407**: 321-326.

Cosgrove, D. J. and Cleland, R. E. (1983) Solutes in the free space of growing stem tissues. *Plant Physiol* **72**: 326-331.

Delrot, S., Picaud, S. and GaudillPre, J. P. (2001) Water transport and aquaporins in grapevine. In *Molecular Biology and Biotechnology of the Grapevine* (K. A. Roubelakis-Angelakis, ed.) pp. 241-262. Kluwer Acad. Pub. Dordrecht/Boston/London.

Düring, H., Lang, A. and Oggionni, F. (1987) Patterns of water flow in Riesling berries in relation to developmental changes in their xylem morphology. *Vitis* **26**: 123-131.

Elfving, D. C. and Kaufmann, M. R. (1972) Diurnal and seasonal effects of environment on plant water relations and fruit diameter of citrus. *J Am Soc Horticultural Sci* **97**: 566-570.

Findlay, N., Oliver, K. J., Nii, N. and Coombe, B. G. (1987) Solute accumulation by grape pericarp cells. IV. Perfusion of pericarp apoplast via the pedicel and evidence for xylem malfunction in ripening berries. *J Exp Bot* **38**: 668-679.

Fishman, S., Genard, M. and Huguet, J. (2001) Theoretical analysis of systematic errors introduced by a pedicel-girdling technique used to estimate separately the xylem and phloem flows. *J Theoret Biol* **213**: 435-446.

Grange, R. I. and Andrews, J. A. (1994) Expansion rate of young tomato fruit growing on plants at positive water potential. *Plant Cell Environ* **17**: 181-187.

Greenspan, M. D., Schultz, H. R. and Matthews, M. A. (1996) Field evaluation of water transport in grape berries during water deficit. *Physiol Plantarum* **97**: 55-62.

Greenspan, M. D., Shackel, K. A. and Matthews, M. A. (1994) Developmental changes in the diurnal water budget of the grape berry exposed to water deficits. *Plant Cell Environ* **17**: 811-820.

Guichard, S., Gary, C., Longuenesse, J. J. and Leonardi, C. (1999) Water fluxes and growth of greenhouse tomato fruits under summer conditions. *Acta Horticulturae* **507**: 223-230.

Ho, L. C., Grange, R. I. and Picken, A. J. (1987) An analysis of the accumulation of water and dry matter in tomato fruit. *Plant Cell Environ* **10**: 157-162.

Johnson, R. W., Dixon, M. A. and Lee, D. R. (1992) Water relations of the tomato during fruit growth. *Plant Cell Environ* **15**: 947-953.

Klepper, B. (1968) Diurnal pattern of water potential in woody plants. *Plant Physiol.* **43**: 1931-1934.

Lang, A. (1990) Xylem, phloem and transpiration flows in developing apple fruits. *J Exp Bot* **41:** 645-651.

Lang, A. and Düring, H. (1991) Partitioning control by water potential gradient: Evidence for compartmentation breakdown in grape berries. *J Exp Bot* **42:** 1117-1122.

Lang, A.. and Ryan, K. G. (1994) Vascular development and sap flow in apple pedicles. *Ann Bot* **74:** 381-388.

Lang, A. and Thorpe, M. R. (1989) Xylem, phloem and transpiration flows in a grape: Application of a technique for measuring the volume of attached fruits using Archimedes' principle. *J Exp Bot* **40:** 1069-1078.

Lee, D. R. (1990) A unidirectional water flux model of fruit growth. *Can J Biol* **68:** 1286-1290.

Lee, D. R., Dixon, M. A. and Johnson, R. W. (1989) Simultaneous measurements of tomato fruit and stem water potential using in situ stem hygrometers. *Can J Biol* **67:** 2352-2355.

Malone, M. and Andrews, J. (2001) The distribution of xylem hydraulic resistance in the fruiting truss of tomato. *Plant Cell Environ* **24:** 565-570.

Maotani, T., Machida, Y. and Yamatsu, K. (1977) Studies on leaf water stress in fruit trees. VI. Effect of leaf water potential on growth of satsuma mandarin (*Citrus unshiu* Marc.) trees. *J Japanese Soc Horticultural Sci* **45:** 329-334.

Matthews, M. A., Matthews, M. A., Cheng, G. and Weinbaum, S. A. (1987) Changes in water potential and dermal extensibility during grape berry development. *J Am Soc Horticultural Sci* **112:** 314-319.

Nakano, M. (1989) Characteristics of berry growth during stage II, and the components of "speck" disorder in 'Muscat of Alexandria' grape. *J Japanese Soc Horticultural Sci* **58:** 529-536.

Ojeda, H., Deloire, A., Carbonneau, A., Ageorges, A. and Romieu, C. (1999) Berry development of grapevines: Relations between the growth of berries and their DNA content indicate cell multiplication and enlargement. *Vitis* **38:** 145-150.

Pate, J. S. (1988) Water economy of fruits and fruiting plants: Case studies of grain legumes. In *Senescence and Aging in Plants* (L. D. Nooden and A. C. Leopold, eds.) pp 219-239. Academic Press, San Diego.

Pate, J. S., Peoples, M. B., van Bel, A. J. E., Kuo, J. and Atkins, C. A. (1985) Diurnal water balance of the cowpea fruit. *Plant Physiol* **77:** 148-156.

Pate, J. S., Sharkey, P. J. and Atkins, C. A. (1977) Nutrition of a developing legume fruit. Functional economy in terms of carbon, nitrogen, and water. *Plant Physiol* **59:** 506-510.

Peoples, M. B., Pate, J. S., Atkins, C. A. and Murray, D. R. (1985) Economy of water, carbon, and nitrogen in the developing cowpea fruit. *Plant Physiol* **77:** 142-147.

Rogiers, S. Y., Smith, J. A., White, R., Keller, M., Holzapfel, B. P. and Virgona, J. M. (2001) Vascular function in berries of *Vitis vinifera* (L) cv Shiraz. *Aust J Grape Wine Res* **7:** 46-51.

Serpe, M. and Matthews, M. (2000) Turgor and cell wall yielding in dicot leaf growth in response to changes in relative humidity. *Aust J Plant Physiol* **27:** 1131-1140.

Shackel, K. A., Greve, C. G., Labavitch, J. M. and Ahmadi, H. (1991) Cell turgor changes associated with ripening in tomato preicarp tissue. *Plant Physiol* **97:** 814-816.

Shackel, K. A., Matthews, M. A. and Morrison, J. C. (1987) Dynamic relation between expansion and cellular turgor in growing grape (*Vitis vinifera*) leaves. *Plant Physiol* **84:** 1166-1171.

Shimomura, K. (1967) Effects of soil moisture on the growth and nutrient absorption of grapes. *Acta Agronomie Hungary* **16:** 209-216.

Syvertsen, J. P. and Albrigo, L. G. (1980) Seasonal and diurnal citrus leaf and fruit water relations. *Bot Gaz* **141:** 440-446.

Tromp, J. (1984) Diurnal fruit shrinkage in apple as affected by leaf water potential and vapour pressure deficit of the air. *Scientia Horticulturae* **22:** 81-87.

Tyvergyak, P. J. and Richardson, D. G. (1979) Diurnal changes of leaf and fruit water potentials of sweet cherries during the harvest period. *Hort Sci* **14:** 520-521.

van Zyl, J. L. (1987) Diurnal variation in grapevine water stress as a function of changing soil water status and meteorological conditions. *South Afr J Enol Viticulture* **8:** 45-52.

Wei, C., Steudle, E., Tyree, M. T. and Lintilhac, P. M. (2001) The essentials of direct xylem pressure measurement. *Plant Cell Environ* **24:** 549-555.

Welbaum, G. E. and Meinzer, F. C. (1990) Compartmentation of solutes and water in developing sugarcane stalk tissue. *Plant Physiol* **93:** 1147-1153.

Yamamoto, T. (1983) Models of water competition between fruits and leaves on spurs of "Bartlett" pear trees and its measurement by a heat-pulse method. *Scientia Horticulturae* **20:** 241-250.

# Part III

## Integration of Xylem and Phloem

# 10

# The Stem Apoplast: A Potential Communication Channel in Plant Growth Regulation

*Michael Thorpe, Peter Minchin, Nick Gould,*
*and Joanna McQueen*

Plant growth and survival depend on the coordinated synthesis and use of carbohydrate. Both sources and sinks of carbohydrate are influenced by processes on a wide range of time scales and organizational levels throughout the plant, but we do not understand how these processes are coordinated. Carbohydrate synthesis varies enormously as a result of environmental influences over a wide range of time scales: very short, owing to sun flecks; medium term, such as diurnal; and longer term owing to seasonal and yearly variation. But utilization rates also vary and not usually in synchrony with production, and so a wide range of storage and remobilization processes and associated rhythms have evolved that help to uncouple carbohydrate production and use. In leaves, for example, storage and remobilization of carbohydrates such as starch allow diurnal fluctuations in synthesis to be much higher in amplitude than fluctuations in export. In barley, a large fraction of grain fill comes from remobilized carbohydrate that was fixed before anthesis (Biscoe *et al.*, 1975). In many tree species, the carbohydrate stored in stems during the autumn is essential for spring development of new leaves, and stored carbohydrate can also be required for flowers and early fruit development. As well as coping with predictable rhythms in the environment, plants must also deal with random events such as fire, flood, and drought. One strategy is to maintain carbohydrate storage in roots or lignotubers and use it for re-growth (e.g., Bowen and Pate, 1993).

Sources and sinks of carbohydrate are distributed throughout a plant. Sinks are everywhere, because all living tissues contain cells with mitochondria, and some tissues need a large carbon influx for growth or storage. Sources are the photosynthetic tissues in the shoot, and tissues from where remobilization can occur. The coordination of these sources and sinks entails the transfer of information as well as material via the long distance transport systems. Transport phloem—the long distance pathway from the minor veins in exporting leaves to petioles, branches, the main stem, and roots—therefore has two important functions: long-distance axial transport and radial exchange. The same is true of the xylem. Early consideration of the mechanisms of the long-distance transport systems usually envisaged a single source and a single sink, but attention is turning to the coordination of transport in a more realistic plant with multiple sources and sinks, and where radial transfers are necessary. Discussions of radial exchange have mostly concerned xylem-to-phloem transport, with attention to nutrient cycling (Atkins, 2000; Sattelmacher, 2001), but with growing attention to radial exchanges of carbohydrate (van Bel, 1990). To understand the processes of loss and retrieval, they must be measured, and to understand their regulation, it is necessary to measure them noninvasively so that their time-dependence can be observed in relation to diagnostic treatments. However, there are very few actual measurements of radial fluxes; observations have been rather of the *potential* for such fluxes. For example, observing the movement of membrane-impermeant fluorochromes shows the possibility of symplastic transport via intercellular connections; observing sucrose carrier proteins shows the possibility for apoplastic/symplastic transport; and the activity of appropriate enzymes indicates that specific metabolic processes are possible. Nevertheless, *in vivo* measurement of radial fluxes is possible, because the radial flows affect the axial flow of photoassimilate that can be observed continuously *in vivo* using radioactive $^{11}C$. Due to the tracer's short 20-minute half-life, the radial flow itself is often not radiolabeled.

In this chapter we discuss radial exchange processes as measured with radiocarbon, and the concept of buffering is central. We use the term *buffering* to refer to the operation of lateral carbohydrate flows to and from the transport phloem that help to maintain a constant flow from source to sink during times of changing supply and demand. The discussion focuses on pathways of these flows and their short-term regulation. We discuss these processes of buffering in two quite different species (bean - *Phaseolus vulgaris* L. and apple - *Malus × domestica* Borkh.). Buffering of phloem transport was shown to vary diurnally, respond to source/sink properties, and involve the apoplast. This involvement leads us to discuss how phenomena in the apoplast, including the xylem stream, can influence phloem transport. Further, we discuss how the physicochemical properties of the apoplast can be influenced by the phloem, giving the opportunity for the phloem to affect xylem transport and plant water relations.

# The Short-Term Buffer

## Evidence for Short-Term Buffering of Long-Distance Phloem Transport

Short-term buffering of long-distance phloem transport is probably a direct consequence of the lateral unloading and reloading associated with stem growth and longer term storage. The marked seasonality in stem storage raises the question of whether the capacity for buffering also follows a seasonal pattern or develops when it is needed. Our experience using radiolabeled photosynthate has been that continuous unloading of carbohydrate from the long-distance pathway is always observed. After pulse labeling, radiocarbon moves relatively slowly through the leaf tissues to the phloem where it enters the phloem pathway for rapid long-distance transport. Tracer levels in the long-distance pathway (e.g., stem) rise rapidly to a maximum, and then fall slowly due to the dynamics of phloem loading in the leaf and unloading in the stem. If we analyze the changes in the tracer profile as it moves along the phloem pathway, the data always show that there is continuous unloading of the tracer, and hence of photosynthate (i.e., the pathway loss is finite: Minchin and Troughton, 1980; Minchin *et al.*, 2002). Unloading within the transport phloem of a mature exporting leaf is also detectable (Minchin *et al.*, 2002). In addition to continuous unloading, there is also continuous reloading of carbohydrate. For example, Ho and Peel (1969) measured $^{14}$C in exudate from aphids feeding in stem sieve tubes before and after phloem transport from a $^{14}CO_2$-labeled leaf was stopped, calculating that 25% of the carbohydrate in the sieve tube sap had been remobilized. Concurrent loss and retrieval of photoassimilate were reported for young bean stems (Minchin and Thorpe, 1987), and the separate components were quantified. Their ratio varied among the plants: Retrieval was between 10% and 50% of the gross loss, with the net loss up to 4% per centimeter.

With these concurrent unloading and reloading flows, it is not surprising to find that their balance can alter and serve to buffer changes in source or sink activity. The ability of this balance to respond quickly was demonstrated by measuring the time course of $^{11}$C-labeled photoassimilate in various parts of a *Phaseolus* stem. The radiolabel was observed continuously, and transport was suddenly stopped at one position by chilling a short length of the stem (Fig. 10.1A; Minchin *et al.*, 1983). Downstream of the block, axial flow toward the stem apex continued for about 90 minutes before finally stopping. Upstream of the block, flow from the labeled leaf and in the stem continued for a similar time. The interpretation was that pressure gradients in the sieve tubes, on either side of the cold block, were maintained by radial flows of water associated with net radial flows of carbohydrate. The net radial flow on either side of the cold block responded in opposite directions. There was increased net unloading upstream of the block and increased net reloading downstream. These data demonstrated that there was a capability for short-term buffering in the stems. We have seen similar behavior in a variety of species, including apple, moonflower (*Ipomoea alba* L.), a water lily (*Nymphoides peltata* L.), cowpea (*Vigna unguiculata* L.), strawberry (*Fragaria ananassa* Duschesne), sow thistle (*Sonchus oleraceus* L.), and Arabidopsis (*Arabidopsis thaliana* L.). The time course of hydrostatic pressure within the sieve element near such a chill, measured with a pressure sensor glued to the excised stylet of an aphid, was consistent with the buffering dynamics seen with $^{11}$C. The pressure downstream from the chilled region fell within minutes after the chill to a lower value which was then maintained as long as the blockage remained (Fig. 10.1B); upstream of the blockage the pressure increased (see Gould *et al.*, 2004b). Further, the sap's osmotic pressure and sucrose concentration showed no change. The interpretation (Gould *et al.*, 2004b) was that radial exchanges of both water and solute maintained pressure and solute levels after the blockage, just as we had inferred by observing axial phloem flow with $^{11}$C (Minchin *et al.*, 1983). The mechanism that stops axial flow, at least in legumes, is probably due to the explosive change in the conformation of a "forisome," a crystalloid protein body discovered by Knoblauch *et al.* (2001) within the sieve elements of legumes. That report showed that forisome explosion is caused by $Ca^{2+}$, and Plieth *et al.* (1999) reviewed strong evidence that a drop in temperature initiates an increase in cytosolic $Ca^{2+}$.

### Pathway Between Sieve Elements and the Short-Term Buffering Pool

The nature of the cellular pathway for transport of carbohydrate between sieve elements and the short-term buffering pool determines which transport mechanisms can operate. Therefore it is critical to know the pathway before we can discover which mechanisms might regulate lateral exchange and thus coordinate flows between sources and sinks. The majority of carbohydrate in plants is located in the symplast, being starch in chloroplasts of green tissues or in plastids of storage parenchyma and ray cells, or sucrose and soluble polymers of glucose within the vacuoles. The apoplast is a small fraction of those tissues (approximately 10% to 20%), and the carbohydrate concentrations there are low. The pathway between sieve elements and storage could be apoplastic, symplastic, or a combination. Structural and functional evidence suggests that both routes are available, although the majority of evidence points to the apoplast as the main route in many species (van Bel, 1990). The unloading flux in young cells usually takes a symplastic route, but, as they mature, the cells become symplastically isolated leaving only an apoplastic path. This development of symplastic isolation is true for transport phloem, as shown for cambial cells of the stem (van Bel and van Rijen, 1994; Kempers *et al.*, 1998), and other tis-

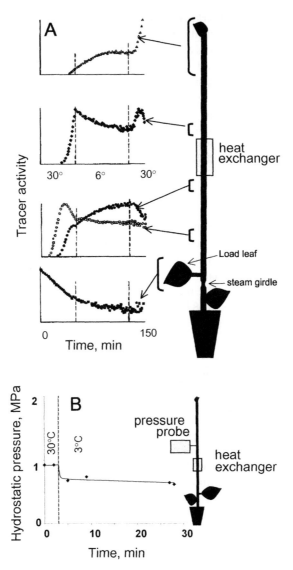

**Figure 10.1** Dynamic buffering responses of the phloem transport system after a cold block. (A) Variation of 11C-photoassimilate, corrected for isotope decay, in several locations in a Phaseolus vulgaris plant after pulse-labeling the load leaf. Phloem transport at one position on the stem was stopped suddenly by chilling but did not stop upstream or downstream of the chill until about 90 minutes later; transport resumed when the heat exchanger was rewarmed (Minchin et al., 1983). The stem was girdled using steam 60 minutes before labeling the leaf to prevent basal transport. (B) Sieve tube hydrostatic pressure measured with the pressure probe which was glued to an aphid stylet 50mm downstream from a heat exchanger. When the stem was chilled suddenly and blocked flow, the pressure declined immediately from its initial value of 0.9 MPa to 0.7 MPa, and remained at that level (Gould *et al.*, 2004a).

sues such as root tips (Oparka and Turgeon, 1999; Imlau *et al.*, 1999; Hukin *et al.*, 2002). The carbohydrate fluxes between these symplastically isolated tissues require sugar carriers at each intervening membrane. Sucrose transporters have been localized to both sieve elements and companion cells within petioles and stems (Stadler *et al.*, 1995; Kühn *et al.*, 1997), indicating that there are retrieval systems for apoplastic sucrose in these tissues, and concurrent unloading and reloading along the length of the transport phloem is now thought to be widespread, if not universal (van Bel, 2003b; Oparka and Turgeon, 1999). The few plasmodesmata between companion cells and phloem parenchyma cells in transport phloem are usually closed but can open under appropriate conditions, and these plasmodesmata may be opened by movement proteins (van Bel, 2003b).

It is inevitable that there will be an apoplastic component of sieve element carbohydrate unloading, because leakage from the high concentration in sieve elements is to be expected (Patrick, 1990). Where the apoplast is involved in phloem unloading from the axial pathway, we might expect to find sucrose in that space. Indeed, apoplastic sucrose in *Phaseolus* bean stems has been found and measured at concentrations around 50 mM (Minchin and Thorpe, 1984). In addition, when photoassimilate was radioactively labeled, tracer appeared in a solution bathing the apoplastic space of the stem soon after tracer first appeared in the stem, and at a greater rate if axial transport was stopped downstream, demonstrating that this apoplastic tracer was closely linked to the phloem. The tracer was non-volatile, so it was not due to respiration of labeled photosynthate, and a "tracer trapping" technique was used to show that labeled sucrose was being taken up across plasma membranes from the stem apoplast. Tracer trapping involves applying appropriate unlabeled solute to compete with the labeled solute for saturable membrane carriers. If there is competition for these binding sites then there is an increase in the escape of labeled solute into the bathing solution. We found that the apoplast can be a pathway of phloem unloading that enhances short-term buffering in bean stems when a cold block was imposed downstream, reflecting an increase in apoplastic photosynthate when net unloading increased (Minchin and Thorpe, 1984).

If photoassimilate is present in the apoplast, then it should be possible to detect some of it moving in the xylem, and indeed this has been observed using very sensitive tracer methods (Minchin and McNaughton, 1987). The quantity involved was not nutritionally significant, but there is a potential for a signaling role. Significant xylem transport of sugars does occur in spring when root reserves are remobilized before bud break. This process generates considerable root pressures, and even leads to bleeding from damaged stems. The consequences for horticulture can be detrimental, as valuable resources are lost if some perennial crops are pruned late in the winter. Such examples include sugar maple (Sauter *et al.*, 1973), *Actinidia* (Ferguson, 1980), and grape (Andersen and Brodbeck, 1989). While this is recognized within the horticultural field, it has hardly been considered in the plant physiology literature. Also, apoplastic sugars increase dramatically during cold hardening (e.g., of winter oat: Livingston and Henson, 1998), which has been attributed to a role in cryoprotection. Photosynthate in the stem apoplast is the substrate for microorganisms and parasitic angiosperms (Patrick, 1989; Hall and Williams, 2000).

Both apoplastic and symplastic pathways have been shown to play a role in the movement of solutes between sieve elements and the short-term buffering pool in mature *Phaseolus* stems. The Newcastle group (Hayes *et al.*, 1987; Patrick and Offler, 1996) reported several strands of evidence for a symplastic route in plants that had a high source/sink ratio (winter grown). Plasmodesmatal connections were visible in micrographs, there was radial movement from the sieve elements of a membrane-impermeant fluorochrome, and radial movement of $^{14}$C-photosynthate was strongly inhibited by plasmolysis (which causes cellular isolation). They also reported evidence using p-chlorobenzene sulfonic acid (PCMBS), a very useful inhibitor to demonstrate that there is trans-membrane transport of sugar. It inhibits all sucrose transporters, but its membrane permeability is low, so little enters the symplast from the apoplast (Delrot *et al.*, 1980). PCMBS had no effect on radial movement of $^{14}$C-photosynthate in the stem of the winter-grown bean, showing that symplastic transport dominated radial exchange in those sink-limited plants. However, pruned or shaded plants, with low source/sink ratio, showed no symplastic unloading (Hayes *et al.*, 1987), and they concluded that the plasmodesmata were closed in response to a high pressure difference (Oparka and Prior, 1992). In fact, gating of plasmodesmata in transport phloem has been suggested as a means of directing photosynthate flow (van Bel, 2003a). Grimm *et al.* (1997) found that axial transport was unaffected by PCMBS in the apoplast of *Cyclamen* and *Primula* petioles and concluded that retrieval from the apoplast was not important for axial transport in that tissue.

The apoplast contains only a small amount of carbohydrate compared to that available from the symplast, the source for long-term buffering. Most reserve carbohydrate is stored within the parenchyma tissue and in ray cells (van Bel, 1990). Little work has been done to investigate mechanisms or pathways involved in the formation and remobilization of these storage systems. van Bel (1990) reviewed the role of rays in two-way xylem-phloem transfers and argued for the existence of both a symplasmic and apoplastic route involving the rays. In a plausible scenario, these reserves are derived from the short-term buffer pool and remobilize via the same apoplastic route (Fig. 10.2). Short-term buffering would depend on soluble carbohydrate within the stem apoplast, and when this was either well sup-

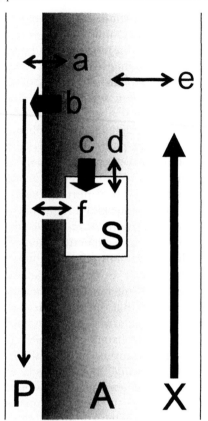

**Figure 10.2** Schematic of pathways in the plant stem for transport of carbohydrates between sieve tubes and storage compartments. From the sieve elements $P$ carbohydrate can (a, d) diffuse through the membrane into the stem apoplast $A$, or (f) move by diffusion or convection via plasmodesmata into a symplastic compartment $S$ (e.g., ray cells or phloem parenchyma). From the apoplast $A$ adjacent to the phloem, carbohydrate can (b) be actively re-loaded into the sieve tubes, (c) be loaded into storage cells $S$, or (e) diffuse further through the apoplast into the xylem stream $X$, where flow is usually in the opposite direction to that in sieve tubes $P$.

plied or depleted by the phloem, the longer term symplastic pool would be replenished or depleted. The apoplastic pool's sugar concentration could then act as the control variable for this buffering cascade.

### Mechanisms That Regulate Lateral Flow Within the Transport Phloem

Short-term buffering of sources and sinks of carbohydrate is due to a change in the net loss or retrieval of stored carbohydrate. Therefore, there are two components of the flow that could change, loss and retrieval, and they are likely to operate with very different mechanisms. The *in vivo* technique to measure stem buffering was used to unravel the responses of each component by using the membrane transport inhibitor PCMBS to inhibit retrieval (as discussed previously) and then perturb axial flow. First, in a situation in which buffering required increased net loading of carbohydrate (i.e., remobilization), the buffering of phloem flow was reduced if the apoplast of that region contained PCMBS (Fig. 10.3A; see also Minchin *et al.*, 1984). The increased net retrieval in this situation was therefore due to increased reloading and not reduced unloading. Second, where there was increased net loss, PCMBS was used to test whether the response was a decline in retrieval or an increase in loss. The observation was that net unloading increased even in the presence of PCMBS (i.e., when retrieval was inhibited) (Minchin *et al.*, 1984). Therefore, we concluded that the unloading component of the flow had increased, rather than that the retrieval declined, demonstrating that both components of exchange with the apoplast can vary.

The mechanism of retrieval is believed to be proton-symport, and such sucrose transporters have been immunolocalized in transport phloem throughout the plant, sometimes in sieve element membranes and sometimes companion cells (Kühn *et al.*, 1997; Stadler *et al.*, 1995). Less attention has been paid to the mechanisms of leakage. Patrick (1990) stressed that a diffusive unloading flow will be relatively insensitive to apoplastic concentration, as the symplastic concentration is high and the apoplastic very low, so plausible changes in the apoplastic concentration will have little effect on the concentration difference across the membrane. Thus, unloading flow will be linearly related to sieve tube concentration. Reloading via a sucrose symporter will be very sensitive to apoplastic concentration and probably not to symplastic concentration. For example, phloem loading in the leaves of barley, where there is a similar symporter to that involved in stem reloading (SUC1 or SUC2), has been shown not to increase when symplastic solute concentration declines (Minchin *et al.*, 2002). However, there is a body of evidence that active uptake of sucrose is turgor sensitive (Grimm *et al.*, 1990; Daie, 1987; Smith and Milburn, 1980), providing a mechanism for short-term responses. We have shown that sieve tube hydrostatic pressure changes after axial transport is stopped by a cold block, showing evidence of turgor regulation (Fig. 10.1B). Longer term control arises from changes in carrier expression. Expression of sucrose transporters as well as enzymes associated with remobilization of symplastic reserves has been shown to be dependent on the apoplastic sucrose concentration (Stadler *et al.*, 1995; Kühn *et al.*, 1997; Li *et al.*, 2003).

A useful working hypothesis for plant regulation of unloading or reloading of carbohydrate is that the apoplastic sucrose concentration is held constant. Any change in apoplastic sucrose concentration will have little effect on diffusive unloading (see previously) but is expected to have a big

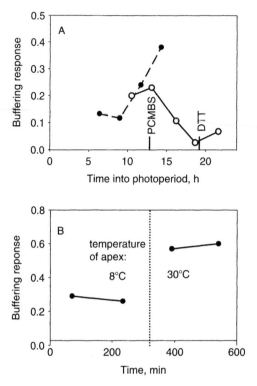

**Figure 10.3**  The buffering response in stems of apple seedlings. (A) Buffering increased during the photoperiod (*closed circles*), suggesting that temporary storage had been depleted during the previous dark period, and was filling through the light period. Buffering reduced after apoplastic retrieval was reduced on application of PCMBS (*open circles*). The effect of PCMBS was partially reversed by DTT, confirming that the decline in buffering was due to the PCMBS, and therefore that retrieval followed an apoplastic pathway. (B) Effect of warming the terminal sink. Buffering was measured for a 65 mm length of stem, from the change in axial transport of [11]C-photosynthate 65 mm downstream from a cold block. The buffering response is defined as the ratio of that axial flow just after a cold block to that just before the block. With a buffering response equal to 0, the axial flow would stop completely, and, if equal to 1, the axial flow would be fully buffered by increased retrieval along the 65 mm of stem. Cold blocks were imposed for 20 minutes at roughly 120-minute intervals. The stem that was treated with PCMBS was prepared on the day before the experiment with a wick (one strand from the wick for an oil lamp) threaded twice through the stem using a darning needle and all ends placed in an apoplastic bathing solution. The times when we applied 2 mM PCMBS and 2 mM DTT are indicated on the abscissa.

effect on active loading. A reduction in apoplastic sucrose concentration will cause a reduction in loading, and vice versa. The apoplastic sucrose concentration can also be expected to affect sucrose uptake or release from the phloem parenchyma and ray tissues involved in longer term storage. Similar uptake kinetics to phloem loading seems likely, both being active, and sucrose release from storage could be induced by reduced uptake. Release of storage carbohydrate into the apoplast becomes important when phloem delivery cannot maintain the target apoplastic sucrose level, and phloem transport will then be maintained from stem reserves. In this model the apoplastic volume acts as both the controller of flows and as the short-term pool for buffering the transport phloem.

Early investigators of the mechanism of phloem transport found the metabolic activity of the pathway to be high (Willenbrink, 1957) and took this to show that the transport process consumed energy along the path. Recent work has shown that the phloem operates at a low oxygen level because of its high respiration rate and poor connection to the atmosphere (van Dongen *et al.*, 2003). This metabolic cost is a necessary part of the leak-retrieval nature of the pathway. Energy for longitudinal transport is provided primarily in the source regions, but the occurrence of remobilization creates an additional energy requirement that must be supplied along the entire length of the transport phloem. There is evidence that symplastic isolation, by means of closed plasmodesmata, may also need metabolic support. In the root tips of *Arabidopsis*, the symplastic isolation of the mature cortical cells requires energy (Wright and Oparka, 1997). Inhibition of adenosine triphosphate production within the transport phloem induced massive assimilate loss (Geigenberger *et al.*, 1993), presumably as a result of a decline in reloading.

The importance of reloading was vividly demonstrated when we treated 20 cm of immature bean stem with PCMBS and found that axial phloem transport to the shoot apex was stopped (unpublished). Sugar retrieval would have been inhibited by PCMBS over this 20-cm length (or more, as PCMBS can move in the xylem), and unloading flow had apparently reduced the sieve-tube solute concentration to such a low level that there was no flow beyond the treated region. An alternative interpretation is the relay hypothesis (Lang, 1979), where the carbohydrate path from source to sink is not just one file of sieve tubes, but a succession of relays where there is unloading and reloading into another file (overcoming problems associated with Münch flow over long distances: Thompson and Holbrook, 2003). If all files of sieve tubes terminated with relay stations were exposed to PCMBS in our experiment, axial transport of photoassimilate would have stopped.

## Quantification of the Amount of Buffering Available

The amount of buffering available to make up for a shortfall from the sources is determined by both the flow rate and amount of carbohydrate available, which together determine the total buffering capacity. In the short term, when the apoplast alone is the buffer volume, the buffer capacity is the total apoplast sugar content; but in the longer term, the capacity will probably be greater depending on the amount of carbohydrate in stem storage, and whether the necessary machinery can be developed to access it. Sauter (1982) showed that there are seasonal changes in sugar uptake from xylem in *Salix*, and van Bel (1990) discussed evidence that the mechanism of uptake varies dramatically by season in *Populus*. In the converse situation when buffering occurs because of excess production, there will be a trade-off between buffering capacity (i.e., capability for storage) and the downregulation of production. Long-term storage requires investment in the facilities, such as storage tissues with appropriate anabolic machinery. If they are not present, storage cannot occur and the source must downregulate. Storage capability could be preprogrammed for predictable imbalances of supply and demand, such as in diurnal or seasonal trends, or it could develop as the need arises.

The only method currently available to quantify the buffering capacity is by observing movement of $^{11}$C-labeled photosynthate. An indicator of the short-term buffer capacity is the response of axial flow at a standard length downstream from where the stem has been blocked. If the axial flow stops, there has been no radial influx anywhere along that length to buffer the flow; if axial flow continues unaffected, there has been complete buffering (until the available reserves are exhausted). Therefore, we have quantified the available short-term buffering as the axial flow just after a cold block, expressed as a fraction of the flow just before the blockage. In apple seedlings we discovered a diurnal variation in this buffering response (Fig. 10.3A). The buffering downstream from a cold block provided by 65 mm of stem was only about 10% of the axial flow early in the photoperiod, but it increased to about 40% after 14 hours of light. It appeared that early in the day little carbon was available for buffering, suggesting that the stem reserves had been depleted overnight. In a corresponding situation, buffering in the very long (30 cm) petiole of the water lily was prevented when source leaves on the plant were blocked for a few hours and presumably depleted the buffers (Thorpe *et al.*, 1983). The importance of the demand for carbon on the size (magnitude) of the buffer response has been demonstrated by warming the major sink, the shoot apex. With a warmer apex, its higher demand for carbohydrate resulted in much more buffering in the stem (Fig. 10.3B).

## What Are the Opportunities to Affect Leakage and Retrieval?

The observations just reviewed show that the net radial flow of carbohydrate can respond rapidly to carbon availability, that this involves an apoplastic pathway, and that it can be affected by the composition of the apoplast. Both unloading and reloading components of radial flows are able to respond, and short-term buffering uses carbohydrate within the apoplast pool. Also, carbohydrate remobilized from storage parenchyma takes an apoplastic path to the phloem, and we have suggested that the requirement for remobilization or storage is communicated through apoplast carbohydrate level. The apoplast is a poorly mixed compartment, so large gradients are likely, especially where there are diffusive fluxes, or diffusion barriers (Canny, 1995). The flow between apoplast and symplast will be related to sugar concentrations at the membrane, which will not be simply the average in the tissue apoplast. The influence of the apoplast on radial flows can therefore be expected to be much stronger than average apoplast concentrations may suggest. Various specialized regions in the apoplast have been reported, such as in the vicinity of guard cells where evaporation occurs and sucrose can reach osmotically significant levels, so that transpiration rate can affect stomatal opening (Ewert *et al.*, 2000). There are sugar storage regions in sugar cane (Welbaum *et al.*, 1992), and there are apoplastic domains associated with haustorial membranes of fungal biotrophs (Hall and Williams, 2000). The close interaction of apoplast and symplast gives opportunities for xylem influence on phloem transport, as well as for phloem influence on xylem transport, and thus a role in regulating both carbon partitioning and plant water status.

### Involvement of Apoplast in Leakage and Retrieval

*Sucrose* Gene regulation by sucrose is widespread, some being induced by low sucrose (the famine response), some by high sucrose (the feast response) (Lalonde *et al.*, 1999). The response for storage or remobilization can be affected by the appropriate genes. For example, apoplastic sucrose down-regulates the expression of the SUT1 sucrose symporter in leaves of sugar beet (Vaughn *et al.*, 2002; Chiou and Bush, 1998), a species with apoplastic phloem loading. This modulation of expression gives an opportunity for coarse control of phloem loading rate in the leaves, because sucrose washed there in the xylem stream will reflect the amount of sucrose in the apoplast of the stem. The latter will be low if sink activity in the plant is high. Li *et al.* (2003) found that in citrus trees with a high fruit load, soluble sugar levels in the roots were low, and two proteins with a role in remobilization of storage carbohydrate (SUT1 and α-amylase) had increased expression.

*Micro-organisms and Parasites* Biotrophic organisms use apoplastic carbohydrates and can severely modify partitioning and metabolism in the plant, by depriving the plant of carbohydrate (Wolswinkel *et al.*, 1974) and by modification of membrane transport (Patrick, 1989). In all cases of infection by biotrophic organisms elevated cytosolic solute levels of the host plant have been observed, resulting in increased efflux from the host

(Patrick, 1989). The cause of these elevated solute levels is unknown, but it appears to be induced by the infection and thus implies a modification of the host by the invader. In the case of fungal infection, the associated apoplastic space has high invertase levels (resulting in cleavage of the unloaded sucrose) and the fungal membrane has a strong expression of a hexose transporter (Voegele *et al.*, 2001; Roitsch *et al.*, 2003). Sucrose cleavage increases osmotic pressure in the apoplast, helping to direct phloem flow to this region. Host-parasite interaction is a worthwhile field for investigating how solute release from host cells can be regulated.

***Carbon Partitioning***   It has been demonstrated that when the availability of carbohydrate is modified, the distribution between sinks also changes. This can be understood in the context of competing sinks, where a change in available supply of carbohydrate has been shown to affect partitioning between sinks as a direct consequence of Münch flow when unloading into sinks follows saturable kinetics (Minchin *et al.*, 1993). Any modification of the radial flow of carbohydrate along the long-distance pathway affects carbohydrate availability and therefore has the potential to alter carbon partitioning within the entire plant. We have noted that the balance of unloading and retrieval can be modified by symporter activity and apoplastic sugar levels. Other properties of the apoplastic solution that affect membrane transport will also be important, such as $K^+$, $Ca^{2+}$, and pH. Modulation of phloem osmotic pressure by solutes other than carbohydrate also alter carbon partitioning. For example, potassium is a major component of phloem sap, and its strong exchange with the xylem is likely to affect phloem partitioning (Gould *et al.*, 2004a).

***Plant Water Relations***   Because hydrostatic pressure in the phloem is affected by local water potential, xylem water potential affects carbon partitioning. The xylem water potential can be affected by transpiration rate, hydraulic conductance, or apoplastic osmolytes, having a major effect on pressure gradients in the phloem with consequences for carbon partitioning (Daudet *et al.*, 2002). In fruits and seeds, apoplastic solutes have a major effect on phloem pressure gradients, usually at late stages of development, assisting transport into those sinks by maintaining a low turgor in sieve elements despite high solute levels (Patrick, 1997; Lang and Düring, 1991). We suggested above that a similar phenomenon helps fungi acquire solutes from their host plant.

### Involvement of Phloem-Borne Signals

***Macro Molecular Signaling***   A likely function for some of the increasing range of RNA fragments or proteins now being found within phloem sap is in information transport and signaling. A possible way for such signals to affect partitioning is via the modulation of sink or retrieval activity through induction or repression of key unloading/utilization transporters or enzymes such as invertase.

***Plant Water Relations*** There is the intriguing opportunity for the phloem itself to influence water relations, as xylem conductance can be influenced by the ionic constitution of the xylem sap (Zwieniecki *et al.*, 2001), which would be modified by appropriate secretion of ions from the phloem. Phloem sap contains a high concentration of potassium. Unloading of potassium into the stem apoplast (Gaymard *et al.*, 1998) has the potential to alter xylem conductance and the local water potential gradient, affecting water potentials throughout the whole plant. The information flow is therefore potentially very subtle, as the exchange of ions between phloem and xylem could affect water potential gradients, with consequences for carbon partitioning as discussed previously for ion circulation and uptake, as well as for all the other phenomena that are influenced by plant water relations.

## Involvement of Carbon-Nitrogen Interactions

The study of the interactions between carbon and nitrate assimilation has a long history linked to the importance of these elements to crop production. Nitrogen deficiency leads to increased carbon partitioning to the root to provide for more root growth, while reduced $CO_2$ assimilation leads quickly to an increase in shoot partitioning. We have observed that changes in nitrate uptake can affect carbon partitioning within an hour, without affecting photosynthesis (Minchin *et al.*, 1994). Whole plant growth models must incorporate these carbon/nitrogen (C/N) interactions. Thornley's (1972) transport-resistance model does this by making both carbon and nitrate assimilation dependent on the amount of both substrates, forcing a C/N balance. More recent whole plant growth modeling incorporates several metabolic signals linking organic carbon and nitrogen pools (Bijlsma and Lambers, 2000).

Microarray analysis has found more than 40 genes in *Arabidopsis* that are induced by the nitrate ion, encoding enzymes of both carbon and nitrogen metabolism (Coruzzi and Bush, 2001). A role for sucrose as a signaling molecule is now well established (Koch, 1997), and apoplastic sucrose has a demonstrated effect on expression of nitrate uptake channels (Forde, 2002). In addition to C/N cross-talk, C-metabolites have recently been linked to pathways for response to ethylene (Zhou *et al.*, 1998) and abscisic acid (Laby *et al.*, 2000). Further, Parsons *et al.* (1993) demonstrated that phloem-borne nitrogen affects nodule growth and activity in legumes. So both nitrate and sucrose, although important metabolites, are also involved in inducing gene expression. There is a complex network of signals regulating growth, with a potential for both phloem and xylem transport systems to alter each other's behavior, and emphasizing the highly integrated nature of plant growth regulation.

## Summary

Our understanding of phloem transport has moved on a long way since Münch first proposed his hypothesis as to what drives phloem flow. Although his hypothesis has come in and out of acceptance, it now forms the basis of most interpretations of carbohydrate movement within vascular angiosperm plants, while phloem transport in gymnosperms remains an almost untouched field. The hypothesis has required remarkably little modification. Three extensions are currently accepted: (1) the route of phloem loading and unloading can be apoplastic or symplastic; (2) a high solute concentration in the sink apoplast affects bulk flow through the phloem, an instance of the inevitable consequence of water relations; and (3) there is unloading and reloading throughout the transport phloem. These three extensions are fundamental to our scenario in which the coordination of sources and sinks of carbohydrate, at a wide range of time scales, is mediated through the unloading and retrieval of carbohydrate throughout the transport phloem, with the apoplast often playing a key role.

The existence of continuous unloading and retrieval of carbohydrate was demonstrated by the use of radiotracers, which also showed that these lateral fluxes could change at any time and uncouple sources and sinks. This changing distribution of carbohydrate within the plant is an essential component in the regulation of plant growth. The dominant pathway for these exchanges was the apoplast in the species investigated, which gives opportunity for the regulation of lateral fluxes by membrane transporters, and from the composition of the apoplastic spaces bordering the phloem tissues where retrieval and unloading occur. There is also evidence for symplastic pathways, and the nature of this pathway, subject to modulation by sundry proteins, may also allow lateral exchange of larger molecules.

A working hypothesis for regulation of unloading or reloading of carbohydrate in the transport phloem is that the apoplastic sucrose concentration is held constant. This is achieved by the balance of unloading and retrieval by the sieve elements and of remobilization or storage in symplastic compartments. The apoplastic volume acts both as the controller of flows and also as the short-term pool for buffering the transport phloem. Short-term regulation will occur via the kinetics of transport, and longer term changes in transporter gene expression responding to such factors as sugar and nitrate.

## Acknowledgments

Support was provided in part by the Royal Society of New Zealand (ISAT and Marsden Funds) and an LDRD grant from DOE/BNL. Joanna McQueen received an AGMARDT Doctoral Scholarship.

## References

Andersen, P. and Brodbeck, B. (1989) Diurnal and temporal changes in the chemical profile of xylem exudate from *Vitis rotundifolia. Physiol Plant* **75:** 63-70.

Atkins, C. (2000) Biochemical aspects of assimilate transfers along the phloem path: N-solutes in lupins. *Aust J Plant Physiol* **27:** 531-537.

Bijlsma, R. J. and Lambers, H. (2000) A dynamic whole-plant model of integrated metabolism of nitrogen and carbon. 2. Balanced growth driven by C fluxes and regulated by signals from C and N substrate. *Plant Soil* **220:** 71-87.

Biscoe, P. V., Gallagher, J. N., Littleton, E. J., Monteith, J. L. and Scott, R. K. (1975) Barley and its environment. IV. Sources of assimilate for the grain. *J Appl Ecol* **12:** 295-318.

Bowen, B. J. and Pate, J. S. (1993) The significance of root starch in post-fire shoot recovery of the resprouter *Stirlingia latifolia* R. Br.(Proteaceae). *Ann Bot* **72:** 7-16.

Canny, M. J. (1995) Apoplastic water and solute measurement: New rules for an old space. *Annu Rev Plant Physiol Plant Mol Biol* **46:** 513-542.

Chiou, T.-J. and Bush, D. R. (1998) Sucrose is a signal molecule in assimilate partitioning. *Proc Natl Acad Sci U S A* **95:** 4784-4788.

Coruzzi, G. and Bush, D. R. (2001) Nitrogen and carbon nutrient and metabolite signaling in plants. *Plant Physiol* **125:** 61-64.

Daie, J. (1987) Interaction of cell turgor and hormones on sucrose uptake in isolated phloem of celery. *Plant Physiol* **84:** 1033-1037.

Daudet, F., Lacointe, A., Gaudillere, J. P. and Cruizat, P. (2002) Generalized Münch coupling between sugar and water fluxes for modeling carbon allocation as affected by water status. *J Theor Biol* **214:** 481-498.

Delrot, S., Despeghel, J.-P. and Bonnemain, J.-L. (1980) Phloem loading in *Vicia faba* leaves: Effect of n-ethylmaleimide and parachloromercuribenzenesulfonic acid on $H^+$ extrusion, $K^+$ and sucrose uptake. *Planta* **149:** 144-148.

Ewert, M. S., Outlaw, W. H., Zhang, S., Aghoram, K. and Riddle, K. (2000) Accumulation of an apoplastic solute in the guard-cell wall is sufficient to exert an influence on transpiration in *Vicia faba* leaves. *Plant Cell Environ* **23:** 195-203.

Ferguson, A. (1980) Xylem sap from *Actinidia chinensis*: Apparent differences in sap composition arising from the method of collection. *Ann Bot* **46:** 791-801.

Forde, B. G. (2002) Local and long-range signaling pathways regulating plant responses to nitrate. *Annu Rev Plant Biol* **53:** 203-224.

Gaymard, F., Pilot, G., Lacombe, B., Bouchez, D., Bruneau, D., Boucherez, J., Michaux-Ferriere, N., Thibaud, J. B. and Sentenac, H. (1998) Identification and disruption of a plant shaker-like outward channel involved in K+ release into the xylem sap. *Cell* **94:** 647-655.

Geigenberger, P., Langerberger, S., Wilke, I., Heineke, D., Heldt, H. W. and Stitt, M. (1993) Sucrose is metabolised by sucrose synthase and glycolysis within the phloem complex of *Ricinus communis* L. seedlings. *Planta* **190:** 446-453.

Gould, N., Minchin, P. E. H. and Thorpe, M. R. (2004b) Direct measurements of sieve element hydrostatic pressure reveal strong regulation of sieve element hydrostatic pressure after pathway blockage. *Funct Plant Biol* **31**: 987-993.

Gould, N., Thorpe, M. R., Minchin, P. E. H., Pritchard, J. and White, P. J. (2004a) Solute is imported to elongating root cells of barley as a pressure-driven flow of solution. *Funct Plant Biol* **31**: 391-397.

Grimm, E., Bernhardt, G., Rothe, K., and Jacob, F. (1990) Mechanism of sucrose retrieval along the phloem path—a kinetic approach. *Planta* **182**: 480-485.

Grimm, E., Jahnke, S. and Rothe, K. (1997) Photoassimilate translocation in the petiole of *Cyclamen* and *Primula* is independent of lateral retrieval. *J Exp Bot* **48**: 1087-1094.

Hall, J. L. and Williams, L. E. (2000) Assimilate transport and partitioning in fungal biotrophic interactions. *Aust J Plant Physiol* **27**: 549-560.

Hayes, P. M., Patrick, J. W. and Offler, C. E. (1987) The cellular pathway of radial transfers of photosynthates in stems of *Phaseolus vulgaris* L.: Effects of cellular plasmolysis and *p*-chloromercuribenzene sulphonic acid. *Ann Bot* **59**: 635-642.

Ho, L. C. and Peel, A. J. (1969) The relative contributions of sugars from assimilating leaves and stem storage cells to the sieve tube sap in willow cuttings. *Physiol Plant* **22**: 379-385.

Hukin, D., Doering-Saad, C., Thomas, C. R. and Pritchard, J. (2002) Sensitivity of cell hydraulic conductivity to mercury is coincident with symplasmic isolation and expression of plasmalemma aquaporin genes in growing maize roots. *Planta* **215**: 1047-1056.

Imlau, A., Truernit, E. and Sauer, N. (1999) Cell-to-cell and long-distance trafficking of the green fluorescent protein in the phloem and symplastic unloading of the protein into sink tissues. *Plant Cell* **11**: 309-322.

Kempers, R., Ammerlaan, A. and van Bel, A. J. E. (1998) Symplasmic constriction and ultrastructural features of the sieve element companion cell complex in the transport phloem of apoplasmically and symplasmically phloem-loading species. *Plant Physiol* **116**: 271-278.

Knoblauch, M., Peters, W. S., Ehlers, K. and van Bel, A. J. E. (2001) Reversible calcium-regulated stopcocks in legume sieve tubes. *The Plant Cell* **13**: 1221-1230.

Koch, K. E. (1997) Molecular crosstalk and regulation of C- and N-responsive genes. In *A Molecular Approach to Primary Metabolism in Higher Plants* (C. H. Foyer and W. P. Quick, eds.) pp. 105-124. Taylor & Francis, London.

Kühn, C., Franceschi, V. R., Schulz, A., Lemoine, R., and Frommer, W. B. (1997) Macromolecular trafficking indicated by localization and turnover of sucrose transporters in enucleate sieve elements. *Science* **275**, 1298-1300.

Laby, R., Kincaid, M., Kim, D. and Gibson, S. (2000). The *Arabidopsis* sugar-insensitive mutants *sis4* and *sis5* are defective in abscisic acid synthesis and response. *Plant J* **23**: 587-596.

Lalonde, S., Boles, E., Hellmann, H., Barker, L., Patrick, J. W., Frommer, W. B. and Ward, J. M. (1999) The dual function of sugar carriers: Transport and sugar sensing. *Plant Cell* **11**, 707-726.

Lang, A. (1979) A relay mechanism for phloem translocation. *Ann Bot* **44**: 141-145.

Lang, A. and Düring, H. (1991) Partitioning control by water potential gradient: Evidence for compartmentation breakdown in grape berries. *J Exp Bot* **42**: 1117-1122.

Li, C. Y., Weiss, D. and Goldschmidt, E. E. (2003) Girdling affects carbohydrate-related gene expression in leaves, bark and roots of alternate-bearing citrus trees. *Ann Bot* **92**: 137-143.

Livingston, D. P. and Henson, C. A. (1998) Apoplastic sugars, fructans, fructan exohydrolase, and invertase in winter oat. Responses to second-phase cold hardening. *Plant Physiol* **116**: 403-408.

Minchin, P. E. H., Cram, W. J. and Thorpe, M. R. (1994) Carbohydrate source-sink interactions. *Biologia Plantarum* **36**: S251.

Minchin, P. E. H., Lang, A. and Thorpe, M. R. (1983) Dynamics of cold induced inhibition of phloem transport. *J Exp Bot* **34**: 156-162.

Minchin, P. E. H. and McNaughton, G. S. (1987) Xylem transport of recently fixed carbon with lupin. *Aust J Plant Physiol* **14**: 325-329.

Minchin, P. E. H., Ryan, K. G. and Thorpe, M. R. (1984) Further evidence of apoplastic unloading in the stem of bean: Identification of the phloem buffering pool. *J Exp Bot* **35:** 1744-1753.

Minchin, P. E. H., and Thorpe, M. R. (1984) Apoplastic phloem unloading in the stem of bean. *J Exp Bot* **35:** 538-550.

Minchin, P. E. H. and Thorpe, M. R. (1987) Measurement of unloading and reloading of photo-assimilate within the stem of bean. *J Exp Bot* **38:** 211-220.

Minchin, P. E. H., Thorpe, M. R., and Farrar, J. (1993) A simple mechanistic model of phloem transport which explains sink priority. *J Exp Bot* **44:** 947-955.

Minchin, P. E. H., Thorpe, M. R., Farrar, J. and Koroleva, O. (2002) Source-sink coupling in young barley plants and control of phloem loading. *J Exp Bot* **53:** 1671-1676.

Minchin, P. E. H. and Troughton, J. H. (1980) Quantitative interpretation of phloem translocation data. *Annu Rev Plant Physiol* **31:** 191-215.

Oparka, K. J. and Prior, D. A. M. (1992) Direct evidence for pressure-generated closure of plasmodesmata. *Plant J* **2:** 741-750.

Oparka, K. J. and Turgeon, R. (1999) Sieve elements and companion cells—traffic control centers of the phloem. *Plant Cell* **11:** 739-750.

Parsons, R., Stanforth, A., Raven, J. A., and Sprent, J. (1993) Nodule growth and activity may be regulated by a feedback mechanism involving phloem nitrogen. *Plant Cell Environ* **16:** 125-136.

Patrick, J. W. (1990) Sieve element unloading: Cellular pathway, mechanism and control. *Physiol Plant* **78:** 298-308.

Patrick, J. W. (1989) Solute efflux from the host at plant-microorganism interfaces. *Aust J Plant Physiol* **16:** 53-67.

Patrick, J. W. (1997) Phloem unloading: Sieve element unloading and post-sieve element transport. *Annu Rev Plant Physiol Plant Mol Biol* **48:** 191-222.

Patrick, J. W., and Offler, C. E. (1996) Post-sieve element transport of photoassimilates in sink regions. *J Exp Bot* **47:** 1165-1177.

Plieth, C., Hansen, U. P., Knight, H. and Knight, M. R. (1999) Temperature sensing by plants: The primary characteristics of signal perception and calcium response. *Plant J* **18:** 491-297.

Roitsch, T., Balibrea, M. E., Hofmann, M., Proels, R. and Sinha, A. K. (2003) Extracellular invertase: Key metabolic enzyme and PR protein. *J Exp Bot* **54:** 513-524.

Sattelmacher, B. (2001) Tansley review no. 22. The apoplast and its significance for plant mineral nutrition. *New Phytologist* **149:** 167-192.

Sauter, J. (1982) Efflux and reabsorption of sugars in the xylem. I. Seasonal changes in sucrose efflux in *Salix. Z Pflanzenphysiol* **106:** 325-336.

Sauter, J. J., Iten, W. and Zimmermann, M. H. (1973) Studies on the release of sugar into the vessels of sugar maple (*Acer saccharum*). *Can J Bot* **51:** 1-8.

Smith, J. A. C. and Milburn, J. A. (1980) Phloem turgor and the regulation of sucrose loading in *Ricinus communis* L. *Planta* **148:** 42-48.

Stadler, R., Brandner, J., Schulz, A., Gahrtz, M. and Sauer, N. (1995) Phloem loading by the PmSUC2 sucrose carrier from *Plantago major* occurs into companion cells. *The Plant Cell* **7:** 1545-1554.

Thompson, M. V. and Holbrook, N. M. (2003) Application of a single-solute non-steady-state phloem model to the study of long-distance assimilate transport. *J. Theor. Biol.* **220:** 419-455.

Thornley, J. H. M. (1972) A balanced quantitative model for root:shoot ratios in vegetative plants. *Ann Bot* **36:** 431-441.

Thorpe, M. R., Lang, A. and Minchin, P. E. H. (1983) Short term interactions between flows of photosynthate. *J Exp Bot* **34:** 10-19.

van Bel, A. J. E. (1990) Xylem-phloem exchange via the rays: The undervalued route of transport. *J Exp Bot* **41:** 631-644.

van Bel, A. J. E. (2003a) The phloem, a miracle of ingenuity. *Plant Cell Environ* **26:** 125-149.

van Bel, A. J. E. (2003b) Transport phloem: Low profile, high impact. *Plant Physiol.* **131:** 1509-1510.

van Bel, A. J. E. and van Rijen, H. (1994) Microelectrode-recorded development of the symplastic autonomy of the sieve element/companion cell complex in the stem phloem of *Lupinus luteus* L. *Planta* **192:** 165-175.

van Dongen, J. T., Schurr, U., Pfister, M. and Geigenberger, P. (2003) Phloem metabolism and function have to cope with low internal oxygen. *Plant Physiol* **131:** 1529-1543.

Vaughn, M. W., Harrington, G. N. and Bush, D. R. (2002) Sucrose-mediated transcriptional regulation of sucrose symporter activity in the phloem. *Proc Natl Acad Sci U S A* **99:** 10876-10880.

Voegele, R. T., Struck, C., Hahn, M. and Mendgen, K. (2001) The role of haustoria in sugar supply during infection of broad bean by the rust fungus Uromycesfabae. *Proc Natl Acad Sci U S A* **98:** 8133-8138.

Welbaum, G. E., Meinzer, F. C., Grayson, R. L. and Thornham, K. T. (1992) Evidence for and consequences of a barrier to solute diffusion between the apoplast and vascular bundles in sugarcane stalk tissue. *Aust J Plant Physiol* **19:** 611-623.

Willenbrink, J. (1957) Über die Hemmung des Stufftransports in den Siebröhren durch lokale Inaktivierung verschiedener Atmungsenzyme. *Planta* **48:** 269-342.

Wolswinkel, P. (1974) Complete inhibition of setting and growth of fruits of *Vicia faba* L., resulting from the draining of the phloem system by *Cuscuta* species. *Acta Bot Neer* **23:** 48-60.

Wright, K. M. and Oparka, K. J. (1997) Metabolic inhibitors induce symplastic movement of solutes from the transport phloem of *Arabidopsis* roots. *J Exp Bot* **48:** 1807-1814.

Zhou L., Jang J. C., Jones T. and Sheen J. (1998) Glucose and ethylene signal transduction cross talk revealed by an *Arabidopsis* glucose-insensitive mutant. *Proc Natl Acad Sci U S A* **95:** 10294-10209.

Zwieniecki, M. A., Melcher, P. J. and Holbrook, N. M. (2001) Hydrogel control of xylem hydraulic resistance in plants. *Science* **291:** 1059-1062.

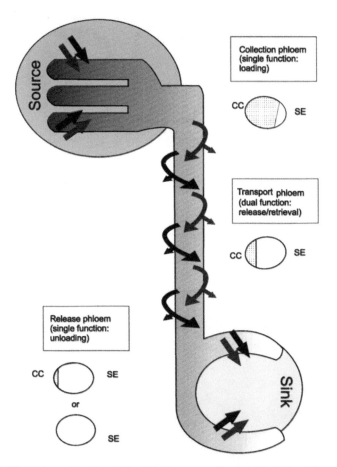

**Figure 2.2** A dynamic version of the Münch pressure flow model, the local fluxes of photoassimilates (violet arrows) and water (blue arrows) and the relative proportion of sieve elements (SE) and companion cells (CC) in the respective phloem zones. Photoassimilates are translocated via the phloem through essentially leaky instead of hermetically sealed pipes (sieve tubes). The solute content, and implicitly the turgor, are controlled by release/retrieval systems located on the plasma membrane of the sieve element/companion cell complexes (SE/CC-complexes). The retrieval mechanisms are energized by the proton-motive force. Differential release/retrieval balances along the pathway control the influx/efflux of sugars and water in the respective phloem zones. In collection phloem (where phloem loading takes place), the retrieval or uptake dominates, in the release phloem (where phloem unloading takes place) the release dominates. In transport phloem having a dual task (nourishment of axial sinks along the pathway and terminal sinks), the balance between release and retrieval varies with the requirements of the plant. The gradual loss of solute and the commensurate amounts of water has been ascribed to a slightly decreasing proton-motive gradient along the phloem pathway. Alternatively, the relative size reduction of companion cells along the source-to-sink path may explain a decreasing retrieval capacity of the SE/CC-complexes in the direction of the sink. The massive photo-assimilate delivery in the sinks is assigned to symplasmic and/or apoplasmic loss of photo-assimilates from the SEs driven by the high consumption and/or storage rates in the sink tissues.

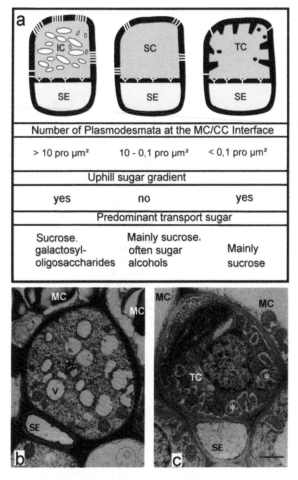

| Number of Plasmodesmata at the MC/CC Interface | | |
|---|---|---|
| > 10 pro μm² | 10 - 0,1 pro μm² | < 0,1 pro μm² |
| Uphill sugar gradient | | |
| yes | no | yes |
| Predominant transport sugar | | |
| Sucrose. galactosyl-oligosaccharides | Mainly sucrose, often sugar alcohols | Mainly sucrose |

**Figure 2.3** Relationship between ultrastructure of the companion cell (CC) in the minor veins and the mode of phloem loading in dicotyledons. Three types of companion cells are presented here. (A) - Intermediary cells (ICs) characterized by numerous cytoplasmic vesicles and many plasmodesmata at the wall interface between mesophyll and companion cells (MC/CC interface). Simple companion cells (SCs), without particular properties, have a moderate plasmodesmal density at the MC/CC-interface. Transfer cells (TC) possess cell wall invaginations to increase plasma membrane surface and have virtually no plasmodesmata at the MC/CC-interface. The structural differences between ICs (B) and TCs (C) are clearly visible in electron-micrographs (courtesy of S. Dimitrovska). V vesicles, * wall invaginations.

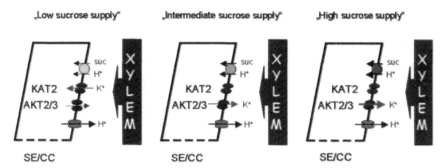

**Figure 2.5** Schematic drawing of the interaction between sucrose uptake and potassium channel gating in the plasma membrane of SE/CCs in the collection phloem of apoplasmically loading species (collective data of Ache *et al.* 2001, Deeken *et al.* 2002, Philippar *et al.* 2003). Sucrose loading is mediated by sucrose/H⁺-symport and energized by plasma membrane H⁺-ATPase. The voltage drop as result of sucrose symport is counteracted by a commensurate potassium efflux from the SE/CCs through the weakly voltage-dependent shaker-like AKT2/3 channel (Ache *et al.* 2001, Deeken *et al.* 2002). Thus, AKT2/3 stabilizes the membrane potential, an importance feature for maintenance of driving forces. As the sucrose supply increases (left to right) membrane depolarization will increase. Gating of the AKT2/3 channels therefore increases with the supply of sucrose. At low sucrose supply (left), voltage-dependent KAT2 channels open up due to the negative membrane potential. The putative role of KAT2 is the involvement in osmotic homeostasis of SEs. KAT2 channels are probably also engaged in "potassium recycling" from the xylem and act as a functional link between xylem and phloem.

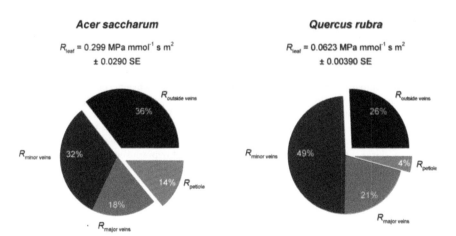

**Figure 5.2** The distribution of hydraulic resistance (1/hydraulic conductance) in the leaves of sugar maple (*Acer saccharum*) and red oak (*Quercus rubra*). From Sack *et al.*, (2004).

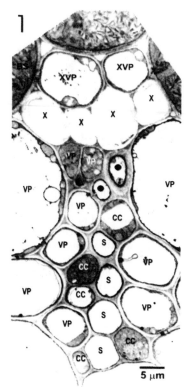

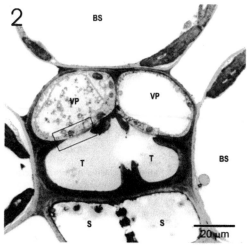

**Figure 6.2** Ultrastructure of a small vein in a wheat leaf, which contains two vascular parenchyma cells (VP) associated with two tracheary elements (T), below which are two sieve tubes (S). Note that no companion or vascular parenchyma cells are visible in this micrograph.

**Figure 6.1** Intermediate vascular bundle in *Eragrostis plana*, a C₄ grass, showing the phloem, which is composed of two thick-walled (*solid dots*) and four functional thin-walled sieve tubes. Thick-walled sieve tubes are close to xylem. The thin-walled sieve tubes (S) are associated with companion cells (CC) as well as vascular parenchyma (VP) cells. The thick-walled sieve tubes (*solid circles*) are associated with vascular parenchyma only. Note the very large lateral vascular parenchyma cells situated on either side of the central phloem core in this bundle.

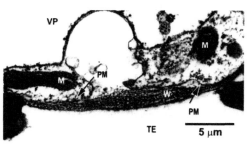

**Figure 6.3** Details of the highly fenestrated wall between vascular parenchyma cells and tracheary elements in wheat (boxed region in Fig. 2). The plasmamembrane (PM) is evident at the boundary between the pit and the cytoplasm of the parenchyma cell. Parenchyma cell contains several mitochondria (M). Note the highly porous wall in the pit region, which could allow the transfer of water and solutes from the vascular parenchyma cell to the tracheary element and vice versa.

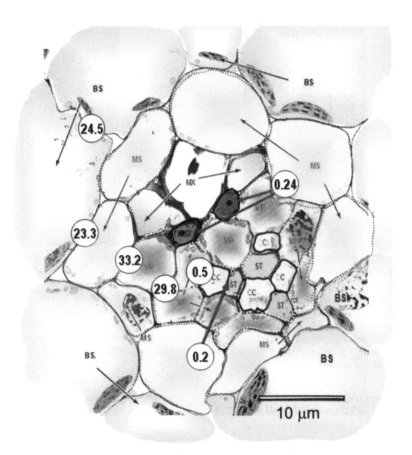

**Figure 6.4** Plasmodesmal frequency, calculated on a per fm common cell wall interface basis for a mature intermediate leaf vascular bundle from barley (*Hordeum vulgare* L. cv. Dyan). Note the close spatial association between the thick-walled sieve tubes (solid dots) and the metaxylem vessels (MX). Thick-walled sieve tubes are separated from the earlier thin-walled metaphloem, which comprises parenchymatic elements (VP), sieve tubes (ST) and associated companion cells (CC). MS = mestome sheath cells; BS = bundle sheath cells.

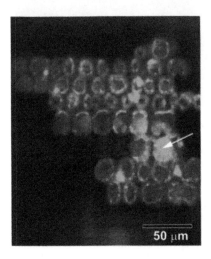

**Figure 6.5** Surface view of a barley leaf 10 minutes after gentle scraping and application of 5,6-CF diacetate to the abraded portion of the leaf (*arrow*). Note that the fluorescence has spread to adjacent cells, and that those further away from the point of application show 5,6-carboxyfluorescine within the cytosol and less evidently associated with the vacuole.

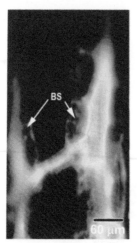

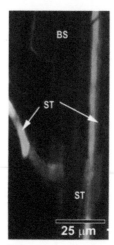

**Figure 6.6** Epifluorescence micrograph showing uptake of 5,6-CF and symplasmic transport into mesophyll cells in *Hordeum vulgare* leaf 10 minutes after application to a lightly-abraded region of the leaf, unlabelled arrow points to abraded cell to which 5,6-carboxyfluorescein diacetate was applied. Note that 5,6-CF has spread to a large number of cells during this short period, but that the bundle sheath cells (BS) are essentially devoid of 5,6-CF. Thin strands of fluorescence are cytoplasm-related; no evidence of vacuolar uptake is evident in this or other similarly-treated tissue.

**Figure 6.7** Confocal image showing localisation of 5,6-CF within the single sieve tube in a cross-vein (*left*) and a sieve tube in a small intermediate longitudinal leaf blade vascular bundle (*right*) in wheat.

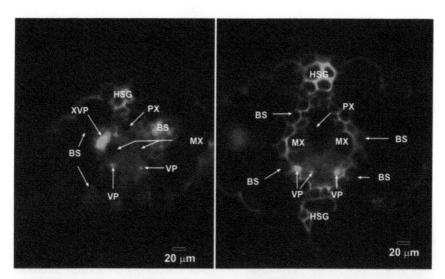

**Figure 6.8 and 6.9** Distribution of 5,6-CF in small and intermediate respectively, vascular bundles in *Zea mays* leaf blade tissue, showing recovery of 5,6-CF from the xylem after 120 (Fig. 8, *left*) and 180 (Fig. 9, *right*) minutes uptake of 5,6-CFDA from the cut end of a severed source leaf of *Zea mays*. Sections cut in silicone oil and viewed with a narrow-band filter set specific for 5,6-CF without interference from autofluorescence associated with lignin. Fluorescence is intense in the parenchyma located on the xylem side of the bundle (XVP), as well as in bundle sheath (BS) and vascular parenchyma cells, but is absent from the xylem, suggesting that dissociation of 5,6-CFDA occurred within the xylem parenchyma. One vascular parenchyma and one thin-walled sieve element contain intense label. After 180 minutes (Fig.9, *right*) the fluorescence intensity is greatest in the phloem, with vascular parenchyma and thin-walled sieve elements showing intense label. Hypodermal sclerenchyma girder cells (HSG) contain label as well.

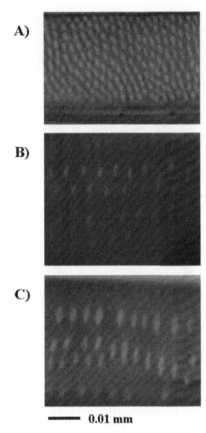

**0.01 mm**

**Figure 17.4** Xylem pit distributions in (A) *Betula papyrifera*, (B) *Acer saccharum*, and (C) *Quercus rubra*. Note the most dense pitting in *B. papyrifera* and least dense pitting in *Q. rubra*, likely contributing to greater integration in *B. papyrifera* and greater sectoriality in *Q. rubra*.

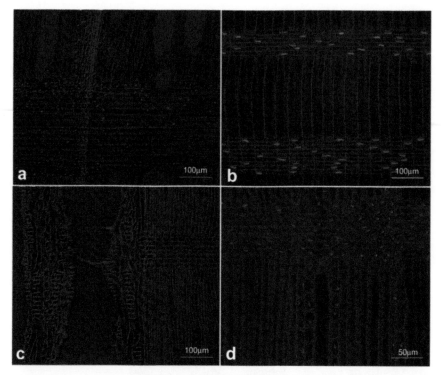

**Figure 22.1** Living ray and axial parenchyma shown by fluorescent staining (DAPI) of nuclei in radial sections of secondary xylem. (A) Ray and marginal (axial) parenchyma in *Liriodendron tulipifera*. Marginal parenchyma refers to axial parenchyma formed at the growth ring boundary and is characteristic of this species. (B) Two rays in the outermost sapwood of *Tsuga canadensis*. The absence of axial parenchyma is typical of many conifers. (C) Ray parenchyma and a network of axial parenchyma and tracheids surrounding an earlywood vessel in *Quercus rubra*. (d) Ray and axial parenchyma surrounding a latewood vessel in *Fraxinus americana*. Marginal (axial) parenchyma is also shown at the growth ring boundary on the far right.

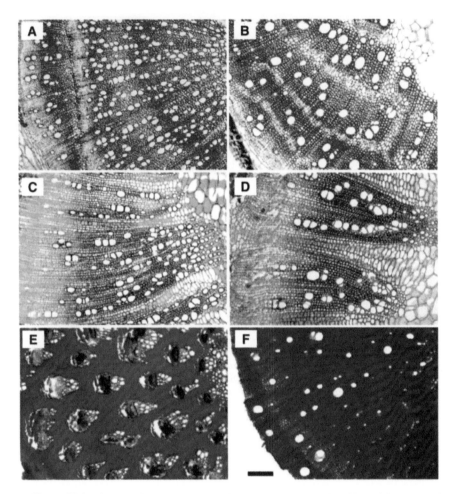

**Figure 25.3** Cross sections of stem xylem from representative pairs of $C_3$ and $C_4$ species of three functional groups. Left panels are $C_4$ and right panels are $C_3$. (A) *Kochia scoparia*; (B) *Chenopodium album*; (C) *Flaveria bidentis*; (D) *Flaveria pringle*; (E) *Atriplex canescens*; (F) *Prosopis glandulosa*. The scale bar in F is 200 μm and applies to all panels.

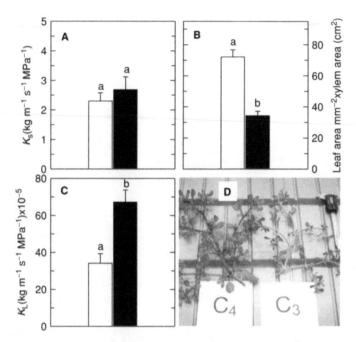

**Figure 25.4**   (A) specific conductivity, $K_s$, (B) leaf area per unit xylem area, and (C) leaf specific conductivity of *Trianthema portulacastrum* ($C_4$) (D, *left*) and *Sesuvium verrucosum* ($C_3$) (D, *right*). Bars are mean (±SE) of 6 plants. Different letters represent significant difference ($P<0.05$). (From data presented in Kocacinar and Sage, 2003.)

# 11

# The Role of Potassium in Long Distance Transport in Plants

*Matthew V. Thompson and Maciej A. Zwieniecki*

Broadly speaking, plant function can be divided into two tasks, the production of energetic and biosynthetic substrates with energy from the sun, and the acquisition of water and mineral nutrients from the soil. The far smaller ancestors of modern land plants could spatially combine these tasks, because diffusion could efficiently distribute resources over the short distances involved. However, the demands of colonizing heterogeneous spatial environments in a dry atmosphere, both in terms of light capture and seed and spore dispersal, led to the competitive selection of complex and highly ramified plant architectures that brought the more "autotrophic" functions of the plant away from the soil (Niklas, 1997). Xylem and phloem—the tissues responsible for the transport of soil water and nutrients to sites of atmospheric gas exchange, and for the translocation of photoassimilates from leaves to other parts of the plant, respectively—evolved to allow this massive increase in plant size by permitting the bulk-flow translocation of resources over great distances.

In all tracheophytes, the flow of both xylem and phloem sap is driven by gradients in hydrostatic pressure generated at the transport endpoints. However, this long-distance transport is founded upon a vast underlying layer of local material exchange between xylem, phloem, and the surrounding tissues (Höll, 1975; Pate, 1975; van Bel, 1990) operating at length scales of no more than a few millimeters. As such, the control of transport in tracheophytes involves a radical juxtaposition of large- and small-scale processes, where the "decision making" done locally by the units that compose the transport pathway is relatively autonomous, but their combined activity is beneficial to the plant as a whole. This strong role for local control of xylem and phloem transport means that the "conduit" portion of the vasculature cannot be viewed as a silent partner subservient to the needs of the transport endpoints.

In this chapter, we discuss the physiology of the xylem and phloem and the interactions between them in relation to their use of potassium to internally regulate transport function in heterogeneous environments. We do not pretend to offer an extended review of potassium transport in plants, but rather an opinion, based on results from our work, of how the modification of sap potassium concentration could act to fine-tune xylem and phloem transport properties at the "decision-making" interface between large- and small-scale physiological processes and anatomical structures.

## Potassium and the Xylem

### Hydrogel Activity in the Xylem

Xylem hydraulic resistance is a function of conduit length and diameter, and of the nature of the connections between adjacent elements. Among angiosperms and some gymnosperms, water passes through vessel junctions composed of a compact cellulose mesh, reinforced by lignin and filled with pectins, called bordered pit membranes (Tyree and Zimmermann, 2002). Pectins are known for their hydrogel-like properties, which allow them to swell/shrink in response to changes in the chemical state or constituents of the surrounding solvent, such as changes in temperature, pH, or solution ionic strength. Pectins can play an important role in governing cell wall hydraulic properties by influencing the size of the very small channels that traverse the pit membranes, and thus the flow of water between vessels after changes in sap potassium concentration (Zwieniecki *et al.*, 2003).

Our physical understanding of hydrogel swelling/shrinking is relatively new (Tanaka, 1981). The volume of a hydrogel depends proximally on the gel's osmotic pressure, itself a function of three gel properties: the elasticity of the pectin strands, polymer-polymer affinity between strands, and the pressure of hydrogen ions in the gel (Tanaka, 1981). Normally, the negatively charged domains in the pectin polymers are neutralized by the positive charges of monovalent cations in solution. In this state, water is expelled and the polymer chains can contract. If deionized water is introduced to the gel, however, these monovalent cations are diluted, leaving hydrogen ions from the water to neutralize the polymer chains' fixed negative charges. Because the dissociability of water is low, these protons will pull a proportional volume of water into the interior of the gel with them, giving rise to a temperature-dependent hydrostatic pressure (Tanaka, 1981) that pushes the pectin strands apart and causes the gel to swell (Annaka and Tanaka, 1992; Shibayama and Tanaka, 1993). Dissolved salts, with high dissociability, shield the negative charges in the polymer network and reduce hydrogen ion pressure. This causes water to flow from the network and the gel shrinks.

Changes in pectin volume caused by ion-gel interactions are expected to be large in the range of cation concentrations (< 20 mM) found in xylem sap (Marschner, 1995; Schurr, 1998). The theoretical limit for gel volume change over this range can be more than an order of magnitude (Tibbits *et al.*, 1998), but the expected volume change will be less for a number of reasons. The presence of cellulose microfibrils restricts the extent to which shrinkage of embedded pectins can increase the size of the pit membrane's microchannels. There is also a limit on gel expansion (and thus microchannel shrinkage) because the xylem sap is never completely deionized. Thus, the negatively charged carboxy groups on the pectin strands may never be fully disassociated. Furthermore, they could be cross-linked with divalent ions such as calcium or boron, thus reducing the number of fixed negative charges (Oosterveld *et al.*, 2000; Ryden *et al.*, 2000). The complications arising from divalent cation cross-linkage, however, do not limit the conductivity response expected after increased concentrations of monovalent cations. A study of the interactive effect of calcium and potassium on water flow through the xylem of *Laurus nobilis* L. suggests that the effects of divalent and monovalent cations are additive (Zwieniecki, unpublished data).

A direct determination of pit membrane pore size and pectin volume change around the pores has not been performed. The opacity of pit membranes in scanning electron microscopy and transmission electron microscopy micrographs (i.e., no visible pores) suggests either that the pores do not follow a straight path across the membrane, or that the fixation required for electron microscopy places the membranes in a "dry" state in which all the pores have contracted. Measurements of the pressure required to force air or microparticles through the membrane, however, indicate that pore diameter in most plants is between 20 and 50 nm (Choat *et al.*, 2003).

Pectin swelling and shrinking appear to affect stem hydraulic conductivity. A survey of conductivity enhancement across a variety of plant species found that stems perfused with a 20-mM solution of potassium salt were anywhere from 0% to 100% more conductive than stems perfused with deionized water (Fig. 11.1). Gymnosperm species and lycopods, which have highly porous margos instead of pectin-hydrogel pit membranes, did not respond to the presence of potassium ions. Ferns were highly responsive, suggesting that hydrogels could play a significant role in their water transport system. In angiosperms, the response was variable, and, in general, was inversely correlated with stem resistance (i.e., with the number of vessel endwalls in the studied segments). The stems of species with long, wide vessels, such as vines and ring porous trees, were less affected by the presence of ions than those with short and narrow vessels (data not shown).

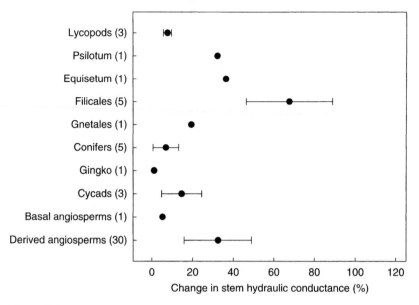

**Figure 11.1**  Conductivity enhancement in vascular plants, as measured by changes in flow rate after perfusion with a 20 mM KCl solution relative to perfusion with deionized water. Enhancement was determined from 10-cm long stem segments collected in a northeastern North American forest (Harvard Forest, MA) or from greenhouse-grown specimens. Changes in conductivity reflect the potential for stem xylem to increase the water potential of the water delivered to leaves at a given flow rate, as well as the potential for the chemical constituents of the sap to govern water redistribution within the plant among branches and vascular sectors. (Circles are the means from each taxon ± SD; numbers in parentheses indicate number of measured species in each taxonomical grouping.)

## Physiological Role of Hydrogel Activity in the Xylem

In the xylem, as shown by Zwieniecki *et al.* (2001), potassium ions can modify the hydration status of bordered pit membranes in many species, significantly decreasing xylem resistance with the addition of only 10 mM potassium chloride to pure water. Because xylem sap flow rates are predominantly controlled by stomatal conductance, ion-induced changes in xylem conductivity cannot directly control sap flow. Instead, those changes in conductivity could change the energetic status of the water on arrival at the leaf. By providing the leaf with water of different water potentials at the same flow rate, the xylem could act to extend or limit the maximum transpiration rate at which stomata could remain open before a drop in water potential led to wilting or a reduction in physiological performance. In other words, an increase in the potassium concentration of xylem sap could extend the capacity of stomata to remain open at a given transpiration rate.

The allocation of water to different parts of the plant could also be affected by the presence of sap cations (see Chapter 10). A recent study of lateral exchange between xylem sectors in tomato showed that ions affect lateral exchange more than longitudinal transport (Zwieniecki *et al.*, 2003). This comes as no surprise, as lateral connections are mostly via bordered pit membranes, the part of xylem conduits most subject to ion-mediated changes in conductivity. The ion-induced enhancement of sap exchange is most pronounced if combined with a large difference in hydrostatic pressure between the adjacent sectors. These results suggest that ion-mediated exchange between independent xylem sectors in branches or stems could be an important process in allowing adjacent leaves or branches to distribute resources based on locally available information, such as ion concentration and water availability. This could be especially important under patchy environmental conditions, where the capacity to acquire resources is not equally distributed among different—even adjacent—roots and leaves.

Xylem sap ionic concentration and composition, two parameters that influence xylem hydraulic conductance, are not constant. They change yearly, monthly, daily, and even over much shorter time periods (Herdel *et al.*, 2001; Peuke *et al.*, 2001; Schurr, 1998). Some of this short time scale variation could be due to the methodological difficulty of collecting xylem sap under pressure, and comprehensive measurements of spatial gradients in ion concentration have yet to be performed. Such information is necessary to assess the potential for ion-controlled redistribution of sap flow to different plant parts.

## Potassium and the Phloem

### "Rules" for Long Distance Transport

Before we can discuss the role of potassium in the phloem, we must first treat the nature of transport along the phloem. At issue are how potassium ions are used along the transport pathway and how their use meets particular global and local physiological needs. Most recent molecular and whole-plant physiological phloem investigations have studied only the source and sink regions, or collection and release phloem, respectively, and although these studies tell us much about the nature of phloem membrane solute flux, they have not been completely integrated into models of long distance translocation. Indeed, there is the common implicit assumption in modeling studies (Goeschl *et al.*, 1976; Thompson and Holbrook, 2003a; Tyree *et al.*, 1974) that the transport phloem is inert or neutral with respect to the control of translocation. However, the involvement of the entire transport pathway in solute and water exchange has been well established

(Lalonde *et al.*, 2003), and although the sieve element/companion cell (SE/CC) complexes (van Bel, 1996) that serially compose the phloem can respond only to locally available signals, they must form a network capable of organizing globally scaled phenomena. As such, it is critical that we identify the cues and signals that govern changes in membrane solute flux, and how those cues are generated at the whole-plant scale.

Direct analysis of the equations used to model axial phloem transport can help in this regard. Theoretical work on the transport phloem (Thompson and Holbrook, 2003b, 2004), using dimensional analysis of the relevant transport scales and parameters, has distilled the functional parameterization of the phloem to just two dimensionless groupings, $\hat{R}$ and $\hat{F}$:

$$\hat{R} = \frac{\text{axial resistance}}{\text{membrane resistance}} = \frac{\mu L}{rk} \Big/ \frac{1}{2LL_p} \tag{1}$$

and

$$\hat{F} = \frac{\text{osmotic pressure}}{\text{axial pressure differential}} = \frac{\Psi_\pi}{\Delta p} = \Psi_\pi \Big/ \frac{\mu LU}{k}, \tag{2}$$

where $\mu$ (MPa s) is viscosity, $L$ (m) is sieve tube length, $r$ (m) is sieve tube radius, $k$ (m$^2$) is axial specific conductivity and is calculated using sieve element and sieve plate geometry following Thompson and Holbrook (2003b), $L_p$ (ms$^{-1}$ MPa$^{-1}$) is membrane permeability to water, $\Psi_\pi$ (MPa) is the osmotic pressure observed in the system at the downstream end of the sieve tube, $\Delta p$ (MPa) is the axial pressure differential or pressure drop between opposing ends of the sieve tube, and $U$ (m s$^{-1}$) is the axial transport flux density at the downstream end of the sieve tube. $\hat{R}$ and $\hat{F}$ govern the behavior of the following system of dimensionless nonlinear partial differential conservation equations:

$$0 = \hat{R}\left[\hat{\psi}_o - \hat{p} + \hat{F}\hat{c}_s\right] + \frac{\partial^2}{\partial \hat{z}^2}\hat{p}, \text{ and}$$
$$\frac{\partial}{\partial \hat{t}}\hat{c}_s = \frac{\partial}{\partial \hat{z}}\left(\hat{c}_s \frac{\partial}{\partial \hat{z}}\hat{p}\right), \tag{3}$$

where $\hat{z}$ is the dimensionless distance, $\hat{t}$ the dimensionless time, $\hat{\psi}_o$ the dimensionless apoplastic water potential, $\hat{p}$ the dimensionless hydrostatic pressure, and $\hat{c}_s$ the dimensionless solute concentration (carets indicate that the variable or grouping is dimensionless). The former conservation equation governs the balance between the divergence of axial sap flow to and from a given point along the pathway, and the rate of membrane water flux at that point. The latter provides a conservation statement for solute concentration (a single solute, in this case, but other solutes can be trivially added by the addition of further solute conservation equations and more concentration terms in the flow divergence equation). These equations provide a general description of the development of gradients in pressure and concentration as a function of time and distance, with $\hat{\psi}_o$ as a forcing function.

The dimensionless form of the equations permits us to describe phloem behavior as a function of change in $\hat{R}$ and $\hat{F}$ alone, without the need to labor through all the possible permutations of the parameters grouped in equations 1 and 2. Any two sieve tubes with the same values of $\hat{R}$ and $\hat{F}$ would be expected to behave in a qualitatively similar manner. Using published data describing the phloem anatomy and transport characteristics of a limited set of species, Thompson and Holbrook (2003b) showed that the product of $\hat{R}$ and $\hat{F}$ is likely to be greater than unity. From this, they demonstrated that the membrane water flux/flow divergence relation in equation 3 reduces to the following dimensionless statement of "water potential equilibrium" between the sieve sap and the apoplast:

$$\hat{\psi}_o = \hat{p} - \hat{F}\hat{c}_s. \tag{4}$$

This reduction implies that it is generally safe to assume that water potential equilibrium exists between the apoplast and phloem symplast—a question of long debate (Goeschl *et al.*, 1976; Lang, 1978)—even in the face of a flow-induced drop in hydrostatic pressure. Furthermore, $\hat{R}$ drops out of the equation, and equation 4 can be substituted into the solute conservation relation in equation 3 to give:

$$\frac{\partial}{\partial \hat{t}} \hat{c}_s = \frac{\partial}{\partial \hat{z}} \left( \hat{c}_s \frac{\partial}{\partial \hat{z}} [\hat{\psi}_o + \hat{F}\hat{c}_s] \right). \tag{5}$$

Equation 5, with only a few caveats and provided that water potential equilibrium is satisfied, represents the whole of expected physicochemical phloem behavior with respect to the axial transport of sap. It is a nonlinear partial differential equation in space and time, but with only one dimensionless grouping ($\hat{F}$), one unknown (concentration), and one forcing function, and is thus considerably easier to analyze. More important, the behavior of equation 5 need only be analyzed with respect to changes in $\hat{F}$ and $\hat{\psi}_o$.

Thompson and Holbrook (2003b) showed that in most plant species and situations for which data are available (primarily crop plants) $\hat{F}$ is expected to be significantly greater than unity. This observation, combined with inspection of equation 2, implies that most sieve tubes operate with a small axial pressure differential, or pressure drop, relative to the osmotic pressure of the sap. Furthermore, analysis of equation 5 (Thompson and Holbrook, 2004) suggests that when $\hat{F}$ is greater than unity, waves of pressure and concentration will be quickly transmitted over the length of the sieve tube, many times faster than the material sap flux rate. A large value of $\hat{F}$, then, permits three important conclusions, or "rules," about the possible nature of phloem transport:

- The drop in pressure between source and sink tissues has been overemphasized. Although it is still true that phloem transport is a pressure-driven process, the pressure drop between the sieve tube's ends is likely to be negligible relative to the capacity of different parts of the sieve tube to resolve small differentials in sap hydrostatic pressure. Therefore, it is unlikely that the plant regulates phloem transport rates by manipulating long-distance pressure drops. Rather, small pressure drops follow inevitably from the viscous axial dissipation of the energy generated by active membrane solute flux. High $\hat{F}$ guarantees a situation similar to that described by the volume-flow hypothesis of Eschrich *et al.* (1972) and Young *et al.* (1973).
- Changes in sieve tube pressure or concentration in one part of the sieve tube will be rapidly transmitted elsewhere in the form of pressure-concentration waves. This rapid transmission of "information" about sieve tube state effectively "folds" the long distances involved in phloem transport by turning changes observed by one part of the transport pathway into changes observed by all.
- Furthermore, because the pressure drop between the sieve tube's ends depends solely on axial sap flux density, sap viscosity, sieve-tube specific conductivity, and the length of the sieve tube, it will be insensitive to axial gradients in apoplastic water potential, a variable intimately tied to xylem performance (Thompson and Holbrook, 2003b). However, to maintain water potential equilibrium along the sieve tube in the presence of a small pressure gradient, apoplastic water potential gradients must be matched by an equally large gradient in sieve sap osmotic potential.

These observations make pressure an excellent signal for regulating sieve tube function: Provided $\hat{F}$ scales at a fairly high value, pressure will vary little throughout the sieve tube; if pressure changes in one part of the sieve tube, that change will be quickly transmitted elsewhere; and the pressure gradient, unlike the osmotic potential gradient, will be insensitive to gradients in apoplastic water potential.

Making local sieve sap hydrostatic pressure the primary control variable for membrane solute exchange fits well with the anatomical organization of the phloem. SE/CC complexes, due to their relatively short length, are individually incapable of measuring axial pressure gradients, as well as incapable of locally altering long distance axial flow rates short of changing local specific conductivity via reversible "stopcocks" (Knoblauch *et al.*, 2001) or the occlusion of sieve plates by P-proteins. However, SE/CC complexes can respond to local changes in pressure. If pressure is everywhere nearly the same, then each SE/CC complex can respond in tandem to the same overall changes in sieve tube state. The tissues in which individual SE/CC complexes find themselves will dictate how they behave in response to those changes. In sites of active photosynthesis, photoassimilates are abundant and will be loaded into the sieve tube according to both the availability of these solutes and drops in the local hydrostatic pressure of the sieve tube. In active sink tissues, increased withdrawal of these solutes from the sieve sap will lower hydrostatic pressure, and signal SE/CC complexes in source regions to load more. SE/CC complexes located along the transport pathway will behave similarly, releasing solutes and reloading them in accordance with demands of the surrounding tissue. If more released solutes are consumed by the surrounding tissue than can be retrieved by the SE/CC complex, then local SE/CC complex sap pressure will drop, and solutes will be loaded from elsewhere to make up the difference (see Chapter 10 for more on the solute buffering potential of the stem apoplast).

The picture that emerges is a translocation system capable of globally integrating SE/CC complex function via changes in hydrostatic pressure, one in which distant demand can signal source tissues with information as to exactly how much photoassimilate is needed, and in which no SE/CC need know exactly where it is located within the plant to properly function. We call this the osmoregulatory flow hypothesis, and it is predicated on the sieve tube having a high value of $\hat{F}$. Under the osmoregulatory flow hypothesis, the collection, transport, and release phloem of van Bel (1996) would be functionally identical, as each can partake, to some extent, in both loading and unloading (Fig. 11.2). Ultimately, the difference between these sectors is anatomical (companion cells are disproportionately large in leaf minor veins and often absent in roots and other sink tissues, van Bel, 2003), which we may take to be a reflection of their different functions with respect to the surrounding tissue. In an ontogenetic and physiological sense, then, anatomical differences between SE/CC complexes are fundamentally context dependent.

Although the pressure drop should remain small at high $\hat{F}$, any gradient in apoplastic water potential must be matched by a gradient in osmotic potential that satisfies water potential equilibrium at each point along the transport pathway. Our current empirical understanding of the phloem is consistent with these expectations. Large pressure gradients have never been conclusively measured, either directly (Hammel, 1968) or by calculating sieve tube pressure as the difference between total and osmotic potential (Fisher, 1978). The only conclusively large gradients to be measured in the phloem are source-to-sink concentration gradients (*cf.* Hocking, 1980), but a gradient in concentration is not the same as a gradient in pressure. A concentration gradient is an additive reflection of both pressure and

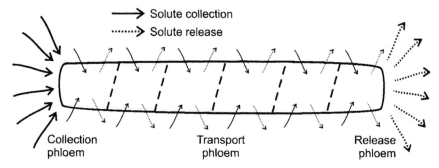

**Figure 11.2**  Osmoregulatory flow hypothesis. The sieve tube is divided into collection, transport, and release phloem, which are distinguished by differences in SE/CC complex anatomy, which is in turn a reflection of their different functional contexts (van Bel, 1996). Solute loading occurs predominantly in the collection phloem, whereas unloading occurs predominantly in the release phloem. Along the transport phloem, solute is both leaked and recovered at a fairly high rate (see especially Minchin and Thorpe, 1987). At a high value of $\hat{F}$ (equation 2), each SE/CC complex in the sieve tube is guaranteed to operate at nearly the same pressure (Thompson and Holbrook, 2003b), with rapid transmission of pressure-concentration waves over the length of the sieve tube (Thompson and Holbrook, 2004). This allows the SE/CC complexes to sense and respond to the same global stimuli. Thus, an increase in solute unloading in a sink organ will lead to a fast drop in pressure throughout the system that can be recovered at all points by an increase in local SE/CC complex solute loading. The majority of this "recovery" would be borne by the source organs, but it could be accommodated, in no small part, by the "transport" phloem if supply from the source organs is limited.

apoplastic water potential gradients, and it arises as a critical, if automatic, part of maintaining water potential equilibrium. In many instances, it could be partially met by controlled loading and unloading of potassium.

## The Role of Potassium

The osmotic potential of potassium and its associated anions constitutes as much as one third to three fourths of the total osmotic potential of phloem sap, and may sometimes exceed that amount (Hoad and Peel, 1965; Lang, 1983; Mengel and Haeder, 1977; Smith and Milburn, 1980a, 1980b, 1980c). In some cases, rates of transport are known to be better correlated with axial gradients in potassium than with gradients in sucrose (Lang, 1983). As noted by Lang (1983), there is "strong reciprocity" between the rates of sucrose and potassium loading. When sugar transport into the phloem sap of *Ricinus communis* is artificially reduced, potassium loading increases (Smith and Milburn, 1980a, 1980b, 1980c); or in willow, when potassium levels are increased, sieve tube sucrose concentration drops (Hoad and Peel, 1965). Even very mild and temporary potassium deficiencies in field grown plants can lead to significant starch accumulation in leaves (Amir and Reinhold, 1971; Ashley and Goodson, 1972), suggesting a reduction in phloem transport capacity (Hartt, 1969; Mengel and Viro, 1974). Potassium also appears to play a critical role in clamping membrane potential (Deeken *et al.*, 2002). It is thus worthwhile to ask how that cation could contribute to turgor regulation and what effect it would have on phloem transport over long distances.

The simplest phloem transport scenario is one in which the apoplastic water potential gradient is negligibly small (Fig. 11.3a). We assume that $\hat{F}$ is greater than unity throughout this discussion (Thompson and Holbrook, 2004), unless otherwise noted. Here, the pressure gradient is accompanied by an equally small matching gradient in osmotic potential (or concentration). Sucrose influx will be greatest near the source and efflux greatest near the sink, but both fluxes will be somewhat distributed throughout the sieve tube.

In another scenario, higher xylem sap flow rates result in a larger gradient in apoplastic water potential (Fig. 11.3b). Steady-state levels of sucrose import and export should remain unchanged from the previous state (i.e., Fig. 11.3a), as these processes are presumably tied to the solute supply and demand of the source and sink tissues, rather than to the solute concentration of the sieve sap. The steady-state gradient in sap pressure should also be largely unaffected. However, the steady-state gradient in sucrose concentration will automatically increase to maintain water potential equilibrium. This will lead to an increase in viscosity and could cause a disproportionately large local increase in the hydrostatic pressure gradient near the source. If $\hat{F}$ is near unity, which could be the case in large plants where transport distances ($L$) are great, this increase in pressure could make it impossible for SE/CC complexes to "measure" pressure locally as if it were a global variable.

The problem of viscosity limitations to phloem transport has been tangentially considered in computer modeling studies (Thompson and Holbrook, 2003a; Tyree *et al.*, 1974), but without adequate attention to how an increase in viscosity could affect the relative magnitude of the pressure gradient and its effect on turgor regulation. If $\hat{F}$ is large, say ~60, as in the case of *R. communis* (Thompson and Holbrook, 2003b), the pressure drop ($\Delta p$) accounts for only a small fraction of the mean osmotic pressure of the sieve sap ($\Psi_\pi$)—that is, a pressure drop of less than 0.02 MPa in a sieve tube with a sap osmotic potential of 1.0 MPa. Even if viscosity were to double, the pressure drop would increase to little more than 0.03 MPa, all else being equal. As such, the range of pressures found in a sieve tube of high $\hat{F}$ is unlikely to be resolvable by different SE/CC complexes and pressure remains a good candidate for whole-sieve tube osmoregulation. However, if $\hat{F} = 5$ at the same osmotic potential, as would be expected for some crop plants and many trees (Thompson and Holbrook, 2003b), a doubling of viscosity would cause an increase in the pressure drop from 0.2 MPa (a marginally high value) to 0.4 MPa. At this point, it is arguable that pressure would no longer be a reliable global control variable for sieve tube function. Thus, because of the asymptotic relationship between

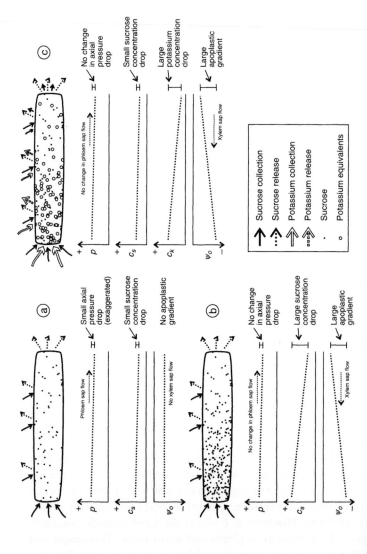

**Figure 11.3** Sucrose and potassium transport under the osmoregulatory flow hypothesis. (A) In a sucrose-only system, and in the absence of a strong apoplastic water potential gradient ($\psi_o$), and at a high value of $\hat{F}$ (that is, a small or negligibly small axial pressure drop relative to the osmotic pressure of the sap), gradients in pressure ($p$) and sucrose concentration ($c_s$, closed circles) will be very small. (B) In the presence of a large gradient in ($\psi_o$), the axial pressure drop will stay small, but the concentration drop will increase considerably to match the new profile in $\psi_o$. Large concentrations of sucrose could lead to high viscosity and elevated pressures if $\hat{F}$ is already near unity. (C) Potassium ($c_k$, open circles) could be used to alleviate the problems associated with high viscosity by its introduction where $\psi_o$ is more negative and it's progressive removal as $\psi_o$ rises. The axial pressure drop, as before, would remain the same, since $\Delta p$ is relatively small when $\hat{F}$ is high.

232

sieve tube behavior and $\hat{F}$, the higher the value of $\hat{F}$, the more the sieve tube's capacity to locally osmoregulate is buffered against changes in variable parameters such as viscosity and sap flow rate. But when apoplastic water potentials are low and concentration and viscosity are high, this buffer may no longer be sufficient.

This problem could be rescued, in our final scenario, by potassium salts (Fig. 11.3c). These have a 10-fold lower aqueous viscosity than sucrose solutions of comparable osmotic potential (Thompson, unpublished data), permitting them a large osmotic benefit relative to their viscous burden. Once loaded, these salts could be progressively unloaded from the sieve tube downstream of the source as demand for them decreased, leaving the flow of sucrose, the pressure gradient, and the sucrose concentration gradient unchanged from our first scenario (Fig. 11.3a), while at the same time maintaining water potential equilibrium and recirculating potassium to lower branch orders. Potassium could be used in sites of active loading where apoplastic water potentials are low and released as apoplastic water potential begins to rise, making potassium into a kind of osmotic "lubricant" for phloem transport, and keeping the pressure drop small.

This hypothetical system (Fig. 11.3) bears some resemblance to the turgor-regulated translocation hypothesis of Lang (1983), especially in its reliance on potassium as a major regulator of sieve tube turgor. However, Lang (1983) proposed that potassium is used to regulate the phloem's axial *pressure gradient*. We would argue as axiomatic that the pressure gradient cannot be controlled explicitly by the plant. No individual unit of the phloem transport pathway (i.e., an SE/CC complex) could possibly measure such a gradient, nor is there any known parallel system in plants that would be capable of meeting such a task. If each SE/CC complex in the phloem regulates local turgor, but the pressure drop changes, then different parts of the sieve tube would have to respond differently to that change in accordance with their position along the pressure gradient. This is especially problematic if the range of pressures found in the sieve tube is already high (i.e., at low $\hat{F}$). Furthermore, it is questionable whether large gradients in pressure exist in any phloem system (Thompson and Holbrook, 2003b), although this assertion has not been adequately tested in larger plants, which, all things being equal, would possess a smaller value of $\hat{F}$ owing to the greater distances involved.

## Vascular Anatomy, Xylem-Phloem Integration, and Potassium

### Xylem-Phloem Integration and Potassium

Here, we focus on potassium as a way to link xylem and phloem function. The synchronous use and recirculation of potassium in such developmentally and anatomically related tissues (Komor *et al.*, 1996; MacRobbie, 1971; Pate, 1975; Patrick *et al.*, 2001; Ziegler, 1975) strongly suggest coordinated activity. Because the accumulation of potassium in plants is largely in stoichiometric proportion to biomass accumulation, and because growth is slow, any change in the export of potassium from leaves via the phloem is likely to be reflected by changes in potassium import via the xylem. Some have asserted that potassium recirculation in the phloem back to the roots may act as a kind of nonallosteric signal, alerting the roots to the fact that either too much or too little potassium has arrived at the leaves (Drew and Saker, 1984). Others have argued that recirculation in the xylem back to the leaves informs the leaves of how much potassium is actually required by phloem sinks, and how much needs to be collected in the future (Komor *et al.*, 1996). Although it is likely that the transfer of potassium between the xylem and phloem offers this kind of signaling between distant parts of the plant, the fact that *so much* potassium is recirculated at an energetic cost suggests that there could be a much deeper and intrinsic role for this solute in the transport process itself, and that it is used not just at the endpoints but along the transport pathway, as well.

In the xylem, sap potassium can enhance xylem conductivity, allowing xylem sap to arrive at the leaves at a higher water potential and possibly extending the range of transpiration rates available to the stomata before they must shut. In the phloem, potassium enables the export of viscous photoassimilates by "lubricating" the sap through regions of low apoplastic water potential. These phenomena could build on one another: (1) low leaf water potentials lead to a drop in phloem sap pressure in the source; (2) solutes, both potassium and photoassimilates, are loaded in response to the pressure drop; (3) potassium is exported from the leaf to points "deeper" in the plant; (4) surplus potassium is transferred from the phloem to the xylem, raising the xylem sap's potassium concentration; (5) hydrogels in the xylem's bordered pits shrink, enhancing xylem conductivity; (6) and xylem sap is delivered to the leaf at a higher water potential, which alleviates the osmotic demand on the phloem and allows the xylem to divert sap toward one branch or another depending on the local rate and source of potassium recirculation.

At the moment, this scheme is largely speculative and is not meant to be exclusive of other modes of potassium use. While the fact of potassium recirculation is not in doubt, the mechanisms behind it and the signals that control it are not well understood. Numerous potassium channels have been identified that are involved in all aspects of potassium physiology, including soil potassium uptake, long distance transport of potassium in the xylem and phloem, and potassium fluxes associated with guard cell regulation (Cherel, 2004), but the characterization of these channels is still in its infancy. More important, the relation of these channels to the local anatomy and global physiological function of whole-plant potassium recirculation is far from certain.

Furthermore, the anatomical associations between the xylem and phloem in all parts of plants have yet to be fully established. Vascular rays have long been proposed as an important point of exchange between the xylem and phloem (Höll, 1975; van Bel, 1990), but we do not yet know how cellular and molecular processes are cued to take advantage of them. Vascular transfer cells also present a tantalizing anatomical opportunity for xylem-phloem exchange (Gunning and Pate, 1974), especially since phloem transfer cells in the leaf insertion points of some plants are segregated to stem vascular bundles while xylem transfer cells are found predominantly in the vascular bundles of the leaf trace. If vascular transfer cells are involved in potassium recirculation, this spatial arrangement could allow a vascular node to "decide" whether to deliver potassium to the stem phloem or the leaf xylem. Unfortunately for the study of whole-plant solute exchange and translocation, transfer cells have received far more attention in leaf minor veins than in the rest of the plant (Turgeon *et al.*, 2001), although what has been learned in leaves can no doubt be applied to other parts of the plant (see Chapter 3).

### The Importance of Detailed Vascular Anatomy

It is true that sieve tubes and xylem conduits form space-filling branched networks, as do the vascular systems of animals, but their function is not centralized. In mammals, a kidney acts on blood that arrives by a discrete renal artery, the heart provides a focused source for the pressure required for blood flow, and the lungs funnel their gas exchange capacity to the heart through a single pulmonary vein. This organ-level focus is lacking in plant vasculature. What we call the xylem and phloem are essentially collections of discrete functional units whose activity is coordinated solely by their developmental history and by the physical connections between them. Despite their decentralization, however, the xylem and phloem are organized such that they are able to produce consistent responses to changing conditions.

The melding of these two scales of organization—cellular and whole-plant—can be best approached with a focus on anatomy. Anatomy provides constraints on how the xylem and phloem could operate at the whole-plant scale, which in turn provides constraints on how the parts should behave to meet whole-plant needs. For example, the dimensionless group $\hat{F}$ provides a convenient diagnostic for our expectations of phloem behavior. If $\hat{F}$ is sufficiently greater than unity, then the sieve tube should have a relatively small axial pressure drop and exhibit rapid propagation of pressure-concentration waves. These "rules" represent significant constraints on whole-plant behavior and lead naturally to certain hypotheses about the function of individual SE/CC complexes, such as their dependence on pressure as a measure of global sieve tube status. But the opposite, that $\hat{F}$ is less than or near unity and that axial pressure drops are relatively large, would entail a different set of hypotheses, such as the need for a mechanism that could allow SE/CC complexes to individually measure pressure gradients and control them at a global level, and that sieve sap pressure and solute concentration would be unreliable indicators of phloem state. $\hat{F}$ is very sensitive to anatomy; therefore, anatomy is the key to developing hypotheses that link cellular and whole-plant translocation.

Concepts of phloem function during the last 40 years underwent considerable change as our understanding of phloem anatomy developed. The observation of sieve plate occlusion by "proteinaceous slime" and callose led to a diverse set of hypotheses, such as electro-osmotic flow (Spanner, 1975), peristaltic flow (Aikman and Anderson, 1971), and cytoplasmic streaming (Thaine, 1969). Each of these required that the energy needed for translocation be generated at a local scale (Canny, 1975; MacRobbie, 1971) rather than at the extremes of the translocation pathway, as called for by the pressure-flow hypothesis. These in turn led to physiological and anatomical investigations of the phloem that eventually led to their termination as viable hypotheses. The physiology of the anatomical parts—as defined by each model—could not be made to fit either empirical evidence or the proper functioning of the whole (see Chapter 2).

Xylem physiology has had a similar history. Because xylem sap is under tension, it is in a metastable state vulnerable to cavitation. The conceptual difficulties arising from cavitation vulnerability led to a number of studies that confirmed the presence of tension in xylem sap (Holbrook *et al.*, 1995; Pockman *et al.*, 1995), as well as to at least one alternative to the canonical tension-cohesion theory of sap ascent (the "compensating pressure theory," Canny, 1995), which attempted to incorporate empirical evidence for xylem tension into a more complex theory with no tension at all. Subsequently, there has been an explosion of research into not only the whole-plant physiological function of the xylem with respect to the cohesion theory and the limitations it places on plant structure (*cf.* Hacke and Sperry, 2001; Koch *et al.*, 2004; McCulloh *et al.*, 2003; Sperry, 2000) but also the anatomical characteristics of the xylem that permit it to endure and repair embolisms (Chapter 18) (Hacke and Sperry, 2003; Holbrook and Zwieniecki, 1999; Zwieniecki and Holbrook, 2000). As with the phloem, the study of xylem anatomy provides a focus for conceptually integrating its function at the whole-plant and cellular scales.

The hypotheses presented here must be worked out by a combined approach in systems that allow equal emphasis on the anatomy, molecular biology, and whole-plant physiology of the organism. This work must continue with a focus on plants with spatially extensive vascular systems, as it is in these plants that the spatial limitations placed on long-distance transport are tested. Chaffey (2002) argued that the time is now for a concerted effort at organizing our understanding of basic questions of secondary vascular systems. We would add that it is vital that an effort be made to integrate larger plants into whole-plant transport studies, especially *Populus*, for which considerable effort has already been made at genetic characterization (Brunner *et al.*, 2004). The importance of viscosity and sap flux density measurements, for instance, pale in comparison to the serious lack of information we currently have for even the simplest of architectural parameters, such as the length of sieve tubes relative to the size of a tree, or sieve pore diameter and sieve plate thickness (Thompson and Holbrook, 2003b), and, as pointed out by van Bel and Hafke (Chapter 2), no consensus has been reached as to how to calculate sieve plate resistance from first principles, although a first approximation has been advanced (Thompson and Holbrook, 2003a). The two parameters included in $\hat{F}$ that relate to phloem anatomy, $k$ and $L$, both vary by several orders of magnitude across even the limited set of taxa for which good data are available, and data for certain critical taxa are still missing. For instance, although the phloem is actively studied in *Arabidopsis*, its quantitative anatomy is largely unknown or unpublished. Solanaceous plants are also poorly characterized. The time is now to refocus our efforts on anatomy, in combination with whole-plant physiological and molecular studies, to effectively advance our understanding of the highly distributed vascular systems of plants.

## Acknowledgments

We wish to thank Michael Thorpe for helpful comments on this manuscript.

## References

Aikman, D. P. and Anderson, W. P. (1971) A quantitative investigation of a peristaltic model of phloem translocation. *Ann Bot London* **35**: 761-772.

Amir, S. and Reinhold, L. (1971) Interaction between K-deficiency and light in $^{14}$C-sucrose translocation in bean plants. *Physiol Plantarum* **24**: 226-231.

Annaka, M. and Tanaka, T. (1992) Multiple phases of polymer gels. *Nature* **355**: 430-432.

Ashley, D. A. and Goodson, R. D. (1972) Effect of time and plant K status on $^{14}$C-labeled photosynthate movement in cotton. *Crop Sci* **12**: 686-690.

Brunner, A. M., Busov, V. B. and Strauss, S. H. (2004) Poplar genome sequence: Functional genomics in an ecologically dominant plant species. *Trends Plant Sci* **9**: 49-56.

Canny, M. J. (1975) Mass transfer. In *Encyclopedia of Plant Physiology, Transport in Plants I, Phloem Transport*. (M. H. Zimmermann and J. A. Milburn, eds.) pp. 139-153. Springer-Verlag, New York.

Canny, M. J. (1995) A new theory for the ascent of sap: Cohesion supported by tissue pressure. *Ann Bot London* **75:** 343-357.

Chaffey, N. (2002) Why is there so little research into the cell biology of the secondary vascular system of trees? *New Phytol* **153:** 213-223.

Cherel, I. (2004) Regulation of K+ channel activities in plants: From physiological to molecular aspects. *J Exp Bot* **55:** 337-351.

Choat, B., Ball, M., Luly, J. and Holtum, J. (2003) Pit membrane porosity and water stress-induced cavitation in four co-existing dry rainforest tree species. *Plant Physiol* **131:** 41-48.

Deeken, R., Geiger, D., Fromm, J., Koroleva, O., Ache, P., Langenfeld-Heyser, R., Sauer, N., May, S. T. and Hedrich, R. (2002) Loss of the AKT2/3 potassium channel affects sugar loading into the phloem of *Arabidopsis*. *Planta* **216:** 334-344.

Drew, M. C. and Saker, L. R. (1984) Uptake and long-distance transport of phosphate, potassium and chloride in relation to internal ion concentration in barley: Evidence for non-allosteric regulation. *Planta* **160:** 500-507.

Eschrich, W., Evert, R. F. and Young, J. H. (1972) Solution flow in tubular semipermeable membranes. *Planta* **107:** 279-300.

Fisher, D. B. (1978) An evaluation of the Münch hypothesis for phloem transport in soybean. *Planta* **139:** 25-28.

Goeschl, J. D., Magnuson, C. E., DeMichele, D. W. and Sharpe, P. J. H. (1976) Concentration-dependent unloading as a necessary assumption for a closed form mathematical model of osmotically driven pressure flow in phloem. *Plant Physiol* **58:** 556-562.

Gunning, B. E. S. and Pate, J. S. (1974) Transfer cells. In *Dynamic Aspects of Plant Ultrastructure*. (A. W. Robards, ed.) pp. 441-480. McGraw-Hill, London.

Hacke, U. G. and Sperry, J. S. (2001) Functional and ecological xylem anatomy. *Perspect Plant Ecol Evol Syst* **4:** 97-115.

Hacke, U. G. and Sperry, J. S. (2003) Limits to xylem refilling under negative pressure in *Laurus nobilis* and *Acer negundo*. *Plant Cell Environ* **26:** 303-311.

Hammel, H. T. (1968) Measurement of turgor pressure and its gradient in the phloem of oak. *Plant Physiol* **43:** 1042-1048.

Hartt, C. E. (1969) Effect of potassium deficiency upon translocation of $^{14}$C in attached blades and entire plants of sugarcane. *Plant Physiol* **44:** 1461-1469.

Herdel, K., Schmidt, P., Feil, R., Mohr, A. and Schurr, U. (2001) Dynamics of concentrations and nutrient fluxes in the xylem of *Ricinus communis*: Diurnal course, impact of nutrient availability and nutrient uptake. *Plant Cell Environ* **24:** 41-52.

Hoad, G. V. and Peel, A. J. (1965) Studies on the movements of solutes between the sieve tubes and surrounding tissues in willow. I. Interference between solutes and rate of translocation measurements. *J Exp Bot* **15:** 433-451.

Hocking, P. J. (1980) The composition of phloem exudate and xylem sap from tree tobacco (*Nicotiana glauca* Grah.). *Ann Bot London* **45:** 633-643.

Holbrook, N. M., Burns, M. J. and Field, C. B. (1995) Negative xylem pressures in plants: A test of the balancing pressure technique. *Science* **270:** 1193-1194.

Holbrook, N. M. and Zwieniecki, M. A. (1999) Xylem refilling under tension. Do we need a miracle? *Plant Physiol* **120:** 7-10.

Höll, W. (1975) Radial transport in rays. In *Encyclopedia of Plant Physiology, Transport in Plants I, Phloem Transport*. (M. H. Zimmermann and J. A. Milburn, eds.) pp. 432-450). Springer-Verlag, New York.

Knoblauch, M., Peters, W. S., Ehlers, K. and van Bel, A. J. E. (2001) Reversible calcium-regulated stopcocks in legume sieve tubes. *Plant Cell* **13:** 1221-1230.

Koch, G. W., Sillett, S. C., Jennings, G. M. and Davis, S. D. (2004) The limits to tree height. *Nature* **428**: 851-854.

Komor, E., Orlich, G., Weig, A. and Köckenberger, W. (1996) Phloem loading -not metaphysical, only complex: Towards a unified model of phloem loading. *J Exp Bot* **47**: 1155-1164.

Lalonde, S., Tegeder, M., Throne-Holst, M., Frommer, W. B. and Patrick, J. W. (2003) Phloem loading and unloading of sugars and amino acids. *Plant Cell Environ* **26**: 37-56.

Lang, A. (1978) A model of mass flow in the phloem. *Aust J Plant Physiol* **5**: 535-546.

Lang, A. (1983) Turgor-regulated translocation. *Plant Cell Environ* **6**: 683-689.

MacRobbie, E. A. C. (1971) Phloem translocation. Facts and mechanisms: A comparative survey. *Biol Rev* **46**: 428-481.

Marschner, H. (1995) *Mineral Nutrition of Higher Plants.* (2nd ed.). Academic Press, San Diego.

McCulloh, K., Sperry, J., and Adler, F. (2003) Water transport in plants obeys Murray's law. *Nature* **421**: 939-942.

Mengel, K. and Haeder, H. E. (1977) Effect of potassium supply on the rate of phloem sap exudation and the composition of the phloem sap of *Ricinus communis. Plant Physiol* **59**: 282-284.

Mengel, K. and Viro, M. (1974) Effect of potassium supply on the transport of photosynthates to the fruits of tomatoes (*Lycopersicon esculentum*). *Physiol Plantarum* **30**: 295-300.

Minchin, P. E. H. and Thorpe, M. R. (1987) Measurement of unloading and reloading of photo-assimilate within the stem of bean. *J Exp Bot* **38**: 211-220.

Niklas, K. J. (1997) *The Evolutionary Biology of Plants.* University of Chicago Press, Chicago.

Oosterveld, A., Beldman, G. and Voragen, A. (2000) Oxidative cross-linking of pectin polysaccharides from sugar beet pulp. *Carbohyd Res* **328**: 199-207.

Pate, J. S. (1975) Exchange of solutes between phloem and xylem and circulation in the whole plant. In *Encyclopedia of Plant Physiology, Transport in Plants I, Phloem Transport.* (M. H. Zimmermann and J. A. Milburn, eds.) pp. 451-473. Springer-Verlag, New York.

Patrick, J. W., Zhang, W., Tyerman, S. D., Offler, C. E. and Walker, N. A. (2001) Role of membrane transport in phloem translocation of assimilates and water. *Aust J Plant Physiol* **28**: 695-707.

Peuke, A. D., Rokitta, M., Zimmermann, U., Schreiber, L. and Haase, A. (2001) Simultaneous measurement of water flow velocity and solute transport in xylem and phloem of adult plants of *Ricinus communis* over a daily time course by nuclear magnetic resonance spectrometry. *Plant Cell Environ* **24**: 491-503.

Pockman, W. T., Sperry, J. S. and O'Leary, J. W. (1995) Sustained and significant negative water pressure in xylem. *Nature* **378**: 715-716.

Ryden, P., MacDougall, A. J., Tibbits, C. W. and Ring, S. G. (2000) Hydration of pectin polysaccharides. *Biopolymers* **54**: 398-405.

Schurr, U. (1998) Xylem sap sampling: New approaches to an old topic. *Trends Plant Sci* **3**: 293-298.

Shibayama, M. and Tanaka, T. (1993) Volume phase transition and related phenomena of polymer gels. *Adv Polymer Sci* **109**: 1-62.

Smith, J. A. C. and Milburn, J. A. (1980a) Osmoregulation and the control of phloem-sap composition in *Ricinus communis* L. *Planta* **148**: 28-34.

Smith, J. A. C. and Milburn, J. A. (1980b) Phloem transport, solute flux and the kinetics of sap exudation in *Ricinus communis* L. *Planta* **148**: 35-41.

Smith, J. A. C. and Milburn, J. A. (1980c) Phloem turgor and the regulation of sucrose loading in *Ricinus communis* L. *Planta* **148**: 42-48.

Spanner, D. C. (1975) Electroosmotic flow. In *Encyclopedia of Plant Physiology, Transport in Plants I, Phloem Transport* (M. H. Zimmermann and J. A. Milburn, eds.), pp. 301-327. Springer-Verlag, New York.

Sperry, J. S. (2000) Hydraulic constraints on plant gas exchange. *Agric For Meteorol* **104**: 13-23.

Tanaka, T. (1981) Gels. *Sci Am* **244**: 124-138.

Thaine, R. (1969) Movement of sugars through plants by cytoplasmic pumping. *Nature* **222:** 873-875.

Thompson, M. V. and Holbrook, N. M. (2003a) Application of a single-solute non-steady-state phloem model to the study of long-distance assimilate transport. *J Theor Biol* **220:** 419-455.

Thompson, M. V. and Holbrook, N. M. (2003b) Scaling phloem transport: Water potential equilibrium and osmoregulatory flow. *Plant Cell Environ* **26:** 1561-1577.

Thompson, M. V. and Holbrook, N. M. (2004) Scaling phloem transport: Information transmission. *Plant Cell Environ* **27:** 509-519.

Tibbits, C. W., MacDougall, A. J. and Ring, S. G. (1998) Calcium binding and swelling behavior of a high methoxyl pectin gel. *Carbohyd Res* **310:** 101-107.

Turgeon, R., Medville, R. and Nixon, K. C. (2001) The evolution of minor vein phloem and phloem loading. *Am J Bot* **88:** 1331-1339.

Tyree, M. T. Christy, A. L. and Ferrier, J. M. (1974) A simpler iterative steady state solution of Münch pressure-flow systems applied to long and short translocation paths. *Plant Physiol* **54:** 589-600.

Tyree, M. T. and Zimmermann, M. H. (2002) *Xylem structure and the ascent of sap.* (2nd ed.). Springer-Verlag, New York.

van Bel, A. J. E. (1990) Xylem-phloem exchange via the rays: The undervalued route of transport. *J Exp Bot* **41:** 631-644.

van Bel, A. J. E. (1996) Interaction between sieve element and companion cell and the consequences for photoassimilate distribution. Two structural hardware frames with associated physiological software packages in dicotyledons. *J Exp Bot* **47:** 1129-1140.

van Bel, A. J. E. (2003) The phloem, a miracle of ingenuity. *Plant Cell Environ* **26:** 125-149.

Young, J. H., Evert, R. F. and Eschrich, W. (1973) On the volume-flow mechanism of phloem transport. *Planta* **113:** 355-366.

Ziegler, H. (1975) Nature of transported substances. In *Encyclopedia of Plant Physiology, Transport in Plants I, Phloem Transport.* (M. H. Zimmermann and J. A. Milburn, eds.) pp. 59-100. Springer-Verlag, New York.

Zwieniecki, M. A. and Holbrook, N. M. (2000). Bordered pit structure and vessel wall surface properties: Implications for embolism repair. *Plant Physiol* **123:** 1015-1020.

Zwieniecki, M. A., Melcher, P. J. and Holbrook, N. M. (2001) Hydrogel control of xylem hydraulic resistance in plants. *Science* **291:** 1059-1062.

Zwieniecki, M. A., Orians, C. M., Melcher, P. J. and Holbrook, N. M. (2003) Ionic control of the lateral exchange of water between vascular bundles in tomato. *J Exp Bot* **54:** 1399-1405.

# 12

## Coordination Between Shoots and Roots

*Arnold Bloom*

In 1900, Johann Hurlinger traveled from Vienna to Paris over a period of 55 days. What distinguishes this journey is that Mr. Hurlinger walked the entire distance on his hands. In contraposition, people who are deprived the use of their hands develop extraordinary dexterity with their feet. These curious facts show that humans, despite an overwhelming preference for remaining upright, have an amazing potential for adjusting to new orientations.

Terrestrial plants do not display such versatility. Few, if any, will survive for more than a day with their shoots underground and their roots in the air. This intolerance derives from the divergence between shoots and roots in their structure and function. Indeed, shoots and roots are like two organisms that spend their lives in entirely different environments, yet form a symbiotic relationship.

The following chapter explores the differences between the shoot and root environment and the relationship that exists between shoots and roots. Communication among partners is paramount for such a relationship to flourish or even survive. The success of plants on land attests to the effectiveness of this shoot/root communication over long distances, and the evolutionary trend is for greater differentiation between shoots and roots. Successful coordination among dissimilar partners is complicated, and many aspects about shoot/root communications remain unresolved. This chapter, nonetheless, glosses over many of the unresolved issues to present a coherent picture.

This chapter concludes with some predictions about changes to shoot/root coordination with changing global climate. Atmospheric levels of $CO_2$ are rising at an unprecedented rate, and shoots are more sensitive than roots to such changes. As a result, the terrestrial flora may experience major shifts.

## Structure and Function

Shoots and roots assume typical, although not exclusive, roles in a plant. Shoots usually acquire light and carbon dioxide while minimizing water loss and promote the dispersal of progeny. Roots normally anchor the plant onto a substrate, acquire water and mineral nutrients from this substrate, and participate in vegetative propagation. These distinct responsibilities require different structures. For instance, shoots are covered with a water-resistant cuticle and optimize their display of photosynthetic organs and their exchange of gases, whereas roots have a water-permeable epidermis and present a large surface area per volume of soil for exchange of mineral nutrients.

Even within a canopy or root system of a single plant, there is specialization of structure/function. The shoot of a single plant contains dissimilar tissues, ranging from reproductive structures to stems to leaves. Among leaves, the dichotomy between sun and shade leaves is striking (Pearcy, 1998). Roots—be they primary, lateral, or adventitious—differ in dimension, orientation, and physiology (Fitter *et al.*, 2002; Wells and Eissenstat, 2002). This chapter, however, will ignore the diversity in structure aboveground or belowground and will treat shoots or roots as monolithic organs.

## Functional Equilibrium

Plants, having limited mobility for most of their lives, must deal with changes in their local environment. Aboveground, light level, temperature, or humidity may fluctuate substantially during the day and across the canopy, but $CO_2$ and $O_2$ concentrations remain relatively uniform. By contrast, soils buffer the roots from temperature extremes, but the belowground concentrations of $CO_2$ and $O_2$, water, and nutrients are extremely heterogeneous, both spatially and temporally. For example, inorganic nitrogen concentrations in a soil may range 1000-fold over a distance of centimeters or the course of hours (Bloom, 1997b). Given such heterogeneity, plants seek the most favorable conditions within their reach.

Shoots branch into light gaps, and leaves follow the sun to the degree permitted by water supply, gravity, and wind. For example, tropical understory plants vary parameters such as leaf orientation, internode length, and specific leaf area to achieve similar efficiencies of light capture and foliage display (Valladares *et al.*, 2002). Sack and Tyree (Chapter 5 in this volume) also detail how hydraulic constraints of leaves may influence shoot resource acquisition.

Roots sense the belowground environment through gravitropism, thigmotropism, chemotropism, and hydrotropism to guide their growth toward soil resources. The extent to which roots proliferate within a soil patch varies with nutrient level (Fig. 12.1). Root growth is minimal in poor soils because the roots become nutrient limited. As soil nutrient availability increases, roots proliferate (Drew, 1975; Jackson and Caldwell, 1989; Samuelson *et al.*, 1991) and form denser architecture (Fitter and Stickland, 1991). Where soil nutrients exceed an optimal level, root growth becomes carbohydrate limited and eventually ceases (Durieux *et al.*, 1994). With high soil nutrients, a few roots—3.5% of the root system in spring wheat (Robinson *et al.*, 1991) and 12% in lettuce (Burns, 1991)—are sufficient to supply all the nutrients required so that a plant may diminish the allocation of resources to roots while increasing its allocation to the shoot and reproductive structures. This is one mechanism through which fertilization stimulates crop yields.

The optimization of root growth with soil nutrient availability is consistent with the hypothesis of a functional equilibrium between shoot and root (Bloom *et al.*, 1993; Brouwer, 1967; Brouwer and DeWit, 1969). This

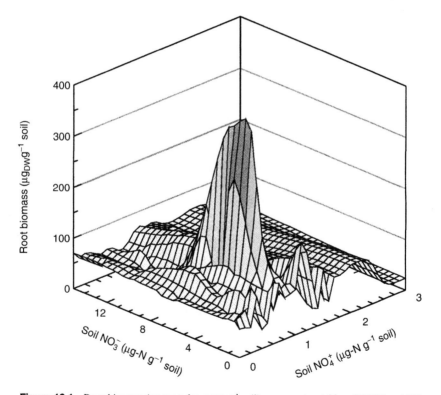

**Figure 12.1**    Root biomass ($\mu$g root dry mass g$^{-1}$ soil) versus extractable soil NH$_4^+$ and NO$_3^-$ ($\mu$g extractable N g$^{-1}$ soil) for tomato, *Lycopersicon esculentum* L. Mill. cv T-5, growing in an irrigated field that had been fallow the previous 2 years. From Epstein & Bloom (2005) *Mineral Nutrition of Plants.* Copyright Sinauer Associates; adapted with permission.

hypothesis proposes that because of the distances between sources and sinks, shoots more readily meet their requirements for carbohydrates than for nutrients, whereas roots more readily meet their requirements for mineral nutrients than for carbohydrates. Under low soil nutrients, the root and shoot growth is nutrient limited and carbohydrate supply in the shoot is relatively high so that carbohydrate translocation to roots is high. When a root encounters a soil patch rich in nutrients, the nutrients absorbed initially remain in that root and enhance growth (Bloom, 2002). As the nutrient levels in that root reach a surfeit, more nutrients are translocated to a shoot. Vascular connections often determine that one part of the shoot receives a significant portion of its nutrients from a particular root and, in return supplies carbohydrates first to that root (Orians *et al.*, 2002; Watson and Casper, 1984). This part of a shoot, now sufficient in nutrients, becomes carbohydrate limited and translocates relatively little carbohydrate to the root; growth of this root then becomes carbohydrate limited. Such a response would explain the hyperbolic response of root production to soil nutrient availability (Fig. 12.1; Durieux *et al.*, 1994). For more information on the sectoriality of vascular connections, see Orians Babst, and Zanne Chapter 17 in this volume.

Functional equilibrium, like any robust hypothesis, has been subject to criticism. Although the hypothesis as originally proposed failed to identify specific physiological mechanisms for the regulation of shoot/root partitioning (Farrar and Jones, 2000), the next section of this chapter forwards several possibilities. The hypothesis also did not address the controversial matter of alternative respiration, in which the electron transport chain in mitochondria uses an alternative oxidase instead of cytochrome *c* oxidase (Lambers, 1983). Nonetheless, the hypothesis of functional equilibrium forms the basis for several successful models of whole plant growth (Hunt *et al.*, 1990; Thornley, 1977) and provides an appropriate framework for the issues discussed here.

## Regulatory Signals

Plants generally use a variety of hormones to orchestrate developmental processes across remote tissues. Coordination of root and shoot development is no exception. Auxins are primarily produced in shoots and transported through the phloem to the roots. By contrast, cytokinins are primarily produced in roots and transported through the xylem to the shoots. Auxins decrease at high nutrient levels, diminishing the proliferation of lateral roots, whereas cytokinins increase with the nutrient status of a plant (Takei *et al.*, 2002) and stimulate leaf expansion, but inhibit root elongation. These responses fit nicely into the scheme for the functional equilibrium between root and shoot outlined in the previous section

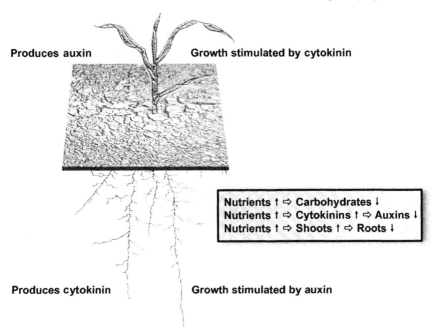

**Produces auxin**     **Growth stimulated by cytokinin**

Nutrients ↑ ⇨ Carbohydrates ↓
Nutrients ↑ ⇨ Cytokinins ↑ ⇨ Auxins ↓
Nutrients ↑ ⇨ Shoots ↑ ⇨ Roots ↓

**Produces cytokinin**     **Growth stimulated by auxin**

**Figure 12.2**   The functional equilibrium between root and shoot growth as determined by resource levels and communicated via auxins and cytokinins, plant hormones. From Epstein & Bloom (2005) *Mineral Nutrition of Plants.* Copyright Sinauer Associates; adapted with permission.

(Bangerth *et al.*, 2000): Low nutrient levels result in low cytokinins and high auxins that favors root over shoot development and vice-versa (Fig. 12.2).

Other hormones such as abscisic acid (ABA) and ethylene coordinate shoot and root responses to water deficits. As soils dry out, roots accumulate additional ABA that enters the transpiration stream (Davies and Zhang, 1991). This ABA and the resultant alkalization of the xylem sap reach the leaves and induce stomatal closure (Wilkinson and Davies, 2002). In addition, the higher ABA levels that occur in a water-stressed plant suppress ethylene production and, thus, promote root growth and stunt shoot growth (Spollen *et al.*, 2000). This results in a substantial increase in root:shoot ratio, which in conjunction with ABA-induced stomatal closure, enables a plant to restore a more favorable water status.

Admittedly, the scenarios outlined above oversimplify a complex network of hormonal and other responses that attune the growth and development of a plant to its environment (Passioura, 2002). For example, stomatal behavior can be independent of a root's ability to produce ABA (Holbrook *et al.*, 2002), and ABA does not mediate stomatal closure that occurs during water deficits resulting from cold soils (Vernieri *et al.*, 2001; Wilkinson *et al.*, 2001).

Certain inorganic nutrients themselves serve as messengers. Most ubiquitous is calcium that acts as a secondary messenger for a broad range of responses (Rudd and Franklin-Tong, 2001). Magnesium is an important cofactor for many biochemical reactions through its ability to bridge nitrogen and phosphorus groups. Most nutrients also regulate plant responses to their own availability. Such regulation acts at the transcription level and beyond. For example, nitrate induces the transcription of mRNA for nitrate reductase as well as stimulates the activity of this enzyme through posttranslational regulation (Forde, 2002). Low phosphate conditions induce genes coding for phosphate transporters, phosphatases, and ribonucleases, products that enhance phosphate acquisition and recycling (Abel *et al.*, 2002). The use of inorganic nutrients as messengers, however, seems limited to short distances within a cell or, at most, within a tissue (Epstein and Bloom, 2005).

Nitrogen is the mineral nutrient that plants require in greatest amounts, and root absorption of nitrogen limits primary productivity in most ecosystems (Bloom, 1997a). The carbon to nitrogen ratio (C:N) in a plant, therefore, reflects the balance between shoot and root functions (Foyer *et al.*, 2001). The amino acids asparagine, glutamine, and glutamate serve as signaling compounds because of their variable nitrogen-to-carbon ratios (2 N to 4 C for asparagine, 2 N to 5 C for glutamine, and 1 N to 5 C for glutamate, Lam *et al.*, 1996). The synthesis of glutamine and glutamate from $NH_4^+$ involves the enzymes glutamine synthetase (GS) and glutamate synthase (GOGAT). The major pathway for asparagine synthesis involves the enzyme asparagine synthetase (AS) that facilitates the transfer of the amide nitrogen from glutamine to aspartate. Conditions of ample energy (i.e., high levels of light and carbohydrates) stimulate GS and GOGAT, inhibit AS, and thus favor nitrogen assimilation into glutamine and glutamate, compounds that are rich in carbon and stimulate root growth. By contrast, energy-limited conditions inhibit GS and GOGAT, stimulate AS, and thus favor nitrogen assimilation into asparagine, a compound that is rich in nitrogen, sufficiently stable for long-distance transport or long-term storage, and favors shoot growth (Lam *et al.*, 1996). In summary, low nitrogen conditions produce glutamine and glutamate that induce root growth, whereas low carbohydrate conditions produce asparagine that induces shoot growth. Regulation of these competing pathways thus helps balance shoot and root growth as well as carbon and nitrogen metabolism.

## Global Change and Shoot/Root Coordination

Humans have become participants in a huge, uncontrolled experiment, one that will dramatically alter the C:N balance of plants and, thereby, change shoot/root coordination. The experiment is global climate changes that result from anthropogenic release of greenhouse gases, particularly the release of $CO_2$ from the burning of fossil fuels. Atmospheric $CO_2$ concentrations have increased from about 280 to 370 $\mu mol\ mol^{-1}$ since 1800 (Whorf and Keeling, 1998) and may reach between 500 and 900 $\mu mol\ mol^{-1}$ by the end of the century (Joos *et al.*, 1999).

Several responses of higher plants to such $CO_2$ enrichment have been surprising (Bazzaz, 1990). For instance, a doubling of $CO_2$ level initially accelerates carbon fixation in $C_3$ plants by over 30%, yet after days to weeks of exposure to elevated $CO_2$ concentrations, depending on species and conditions, carbon fixation and plant growth decline until they stabilize at rates that average 8% above ambient controls (Poorter and Navas, 2003). This general phenomenon, known as $CO_2$ acclimation, is correlated with a decline in the activities of Rubisco and other enzymes in the Calvin cycle (Moore *et al.*, 1998). These changes in the activities of Calvin cycle enzymes are not necessarily specific; rather, they often follow a decline in overall shoot protein and nitrogen contents (Makino and Mae, 1999). Shoot nitrogen contents diminish by an average of 14% with a doubling of $CO_2$ (Cotrufo *et al.*, 1998), a difference that is nearly double what would be expected if a given amount of nitrogen were diluted by additional biomass (Makino and Mae, 1999).

Root responses to elevated $CO_2$ fail to follow any pattern (Norby and Jackson, 2000). Root-to-shoot ratios do not vary significantly (Rogers *et al.*, 1996). Changes in root mineral nitrogen absorption are inconsistent (BassiriRad *et al.*, 2001). The number and activity of nitrogen fixing nodules per unit root remain unaffected (Vitousek and Field, 1999), as do the extent and activity of mycorrhizae (Fitter *et al.*, 2000).

We have recently discovered a phenomenon that helps explain the variability in shoot and root responses to elevated $CO_2$: Elevated $CO_2$ inhibits $NO_3^-$ assimilation in the shoots of $C_3$ plants (Figs. 12.3 & 12.4, barley, Bloom *et al.*, 1989; wheat, Bloom *et al.*, 2002b; tomato, Searles and Bloom, 2003; and *Arabidopsis*, Rachmilevitch *et al.*, 2004). By contrast, $NO_3^-$ assimilation in the roots of a $C_3$ plant (Bloom *et al.*, 2002b) and in the shoots of maize, a $C_4$ plant (Fig. 12.3, Cousins and Bloom, 2003) are insensitive to atmospheric $CO_2$ levels. Because $NO_3^-$ is the prominent form of inorganic nitrogen available to plants from temperate soils (Bloom, 1997b), diminished $NO_3^-$ assimilation dramatically alters the nitrogen balance in $C_3$ plants (Bloom *et al.*, 2002b).

Much of our evidence is based on estimates of shoot $NO_3^-$ assimilation derived from calculations of $\Delta AQ$—the difference in the *assimilatory quotient* (ratio of net $CO_2$ consumption to net $O_2$ evolution) when a plant's nitrogen source was shifted from $NO_3^-$ to $NH_4^+$. The $\Delta AQ$ depends on $NO_3^-$ assimilation because transfer of electrons to $NO_3^-$ and then to $NO_2^-$ increases net $O_2$ evolution from the light-dependent reactions of photosynthesis, while net

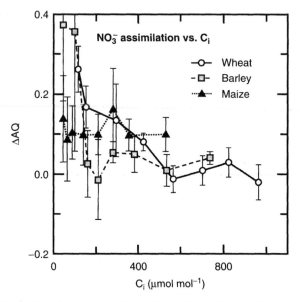

**Figure 12.3**   In wheat, barley, and maize seedlings, shoot $NO_3^-$ assimilation—as monitored by the change in assimilatory quotient ($AQ = CO_2$ consumed/$O_2$ evolved) with a shift in N source from $NO_3^-$ to $NH_4^+$—as a function of internal $CO_2$ concentration, $C_i$. Shown are mean ± SE for 5-8 replicate plants (Epstein and Bloom, 2005).

$CO_2$ consumption remains unchanged (Bloom *et al.*, 1989; Bloom *et al.*, 2002b; Cen *et al.*, 2001; Myers, 1949). Several other biochemical processes such as lipid metabolism might influence $\Delta AQ$, but $NO_3^-$ assimilation has a much greater effect than any of these (Cen *et al.*, 2001).

Three mechanisms may be responsible for $CO_2$ inhibition of $NO_3^-$ assimilation. First, photorespiration serves as an important redox transfer mechanism that increases the cytosolic NADH/NAD ratio via the export of malate from the chloroplast (Backhausen *et al.*, 1994). The initial step of $NO_3^-$ assimilation (i.e., the reduction of $NO_3^-$ to $NO_2^-$) occurs in the cytosol (Rufty *et al.*, 1986; Vaughn and Campbell, 1988) and uses NADH from the malate shuttle (Robinson, 1987). This may explain why greater $NO_3^-$ assimilation is observed under ambient $CO_2$ and $O_2$ (Fig. 12.4A), conditions under which photorespiration would be highest.

Second, the reduction of $NO_2^-$ to $NH_4^+$, the incorporation of $NH_4^+$ into amino acids, and the Calvin cycle all occur in the stroma of a chloroplast and require ferredoxin that is reduced via photosynthetic electron transport (Sivasankar and Oaks, 1996). The $K_m$s for reduced ferredoxin of the key enzymes—ferredoxin-NADP reductase (FNR), nitrite reductase (NiR), and glutamate synthase (GOGAT)—are 0.1, 0.6, and 60 μM, respectively

(D. B. Knaff, personal communication). As a result, $NO_3^-$ assimilation proceeds only if the availability of reduced ferredoxin exceeds that needed for NADPH formation (Baysdorfer and Robinson, 1985; Peirson and Elliott, 1988). This occurs only at high light under ambient $CO_2$ and $O_2$ (Fig. 12.4), when the availability of $CO_2$ limits photosynthesis. This seems reasonable in that plants can store $NO_3^-$ in large quantities and even use it as a metabolically-benign osmoticant (Bloom *et al.*, 2002a) until excess reductant becomes available, but cannot store quantities of $CO_2$ for extended periods.

Third, $NO_2^-$ transport from the cytosol into the chloroplast involves the diffusion of $HNO_2$ across chloroplast membranes and, therefore, requires the stroma to be more alkaline than the cytosol (Shingles *et al.*, 1996). Carbon dioxide at elevated concentrations can dissipate this pH gradient because additional $CO_2$ movement into the chloroplast acidifies the stroma and because enhanced carbon fixation hydrolyzes ATP faster and requires supplementary proton exchange across the thylakoid membrane to regenerate this ATP. As a result, $NO_2^-$ transport into the chloroplast is inhibited by elevated $CO_2$ (Fig. 12.5).

Because $CO_2$ at elevated concentrations inhibits $NO_3^-$ photoassimilation in $C_3$ plants, plants receiving $NH_4^+$ as a N source prove more responsive to $CO_2$ enrichment than those receiving $NO_3^-$. The form in which N was supplied (0.2 mM $NH_4^+$ or 0.2 mM $NO_3^-$) did not influence the growth of wheat seedlings at 360 $\mu$mol mol$^{-1}$ (ambient) $CO_2$, but had a dramatic effect at 700 $\mu$mol mol$^{-1}$ (elevated) $CO_2$ (Fig. 12.6). Leaf area in the ele-

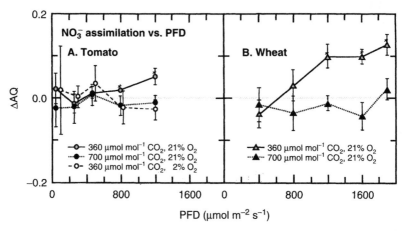

**Figure 12.4** In tomato (A) and wheat (B), shoot $NO_3^-$ assimilation—as monitored by the change in assimilatory quotient (AQ = $CO_2$ consumed/$O_2$ evolved) with a shift in N source from $NO_3^-$ to $NH_4^+$—as a function of photosynthetic photon flux density, *PFD*. Shown are mean $\pm$ SE for 5-8 replicate plants. Asterisks mark the means that were significantly different from zero ($P < 0.05$, a student's *t*-test) (Epstein and Bloom, 2005).

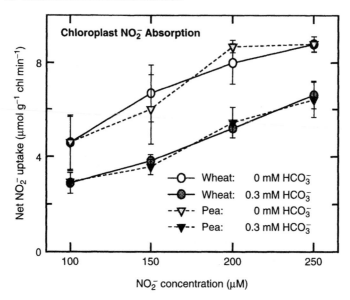

**Figure 12.5**    Net $NO_2^-$ uptake ($\mu$mol mg$^{-1}$ chlorophyll min$^{-1}$) by isolated chloroplasts as a function of $NO_2^-$ concentration when the medium contained 0 (light symbols) or 0.3 (dark symbols) mM $HCO_3^-$. Shown are the mean $\pm$ SE (n = 3) for wheat (circles) and pea (inverted triangles). From Bloom *et al.* (2002) *PNAS* 99:1730-1735. Copyright PNAS; adapted with permission.

vated $CO_2$ treatment relative to the ambient $CO_2$ treatment increased 49% under $NH_4^+$ nutrition, but only 24% under $NO_3^-$ nutrition (Fig. 12.6). Total plant biomass increased 78% under $NH_4^+$ nutrition, but only 44% under $NO_3^-$ nutrition (Fig. 12.6).

Shoot and root nitrogen concentrations were similar under the two $CO_2$ regimes, indicating that nitrogen absorption per unit plant mass remained unchanged (Fig. 12.7). The fate of this nitrogen after it was absorbed, however, differed under ambient and elevated $CO_2$ as demonstrated by the balance between inorganic and organic nitrogen (Fig. 12.7). In the elevated $CO_2$ treatment relative to the ambient $CO_2$ treatment, shoot protein concentrations decreased 6% under $NH_4^+$ nutrition, as might be expected given the dilution by additional biomass, but decreased 13% under $NO_3^-$ nutrition despite less additional biomass (Fig. 12.6). Thus, shoot protein per plant increased 73% and 32% under $NH_4^+$ and $NO_3^-$, respectively. Shoot $NO_3^-$ concentrations were undetectable in plants receiving $NH_4^+$, but increased 62% at elevated $CO_2$ in those receiving $NO_3^-$ (Fig. 12.7). *In vitro* shoot activities of $NO_3^-$ reductase and $NO_2^-$ reductase decreased 12% and 27% from ambient to elevated $CO_2$, respectively, on a total protein basis (Fig. 12.7) and decreased 33% and 30%, respectively, on a fresh mass basis. Root protein,

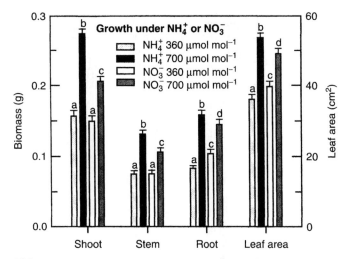

**Figure 12.6**   Biomass (g dry mass) and leaf area (cm²) per plant of wheat seedlings grown for 14 days in controlled environment chambers at 360 or 700 μmol mol⁻¹ $CO_2$ and under $NH_4^+$ or $NO_3^-$ nutrition. Shown are mean ± SE for 4 replicate experiments, each with 8 to 10 plants per treatment. Treatments labeled with different letters differ significantly ($P < 0.05$). From Bloom *et al.* (2002) *PNAS* 99:1730-1735. Copyright PNAS; adapted with permission.

$NO_3^-$, and enzyme activities were similar under both $CO_2$ treatments (Fig. 12.7). In summary, shoot organic nitrogen parameters under $NO_3^-$ nutrition declined at elevated $CO_2$, but root parameters did not.

If $CO_2$ inhibition of shoot $NO_3^-$ assimilation were a general phenomenon among $C_3$ plants, it would contribute to the response of natural ecosystems to rising $CO_2$ levels. Ecosystems vary in the relative availabilities of $NH_4^+$ and $NO_3^-$ in their soils. Plants vary in their relative dependence upon $NH_4^+$ and $NO_3^-$ as nitrogen sources (Bloom, 1997b) and in their balance between shoot and root $NO_3^-$ assimilation (Andrews, 1986). Our results suggest that rising atmospheric $CO_2$ will favor taxa that prefer $NH_4^+$ as a nitrogen source or that assimilate $NO_3^-$ primarily in their roots. This might produce major changes in the terrestrial flora.

Consistent with the theme of this chapter, $CO_2$ inhibition of shoot $NO_3^-$ assimilation would severely alter shoot and root coordination. To avoid $NH_4^+$ toxicity, roots rapidly assimilate $NH_4^+$ as they absorb it. A shift from shoot $NO_3^-$ assimilation to root assimilation of $NH_4^+$ and $NO_3^-$ would require translocation of supplementary carbohydrate to the roots in order to provide sufficient energy and carbon skeletons for these processes. With less nitrogen in the shoot as well as more carbohydrate going belowground, plant priorities would turn upside down. The functional equilibrium between shoot and root shift to a new set point.

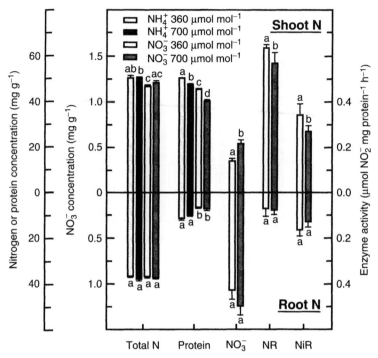

**Figure 12.7** Total N concentration (mg g$^{-1}$ dry mass), protein concentration (mg g$^{-1}$ fresh mass), NO$_3^-$ concentration (mg g$^{-1}$ fresh mass), NO$_3^-$ reductase activity (μmol NO$_2^-$ generated mg$^{-1}$ protein h$^{-1}$), and NO$_2^-$ reductase activity (μmol NO$_2^-$ consumed mg$^{-1}$ protein h$^{-1}$) in the shoot (top panel) or root (bottom panel) of wheat grown for 14 days in controlled environment chambers at 360 or 700 μmol mol$^{-1}$ CO$_2$ and under NH$_4^+$ or NO$_3^-$ nutrition. Shown are mean ± SE for 4 replicate experiments, each with 8 to 10 plants per treatment. Treatments labeled with different letters differ significantly ($P < 0.05$). Chlorophyll concentrations were 0.32 ± 0.02 and 0.34 ± 0.07 g liter$^{-1}$ (mean ± SE, n = 2) for the NH$_4^+$ treatment at 360 and 700 μmol mol$^{-1}$, respectively, and 0.30 ± 0.01 and 0.26 ± 0.04 g liter$^{-1}$ (mean ± SE, n = 2) for the NO$_3^-$ treatment at 360 and 700 μmol mol$^{-1}$, respectively. From Bloom *et al.* (2002) *PNAS* 99:1730-1735. Copyright PNAS; adapted with permission.

The implications of this phenomenon resonate up trophic levels. Under elevated CO$_2$, a greater percentage of the nitrogen in a plant would remain as free NO$_3^-$, a nitrogen form preferred less by plant pests from bacteria to humans. For example, protein content and bread quality of flour tend to decline when wheat is grown at elevated CO$_2$ (Kimball *et al.*, 2001). Because of the transfer of additional resources belowground under elevated CO$_2$, herbivores that graze on shoots might be at a disadvantage with respect to those that target roots. The rapid rise of atmospheric CO$_2$ concentrations, unfortunately, leaves less time for evolutionary adjustments.

## Conclusion

Shoots and roots of higher plants are distinct in form and function. They coordinate their activities achieving a functional equilibrium between the two organs through several mechanisms. Translocation of carbohydrates from shoots to roots and water and nutrients from roots to shoots are balanced according to the resource needs of the exporting organ. Phytohormones such as cytokinins and auxins and amino acids such as asparagine, glutamine, and glutamate act as messengers between the two organs and regulate shoot versus root growth. Rising atmospheric $CO_2$ concentrations will inhibit shoot $NO_3^-$ assimilation and thus will perturb these regulatory mechanisms. Under elevated $CO_2$, additional carbohydrate may be transferred to roots and additional organic nitrogen generated in roots.

## Acknowledgments

Funded in part by NSF IBN-99-74927, NSF IBN-03-43127, and USDA NRI-CGP 2000-00647

## References

Abel, S., Ticconi, C. A. and Delatorre, C. A. (2002). Phosphate sensing in higher plants. *Physiol. Plant.* **115:** 1-8.

Andrews, M. (1986). The partitioning of nitrate assimilation between root and shoot of higher plants. *Plant Cell Environ.* **9:** 511-519.

Backhausen, J. E., Kitzmann, C. and Sheibe, R. (1994). Competition between electron acceptors in photosynthesis: Regulation of the malate valve during $CO_2$ fixation and nitrite reduction. *Photosyn. Res.* **42:** 75-86.

Bangerth, F., Li, C.-J. and Gruber, J. (2000). Mutual interaction of auxin and cytokinins in regulating correlative dominance. *Plant Growth Regulation* **32:** 205-217.

BassiriRad, H., Gutschick, V. P. and Lussenhop, J. (2001). Root system adjustments: Regulation of plant nutrient uptake and growth responses to elevated $CO_2$. *Oecologia* **126:** 305-320.

Baysdorfer, C. and Robinson, M. J. (1985). Metabolic interactions between spinach leaf nitrite reductase and ferredoxin-NADP reductase. *Plant Physiol.* **77:** 318-320.

Bazzaz, F. A. (1990). The response of natural ecosystems to the rising global $CO_2$ levels. *Annu. Rev. Ecol. Syst.* **21:** 167-196.

Bloom, A. J. (1997a). Interactions between inorganic nitrogen nutrition and root development. *Z. Pflanzenernähr. Bodenk.* **160:** 253-259.

Bloom, A. J. (1997b). Nitrogen as a limiting factor: Crop acquisition of ammonium and nitrate. In *Ecology in Agriculture* (L. E. Jackson, ed.) pp. 145-172. Academic Press, San Diego.

Bloom, A. J. (2002). Mineral Nutrition, Chapter 5. In *Plant Physiology* (L. Taiz and E. Zeiger, eds.) pp. 67-86. 3rd Ed. Sinauer Associates, Sunderland, MA.

Bloom, A. J., Caldwell, R. M., Finazzo, J., Warner, R. L. and Weissbart, J. (1989). Oxygen and carbon dioxide fluxes from barley shoots depend on nitrate assimilation. *Plant Physiol.* **91:** 352-356.

Bloom, A. J., Jackson, L. E. and Smart, D. R. (1993). Root growth as a function of ammonium and nitrate in the root zone. *Plant Cell Environ.* **16:** 199-206.

Bloom, A. J., Meyerhoff, P. A., Taylor, A. R. and Rost, T. L. (2002a). Root development and absorption of ammonium and nitrate from the rhizosphere. *J. Plant Growth Regulation* **21:** 416-431.

Bloom, A. J., Smart, D. R., Nguyen, D. T. and Searles, P. S. (2002b). Nitrogen assimilation and growth of wheat under elevated carbon dioxide. *Proc. Natl. Acad. Sci. USA* **99:** 1730-1735.

Brouwer, R. (1967). Beziehungen zwischen Spross-und Wurzelwachstum. *Angewandte Botanik* **41:** 244-250.

Brouwer, R. and DeWit, C. T. (1969). A simulation model of plant growth with special attention to root growth and its consequences. In *Root Growth* (W. J. Whittington, ed.) pp. 224-244. Butterworths, London.

Burns, I. G. (1991). Short- and long-term effects of a change in the spatial distribution of nitrate in the root zone on N uptake, growth and root development of young lettuce plants. *Plant Cell Environ.* **14:** 21-33.

Cen, Y.-P., Turpin, D. H. and Layzell, D. B. (2001). Whole-plant gas exchange and reductive biosynthesis in white lupin. *Plant Physiol.* **126:** 1555-1565.

Cotrufo, M. F., Ineson, P. and Scott, A. (1998). Elevated $CO_2$ reduces the nitrogen concentration of plant tissues. *Global Change Biol.* **4:** 43-54.

Cousins, A. B. and Bloom A. J. (2003). Influence of elevated $CO_2$ and nitrogen nutrition on photosynthesis and nitrate photoassimilation in maize (*Zea mays* L.). *Plant Cell Environ.*, **26:** 1525-1530.

Davies, W. J. and Zhang, J. H. (1991). Root signal and the regulation of growth and development of plants in drying soil. *Annu. Rev. Plant Physiol.* **42:** 55-76.

Drew, M. C. (1975). The effect of the supply of mineral nutrients on root morphology, nutrient uptake, and shoot growth in cereals, pp. 63-73. Agricultural Research Council Letcombe Laboratory Annual Report. Wantage, Berkshire, UK.

Durieux, R. P., Kamprath, E. J., Jackson, W. A. and Moll, R. H. (1994). Root distribution of corn-the effect of nitrogen fertilization. *Agron. J.* **86:** 958-962.

Epstein, E. and Bloom, A. J. (2005). *Mineral Nutrition of Plants: Principles and Perspectives.* 2nd Ed. Sinauer Associates, Sunderland, MA.

Farrar, J. F. and Jones D. L. (2000). The control of carbon acquisition by roots. *New Phytol.* **147:** 43-53.

Fitter, A., Williamson, L., Linkohr, B. and Leyser, O. (2002). Root system architecture determines fitness in an *Arabidopsis* mutant in competition for immobile phosphate ions but not for nitrate ions. *Proc. R. Soc. London Ser. B* **269:** 2017-2022.

Fitter, A. H., Heinemeyer, A. and Staddon, P. L. (2000). The impact of elevated $CO_2$ and global climate change on arbuscular mycorrhizas: A mycocentric approach. *New Phytol.* **147:** 179-187.

Fitter, A. H. and Stickland, T. R. (1991). Architectural analysis of plant root systems: 2. Influence of nutrient supply on architecture in contrasting plant species. *New Phytol.* **118:** 383-389.

Forde, B. G. (2002). Local and long-range signaling pathways regulating plant responses to nitrate. *Annu. Rev. Plant Physiol. Plant Mol. Biol.* **53:** 203-224.

Foyer, C. H., Ferrario-Mery, S. and Noctor, G. (2001). Interactions between carbon and nitrogen metabolism. In *Plant Nitrogen* (P. J. Lea and J.-F. Morot-Gaudry, eds.) pp. 237-254. Springer-Verlag, Berlin.

Holbrook, N. M., Shashidhar, V. R., James, R. A. and Munns, R. (2002). Stomatal control in tomato with ABA-deficient roots: Response of grafted plants to soil drying. *J. Exp. Bot.* **53:** 1503-1514.

Hunt, R., Wilson, J. W. and Hand, D. W. (1990). Integrated analysis of resource capture and utilization. *Ann. Bot. (London)* **65:** 643-648.

Jackson, R. B. and Caldwell, M. M. (1989). The timing and degree of root proliferation in fertile-soil microsites for 3 cold-desert perennials. *Oecologia* **81:** 149-153.

Joos, F., Plattner, G. K., Stocker, T. F., Marchal, O. and Schmittner, A. (1999). Global warming and marine carbon cycle feedbacks on future atmospheric $CO_2$. *Science* **284:** 464-467.

Kimball, B. A., Morris, C. F., Pinter, P. J., Wall, G. W., Hunsaker, D. J., Adamsen, F. J., LaMorte, R. L., Leavitt, S. W., Thompson, T. L., Matthias, A. D. and Brooks, T. J. (2001). Elevated $CO_2$, drought and soil nitrogen effects on wheat grain quality. *New Phytol.* **150:** 295-303.

Lam, H.-M., Coschigano, K. T., Oliveira, I. C., Melo-Oliveira, R. and Coruzzi, G. M. (1996). The molecular-genetics of nitrogen assimilation into amino acids in higher plants. *Annu. Rev. Plant Physiol. Plant Mol. Biol.* **47:** 569-593.

Lambers, H. (1983). 'The functional equilibrium', nibbling on the edges of a paradigm. *Neth. J. Agric. Sci.,* **31:** 305-311.

Makino, A. and Mae, T. (1999). Photosynthesis and plant growth at elevated levels of $CO_2$. *Plant Cell Physiol.* **40:** 999-1006.

Moore, B. D., Cheng, S. H., Rice, J. and Seemann, J. R. (1998). Sucrose cycling, Rubisco expression, and prediction of photosynthetic acclimation to elevated atmospheric $CO_2$. *Plant Cell Environ.* **21:** 905-915.

Myers, J. (1949). The pattern of photosynthesis in Chlorella. In *Photosynthesis in Plants* (J. Franck and W. E. Loomis, eds.) pp. 349-364. Iowa State College Press, Ames, Iowa.

Norby, R. J. and Jackson, R. B. (2000). Root dynamics and global change: Seeking an ecosystem perspective. *New Phytol.,* **147:** 3-12.

Orians, C. M., Ardon, M. and Mohammad, B. A. (2002). Vascular architecture and patchy nutrient availability generate within-plant heterogeneity in plant traits important to herbivores. *Am. J. Bot.* **89:** 270-278.

Passioura, J. B. (2002). Soil conditions and plant growth. *Plant Cell Environ.* **25:** 311-318.

Pearcy, R. W. (1998). Acclimation to sun and shade. In *Photosynthesis: A Comprehensive Treatise* (A. S. Raghavendra, ed.) pp. 250-263. Cambridge University Press, Cambridge.

Peirson, D. R. and Elliott, J. R. (1988). Effect of nitrite and bicarbonate on nitrite utilization in leaf tissue of bush bean (*Phaseolus vulgaris*). *J. Plant Physiol.* **133:** 425-429.

Poorter, H. and Navas, M. L. (2003). Plant growth and competition at elevated $CO_2$: On winners, losers and functional groups. *New Phytol.* **157:** 175-198.

Rachmilevitch S., Cousins A. B., and Bloom A. J. (2004). Nitrate assimilation in plant shoots depends on photo respiration. *Proc. Natl. Acad. Sci. USA,* **101:** 11506-11510.

Robinson, D., Linehan, D. J. and Caul, S. (1991). What limits nitrate uptake from soil? *Plant Cell Environ.* **14:** 77-85.

Robinson, J. M. (1987). Interactions of carbon and nitrogen metabolism in photosynthetic and non-photosynthetic tissues of higher plants: Metabolic pathways and controls. In *Models in Plant Physiology and Biochemistry* (D. W. Newman and K. G. Stuart, eds.) Vol. 1, pp. 25-35. CRC Press, Boca Raton, FL.

Rogers, H. H., Prior, S. A., Runion, G. B. and Mitchell, R. J. (1996). Root to shoot ratio of crops as influenced by $CO_2$. *Plant Soil* **187:** 229-248.

Rudd, J. J. and Franklin-Tong, V. E. (2001). Unravelling response-specificity in $Ca^{2+}$ signalling pathways in plant cells. *New Phytol.* **151:** 7-33.

Rufty, T. W., Thomas, J. F., Remmler, J. L., Campbell, W. H. and Volk, R. J. (1986). Intercellular localization of nitrate reductase in roots. *Plant Physiol.* **82:** 675-680.

Samuelson, M. E., Eliasson, L. and Larsson, C.-M. (1991). Nitrate-regulated growth and cytokinin responses in seminal roots of barley. *Plant Physiol.* **98:** 309-315.

Searles, P. S. and Bloom, A. J. (2003). Nitrate photoassimilation in tomato leaves under short-term exposure to elevated carbon dioxide and low oxygen. *Plant Cell Environ.* **26:** 1247-1255.

Shingles, R., Roh, M. H. and McCarty, R. E. (1996). Nitrite transport in chloroplast inner envelope vesicles. *Plant Physiol.* **112:** 1375-1381.

Sivasankar, S. and Oaks, A. (1996). Nitrate assimilation in higher plants—the effect of metabolites and light. *Plant Physiol. Biochem.* **34:** 609-620.

Spollen, W. G., LeNoble, M. E., Samuels, T. D., Bernstein, N. and Sharp, R. E. (2000). Abscisic acid accumulation maintains maize primary root elongation at low water potentials by restricting ethylene production. *Plant Physiol.* **122:** 967-976.

Takei, K., Takahashi, T., Sugiyama, T., Yamaya, T. and Sakakibara, H. (2002). Multiple routes communicating nitrogen availability from roots to shoots: A signal transduction pathway mediated by cytokinin. *J. Exp. Bot.* **53:** 971-977.

Thornley, J. H. M. (1977). Root:shoot interactions. In *Integration of Activity in the Higher Plant* (D. H. Jennings, ed.) pp. 367-389. Cambridge University Press, Cambridge.

Valladares, F., Skillman, J. B. and Pearcy, R. W. (2002). Convergence in light capture efficiencies among tropical forest understory plants with contrasting crown architectures: A case of morphological compensation. *Am. J. Bot.* **89:** 1275-1284.

Vaughn, K. C. and Campbell, W. H. (1988). Immunogold localization of nitrate reductase in maize leaves. *Plant Physiol.* **88:** 1354-1357.

Vernieri, P., Lenzi, A., Figaro, M., Tognoni, F., and Pardossi, A. (2001). How the roots contribute to the ability of *Phaseolus vulgaris* L. to cope with chilling-induced water stress. *J. Exp. Bot.* **52:** 2199-2206.

Vitousek, P. M. and Field, C. B. (1999). Ecosystem constraints to symbiotic nitrogen fixers: A simple model and its implications. *Biogeochemistry (Dordrecht)* **46:** 179-202.

Watson, M. A. and Casper, B. B. (1984). Constraints on the expression of plant phenotypic plasticity. *Annu. Rev. Ecol. Syst.* **15:** 233-258.

Wells, C. E. and Eissenstat, D. M. (2002). Beyond the roots of young seedlings: The influence of age and order on fine root physiology. *J. Plant Growth Regulation* **21:** 324-334.

Whorf, T. and Keeling, C. D. (1998). Rising carbon. *New Scientist* **157:** 54.

Wilkinson, S., Clephan, A. L. and Davies, W. J. (2001). Rapid low temperature-induced stomatal closure occurs in cold-tolerant *Commelina communis* leaves but not in cold-sensitive tobacco leaves, via a mechanism that involves apoplastic calcium but not abscisic acid. *Plant Physiol.* **126:** 1566-1578.

Wilkinson, S. and Davies, W. J. (2002). ABA-based chemical signalling: The co-ordination of responses to stress in plants. *Plant Cell Environ.* **25:** 195-210.

# 13

# Sweeping Water, Oozing Carbon: Long Distance Transport and Patterns of Rhizosphere Resource Exchange

*Zoë G. Cardon and Patrick M. Herron*

Effects of long distance transport of water and carbon within plants extend beyond the boundaries of the plants themselves. Focusing belowground exactly 100 years ago, Lorenz Hiltner coined the term *rhizosphere* to identify the zone in which legume roots affected soil bacteria (Lynch, 1990). The definition has since expanded to include the volume of soil around any root that is influenced physically or chemically by the root. Long distance transport of water and organic carbon within plants contributes to creating rhizosphere soil conditions where a particularly active microbial, mycorrhizal, and microfaunal community thrives.

This vibrant rhizosphere microbial community has long been recognized to be supported by rhizodeposition of organic carbon by roots to soil, including exudation (diffusion) of low molecular weight compounds to soil (e.g., sugars, organic acids, and amino acids), secretion of enzymes and mucilage, and sloughing of root cap cells (Grayston *et al.*, 1996; Whipps, 1990). However, active microbes (both bacteria and fungi) in the rhizosphere require more than just root-derived, organic carbon to grow and reproduce, and the sources of other essential elements that the microbes tap influences nutrient availability to plants. Nitrogen and phosphorus (among other nutrients) are essential for construction of microbial biomass, and they can be gathered either from the soil solution or from decomposition of soil organic matter (SOM) catalyzed by microbial extracellular enzymes (e.g., Moorhead and Linkins, 1997; Schimel and Weintraub, 2003). Which of these nutrient sources the microbes tap to make new biomass sets the stage for whether (1) roots and microbes compete for the dissolved

Copyright © 2005 by Elsevier, Inc.
All rights of reproduction in any form reserved.

mineral nutrients in soil solution (which both plants and microbes can take up (e.g., Diaz *et al.*, 1993), or (2) microbes move nutrients from the SOM pool (otherwise largely unavailable to plant roots) into new, active microbial biomass. (The nutrients in SOM are largely unavailable to plant roots because, in general, roots lack the capacity to produce the suite of extracellular enzymes necessary to break common chemical bonds in SOM and release monomers or mineral nutrients.) The microbial incorporation (immobilization) of nutrients newly acquired from SOM into new microbial biomass creates the potential for those nutrients to be released ultimately to roots, if the microbial biomass turns over (i.e., dies) through protozoal grazing or physical stress (e.g., Clarholm, 1985). Thus the dynamics with which SOM-derived nutrients within microbial biomass can become available to plant roots depends on microbial turnover. Taking all this information together, the rhizosphere can be considered a commodities exchange in terrestrial ecosystems, where organic carbon lost from roots fuels a rhizosphere community that, on one hand, can compete with roots for dissolved mineral nutrients in soil solution, but that also can decompose SOM, mining it for essential nutrients that otherwise would be inaccessible to plant roots.

The role of mycorrhizae in the nutrition of the vast majority of plant species also requires mention, though we do not focus on mycorrhizae in this chapter. Recognizing the important role of mycorrhizae in the mineral nutrition of plants does not negate the potential significance of other rhizosphere microbes for plant nutrition. Many mycorrhizal fungi (particularly arbuscular mycorrhizae) do not have the ability to decompose organic matter with nearly the efficiency of saprotrophic microbes, though it appears that ericoid and ectomycorrhizae are more capable in their saprotrophic function than originally realized (see Read and Perez-Moreno, 2003, for a review). From the point of view of mineral nutrient availability to plants, the complementarity of mycorrhizae and the saprotrophic rhizosphere microbial community deserves greater study (Bonkowski *et al.*, 2001).

Though the emphasis on carbon loss to the rhizosphere through rhizodeposition has dominated the thinking of ecological rhizosphere researchers (e.g., Newman, 1985), another aspect of plant root biology has the potential to dramatically influence rhizosphere community composition and activity. The rhizosphere soil directly around roots is also the portal for vast water flux from bulk soil to plants. Of the 60 trillion tons of water that cycle from soils to the atmosphere each year, nearly two thirds passes through the bodies of plants and is transpired from leaves (Holbrook and Zwieniecki, 2003). All of that water was first drawn into plants through the rhizospheres of innumerable plant roots. When transpiration is occurring, the bulk sweep of water toward the root carries dissolved nutrients through the rhizosphere toward the root surface, while likely inhibiting the diffusional flux of soluble carbon compounds out of the root to surrounding rhizosphere soil. An enormous literature exists describing water and nutrient fluxes to plant roots (see Tinker and Nye, 2000, for details), and an essential next step in rhizosphere research is knitting this literature with the organismal and ecological literature focusing on the rhizosphere community. Soil water content and films of water on soil particles change with the diurnal ebb and flow of the transpiration stream, likely influencing the activity and interactions of soil bacteria, fungi, protozoa, nematodes, and other organisms (Or and Tuller, 1999; Parr *et al.*, 1981). Overall, the transpiration stream draws down soil water content until soil water is recharged by precipitation, by capillary movement from wetter soil layers, or by passive hydraulic redistribution via plant root conduits. The spatial heterogeneity and dynamics of water potential and water movement in zones around roots add a fascinating layer of complexity to the story of carbon and nutrient cycling in the rhizosphere, a layer that is explored later.

Such exchanges of essential energy and nutrients in the rhizosphere system, but with costs, benefits, competition, and interdependence, suggest there is potential for evolutionary innovation and natural selection associated with systemic feedback loops. Roots have been surrounded by a soil microbial and faunal community throughout the 400 million years of their evolution (Algeo and Scheckler, 1998; Raven and Edwards, 2001). This inevitable juxtaposition has set the stage for a dynamic interaction, particularly since mineral nutrients often limit the primary production and growth of plants (Schlesinger, 1997), and the availability of labile organic carbon in bulk soil (distant from roots) limits microbial growth (Cheng *et al.*, 1996). A deeper understanding of rhizosphere function requires probing the essence of rhizosphere system behavior with mathematical modeling and particularly with nondestructive measures that can be made continuously, or at least repeatedly, *in situ* in soil (e.g., Bringhurst *et al.*, 2001). Such basic questions as, "how plastic a trait is rhizodeposition?" still remain unanswered today, largely because of the difficulty in studying rhizosphere processes belowground without digging up and thus destroying the system.

In this chapter, we consider mechanisms by which long distance transport processes in plants may affect resource exchange in the rhizosphere. First, we briefly discuss whether known phenological patterns of carbon allocation from shoots to roots within plants translate into temporal patterns of carbon availability and/or microbial and microfaunal activity in the rhizosphere. Second, we consider potential effects of spatial and temporal patterns of root water uptake on rhizosphere community function, via effects on dissolved nutrients delivered to the rhizosphere community by bulk flow, effects on soil water content, or effects on diffusion of carbon from roots to the rhizosphere. Together, these ideas naturally spur a view of roots as part of a larger belowground nutrient- and water-acquisition system, a system with varying membership depending on the rhizosphere community. Shoots deliver carbon to the roots and thus to this system via long distance transport; the strength of the carbon sink belowground can potentially vary with root control over components of rhizodeposition. Long distance transport of water initiated by stomatal opening, and potentially modulated further by regulation of aquaporins (Tournaire-Roux *et al.*, 2003), can likely affect rhizosphere water and nutrient content. It is not known to what extent rhizodeposition and water fluxes are influenced by feedback loops from mineral nutrient availability in the rhizosphere, though intriguing information is building. Clarkson *et al.* (2000), for example, review experiments with isolated roots indicating that hydraulic conductivity of root segments is sensitive to the concentration of particular mineral nutrients in surrounding solution. Over evolutionary time, environmental pressures may have driven plants toward various allocation and uptake strategies belowground that should be viewed not simply as strategies of carbon allocation and water uptake *within* plants, but also as strategies affecting rhizosphere function with potential feedbacks from nutrient availability. We end the chapter with suggestions for future rhizosphere research.

## Long Distance Transport of Carbon

Allocation of carbon belowground by plants can be a substantial portion of the plant carbon budget. Whipps (1990) and Grayston *et al.* (1996), for example, review estimates of carbon allocation belowground in systems ranging from hydroponic plant culture to mature forests. They report that large allocations of 50% or more of assimilated carbon to roots and rhizodeposition are common, with values as high as 73% being listed for Douglas fir roots and mycorrhizae (Whipps, 1990). Within that allocated carbon flux, a substantial proportion can be associated with rhizodeposition to soil alone (e.g., 2% to 70%, for a range of species in Whipps, 1990). Even within single species, estimates of the allocation of carbon belowground vary experiment to experiment; for wheat, Whipps lists values of 32% to 59% of net fixed carbon being transferred to roots in different experiments. The timing of carbon allocation also varies dramatically with season and phenology. A clear challenge for rhizosphere research is discerning predictable patterns in this variation, and considering whether the variation has implications for rhizosphere processes.

### Phenology of Carbon Allocation Belowground

An extensive literature describes patterns of carbon allocation belowground in annual and perennial plants, particularly in important agricultural and silvicultural species (for reviews, see Borchert, 1991; Dickson, 1991a; Grayston *et al.*, 1996; Kuzyakov, 2001; Kuzyakov and Domanski, 2000; Whipps, 1990). Seasonal changes in root growth, root activity, and long distance transport of sugars to roots, discerned using carbon isotopic techniques, frequently accompany phenological patterns of vegetative growth, flowering, fruiting, and senescence. For example, in recent work with annual plants, Fu *et al.* (2002) used the natural carbon isotopic labeling provided by the $C_3$ and $C_4$ photosynthetic pathways to explore rhizosphere respiration (defined as root respiration plus microbial respiration of root-derived organic compounds) as a function of plant species (soybean, sunflower, sorghum, and amaranthus) and seasonal phenology. By growing $C_3$ plants in soils with SOM derived from $C_4$ plants, and growing $C_4$ plants in soils with SOM derived from $C_3$ plants, contributions of newly-fixed carbon and SOM-derived carbon to soil respiration and soil carbon pools could be quantified. Rhizosphere respiration (by definition derived from newly fixed carbon), expressed per gram living root, decreased as fruits developed and senescence approached, suggesting that allocation of carbon belowground to roots decreased once reproduction was initiated.

Experiments relying on isotopically labeled photosynthate have also been used to examine patterns of carbon allocation belowground in perennial plants, as well as potential rhizosphere responses. For example, oaks (and a number of other trees, see Borchert, 1991) have a semideterminate growth pattern, meaning that shoot growth occurs in pulses (flushes) over the course of the growing season. (In oaks, root growth also proceeds in flushes out of phase with shoot growth, but this is not the case for all semi-determinate trees; Borchert, 1991.) Dickson (1991b) showed that mature first flush leaves in northern red oaks allocate 95% of newly fixed carbon belowground when roots are actively growing, and then switch to allocate 90% to new shoots when new shoots and leaves are growing. This switching of allocation of newly fixed photosynthate to belowground and aboveground sinks continues as roots and shoots alternate growth. Does the enhanced allocation of carbon belowground when roots are growing necessarily mean that carbon movement to the rhizosphere is also enhanced? Stating the question more broadly, do known temporal patterns in carbon allocation to roots within plants translate necessarily into parallel patterns of carbon loss from roots to the rhizosphere? We do not know, but it seems likely that at the minimum, higher root growth rates will enhance rhizodeposition of root cap cells to the surrounding rhizosphere soil.

Semideterminate tree species such as oaks offer an unusual opportunity for exploring effects of varying plant carbon allocation belowground on rhizosphere microbial biomass and activity. Several episodes of high carbon allocation belowground occur during a single growing season at easily discerned times, when shoots are not growing. This natural, pulsed allocation pattern allows separation of the effects of variation in carbon allocation belowground from effects of seasonal environmental variation on rhizosphere microbial biomass and activity. In a simple first experiment to explore this idea using northern red oaks, Cardon *et al.* (2002) found that soil respiration rates (including root, microbial, and soil faunal respiration) varied inversely with pulsed shoot growth rates and in parallel with pulsed root growth, but net mineralization of nitrogen was unaffected by shoot and root flush stage. Rhizosphere microbial biomass was higher around roots of oaks with mature shoots (growing roots) than around roots of oaks with expanding shoots (and nongrowing roots) at a final harvest, but it was not possible to discern whether this stimulation was caused simply by enhanced root growth (and, for example, root cap sloughing), by enhanced exudation or secretion of organic compounds, or a combination of these fluxes.

It remains to be seen whether exudation and secretion of compounds are generally enhanced by increased allocation from shoots to roots. If the carbon allocation patterns within plants do affect fluxes of carbon to the rhizosphere, then the extensive physiological, silvicultural, and agricultural literatures describing known carbon allocation strategies are potentially a goldmine of information. A reexamination of correlations between well-known allocation strategies and soil environmental conditions under which these strategies are found could lead to an interesting reinterpretation of potential consequences of plant carbon allocation strategies for rhizosphere processes and nutrient availability.

### Microbial Response to Labile Carbon Availability in the Rhizosphere

As noted in the introduction to this chapter, the availability of labile carbon in the rhizosphere increases the microbial demand for nutrients as the microbial community grows (Cheng *et al.*, 1996; Chen and Stark, 2000) and that increased microbial demand may stimulate decomposition of SOM. However, rhizodeposition does not *necessarily* have a positive effect on decomposition of SOM by microbes in the rhizosphere, or on availability of limiting mineral nutrients to plants. Kuzyakov (2002) suggested that the relative availability of mineral nutrients in soil solution strongly influences whether release of labile (easily utilized) carbon flux from roots ultimately increases or decreases nutrient availability to plants. If mineral forms of nutrients (e.g., $NO_3^-$, $NH_4^+$, $PO_4^{3-}$) are available for uptake from soil solution, microbes have no reason to attack SOM for nutrients (Fig. 13.1). In this situation, microbes and plant roots may compete for dissolved mineral nutrients in soil solution; even with eventual turnover of new microbial biomass built using nutrients from soil solution, no nutrients will have been moved from a pool unavailable to roots (e.g., SOM) to a microbial biomass pool ultimately more readily available to roots. In this situation, the plant does not benefit nutritionally from having spurred microbial proliferation. On the contrary, enhancing microbial growth may well have increased plant-microbe competition for dissolved mineral nutrients in soil solution. Further, microbes with access to abundant labile carbon and dissolved mineral nutrients might turn away from breaking down SOM for carbon building blocks and preferentially use labile carbon substrate, decreasing mineralization of SOM in the presence of plant roots. Cardon *et al.* (2001) suggest that this may explain why decomposition of older pools of SOM carbon is depressed when exposure of California grassland communities to elevated $CO_2$ increases root growth and, likely, rhizodeposition in microcosms.

If labile carbon is available in the rhizosphere, but dissolved nutrients are not readily available in soil solution, then microbes may attack SOM with extracellular enzymes, and move nutrients into the active microbial pool by immobilizing them into biomass (e.g., Clarholm, 1985; Fig. 13.1). Protozoal grazing on that microbial biomass can result in protozoa eliminating excess prey N as ammonium (e.g., Kuikman *et al.*, 1989), making it available to roots. (Turnover of microbial biomass under physical stresses, too, could encourage release of nutrients trapped in microbial biomass, Kuikman *et al.*, 1989.) When protozoal grazing on microbes releases ammonium, carbon loss by roots could be viewed as having fueled a multitrophic interaction in the rhizosphere that led to increased mineral nutrient (ammonium) availability to plants (e.g., Kuzyakov, 2002). This mechanism clearly can operate in microcosms (e.g., Elliott *et al.*, 1979; Ingham *et al.*, 1985), but its importance in nature is unclear. Expanding on this conceptual model, it is important to note that microbes attack SOM using extracellular enzymes,

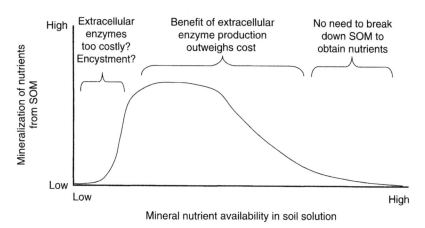

**Figure 13.1**   Enhancement of microbial breakdown of soil organic matter (SOM), caused by labile carbon substrates from roots, along a gradient of mineral nutrient availability in soil solution.

which have a nitrogen cost. If mineral nutrient concentrations are extremely low in the environment of microbes, production of extracellular enzymes to attack SOM for nutrients may be too costly. (For a model of microbial allocation to extracellular enzymes, see Sinsabaugh and Moorhead, 1994.) Stimulation of SOM breakdown by microbes with increased labile rhizosphere carbon, then, may be most likely to occur when mineral nutrient availability is in an intermediate range, low enough that soil solution ions cannot supply sufficient nutrients to produce microbial biomass, but high enough that the costs of extracellular enzyme production are not too high (Fig. 13.1).

Considering plant communities, Kuzyakov (2001) hypothesized that annual plants transport more carbon belowground specifically for labile rhizodeposits than do perennial woody plants (per unit root). Kuzyakov argued that because perennials annually retranslocate and store mineral nutrients from senescent structures before dropping them, this storage ensures that perennials do not rely as completely on soil supplies of mineral nutrients (and thus, potentially, on rhizosphere microbial activity that may be key for releasing those nutrients) to support future biomass production. The implicit assumption behind this hypothesis is that enhanced rhizodeposition drives increased nutrient availability; this may or may not be true depending on environmental conditions. Interestingly, Grayston *et al.* (1996) suggest the opposite broad pattern, that the proportion of assimilated carbon ending up as rhizodeposition is larger in perennials than annuals. However, Grayston *et al.* (1996) then cites percentages of carbon allocated by trees to rhizodeposition *and roots*, which can indeed be extraordinarily high (e.g., 73% of assimilate), so their argument appears to focus on total carbon allocation belowground, rather than carbon allocated specifically to rhizodeposition. Likely, the nutrient demand within the plant, root/shoot ratios, and the proportion of root mass that is active in nutrient uptake and rhizodeposition will all enter into determining whether annuals and perennials have different rhizodeposition strategies potentially influencing rhizosphere nutrient cycling.

This scheme suggests several more questions, as rhizodeposition, or, more specifically, exudation or secretion of labile carbon compounds, could be under different selective pressures in different soil environments. Do plants that tend to be associated with nutrient-rich environments tend to have lower labile carbon loss per unit root than plants that tend to be associated with environments poor in mineral nutrients? Can rhizodeposition be adjusted on physiological time scales in response to perceived nutrient availability around roots or plant demand for nutrients? There is evidence in the mycorrhizal literature for plasticity in the extent of labile carbon loss from roots (e.g., Johnson, 1993); enhanced carbon flux encourages establishment of the mycorrhizal symbiosis. Beyond such obvious symbioses, has the general microbial milieu in which roots have evolved influenced traits associated with the quality, quantity, timing, or plasticity of rhizodeposition? Answers to these questions await rhizosphere research with an evolutionary perspective.

## Spatial Heterogeneity of Carbon Fluxes Along Roots

Many of the experiments described here relied on tracking carbon fixed by plants and transferred to soils by using isotopically labeled carbon dioxide ($CO_2$). Either labeled $CO_2$ is applied to photosynthesizing shoots, or researchers have taken advantage of natural labeling of new photosynthate associated with the $C_3$ and $C_4$ photosynthetic pathways (e.g., Kuzyakov, 2001; Kuzyakov and Domanski, 2000). These isotopic approaches have yielded important information about gross carbon fluxes from roots to soils and rhizosphere organisms, but their focus has not been the spatial and temporal heterogeneity in rhizodeposition along roots. From the microbial perspective, however, this heterogeneity may be extremely important in determining where and when microbes grow (e.g., Darrah, 1991) and thus where SOM is mineralized, and where protists graze. Numerous conceptual and mathematical models have been used to consider whether heterogeneity in the spatial distribution of carbon loss along young roots could affect microbial growth and, potentially, mineralization of organic matter (e.g., see Toal *et al.*, 2000, for a review). Kuzyakov (2001) presented a conceptual model, illustrating the root tip, elongation, and maturation zones, as well as the zone of suberization, and suggested spatial patterns in microbial processes (including mineralization of SOM) associated with these root zones. Maximal microbial growth and priming of decomposition of SOM are predicted in the zone of elongation, between the tip and the maturation zone where root hairs are produced.

Currently, the basic data needed to parameterize or test such models for roots growing in nonsterile soils are largely lacking. Ideally, measurements of processes in the rhizosphere should be made nondestructively, continuously, and at spatial and temporal scales appropriate to the questions being asked. This is difficult to achieve, as the processes are occurring out of view in soil belowground, labile carbon substrates lost from roots can be rapidly used by rhizosphere microbes, and important spatial heterogeneity in carbon, mineral nutrient, and water availability may occur over micrometer scales. One experimental approach is to use living "reporter" microorganisms seeded into nonsterile soil to report the physical or chemical conditions in their soil environment, or report their own biological responses to those physical or chemical conditions. These genetically engineered microbes can be designed so that production of any one of a number of reporter molecules or enzymes is under the control of an inducible promoter of choice. Engineered microbiosensors commonly produce green fluorescent protein (GFP, encoded by the *gfp* gene), β-galactosidase (encoded by *lacZ*), ice nucleation protein (encoded by *inaZ*), or luciferase (*lux* genes), among other molecules (for reviews of microbial reports and reporters, see Hansen and Sorensen, 2001; Leveau and Lindow, 2002). Each of these reporting mechanisms has advantages and disadvantages. For example, the ice nucleation report can be extraordinarily sensitive because of a built-in amplification system (Jaeger *et al.*, 1999); however, assay of the ice nucleation report requires destructive harvest of soils, making repeated measures impossible. In contrast, GFP production in rhizosphere microorganisms can be detected through clear-sided microcosms nondestructively by using epifluorescence or confocal fluorescence microscopy (e.g., Bringhurst *et al.*, 2001). The GFP report, though not as sensitive as the ice nucleation report, allows spatial information about the rhizosphere to be gathered nondestructively at micrometer scales over time. Interpretation of data from all microbiosensors, no matter the form of the report, requires recognition that a number of microbe- and reporter-specific properties beyond promoter activity can influence the report including (1) the time required to make and fold the reporter protein, (2) the turnover time of the reporter molecule in the cell, and (3) the possibility that differential bacterial growth rates could dilute the report to different extents independent of activity of the promoter (see Leveau and Lindow, 2001).

With proper care in interpretation, microbiosensors can provide a new glimpse of the rhizosphere environment, as perceived by microbes, and can report the microbial metabolic response to that environment. In a pioneering study, Jaeger *et al.* (1999) used genetically engineered bacteria seeded into nonsterile soils to map tryptophan and sucrose pools around young *Avena barbata* roots. Jaeger *et al.* (1999) found highest sucrose presence in the soil harvested 0 to 4 cm from the root tip, and sucrose declined to very low levels (near those found in bulk soil) approximately 8 to 12 cm from the root tip, where lateral root primordia were beginning to appear. Tryptophan, in contrast, was at low levels around the root tip, and increased in concentration from 8 to 12 and 12 to 16 cm from the tip, where lateral roots were emerging. The number of engineered bacteria per unit weight soil was highest at the root tip (0 to 4 cm) and where lateral roots had emerged (~16 to 20 cm from the tip), and lowest where both sucrose and tryptophan pools were low (~8 to 12 cm from the tip). This microbial biomass pattern is in contrast to predictions that might emerge from conceptual models such as Kuzyakov's, where a single maximum biomass is predicted behind the root tip. Zelenev *et al.* (2000), in contrast, predict distributions of bacterial biomass along roots that take the form of waves, with population booms and busts generating the pattern as a function of high carbon loss at the tip and root growth. In natural, unsterilized soil, patterns of bacterial biomass along roots undoubtedly reflect the integration of abiotic and biotic environmental factors influencing microbial growth, division, and death, including, but not limited to, patterns of fluxes of labile carbon from roots and grazing of microbial biomass by protozoa.

Several research groups have now developed microbiosensors for use exploring microbial ecology in soils (e.g., Hansen and Sorensen, 2001; Killham and Yeomans, 2001; Leveau and Lindow, 2002; Ramos *et al.*, 2000). In collaboration with Dr. Daniel Gage and his laboratory group, we are developing such tools, and using them to begin exploring rhizosphere interactions nondestructively at high spatial resolution in nonsterile soils. The genetically engineered *Sinorhizobium meliloti* we have developed are useful for exploring dynamic resource availability and microbial growth. Galactoside-sensing microbes report, with green fluorescence, the presence of galactosides (common in phloem sap) around roots in nonsterile soils by using the galactoside-inducible *melA* promoter to drive transcription of the *gfp* gene (Bringhurst *et al.*, 2001). Another biosensor strain we have developed contains a *S. meliloti* rRNA promoter linked to *gfp(mut3)*; this biosensor produces GFP in proportion to microbial growth rate under steady state conditions (Rosado and Gage, 2002). Still other sensors, for example those engineered to constitutively express red fluorescent protein, fluoresce for hours from inside vacuoles after being engulfed by protozoa in the rhizosphere, thus flagging predators in zones of intense protozoal grazing (Bringhurst *et al.*, 2001). All these microbiosensors can be seeded into nonsterile soils and viewed nondestructively over time through glass-sided microcosms using epifluorescence or confocal microscopy.

Another important step toward understanding spatial heterogeneity in the flux of labile carbon from roots to soils lies in understanding how sugars move from mature phloem, through immature, growing primary root tissue, to the rapidly dividing apical meristem. Using empirical and modeling techniques, Bret-Harte and Silk (1994) concluded that the symplastic route for carbon movement to the root apical meristem could not support the carbon fluxes necessary for root growth. Some transport was likely apoplastic. Once sugars are in the apoplasm, however, they can diffuse into the surrounding soil. Taking a rhizosphere perspective, are such zones of apoplastic transport hot-spots for availability of sugars to microbes? Is this part of the reason that labile carbon availability at the root tip is particularly high?

Pritchard and Newbury (Chapter 8) provide a detailed model of apoplastic and symplastic sugar movement from sieve elements to surrounding tissue in growing roots, describing three major zones behind the root tip where sieve elements are functional and the surrounding root tissue is maturing. These zones (called symplastic growing, apoplastic growing, and apoplastic mature, from the tip back) vary in the dominance of apoplastic and symplastic routes of sucrose delivery from phloem to cells; distributions of aquaporins, cell wall invertases, plasmalemma hexose transporters, and tonoplast transporters in root tissue reflect the balance of these different routes. Within a small root length (~ millimeter scale, depending on the species), widely differing functional zones clearly exist, each with its own characteristics that could strongly affect diffusional flux of sugar from the root to the rhizosphere. When sugars are localized in the apoplast, there is no membrane to oppose their diffusing to soil, but the activity of invertase and hexose transporters strongly influences the size of the pool of sugar that might diffuse to soil. An immobile rhizosphere community will be subjected to several root zones sequentially as a root grows past and matures, first encountering the root cap with its mucilage secretions, second the meristem and undifferentiated early tissue, and then the three zones defined by Pritchard and Newbury (Chapter 8) where some phloem has matured. The size of these zones and the speed of root growth will interact to expose the rhizosphere community to varying resource availability through time. Measurement of these variations *in situ*, and of microbial response, would provide useful clues to understanding links between characteristics of carbon fluxes from roots and associated rhizosphere microbial activity.

## Long Distance Transport of Water

As noted in the introduction, high rhizosphere microbial biomass and activity has long been thought to be associated with readily available carbon from roots, and extensive empirical rhizosphere research and mathematical modeling have focused on exchanges of carbon and nutrients between plants and microbes. Effects of plant roots on the rhizosphere, however, extend beyond the delivery of newly fixed carbon to the soil. Transpirational flux drives fluctuations in rhizosphere water potential day to day during seasons when plants are active. Under the influence of the transpiration stream, water flows through the rhizosphere toward roots, changing soil water content and water film thicknesses diurnally, and exposing the rhizosphere community to a sweep of soil solution with dissolved organic and inorganic nutrients. Literature focusing on water fluxes (and associated dissolved nutrient delivery) to plant roots has remained substantially separate from research focusing on the rhizosphere soil community. Numerous mathematical models and empirical experiments in the agricultural, soil hydrological, and ecological literatures (see Tinker and Nye, 2000, and Hillel, 1998, for broad reviews) have examined diffusion of mineral nutrients to roots and delivery of nutrients to roots by bulk flow of soil solution, and these ideas and models need to be linked with the many organismal and ecological models focusing on rhizosphere resource exchange. Here we briefly consider how water fluxes potentially affect the rhizosphere microbial and faunal community biomass and activity, and associated carbon and nutrient cycling.

### Spatial and Temporal Heterogeneity of Water Fluxes Along Roots

At the scale of individual roots, spatial heterogeneity in water uptake along the root length has striking implications for activity in the rhizosphere community. Vasculature in the root matures some distance behind the root tip, and that distance varies with plant species and rate of root growth (e.g., centimeters behind the tip in corn, North and Peterson [Chapter 7]; millimeters behind the tip in pea, Kramer and Boyer, 1995). A hydraulic "dead zone" at the root tip results that is spared the dramatic, transpiration-driven fluctuations in water potential experienced by mature portions of the root (Zwieniecki *et al.*, 2002). Hsiao and Xu (2000) suggested that the delayed xylem maturation, coupled with the "leaky" nature of xylem conduits, thus ensures that the growing tip of the root has essential first access to water available in the newly explored soil volume. This characteristic hydraulic dead zone, which for example in corn may extend 2 cm back from the tip (Chapter 7), also potentially spares the rhizosphere community at the root tip from large, diurnal fluctuations in water potential. Depending on the speed of root tip elongation, rhizosphere organisms near the root tip are bathed for a time in mucilage, sloughed root cap cells, and, likely some labile carbon substrates diffusing into the soil, without enormous diurnal rhythms in water flux and water potential influencing rhizosphere community behavior. This period of relative hydraulic calm in the rhizosphere, however, does not last.

Behind the root tip, in root zones where xylem has matured, we speculate that bulk flow of water into the root under control of the transpiration stream likely opposes any diffusion of labile carbon from the root to the rhizosphere during the day. At night, as the water potential gradient between the soil and plant relaxes, bulk flow into the root ceases, and diffusion of labile carbon from root to rhizosphere may occur more readily. Beyond this carbon flux, Caldwell *et al.* (1998) suggest that in dry environments, hydraulic redistribution of water *out* of roots into drier rhizosphere soil at night may contribute to sustaining rhizosphere activity and biogeochemical cycling in soils that otherwise would be too dry for microbial activity to continue. Not only might delivery of water at night sustain rhizosphere microbes in such environments, but also the water flow out of roots to dry soils could flush labile, apoplastic carbon out of roots and into the surrounding rhizosphere. (The availability of labile carbon within the root, however, may decrease substantially at night because photosynthesis has ceased). Diurnal cycling of the transpiration stream could thus strongly affect rhizosphere microbial activity through changing water content and/or pulsed opposition to diffusion of carbon out of the root; these possibilities need to be explored.

Adding further complexity to the picture, the rhizosphere community around root zones that are actively taking up water is cyclically bathed in a transpirationally driven stream of soil solution containing dissolved organic and inorganic mineral nutrients. When the transpiration stream stops, immobilization by microbes and roots continues, but in a standing pool of soil solution around the roots. How does the microbial community respond to these different conditions of water and nutrient flux that vary on a diurnal cycle? And, as aquaporin activity and distribution in root cell membranes can also vary with environmental conditions (e.g., see Chapter 7), what are the implications of this root-controlled variation, and associated altered water flux, for resource exchange in the rhizosphere? A major challenge for future rhizosphere research is to embrace the biogeochemical implications of heterogeneous water fluxes in space and time along roots; these heterogeneities stem from changing radial and longitudinal conductivity of the roots themselves (e.g., Clarkson *et al.*, 2000; Landsberg and Fowkes, 1978; Tournaire-Roux *et al.*, 2003; Zwieniecki *et al.*, 2001), and changing hydraulic conductivity of soils and diffusion of solutes through soils as a function of water content (generally summarized by Hillel, 1998; Moldrup *et al.*, 2001).

### Measurement of Water Potential at Microbial Spatial Scales

In keeping with the idea that resource exchange between roots and the rhizosphere community is likely influenced by spatial and temporal patterns of water uptake by roots, developing a method to measure water potential changes at microbial spatial scales is a high priority. Until now, there has been no method to observe water potential gradients at extremely small scales in soils (Hillel, 1998). Soil psychrometers, time domain reflectometry, and neutron probes all integrate information from much larger volumes than those experienced immediately by microbes and soil fauna. We are developing a soil microbiosensor that reports total soil water potential (matric and osmotic combined) based on work pioneered by Axtell and Beattie (2002). They developed a microbiosensor strain of *Pantoea agglomerans*, a bacterium found on leaves, in which GFP transcription is under control of the promoter from the *Escherichia coli proU* transport system. This system is induced in *E. coli* when the bacterium is in environments with high osmolarity. By exposing the biosensors to concentrations of either sodium chloride (NaCl) or 8000-molecular-weight polyethylene glycol (PEG 8000) known to generate identical water potentials, Axtell and Beattie showed that GFP production was induced in their *P. agglomerans* strain to very similar extents regardless of the source of the osmotic potential outside the cell. (This comparison was made from bacteria growing on agar plates permeated with the known NaCl and PEG8000 concentrations.) Because PEG8000 cannot cross cell membranes, the response of the bacteria to the presence of PEG8000 suggests that the reporters may be sensing total water potential, not simply salt concentration. To test this idea, we constructed vials in which a known water potential of the headspace atmosphere could be generated through equilibration with a solution of known osmotic potential (Fig. 13.2A). We used that atmospheric water potential to control water potential in a small soil sample by supporting a cup containing 36 mg sandy loam on mesh in the atmosphere of each vial. Vials were sealed with rubber septum caps, and soil water potential was allowed to equilibrate at least 3 days with the controlled water potential in the vial atmosphere. Six µl of bacterial suspension in medium were then deposited on the soil in each vial using a syringe needle inserted through the septum cap. After equilibration for 22 hours, bacteria were extracted from soil, fixed, and analyzed by flow cytometry. GFP fluorescence per *P. agglomerans* cell was linearly related to the known, equilibrated total water potential in the vial (Fig. 13.2B). We are now working to engineer the common soil bacterium, *Pseudomonas putida*, to report soil water potential around roots *in situ*.

## Future Directions for Rhizosphere Research

Strategies of carbon allocation belowground in plants, and characteristics of water and nutrient uptake and transport mechanisms within plants, have captured the attention of plant physiologists for decades. Rhizosphere microbial ecologists have been similarly fascinated with the loss of organic compounds from roots, but the pulsing transpiration stream has not caught their attention. In general, the research and literature resulting from all these lines of inquiry have not yet been well integrated, though the interactions of roots, soils, water availability, and soil microbes and fauna are well recognized. Clearly, roots are not straws that passively and nonselectively move carbon to soils and remove water and nutrients from soils. And, soil and its living inhabitants are not floral foam passively delivering water and mineral nutrients to roots. Instead, the root, microbial, soil faunal, and physical soil mix belowground is a remarkably interactive and dynamic system.

How can the mechanisms underlying process and control in this system be uncovered? We believe that one promising approach will be to consider how the essential characters of long distance transport in plants affect, and

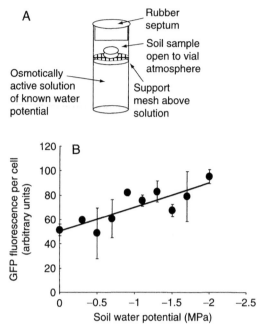

**Figure 13.2**   Dependence of GFP fluorescence per *Pantoea* microbiosensor cell on controlled soil water potential. In this representative experiment, microbiosensors grown in half-strength 21C medium (Halverson and Firestone, 2000) were resuspended in that medium plus 0.5% glucose and 0.08 M NaCl at optical density 0.24 (595 nm). (A) Microbiosensors were added to soil in vials containing osmotic solution at a range of water potentials (n = 3 vials per water potential) as described in the text, harvested after equilibration with atmospheric water potential 22 hours later, and (B) the geometric mean fluorescence per cell for each vial was determined using a Becton-Dickinson Facscalibur (San Jose, CA) dual laser flow cytometer standardized with Inspeck Green 505/515, 6 μm calibration beads. Settings were 488 nm argon ion excitation laser, 530 ± 30 nm bandpass filter for GFP fluorescence, 581 V setting for GFP detector. Data for >3000 bacteria per soil sample were analyzed using WinMDI 2.8. Symbols are mean ± standard error at each water potential. For the regression, $R^2 = 0.33$, $P<0.001$.

are affected by, rhizosphere resource exchange. We speculate, for example, that the rhizosphere community around active, mature portions of young roots may receive diurnally pulsed supplies of mineral nutrients and carbon compounds alternating in time. Water and nutrients flow toward the root during the day when transpiration is occurring, and labile carbon compounds likely diffuse more easily out of the root at night when bulk flow of water toward roots has ceased. Does this pulsing resource availability in the rhizosphere influence the efficiency of rhizosphere processes, including decomposition of soil organic matter? In arid systems where deep water is available, does hydraulic redistribution of water through roots from deep to surface soils at night help maintain microbial activity in otherwise dry surface soils?

Answering these questions with mechanistic understanding will require melding empirical and modeling work that spans microbial and soil faunal ecology, soil physics and chemistry, and physiological ecology of plant roots. Examining rhizosphere resource exchange *in situ* will be particularly challenging, and absolutely essential. On the empirical front, we need to move beyond the destructive indicator assays that have characterized so much of work belowground. *In situ*, continuous, nondestructive monitoring systems are essential for analyzing pools, fluxes, and interactions in soils, and these systems need to monitor at appropriate spatial and temporal scales depending on the question of interest. Because microbial activity is the gateway through which so many transformations within soils proceed, developing techniques to report the microbial perception of, and function within, the highly heterogeneous soil environment around roots is a high priority. Genetically engineered microbial biosensors are a step in the right direction, but of course using such genetically engineered microbes in the field is problematic. What are the next generations of sensors and approaches? Ultimately, the goal should be observing and understanding belowground pattern and process nondestructively, spanning scales important to microbes and fine roots, up to scales capturing patterns and processes across ecosystems.

## Acknowledgments

Supported by a grant from the Andrew W. Mellon Foundation to ZGC. We thank Missy Holbrook, Colin Orians, Jeremy Pritchard, and Maciej Zwieniecki for helpful comments on the manuscript, and John Stark for discussions of microbial limitations in soil.

## References

Algeo, T. J. and Scheckler, S. E. (1998) Terrestrial-marine teleconnections in the Devonian: Links between the evolution of land plants, weathering processes, and marine anoxic events. *Phil Trans R Soc Lond B* **353:** 113-130.

Axtell, C. A. and Beattie, G. A. (2002) Construction and characterization of a proU-gfp transcriptional fusion that measures water availability in a microbial habitat. *Appl Environ Microbiol* **68:** 4604-4612.

Bonkowski, M., Jentschke, G. and Scheu, S. (2001) Contrasting effects of microbial partners in the rhizosphere: Interactions between Norway Spruce seedlings (*Picea abies* Karst.), mycorrhiza (*Paxillus involutus* (Batsch) Fr.) and naked amoebae (protozoa). *Appl Soil Ecol* **18**: 193-204.

Borchert, R. (1991) Growth periodicity and dormancy. In *Physiology of Trees* (A.S. Raghavendra, ed.) pp. 221-245. Wiley, New York.

Bret-Harte, M. S. and Silk, W. K. (1994) Nonvascular, symplasmic diffusion of sucrose cannot satisfy the carbon demands of growth in the primary root tip of *Zea mays* L. *Plant Physiol* **105**: 19-33.

Bringhurst R. M., Cardon, Z. G. and Gage, D. J. (2001) Galactosides in the rhizosphere: Utilization by *Sinorhizobium meliloti* and development of a biosensor. *Proc Natl Acad Sci* **98**: 4540-4545.

Caldwell, M. M., Dawson, T. E. and Richards, J. H. (1998) Hydraulic lift: Consequences of water efflux from the roots of plants. *Oecologia* **113**: 151-161.

Cardon, Z. G., Czaja, A. D., Funk, J. L. and Vitt, P. L. (2002) Periodic carbon flushing to roots of *Quercus* rubra saplings affects soil respiration and rhizosphere microbial biomass. *Oecologia* **133**: 215-223.

Cardon, Z. G., Hungate, B. A., Cambardella, C. A., Chapin III, F. S., Field, C. B., Holland, E. A. and Mooney, H. A. (2001) Contrasting effects of elevated $CO_2$ on old and new soil carbon pools. *Soil Biol Biochem* **33**: 365-373.

Chen, J. and Stark, J. M. (2000) Plant species effects and carbon and nitrogen cycling in a sagebrush-crested wheatgrass soil. *Soil Biol Biochem* **32**: 47-57.

Cheng, W., Zhang, Q., Coleman, D. C., Carroll, C. R. and Hoffman, C. A. (1996) Is available carbon limiting microbial respiration in the rhizosphere? *Soil Biol Biochem* **28**: 1283-1288.

Clarholm, M. (1985) Interactions of bacteria, protozoa and plants leading to mineralization of soil nitrogen. *Soil Biol Biochem* **17**: 181-187.

Clarkson, D. T., Carvajal, M., Henzler, T., Waterhouse, R. N., Smyth, A. J., Cooke, D. T. and Steudle, E. (2000) Root hydraulic conductance: Diurnal aquaporin expression and the effects of nutrient stress. *J Exp Bot* **51**: 61-70.

Darrah, P. R. (1991) Models of the rhizosphere. I. Microbial population dynamics around a root releasing soluble and insoluble carbon. *Plant Soil* **133**: 187-199.

Diaz, S., Grime, J. P., Harris, J. and McPherson, E. (1993) Evidence of a feedback mechanism limiting plant response to elevated carbon dioxide. *Nature* **364**: 616-617.

Dickson, R. E. (1991a) Assimilate distribution and storage. In *Physiology of Trees* (A. S. Raghavendra, ed.) pp. 51-95. Wiley, New York.

Dickson, R. E. (1991b) Episodic growth and carbon physiology in northern red oak. In *The Oak Resource in the Upper Midwest: Implications for Management* (S. B. Laursen and J. F. DeBoe, eds.) pp. 117-124. Minnesota Extension Service, University of Minnesota, St. Paul.

Elliott, E. T., Coleman, D. C. and Cole, C. V. (1979) The influence of amoebae on the uptake of nitrogen in gnotobiotic soil. In *The Soil-Root Interface* (J. L. Harley and R. S. Russell, eds.) pp. 221-230. Academic Press, London.

Fu, S., Cheng, W. and Susfalk, R. (2002) Rhizosphere respiration varies with plant species and phenology: A greenhouse pot experiment. *Plant Soil* **239**: 133-140.

Grayston, S. J., Vaughan, D. and Jones, D. (1996) Rhizosphere carbon flow in trees, in comparison with annual plants: The importance of root exudation and its impact on microbial activity and nutrient availability. *Appl Soil Ecol* **5**: 29-56.

Halverson, L. J. and Firestone, M. K. (2000) Different effects of permeating and nonpermeating solutes on the fatty acid composition of *Pseudomonas putida*. *Appl Environ Microbiol* **66**: 2414-2421.

Hansen, L. H. and Sorensen, S. J. (2001) The use of whole-cell biosensors to detect and quantify compounds or conditions affecting biological systems. *Microb Ecol* **42**: 483-494.

Hillel, D. (1998) *Environmental Soil Physics*. Academic Press, San Diego.

Holbrook, N. M. and Zwieniecki, M. A. (2003) Plant biology: Water gate. *Nature* **425**: 361.

Hsiao, T. C. and Xu, L.-K. (2000) Sensitivity of growth of roots versus leaves to water stress: A biophysical analysis and relation to water transport. *J Exp Bot* **51**: 1595-1616.

Ingham, R. E., Trofymow, J. A., Ingham, E. R. and Coleman, D. A. (1985) Interactions of bacteria, fungi, and their nematode grazers: Effects on nutrient cycling and plant growth. *Ecol Monogr* **55**: 119-140.

Jaeger, C. H., Lindow, S. E., Miller, W., Clark, E. and Firestone, M. K. (1999) Mapping sugar and amino acid availability in soil around roots with bacterial sensors of sucrose and tryptophan. *Appl Environ Microbiol* **65**: 2685-2690.

Johnson, N. C. (1993) Can fertilization of soil select less mutualistic mycorrhizae. *Ecol Appl* **3**: 749-757.

Killham, K. and Yeomans, C. (2001) Rhizosphere carbon flow measurement and implications: From isotopes to reporter genes. *Plant Soil* **232**: 91-96.

Kramer, P. J. and Boyer, J. S. (1995) *Water Relations of Plants and Soils.* Academic Press, San Diego.

Kuikman, P. J., Van Vuuren, M. M. I. and Van Veen, J. A. (1989) Effect of soil moisture regime on predation by protozoa of bacterial biomass and the release of bacterial nitrogen. *Agr Ecosyst Environ* **27**: 271-279.

Kuzyakov, Y. (2002) Review: Factors affecting rhizosphere priming effects. *J Plant Nutr Soil Sci* **165**: 382-396.

Kuzyakov, Y. and Domanski, G. (2000) Carbon input by plants into the soil. Review. *J Plant Nutr Soil Sci* **163**: 421-431.

Kuzyakov, Y. V. (2001) Tracer studies of carbon translocation by plants from the atmosphere in to the soil (a review). *Euras. Soil Sci* **34**: 28-42.

Landsberg, J. J. and Fowkes, N. D. (1978) Water movement through plant roots. *Ann Bot* **42**: 493-508.

Leveau, J. H. J. and Lindow, S. E. (2001) Predictive and interpretive simulation of green fluorescent protein expression in reporter bacteria. *J Bacteriol* **183**: 6752-6762.

Leveau, J. H. J. and Lindow, S. E. (2002) Bioreporters in microbial ecology. *Curr Opin Microbiol* **5**: 259-265.

Lynch, J. M. (1990) Introduction: Some consequences of microbial rhizosphere competence for plant and soil. In *The Rhizosphere* (J. M. Lynch, ed.) pp. 1-10. John Wiley & Sons, Chichester, West Sussex, England.

Moldrup, P., Olesen, T., Komatsu, T., Schjonning, P. and Rolston, D. E. (2001) Tortuosity, diffusivity, and permeability in the soil liquid and gaseous phases. *Soil Sci Soc Am J* **65**: 613-623.

Moorhead, D. L. and Linkins, A. E. (1997) Elevated $CO_2$ alters belowground exoenzyme activities in tussock tundra. *Plant Soil* **189**: 321-329.

Newman, E. I. (1985) The rhizosphere: Carbon sources and microbial populations. In *Ecological Interactions in Soil* (A. H. Fitter, ed.) pp. 107-121. Blackwell Scientific, Oxford.

Or, D. and Tuller, M. (1999) Liquid retention and interfacial area in variably saturated porous media: Upscaling from single-pore to sample-scale model. *Water Resour Res* **35**: 3591-3605.

Parr, J. F., Gardner, W. R. and Elliott, L. F. (1981) *Water Potential Relations in Soil Microbiology.* ASA, Madison, WI.

Ramos, C., Molbak, L. and Molin, S. (2000) Bacterial activity in the rhizosphere analyzed at the single-cell level by monitoring ribosome contents and synthesis rates. *Appl Environ Microbiol* **66**: 801-809.

Raven, J. A. and Edwards, D. (2001) Roots: Evolutionary origins and biogeochemical significance. *J Exp Bot* **52**: 381-401.

Read, D. J. and Perez-Moreno, J. (2003) Mycorrhizas and nutrient cycling in ecosystems: A journey toward relevance? *New Phytol* **157**: 475-492.

Rosado, M. and Gage, D. J. (2002) A *gfp* reporter for monitoring rRNA synthesis and growth rate of the nodulating symbiont *Sinorhizobium meliloti*: Use in the laboratory and in a model complex environment, the rhizosphere. *FEMS Microbiol Lett* **226:** 15-22.

Schimel, J. P. and Weintraub, M. N. (2003) The implications of exoenzyme activity on microbial carbon and nitrogen limitation in soil: A theoretical model. *Soil Biol Biochem* **35:** 549-563.

Schlesinger, W. H. (1997) *Biogeochemistry, an Analysis of Global Change.* Academic Press, San Diego.

Sinsabaugh, R. L. and Moorhead, D. L. (1994) Resource allocation to extracellular enzyme production: A model for nitrogen and phosphorus control of litter decomposition. *Soil Biol Biochem* **26:** 1305-1311.

Tinker, P. B. and Nye, P. H. (2000) *Solute Movement in the Rhizosphere.* Oxford University Press, New York.

Toal, M. E., Yeomans, C., Killham, K., and Meharg, A. A. (2000) A review of rhizosphere carbon flow modelling. *Plant Soil* **222:** 263-281.

Tournaire-Roux, C., Sutka, M., Javot, H., Gout, E., Gerbeau, P., Luu, D.-T., Bligny, R. and Maurel, C. (2003) Cytosolic pH regulates root water transport during anoxic stress through gating of aquaporins. *Nature* **425:** 393-397.

Whipps, J. M. (1990) Carbon economy. In *The Rhizosphere* (J. M. Lynch, ed.) pp. 55-99. John Wiley & Sons, Chichester, West Sussex, England.

Zelenev, V. V., van Bruggen, A. H. C. and Semenov, A. M. (2000) "BACWAVE," a spatial-temporal model for traveling waves of bacterial populations in response to a moving carbon source in soil. *Microb Ecol* **40:** 260-272.

Zwieniecki, M. A., Melcher, P. J. and Holbrook, N. M. (2001) Hydrogel control of xylem hydraulic resistance in plants. *Science* **291:** 1059-1062.

Zwieniecki, M. A., Thompson, M. V. and Holbrook, N. M. (2002) Understanding the hydraulics of porous pipes: Tradeoffs between water uptake and root length utilization. *J Plant Growth Regul* **21:** 315-323.

# Part IV
## Development, Structure, and Function

# 14

## From Cambium to Early Cell Differentiation Within the Secondary Vascular System

*Peter Barlow*

The term *cambium* (Latin for "that which changes") reflects the changing nature of the cells that appear to stream out from the thin-walled cambial zone and differentiate as either wood or bark tissues. Another aspect of the "change" that characterizes the vascular cambium of tree species of temperate regions is its disappearance in autumn and winter, and its reappearance in the following spring. This evanescent nature of cambium is a general feature of the response of meristems to the seasons and, moreover, it indicates the continual opposition between cell reproduction and cell differentiation—first one process dominating, then the other. The only cambial cells that seem able to resist differentiation are the initials, a controversial (see Catesson, 1974) though identifiable (Newman, 1956; Timell, 1980) and conceptually useful (Wilson, 1964) group of cells in which at least one of the daughter cells of a mitosis retains the "stem cell" character of the parental initial cell. As may be seen in cross sections of stem secondarily thickened roots or shoots, a ring of cambial initials, one cell deep, constitutes the cambial perimeter.

Vascular cambium of both roots and shoots contains two types of cells: long, spindle-shaped fusiform cells and smaller, cuboidal ray parenchyma cells (Philipson *et al.*, 1971; Larson, 1994; Lachaud *et al.*, 1999; Savidge, 2000). Ray initials are regularly interspersed with the fusiform initials on the cambial perimeter and the radially elongated files to which they give rise intrude, like the spokes of a bicycle wheel, into both secondary xylem and phloem. Irrespective of whether they are ray or fusiform cells, cambial initial cells are bidirectional in their cell production. Each initial produces alternating sequences of new cells from either its inward- or outward-facing surfaces that pass into the secondary xylem and phloem domains, respectively (Newman, 1956; Barlow *et al.*, 2002). Among the differentiated cells produced by the cambial fusiform cells are those which have become adapted for long-distance vertical transport of solutes (tracheids, xylem vessel elements, and phloem sieve cells) and for the assistance of these processes (e.g., parenchyma cells, vessel accessory cells and companion cells). Other cells (fibers, and also the tracheids) are adapted for the mechanical support of the plant. Ray cells also synthesize and transport radially secondary metabolites into the interior of the wood (Stewart, 1966), as well as storing and transporting trophic materials to the cambium (Sauter and Witt, 1997). From a mechanical point of view, rays physically bolt together the annual rings of xylem, thus preventing shearing of these groups of cells when the stem is bent (Reiterer et al., 2002).

The aims of this chapter are twofold. The first is to highlight features of the cambial meristem, mainly in trees, that bear on the development of the vertical and radial transport systems of stems and roots. The second is to discuss some of the earliest stages of xylem vessel, phloem, and ray development. Because it is often useful to focus on a particular species, data gained from hybrid aspen, Populus tremula L. × P. tremuloides Michx, are mentioned where appropriate. Populus has been advocated as a "model" hardwood tree species (Chaffey, 1999; Bradshaw et al., 2000). However, the genetically advantageous annual weed, Arabidopsis thaliana, is increasingly making a contribution to the analysis of secondary tissue formation (Zhong et al., 2001; Chaffey et al., 2002).

## Vascular Cambium

Development of the vascular cambium and, hence, the initiation of secondary vascular tissue, commences a few centimeters behind the apices of root and shoots. In both tree (Larson, 1976) and herbaceous (Enright and Cumbie, 1973) species, cambial activity and the associated widening growth commence at about the position where primary extension growth ceases. In shoots of hybrid aspen, for example, the cambium becomes established in about the seventh visible internode (Schrader et al., 2003). It is usual in shoots of European and North American trees for primary elongation and secondary widening growth to proceed simultaneously. However, species of other climatic regions, such as the evergreen Michelia champaca of India, show separate periods of shoot elongation that is either accompanied (July-November) or not accompanied (March-May) by cambial activity and radial growth (Chowdhury and Tandan, 1950). Clearly there are important environmental controls over the way in which internal resources are used for growth. A number of publications present anatomical observations that enable the changeover from primary to secondary growth within a shoot axis to be understood (reviewed by Philipson et al. [1971] and Soh [1990]). Although the details of the derivation of fusiform and ray systems from the primary meristematic cells are likely to vary among species (Butterfield, 1976), it is possible, nevertheless, to discuss this alteration to the mode of growth in terms of changes in cell and organ polarities (Barlow *et al.*, 2004).

### Growth Polarity

The primary shoot apex exhibits axial gradients of cellular growth and differentiation (e.g., Maksymowych *et al.*, 1985); the secondary cambial meristems and their derivatives exhibit similar, but radial, gradients. These latter gradients extend centripetally from the vascular cambium into the xylem domain, and centrifugally into the phloem domain. The polarity that accompanies such gradients seems to be quite stable, as reorientation of the cambium by grafting (Thair and Steeves, 1976) reveals that its development and that of the tissues derived from it continue in accordance with their original orientation. Only by deep cutting and partial cambial and phloem girdling can the orientation of cambial growth and cell differentiation be disturbed (Kirschner *et al.*, 1971), the results apparently being attributable to the experimentally redirected flows of auxin.

It is probably auxin, synthesized in the distal primary shoot tissues and transported basipetally, which enables cell division and widening in the cambial zone. There is also evidence for the synthesis of additional auxin in the secondary phloem (Sundberg and Uggla, 1998; Tuominen *et al.*, 2000). Thus, the amount of auxin present at any point in the cambium is accounted for both by long-range transport from the shoot and by short-range transport from a more local site of synthesis. Radial gradients in the amount of auxin transported axially (Schrader *et al.*, 2003) encourage a secondary polarity to develop across the cambial zone, the effects of which are revealed as radial variations in the intensity of cell division and cell differentiation across both the secondary xylem and phloem domains (Uggla *et al.*, 1998; Hellgren *et al.*, 2003). The periclinal orientation of the fusiform cell divisions may also assist in shaping the radial auxin gradient (see Schrader *et al.*, 2003) because such divisions could partition, and thereby dilute, the numbers of auxin-transporter molecules in successive generations of radially disposed daughter cells (see Barlow and Lück, 2004). The high level of auxin in the cambial region may account for its mitotic activity, and its progressive diminution away from the cambial zone may eventually permit the commencement of cell determination and differentiation.

### Storied and Nonstoried Cambia

There are two types of cambia, storied and nonstoried, which differ in the spatial arrangement and growth of their cells (Philipson *et al.*, 1971; Larson, 1994). Angiosperm tree species with storied cambia are often held to be evolutionarily more derived than those whose cambia are nonstoried, a view originally proposed by Bailey (1923) on the basis of a survey of numerous species. In both types of cambium, fusiform and ray cells are capable of dividing in each of three planes: periclinal, anticlinal-radial, and anticlinal-transverse.

As is evident from tangential sections of the cellular predecessors of cambium (e.g., Enright and Cumbie, 1973), the continuing extension of these cells, when all others have ceased extending, leads to the nonstoried cambial condition. Vertically extending corners of procambial cells become the elongating tips of the new fusiform cambial cells. In shoots of *Hoheria angustifolia* this conversion of end walls into pointed intrusive tips occurs over a distance of two internodes (Butterfield, 1976).

Storied cambium is a simplification of the nonstoried state, and the end walls of the nascent fusiform cells do not show a great deal of vertical extension. Each story is a packet of fusiform cells of common descent whose development has been accompanied by the insertion of radial division walls and by growth in the circumferential plane (Beijer, 1927). In a sequence of cell packets from a root of *Aeschynomene elaphroxylon* (Leguminoseae), Beijer (1927) could make out the results of three complete rounds of radial division across a width of 7 to 8 mm. A study by Whalley (1950) of cellular development in a branch of *Thuja occidentalis* revealed four successive anticlinal divisions during 8 mm of radial growth. Here, however, there was a great deal of additional intrusive vertical growth, a characteristic of the nonstoried fusiform cells of this species. Interestingly, the intrusive growth did not seem to disrupt the already existing clones of cells, but took place to one side of them.

The relative degree of tip growth of the fusiform cells is one of the chief features distinguishing nonstoried from storied cambia. In many plant systems, tip growth is accompanied by nuclear movement (Schnepf, 1986) with the nucleus maintaining a constant distance from the growing tip. Similarly, Barghoorn (1940a) proposed that rapid nuclear movements accompanying tip growth in conifer cambium could account for the presence of nuclei at either the apical or basal ends of the fusiform cells (perhaps moving to these sites alternately). Tip growth is supported by actin microfilaments (Baluška *et al.*, 2000), and actin is also required for rapid cell extension of both shoots (Thimann *et al.*, 1992) and roots (Baluška *et al.*, 1997). Because the end walls of storied cambial fusiform cells show limited extension, it is possible that the storied state reflects an inability of the longitudinal actin microfilaments to promote extensive tip growth, perhaps as a result of a particular balance of hormones within the cambium (see Little and Savidge, 1987). For example, Thimann *et al.* (1992) showed that auxin and abscisic acid had powerful shoot-growth promoting and inhibitory effects, respectively, which correlated with the organizational status of the intracellular actin filaments.

Tip growth in nonstoried cambial cells and their derivatives eventually ceases, and the cells reach a final limiting length. In conifers, no further increase occurs in trees older than about 60 years, and in nonstoried dicotyledonous trees after about 20 years (Bailey, 1923). Mean fusiform cell length in storied cambium does not increase with age (Bailey, 1923). In both cases, maximal cell size is probably an expression of a particular limiting nucleocytoplasmic ratio that takes different times to achieve according to the species, and that which may also be related to the mass of the nuclear DNA (Price *et al.*, 1973). Nuclear DNA contents in gymnosperms exceed by × 60-100 those of dicotyledonous trees with storied or nonstoried cambia, and gymnosperm nuclei are correspondingly larger also (Nagl, 1990). Nuclear factors should therefore be taken into account in considering the size of cambial cells, as well as the size of the vascular cells derived from them.

## Cambial Fusiform Cell Divisions

***Periclinal Divisions*** Periclinal divisions of fusiform cells differ in the two types of cambium. In storied cambium, the new division walls run vertically from end to end of the cells, whereas in nonstoried cambium the division walls do not reach completely to the ends of the cells, and their attachments to the mother cell walls are acutely angled (Bailey, 1920a). It is possible that this difference is due to the physical properties of the mother cell walls and their plasma membranes to which the division walls are attached need to attach. For example, the wall accepting a division wall may have to be a nongrowing wall or segment of wall simply because regions of wall that are active in intrusive growth cannot accept a division wall. Brown and Sax (1962) proposed that the pressure exerted on the inner surface of the cambium by the centrifugally advancing front of incompressible secondary xylem, and the presence of a similarly resistant bark at its outer surface, favors periclinal divisions. This idea was supported by cytological observations from experimentally compressed blocks of tobacco callus tissue (Lintilhac and Vesecky, 1984) in which the new cell division walls were oriented perpendicularly to the applied pressure and gave the tissue the appearance of a cambium.

The periclinal divisions in the cambium are the means by which new cells are directed into the secondary xylem and phloem domains. At first, the new cells are nearly as long as the parental cambial cells that produced them. In the cell maturation zone within the xylem domain, however, cells continue to elongate if they are a fiber or tracheid element, or to widen if they are a vessel element. In Bailey's study of a range of conifer species (Bailey, 1920b), tracheids were, on average, 10% longer than the cambial cells that produced them. In the group of less derived dicotyledonous trees that were examined, however, xylem fibers elongated, on average, by an extra 120%, and vessel elements by 14%, whereas in more derived dicots, vessel elements were 15% shorter than their parental cambial cells (Bailey, 1920b). In this last-mentioned class of dicots, the xylem fibers showed a factor of increase of × 4.2 or more (Chalk *et al.*, 1955). The small apparent reduction in vessel length, which has been reproduced was found in many subsequent studies, was also recently reported for the hardwood tree, *Kalopanax pictus*, by Kitin *et al.* (1999). In this study, confocal laser scanning microscopy was used to minimize length measurement artifacts, and this article should be consulted for a discussion of the problems of measuring cell dimensions. Xylem fibers were also observed to be longer in the middle of the annual rings (Chalk *et al.*, 1955), as were the tracheids of conifers (Wilson, 1963), indicating faster cell elongation rates during particularly favorable environmental conditions.

Derivatives of periclinal cambial fusiform cell division within the phloem domain also enlarge, though often the cells become shorter as a result of subsequent transverse divisions. Phloem fibers lengthen, the percentage increase being inversely related to the species-specific length of the cambial fusiform cells (as dictated by their genomic nuclear DNA content) from which they were derived (Esau, 1969, p. 142). This relation also holds for xylem fibers. That is, in species with short fusiform cells, the degree of both xylem and phloem fiber growth is greater than it is in species with inherently longer fusiform cells. Wilson (1963) adduced evidence that, in *Abies concolor*, surface growth of sieve cells in the phloem was just as great as that of tracheids in the xylem, and proceeded at similar rates in both domains. The amount of radial displacement during the development of each of these cells was clearly different, however, as a result of the relatively infrequent production, by periclinal divisions, of new phloem cells compared to xylem cells.

***Radial Divisions***   Distinctive anticlinal divisions are evident in cambial fusiform cells of the two types of cambia. In storied cambia, the anticlinal-radial division walls run from one end of the cell to the other in the same way as do the periclinal division walls, though oriented at right angles to them. In nonstoried fusiform cells, radial divisions are accomplished with pseudotransverse division walls whose angles of orientation, although departing from vertical, are actually quite variable (e.g., Whalley, 1950).

The pseudotransverse division of fusiform cells may be assisted by the rays. In hybrid aspen, an anticlinal division wall was often seen attaching at a site within the mother fusiform cell where it had evidently recently passed though a ray, splitting the ray into two daughter rays in the process (Fig. 14.1). Similar images were also found by Beijer (1927, Fig. 31) in the storied wood of *Alstonia* sp. (Apocyanaceae). Such arrangements of walls could, of course, occur by chance in a system where rays and fusiform cells coexist. However, the suggestion that rays provide platforms on which anticlinal division walls are secured within the fusiform mother cells merits further investigation. Analysis of clones of fusiform cells (Barlow, 2004, unpublished) suggests that it is possible to derive a scheme for their cell division that does indeed involve the participation of the rays.

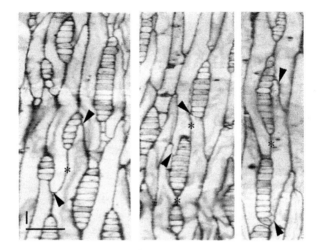

**Figure 14.1**    Images indicating that, after their bisection by a fusiform cell, rays provide sites for the attachment of a division wall in that fusiform cell. Putative division walls are marked (*). After division, the tips of daughter fusiform cells (*arrowheads*) continue to move over the surface of the rays. The photographs are of tangential sections of cambium of hybrid aspen. They have been scanned and compressed vertically by 70%. Horizontal and vertical scale bars: 100 μm. The same magnification applies in each panel. (Sections prepared by Dr. N. J. Chaffey.)

Radial divisions in fusiform cells initiate new radial files of these cells (Fig. 14.2). For a new file to become established, however, the division must occur in the initial zone on the cambial perimeter. Only then can new phloem and xylem mother cells become established. It has been suggested that fusiform initial cells are especially sensitive to the strain imposed on the cambial perimeter by the centrifugal expansion of the xylem tissue, and that each new radial file is evidence of this sensitivity having been transduced, perhaps by actin microfilaments, into a radial orientation of the division wall (Barlow *et al.*, 2004). Another possibility is that the radial divisions are part of a predetermined pattern of cell division.

***Transverse Divisions***    Transverse divisions are found in the fusiform cambial cells of *Tilia* sp. In the study of Zagórska-Marek (1984), their frequency was about 30%, the remainder of the anticlinal divisions being radial, as is characteristic of storied cambium. Although the cambium of *Tilia* is basically storied, it shows features of the nonstoried state in that, after a transverse division in an initial cell, the lower end of one daughter cell and the upper end of the other daughter proceed to grow until each reaches the boundary of the story. The new division wall is mobile, though the limitation to its mobility on upon reaching a story boundary is unexplained. Nevertheless, it is because of this limitation that the stories retain their

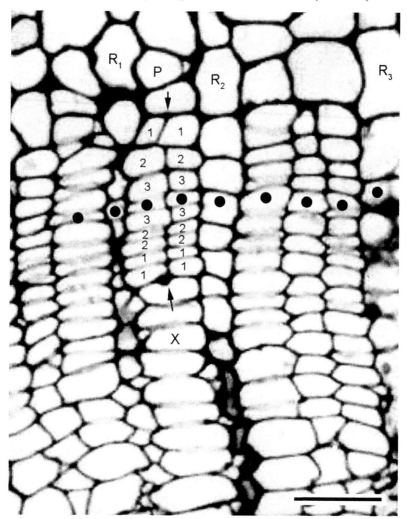

**Figure 14.2**   Establishment of a new radial file of fusiform cells within the cambium of a stem of hybrid aspen, as seen in cross section. Putative initial cells are marked (●). Two cells, both numbered 3 on the phloem (P) and the xylem (X) sides of the initial, are the respective mother cells for phloem- and xylem-directed radial lineages. Cells 1 and 2, which may or may not be subdivided, indicate the first two cell productions from the xylem and phloem mother cells in the radially subdivided lineages. Each file of ray cells (R1-R3) shows a single initial dividing cell, the ray file initial. Scale bar: 50 μm.

identity, though occasionally an overlong fusiform cell can develop, spanning two stories (Zagórska-Marek, 1984).

Transverse divisions of fusiform cells are found in two other situations. First, as a step initiating the differentiation of certain cell types of secondary xylem and phloem, such as parenchyma, companion cells, and sometimes the sieve elements (Esau and Cheadle, 1955; Esau, 1969). In fact, transverse divisions are quite extensive in the fusiform cells on the phloem side of the cambial initials (Esau, 1969), so that, in such cells, there may be as many rounds of mitotic cell division on the phloem side as there are on the xylem side. Divisions are more evident in cross section on the xylem side because of the periclinally oriented division walls, whereas on the phloem side, the division walls are hidden in cross section because they are transverse and so the cambium here has the appearance of being less active. Transverse divisions occur at the conclusion of cambial proliferative activity, during the last part of the growing season, and xylem parenchyma is accordingly differentiated at this time. The second situation involving transverse divisions of fusiform cambial cells is in the development of new rays (Barghoorn, 1940b) (Fig. 14.3). If the new cells are to function as ray initials, however, the transverse divisions must occur within a fusiform initial; otherwise the resulting cell will eventually be lost from the cambium.

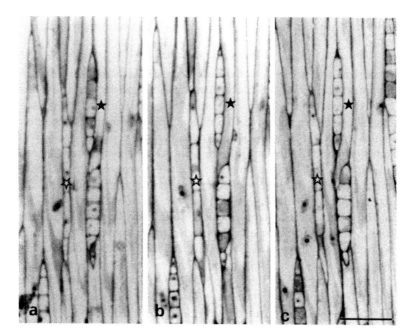

**Figure 14.3** Series of three tangential sections through the cambium of hybrid aspen showing a vertical file of four cells (*open star*), which may be a fusiform cell that has been divided by two rounds of transverse division and is in the process of converting to a set of ray file initials from which a new ray will develop. Also illustrated is the penetration of a ray by the tip of a fusiform cell (*filled star*) and this has created two daughter rays. Scale bar: 100 μm. The same magnification applies in each panel. (Sections prepared by Dr N. J. Chaffey.)

During the transdetermination of fusiform initial cells into rays in various species of conifers (Bannan, 1953; Barghoorn, 1940a), the daughter cells of the accompanying transverse divisions undergo remarkable changes in shape leading to a reorientation of their principal direction of growth. Rays in dicotyledonous trees seem to originate in a similar way (Barghoorn, 1940b), though documentary evidence of the same quality as produced for conifers is lacking. It may be that the intermediate amoeboid fusiform/ray cell form, so clearly illustrated by Bannan (1953) and Barghoorn (1940a), is absent in dicotyledons. The subdivision of fusiform cells and their conversion to ray cells is also convincingly inferred from the method of reconstructing the sequence of division from serial tangential sections, as shown by Beijer (1927, Figs. 7-10) for the secondary xylem of *Aeschynomene elaphroxylon*.

Stress and strain were suggested as stimuli for periclinal and radial divisions, respectively. Transverse mitotic divisions, on the other hand, may occur in the absence of these two factors. In this situation, more simple (default) rules may regulate the division plane (see Barlow *et al.*, 2002, 2004). A decline in overall cellular growth rate could be one condition that favors transverse divisions in fusiform cells as they differentiate into xylem and phloem parenchyma, or during the formation of new ray initials.

### Ray Cell Divisions

Periclinal and transverse cell divisions occur within rays. These divisions contribute to the radial cell files of each ray and to the vertical extent of the ray, respectively. Radial ray cell divisions initiate multiseriate rays in those species where such rays are typical (Chattaway, 1933; Barghoorn, 1941) and may even be a prelude to resin canal formation in conifers. In contrast to the radial files of fusiform cells, where the number of dividing cells along a file varies according to the time of year, each radial file of a ray, at least in hybrid aspen, seems only to have a single dividing cell, the initial cell (Figs. 14.2 and 14.4). When such an initial cell divides periclinally, one of the two daughters continues dividing, thus perpetuating the stem cell state, while the other daughter has no potential for further division, and so can only differentiate. The initial cell is bidirectional in its division producing new cells alternately to the xylem and phloem domains.

Whether a daughter cell of a periclinal division in a ray file initial becomes part of the phloem domain or part of the xylem domain probably depends on the position of the daughter cell with respect to a radially shifting growth center for the ray lineage, as described by Nakielski and Barlow (1995) for the shifting boundary between primary root and root cap tissues. The same general principle of a shifting growth center could also apply in defining the direction of cell production from a fusiform initial cells. The xylem and phloem mother cells (derivatives of the initial), as well as the putative initial cells themselves, could therefore be subject to continual redefinition by this shifting center, as proposed by Newman (1956), and as Barlow and Lück (2004) found would account for certain patterns of secondary phloem development.

The daughters of transverse ray cell divisions are stacked vertically, one upon the other (Fig. 14.4). Usually, but not always, it is the upper and lower cell in a vertical stack of ray initial cells that divides transversely (Braun, 1970), thereby contributing to the growth in height of the ray. In *Picea glauca* and *Abies balsamea* good correlations were found between the production rate of ray cells by transverse divisions and the rate of tracheid

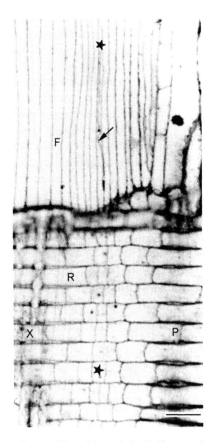

**Figure 14.4**  Portion of a ray (R) and its radial cell files seen in a near-radial section of hybrid aspen stem. Fusiform cells (F) occupy the upper portion of the micrograph. Only one cell, the ray-file initial, divides in each file. The lower star denotes one such cell. The upper star indicates part of a fusiform initial. It is coincident with the vertical stack of ray file initials. An arrow indicates a new division wall within the fusiform initial. Phloem (P) and xylem (X) domains, respectively. Scale bar: 100 μm. (Section prepared by Dr. N. J. Chaffey.)

production by means of periclinal fusiform cell divisions (Gregory and Romberger, 1975), suggesting coordinated rates of cell divisions in these two distinct types of cambial cells.

Ray cell division and concomitant cell growth can resume within the mature portion of the phloem during the second year after ray initiation. These secondary divisions are both periclinal and radial, and occur in what are termed dilatation rays located within the circumferentially expanding bark (Esau, 1965).

## Bisection of Rays

It is generally accepted that groups of ray initials can be formed by the subdivision of fusiform cells. However, this may not be the only way in which new rays form. The observation that rays can be penetrated by the intrusive tips of fusiform cells (Fig. 14.3) suggests that bisection of an already existing ray is another means of increasing the population of rays. The frequency of ray bisection can be assessed from serial tangential sections through many annual growth rings of wood. In *Populus* sp. and *Fagus sylvatica*, for example, rays were found to be bisected, on average, once every two or three years (Braun, 1970). Under steady conditions, rays might therefore also be expected to double their length over a similar period, as was found in hybrid aspen.

If the two daughter portions of a bisected ray are to continue growing both radially and vertically, then each must contain ray file initial cells. The bisection event must therefore take place in a position along the original ray so that both of the rays remain viable and active in cell production. This means that the bisecting fusiform cell penetrates the vertical file of ray cells in which the initials reside (Fig. 14.5).

To bring about bisection of a ray, it is not sufficient for a fusiform cell simply to penetrate a ray so that its tip enters at one side and exits from the other. Nor do images that show this event (Figs. 14.3 and 14.5) give any indication of what happens subsequently. Based on observations from hybrid aspen, it is proposed (Fig. 14.6) that the tip of a fusiform cell slides up (or down) the side of a ray and penetrates the periclinal wall in the region of the ray initials. Perhaps the walls of these cells are thinner, because of their frequent division, than they are in other regions of the ray, and therefore are more susceptible to fusiform penetration. As indicated in Fig. 14.6D, it is as though the penetrative fusiform cell(s) snips the lower portion of the ray, severing its radial continuity. Later, the phloemward portion of the snipped ray initiates an independent existence as a new daughter ray (Fig. 14.6G). Ray bisection is therefore an initial snipping and a subsequent bifurcation (or forking) event (Fig. 14.6).

After bisection and bifurcation, the ray, as seen in cross section, resembles a pitchfork with a handle and two prongs of unequal length (Fig. 14.7). The prongs of the fork portion become prised apart by the radial growth

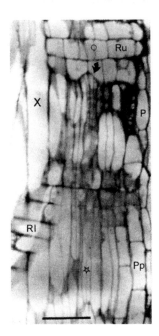

**Figure 14.5**    Radial section through the cambium of hybrid aspen. Fusiform cells (an initial cell is marked by an open star) overlie part of a ray (Rl). In the upper portion of the micrograph, a fusiform initial and its periclinally produced descendents pass under (*arrow*) the upper portion of the ray (Ru). A ray file initial is marked by an open circle (o). P = phloem; Pp = phloem parenchyma; X = xylem. Scale bar: 100 μm. (Section prepared by Dr N. J. Chaffey.)

and divisions of the intervening fusiform cells, and they also lengthen as a result of continuing radial growth and periclinal divisions in each prong of the bisected ray. The new smaller prong, which may also bear a ray terminus as a result of the recent periclinal rupture of the mother ray, becomes the new daughter ray (Fig. 14.6E-G). The other, major prong of the forked ray is the remainder of the preexisting mother ray whose overall length is unchanged. These three groups of cells in the bisected ray—handle and two prongs—remain united and continue to grow within a matrix of fusiform cells and their differentiating derivatives. Vertical growth of each ray is likely to be due to intrusive growth at its upper and lower apices (Fig. 14.6E-G).

In hybrid aspen stems, rays doubled in number during the first year of their observation and seemed likely to double in the second year also (Barlow, unpublished). Thus, over this two-year period the ray system acquired second- and third-order ray branches, the original, first-order rays ("Markstrahlen") being those that run from the outer portion of the phloem all the way to

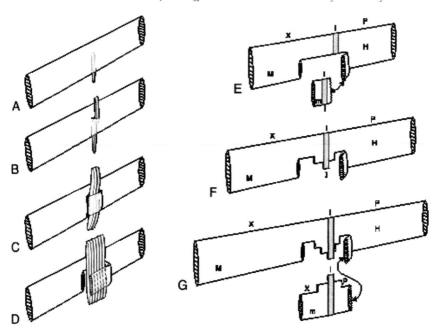

**Figure 14.6** Scheme showing the bisection and radial forking of a ray. (A) A vertical fusiform cell grows up beside and behind a horizontal ray, its tip entering between the periclinal walls of the ray as it extends. (B) The fusiform cell penetrates the ray and its tip emerges on the opposite (*front*) side and continues to grow upwards. (C) The fusiform cell divides periclinally to produce a short radial row of cells (*cf.* Fig. 14.5). The ray is split along the periclinal walls of its cells. (D) The ray has split into major and minor forks connected by a bridge of ray tissue that eventually passes into the phloem. Each fork of the ray grows conformably with the interpolated, radially growing fusiform cells which also begin to divide anticlinally and widen the zone between major and minor forks. The fusiform initial is shaded. (E-G) Cells within the initial zone (*shaded*) contained within the major and minor forks of the ray continue to grow and divide both vertically and radially (minor ray omitted in F). (H) Handle portion of the ray, this being the original end of the ray in the phloem domain. I = initial zone; M = major fork; m = the new, minor fork of a ray; P = phloem domain; X = xylem domain.

the pith. The sites of ray forking could be tentatively recognized within the maturing phloem, appearing in each case as a cellular bridge linking the older, major ray to the shorter, minor ray (Fig. 14.7).

Rays can also fuse as a result of their mobile apices making contact in the vertical plane. In *Tilia cordata*, where rays are regularly bisected by fusiform cells, rays also become reunited. In some years, splitting of rays predominated; in other years, uniting of rays was more frequent (Włoch and Wawrzyniak, 1990). Over an 18-year period of observation, the ratio of splitting to uniting of rays was about 1:0.68, during which time the number of rays increased by 26%.

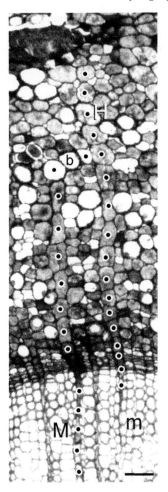

**Figure 14.7**    Early development of the ray system, as seen in cross section of a stem of hybrid aspen. Two rays (*black and white dots*) appear to be linked by a short bridge (b) of three cells (*black dots*) within the phloem domain. This bridge indicates that the ray is forked, as proposed in Fig. 14.6 and, hence, one of the rays, M, is the major fork. M = major ray and is continuous with the handle, H. The other fork, M, is the minor ray. It terminates in the newly formed xylem. Scale bar: 100 μm.

## Vascular Cambium and the Early Stages of Cell Differentiation

The differentiation of secondary vascular tissues has much to do with which of the possible pathways of cell wall synthesis are deployed within the cells produced by the cambium (Mellerowicz *et al.*, 2001). In addition to positive genetic promotion of vascular development (reviewed by Ye, 2002, and by Helariutta and Bhalerao, 2003), there can also be negative effects because of cellular milieux incompatible with certain pathways of cell differentiation. An example of this last-mentioned effect may be inferred from the presence, within the phloem domain, of flavonoids whose properties are antagonistic to peroxidase enzymes that otherwise would enable lignification, as they do within the xylem domain (Ferrer and Ros Barceló, 1994). Also important for the overall structure of secondary xylem is the pattern of cell death. Living cells have precise species-specific locations within an otherwise dead xylem (Wolkinger, 1970).

Many other aspects of cell development are triggered in response to the external, abiotic environment (Creber and Chaloner, 1990; Savidge, 1996; Barlow and Powers, 2005). Details of the route from environment to gene and thence to cell structure are often obscure, though links between, for example, secondary vascular development, day length, gibberellin biosynthesis (Eriksson and Moritz, 2002), and auxin transduction pathways (Moyle et al., 2002; Schrader et al., 2003) are now being uncovered. The abiotic environment also includes nutritional parameters. These can influence cambial development, as indicated by the responses of tomato stems to major nutritional element deficits, estimated by means of the relative areas of secondary xylem and secondary phloem (Venning, 1953). Two extreme effects were elicited by the absences of N (as $NO_3^-$) and $K^+$; the respective xylem:phloem ratios were 15.0 and 0.5. The effect of potassium is of interest because this ion is required for the expansion of cells in the xylem domain (Arend et al., 2002; Langer et al., 2002). In the absence of $K^+$, tomato stems showed a near complete absence of secondary xylem (Venning, 1953).

## Secondary Xylem Vessels

Secondary xylem vessels are conspicuous in cross section on account of their large diameters. In ring-porous woods, the diameters of vessels that develop in spring are larger than those developed later in the growing season. In ash, *Fraxinus excelsior*, the radial diameter of vessels in spring wood ranges from 80 to 170 μm, whereas in summer wood this dimension ranges from 10 to 70 μm (Burggraaf, 1972); in red oak, *Quercus rubra*, the difference is more marked, with spring vessels of 300 μm radial diameter and summer vessels of 60 μm or less (Zasada and Zahner, 1969).

Stacking of serial cross-sectional images of secondary xylem reveals the three-dimensional arrangement of its vessels (Braun, 1959; Burggraaf, 1972). Reconstruction of vessel pathways in wood of ash indicates that, within a segment only 10 mm high, a given vessel may wander up to approximately 700 μm in the tangential direction (that is, by somewhat more than 10 cell widths), but only approximately 70 μm in the radial direction (Burggraaf, 1972). A similar distribution of vessels was shown by Braun (1959) in his study of *Populus* sp. In these two species, as well as in *Fraxinus lanuginosa*, described through the use of confocal microscopy by Kitin et al. (2004), pairs of vessels were shown to come occasionally together, twist around each other while making temporary contact via pit fields, and then move apart during their vertical passage through the wood. It was rare for vessels to terminate blindly. Burggraaf (1972) was puzzled about how the longitudinal alignment of vessels came into being. The best that could be said was that their end-to-end alignment took place by means of an induction process involving a basipetally moving, vessel-forming substance (Burggraaf, 1972).

Information concerning vessel element differentiation makes possible a hypothesis for the determination and subsequent concatenation of secondary xylem vessel elements (summarized in Fig. 14.8). The scheme applies to

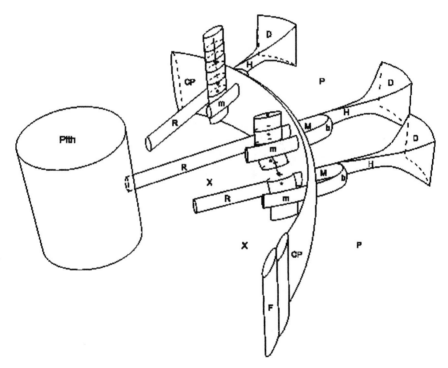

**Figure 14.8**   Scheme showing formation of new vessel elements (∗) in association with new rays, and steps in the vertical basipetal alignment, by homeogenetic induction, of these elements to form a vessel. Vessels formed in this way become linked with other vessel elements, or with vessels of an earlier generation which descend basipetally (*top left, with vertical arrow*) having been initiated in the apices of branches in the shoot system. Three rays are shown with new vessel elements in association with a minor ray. The central chain of three elements is about to join with the single element below it, but this homeogenetic induction event involves a descent that is not quite vertical and gives rise to tangenital displacement of the vessel. One ray contacts the pith; this ray is a Markstrahl. B = bridge between major and minor rays; CP = cambial perimeter; D = dilatation portion of the ray; F = zone of fusiform initials; H = handle portion of ray; M = major ray; m = minor ray; P = phloem domain; R = ray; X = xylem domain.

regions of stem or root remote from apical zones where first-generation vessels form in continuity with leaf petioles (Larson, 1976). Let us commence with the origin of a new ray formed by bisection, as shown in Fig. 14.6. Bisection of a ray leads to the formation of a new minor ray which remains attached to its corresponding parental portion of ray (Fig. 14.7). The site where the major and minor portions of the ray unite is in the phloem. In the xylem, examination of numerous cross sections of hybrid aspen stems prepared throughout the growing season showed that new, expanding vessels were located close to the end of a new minor ray (Fig. 14.9). Each new branch order of rays may therefore be associated with a new generation of vessels.

Contacts between rays and vessels were studied extensively by Braun (1970) and found to consist numerous bordered pits sealed with a pit membrane. Pit membranes are thin areas of primary wall that have been excluded from secondary wall formation. They are regarded as sites of apoplasmic solute transfer between a ray and its associated vessel (Braun, 1984; Chaffey *et al.*, 1997). This ray-vessel contact may occur while the ves-

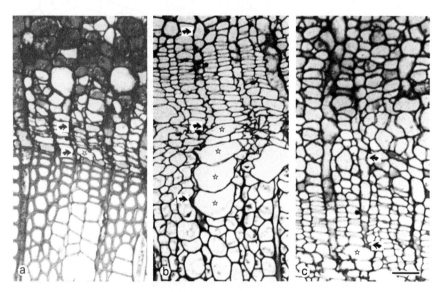

**Figure 14.9**    Relationship between a new ray, possibly formed by bisection, and the development of a new secondary xylem vessel element within a basal region of a stem of hybrid aspen. (A-C) One end of the recently formed ray is in the secondary phloem (*upper area, above the* ●). The other end is in the secondary xylem (*lower area, below the* level of the initial cells) where the ray terminates and associates with a developing vessel element. (A) Earliest recognizable new vessel element at the start of cambial activity (May). (B,C) New enlarging vessels in the secondary xylem domain associated with the end of a new ray (June). Curved arrows mark the two ends of the new ray in the xylem and phloem domains. The initial cells of the arrowed rays are marked by level of the initial cells to the right of those cells. New developing vessel elements are marked by an open star. Scale bar: 50 μm. Same magnification in all cases.

sel initial is still resident within the cambium. Solutes move centripetally along the ray (van Bel, 1990), a process that may be associated with the radially aligned microtubules and actin microfilaments within the ray cells (Chaffey and Barlow, 2000). The subterminal portion of the ray, snipped from the parent ray (Fig. 14.6) and whose cut end is subsequently sealed, is proposed as a site of solute accumulation. If apoplasmic solute transport, especially of sucrose (see Krabel, 2000) and of $K^+$ (Arend *et al.*, 2002; Langer *et al.*, 2002), were to be actively directed at the pit membrane between a contact ray cell and a fusiform derivative of the cambium, which is also a potential vessel element, then turgor pressure could be increased within this vessel initial, its expansion continuing for as long as its wall properties permit (Catesson, 1989). The solutes in the ray are possibly collected from as far away as its dilatation portion within the photosynthetic bark (see also Fig. 14.8). Because auxin is synthesized in the vicinity of the phloem, it may be assumed that auxin, too, is transported to the incipient vessel via the rays, perhaps within the handle portion (Fig. 14.7). The proposed bridge between the major and minor rays (Fig. 14.7) is crucial to ensure that the new ray is also supplied with these two morphogenic solutes, auxin and sucrose.

Although it was proposed by Murakami *et al.* (1999) that contact between a xylem vessel and a ray cell induces a particular pathway of differentiation in the latter, a counterproposal could be that a vessel element is determined as a result of contact between a new ray and a fusiform cambial cell, as outlined previously. Without such contact, the fusiform cell would differentiate as a xylem fiber or a parenchyma cell. After the determination of the nascent vessel element, there follows the induction below it of another, similar potential vessel element (Fig. 14.8). The mediator of this homeogenetic induction event could be the auxin transported from the ray into the first nascent vessel element to be formed. The second cell also receives a complement of vessel-inducing solutes from the nascent vessel element above, to which it responds by enlarging and differentiating, as a vessel element, in a similar manner. Thus, the induction process continues basipetally, resulting in a chain of vessel elements. In this way, it is possible for vessels to cross the boundary between one annual growth ring and the next (Kitin *et al.*, 2004). The leading element of such a chain of vessels may eventually encounter another chain of vessels with which it can establish a link. Two chains of differentiating vessel elements, oriented vertically, one above the other, would then link to form a single vessel (Fig. 14.8). Such linking of vessels would account for the "unifications" recorded by Burggraaf (1972) in his three-dimensional map of vessels.

The preceding scenario, which invokes homeogenetic induction within the xylem domain, derives from the extensive observations of Sachs (Sachs, 1981) concerning the xylogenic effect of auxin, its canalized basipetal transport in undifferentiated tissue, and the resulting differentiation of vessels and tracheids along its flow path. The hypothesis proposed (Fig. 14.8) for vessel formation places all these observations into the context of secondary vascular development in mature portions of stem.

Anatomical observations of young secondary xylem tissue also would be expected to show newly formed vessel elements lacking contact with a ray. Such elements, it is suggested, would have commenced differentiation in response to induction motivated by a new ray outside the area of observation; they may also indicate a continuity of the first-generation vasculature descending from the leaves (Fig. 14.8). The path of the inductive stimuli is basipetal, from cell to cell in the vertical direction. The cells from which the xylem vessel elements arise would not lie precisely above and below one another, but show overlapping of their walls. This inexact alignment of the cell files would account for the wandering of vessels noted by Burggraaf (1972) and Kitin *et al.* (2004). Wandering would be more likely to take place in the tangential plane (as indeed it is) because it is in this plane that fusiform cells (potential vessels) show similar ages and susceptibilities to the basipetal induction process.

It is interesting to speculate briefly on the diversity of development of secondary and tertiary thickening of vessels and tracheary elements. It is well known that these wall thickenings are based on the arrangements of microtubules in the cytoplasmic cortex beneath the plasma membrane (Chaffey *et al.*, 1999) and that these microtubules orient the process of localized cellulose synthesis (Salnikov *et al.*, 2001). Given the enormous variation in the patterning of these wall thickenings throughout the plant kingdom (Bierhorst and Zamora, 1965), and also the variations of this patterning that are seen even within one type of xylem cell, such as a vessel, in different organs at different locations within the body of a single plant (Cheadle, 1944), it seems probable that this overall variation is accounted for by corresponding patterns of cortical microtubules within the respective differentiating cells. Because microtubules are self-organizing structures whose assembly is responsive to environments supporting a biochemical diffusion-reaction system involving tubulin and co-factors (Glade *et al.*, 2002), the diverse, but predictable, patterns of xylem cell wall thickenings throughout the plant are probably due to positionally dependent local variations of the intracellular environment that supports such reactions and within which the diverse lattices of cortical microtubules are consequently formed.

## Secondary Phloem

Secondary phloem tissues of both gymnosperms and dicotyledonous angiosperms are concerned mainly with the basipetal transport of sugars. Just as the structure of secondary xylem tissue shows relative simplicity in conifers but is more complex in dicotyledons, so the same is true of the secondary phloem (Srivastava, 1963; Esau, 1969). The main cell types of the vertical phloem system of conifers are sieve cells, parenchyma cells, and fibers. In dicotyledons, the cell types are sieve tube members, companion cells, parenchyma cells, and fibers. The problem to be addressed now is how the cells of these two vertical phloem systems come to be distinctively distributed within the radial files of phloem. The cell divisions that accompany phloem cell differentiation seem to be quite distinctive (Esau and Cheadle, 1955), suggesting that this feature can be taken as a starting point in a discussion of phloem cell arrangements.

Cell differentiation patterns within radial files of phloem are proposed to reflect the spatiotemporal patterns of cell division within the phloem domain of the vascular cambium (Barlow and Lück, 2004). Here, the cell division system specifies the relative locations of cells within the radial files and the duration for which any location is occupied by a cell. Also important for phloem cell determination and development are radial gradients of morphogens such as auxin (Uggla *et al.*, 1998) and sucrose (Warren Wilson, 1978).

Suggestive of a cell lineage-cell differentiation hypothesis is the regularity of the sequences of cell types within the radial files of secondary phloem of Cupressaceae (Abbe and Crafts, 1939; Bannan, 1955). Denoting the relevant cells as S, sieve cell; P, parenchyma; and F, fiber, the standard radial sequence of differentiated cells in the Cupressaceae is the quartet (F S P S), which is repeatedly produced from the cambium during the development of the secondary phloem. How can this sequence be derived in a way that the proposed solution also has physiological plausibility?

In brief, cellular fates can be approached using a theoretical system that generates files of cells with particular sequences of interdivisional durations (Barlow and Lück, 2004). Using this principle together with positional values specified by a morphogen gradient, the radial quartet (F S P S) described for Cupressaceae can be simulated. The derivation of this standard sequence is shown in Fig. 14.10. Commencing with cell I, the first division at timestep 0 produces a new I cell and a mother cell M. At timestep 1, cell M divides to produce a new M cell and another cell, which, at the conclusion of timestep 2, divides to produce inner and outer daughter cells, which later differentiate as a parenchyma cell P and a sieve cell S, respectively. The precursor of fiber cell F is produced at timestep 4. The meristem extends radially beyond the initial I for one or two cells.

All the radial sequences of differentiated secondary phloem cells mentioned by Bannan (1955) for *Thuja occidentalis* (Cupressaceae) can be generated in this way using, as the criterion for cell determination, the summation of the positional values that occurs as the cells are displaced through the meristem and immediately postmitotic zone (Barlow and Lück, 2004). The standard Cupressaceae-type sequence (F S P S) predominates as long

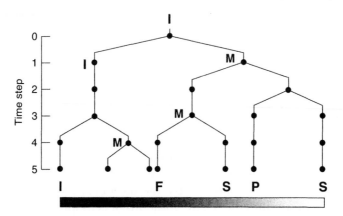

**Figure 14.10** Cell genealogy interpreting the standard recurring quartet of cell types (F S P S) within developing radial files of secondary phloem in the Cupressaceae. All divisions shown are periclinal, though transverse divisions could occur when there is a timestep without a periclinal division. The various cell types are determined according to the positions occupied within a morphogenic gradient across the phloem. This gradient is indicated by shading; the denser the shading, the higher the level of morphogen. F = fiber; P = parenchyma; S = sieve cell; I = initial cell; M = phloem mother cell.

as steady conditions apply. Variant sequences such as (F S P S P S) occur as a result of alteration to the duration of the interdivisional period in the initial cell I with respect to its cell productions into the phloem domain. One perturbation may, for example, be that the initial cell was diverted towards xylem cell production and a new initial cell was derived from a mother cell. But other questions remain. How, for example, do the repeating quartets of cells keep in register across the neighboring files, thereby resulting in a tangentially banded appearance of phloem cells, as is evident in the Cupressaceae? Do groups of initial cells divide in synchrony, or does some additional positional information regulate the outcome of cambial divisions? Similar considerations apply to the secondary phloem of the Pinaceae, whose species display repeating sequences of cells composed of P and S cells only (Barlow and Lück, 2004).

With regard to angiosperms, the types of cells differentiated within the radial files of the secondary phloem are slightly more varied than those of conifers, notably by the inclusion of a companion cell C. Nevertheless, the way in which differentiation in the two groups of trees is regulated in time and space in accordance with morphogen-related positional values is probably no different in principle. For example, the recurrent standard radial cellular sequence (F S P S) characteristic of Cupressaceae is also found in *Robinia pseudoacacia* (Derr and Evert, 1967) but with multiple copies of each of these cell types. In the secondary phloem of *Liriodendron tulipifera*, Cheadle and

Esau (1964) described a consensus recurrent sequence (F F F P S C S C P S C). It, too, can be derived as the consequence of a particular cell division system with a phloem meristem up to four cells wide. The division that produces companion cell C is an innovation, suggesting the presence of supplementary morphogenic information. For example, although C cells can occupy various positions within the phloem, when they are in the vicinity of a ray, they generally make contact with a ray cell (Esau, 1969).

Alternative cell fates might be associated with different cell division patterns in different locations around the cambial perimeter. For example, Zee (1968) deduced two principal sequences of periclinal and radial divisions, in secondary phloem of pea (*Pisum sativum*) epicotyl, as well as an occasional third pathway. Whether a given radial file of the phloem consistently divides according to one or other of the first two pathways, or whether the pathways alternate within a single radial file, is not known.

During the phylogeny of phloem, it seems that there has been a move away from a strict stereotypical division pattern as an accompaniment of histogenesis (viz. conifers) to one where variations of division pattern are permitted (viz. angiosperms). Moreover, transverse divisions in the phloem may also promote diversity of cell types. Whereas it has been proposed (Barlow and Lück, 2004) that phloem fibers in the Cupressaceae are formed as part of a predetermined sequence of divisions within each radial cell file, in angiosperms (as in hybrid aspen) phloem fibers adopt patterns that seem governed by properties inherent to the tissue unit, as well as by those that pertain to the smaller scale of the cell file. The same idea may also be relevant to the question of whether radial cambial cell files have distinct outputs, in terms of the cells differentiated, which in turn relates to the position of the initial cells on the cambial perimeter. Although in many species phloem production precedes that of xylem at the start of the growing season (e.g., the mentioned example of *Robinia* studied by Derr and Evert, 1967), and for which environmental (Wareing and Roberts, 1956; Barlow, 2004) and endogenous hormonal controls (Digby and Wareing, 1966) may play important roles, the question remains of how a preferential direction of cell production from a potentially bidirectional cambial initial could be regulated.

## Final Comments

The cambium is a zone especially adapted to supply cells that are already of nearly the requisite length for fulfilling the functional demands—vertical transport and mechanical strength and flexibility—of the secondary xylem and phloem tissues. Further extension or shortening of cells can be accomplished, respectively, by additional intrusive growth or by cell division, as dictated by the overall pattern of tissue differentiation. The secondary cambial meristem and its postmitotic growth zone contrast in some respects with those features in primary apices where a zone of major cell elongation is interpolated between the primary meristem and the zone where cells achieve their final length. The two meristems do, however, share the feature of formative divisions that create new cell files and proliferative divisions that maintain an optimum output of cells. As indicated by their intrusive growth, cambial cells and their derivatives still retain the tendency to grow vertically at the same time as producing new cells radially. This latter characteristic might even be built into the physically delicate cambial system, once it has been formed, as a result of the mechanical properties associated with, and imposed by, the more robust surrounding transporting tissues. Thus, these surrounding tissues not only supply the whole plant, including the cambium, with essential substrates for growth and cell division but they may also ensure the correct functioning of the cambium with respect to cell production.

## Acknowledgments

I am indebted to Mr. T. Colborn for his help in preparing some of the illustrations. I am also grateful to Dr. N. J. Chaffey (University of College of Bath Spa) for preparing the tissue sections of hybrid aspen.

## References

Abbe, L. B. and Crafts, A. S. (1939) Phloem of white pine and other coniferous species. *Bot Gaz* **100:** 695-722.

Arend, M., Weisenseel, M. H., Brummer, M., Osswald, W. and Fromm, J. H. (2002) Seasonal changes of plasma membrane $H^+$-ATPase and endogenous ion current during cambial growth in poplar plants. *Plant Physiol* **129:** 1651-1663.

Bailey, I.W. (1920a) The significance of the cambium in the study of certain physiological problems. *J. Gen. Physiol.* **2:** 519-534.

Bailey, I. W. (1920b) The cambium and its derivative tissues. II. Size variation of cambial initials in gymnosperms and angiosperms. *Am J Bot* **7:** 355-367.

Bailey, I. W. (1923) The cambium and its derivative tissues. IV. The increase in girth of the cambium. *Am J Bot* **10:** 499-509.

Baluška, F., Barlow, P. W. and Volkmann, D. (2000) Actin and myosin VIII in developing root apex cells. In *Actin: A Dynamic Framework for Multiple Plant Cell Functions* (C. J. Staiger, F. Baluška, D. Volkmann and P. W. Barlow, eds.) pp. 457-476. Kluwer Academic Publishers, Dordrecht.

Baluška, F., Vitha, S., Barlow, P. W. and Volkmann, D. (1997) Rearrangements of F-actin arrays in growing cells of intact maize root apex tissues: A major developmental switch occurs in the postmitotic transition zone. *Eur J Cell Biol* **72:** 113-121.

Bannan, M. W. (1953) Further observations on the reduction of fusiform cambial cells in *Thuja occidentalis* L. *Can J Bot* **31:** 63-74.

Bannan, M. W. (1955) The vascular cambium and radial growth in *Thuja occidentalis* L. *Can J Bot* **33:** 113-138.

Barghoorn, E. S., Jr. (1940a) Origin and development of the uniseriate ray in the Coniferae. *Bull Torrey Bot Club* **67:** 303-328.

Barghoorn, E. S., Jr. (1940b) The ontogenetic development and phylogenetic specialization of rays in the xylem of dicotyledons. I. The primitive ray structure. *Am J Bot* **27:** 918-928.

Barghoorn, E. S., Jr. (1941) The ontogenetic development and phylogenetic specialization of rays in the xylem of dicotyledons. II. Modification of the multiseriate and uniseriate rays. *Am J Bot* **28:** 273-282.

Barlow, P. W. and Powers, S. J. (2005) Predicting the environmental thresholds for cambial and secondary vascular tissue growth and development in stems of hybrid aspen. *Ann Forest Sci* (in press).

Barlow, P. W., Brain, P. and Powers, S. J. (2002) Estimation of directional division frequencies in vascular cambium and in marginal meristematic cells of plants. *Cell Prolif* **35:** 49-68.

Barlow, P. W. and Lück, J. (2004) Cell division systems that account for the various arrangements and types of differentiated cells within the secondary phloem of conifers. *Plant Biosystems* **25:** 179-202.

Barlow, P. W., Volkmann, D. and Baluška, F. (2004) Polarity in roots. In *Polarity in Plants* (K. Lindsey, ed.) pp. 192-241. Blackwells Scientific Publications, Oxford.

Beijer, J. J. (1927) Die Vermehrung der radialen Reihen im Cambium. *Rec Trav Bot Neerl* **24:** 631-786.

Bierhorst, D. W. and Zamora, P. M. (1965) Primary xylem elements and element associations of angiosperms. *Am J Bot* **52:** 657-710.

Bradshaw, H. D., Jr., Ceulemans, R., Davis, J. and Stettler, R. (2000) Emerging model systems in plant biology: Poplar (*Populus*) as a model forest tree. *J Plant Growth Regul* **19:** 306-313.

Braun, H. J. (1959) Die Vernetzung der Gefäße bei *Populus. Z Bot* **47:** 421-435.

Braun, H. J. (1970) Funktionelle Histologie der Sekundären Sprossachse I. Das Holz. In *Encyclopedia of Plant Anatomy* (W. Zimmermann, P. Ozenda and H. D. Wulff, eds.) Vol. 10, part 1. Gebrüder Borntraeger, Berlin.

Braun, H. J. (1984) The significance of the accessory tissues of the hydrosystem for osmotic water shifting as the second principle of water ascent, with some thoughts concerning the evolution of trees. *IAWA Bull* **5:** 275-294.

Brown, C. L. and Sax, K. (1962) The influence of pressure on the differentiation of secondary tissues. *Am J Bot* **49:** 683-691.

Burggraaf, P. D. (1972) Some observations on the course of the vessels in the wood of *Fraxinus excelsior* L. *Acta Bot Neerl* **21:** 32-47.

Butterfield, B. G. (1976) The ontogeny of the vascular cambium in *Hoheria angustifolia* Raoul. *New Phytol* **77:** 409-420.

Catesson, A.-M. (1974) Cambial cells. In *Dynamic Aspects of Plant Ultrastructure* (A. W. Robards, ed.) pp. 358-390. McGraw-Hill, New York.

Catesson, A.-M. (1989) Specific characters of vessel primary walls during the early stages of wood differentiation. *Biol Cell* **67:** 221-226.

Chaffey, N. (1999). Cambium: old challenges – new opportunities. *Trees* **13,** 138-151.

Chaffey, N., Barnett, J. and Barlow, P. (1999) A cytoskeletal basis for wood formation in angiosperm trees: The involvement of cortical microtubules. *Planta* **208:** 19-30.

Chaffey, N. and Barlow, P. W. (2000) Actin in the secondary vascular system of woody plants. In *Actin: A Dynamic Framework for Multiple Plant Cell Functions* (C. J. Staiger, F. Baluška, D. Volkmann and P. W. Barlow, eds.) pp. 587-600. Kluwer Academic Publishers, Dordrecht, The Netherlands.

Chaffey, N., Cholewa, E., Regan, S. and Sundberg, B. (2002) Secondary xylem development in *Arabidopsis:* A model for wood formation. *Physiol Plant* **114:** 594-600.

Chaffey, N. J., Barnett, J. R. and Barlow, P. W. (1997) Cortical microtubule involvement in bordered pit formation in secondary xylem vessel elements of *Aesculus hippocastanum* L. (Hippocastanaceae): A correlative study using electron microscopy and indirect immuno-fluorescence microscopy. *Protoplasma* **197:** 64-75.

Chalk, L., Marstrand, E. B. and Walsh, J. P. de C. (1955) Fibre length in storeyed hardwoods. *Acta Bot Neerl* **4:** 339-347.

Chattaway, M. M. (1933) Ray development in the Sterculiaceae. *Forestry* **7:** 93-108.

Cheadle, V. I. (1944) Specialization of vessels within the xylem of each organ in the Monocotyledoneae. *Am J Bot* **31:** 81-92.

Cheadle, V. I. and Esau, K. (1964) Secondary phloem of *Liriodendron tulipifera. Univ Calif Publ Bot* **36:** 143-252.

Chowdhury, K. A. and Tandan, K. N. (1950) Extension and radial growth in trees. *Nature* **165:** 732-733.

Creber, G. T. and Chaloner, W. G. (1990) Environmental influences on cambial activity. In *The Vascular Cambium* (M. Iqbal, ed.) pp.159-199. Research Studies Press, Taunton / John Wiley, New York

Derr, W. F. and Evert, R. F. (1967) The cambium and seasonal development of the phloem in *Robinia pseudoacacia. Am J Bot* **54:** 147-153.

Digby, J. and Wareing, P. F. (1966) The relationship between endogenous hormone levels in the plant and seasonal aspects of cambial activity. *Ann Bot* **30:** 607-622.

Enright, A. M. and Cumbie, B. G. (1973) Stem anatomy and internodal development in *Phaseolus vulgaris. Am J Bot* **60:** 915-922.

Eriksson, M. A. and Moritz, T. (2002) Daylength and spatial expression of a gibberellin 20-oxidase isolated from hybrid aspen (*Populus tremula* L. × *P. tremuloides* Michx.). *Planta* **214:** 920-930.

Esau, K. (1965) On the anatomy of the woody plant. In *Cellular Ultrastructure of Woody Plants* (W.A. Côté, ed.) pp. 35-50. Syracuse University Press, Syracuse, New York.

Esau, K. (1969) The phloem. In *Encyclopedia of Plant Anatomy* (W. Zimmermann, P. Ozenda and H. D. Wulff, eds.) Vol. 5, part 2. Gebrüder Borntraeger, Berlin.

Esau, K. and Cheadle, V. I. (1955) Significance of cell divisions in differentiating secondary phloem. *Acta Bot Neerl* **4:** 348-357.

Ferrer, M. A. and Ros Barceló, A. (1994) Control of the lignification in *Lupinus* by genistein acting as superoxide scavenger and inhibitor of the peroxidase-catalyzed oxidation of coniferyl alcohol. *J Plant Physiol* **144:** 64-67.

Glade, N., Demongeot, J. and Tabony, J. (2002) Comparison of reaction–diffusion simulations with experiment in self-organised microtubule solutions. *C R Biol* **325:** 283-294.

Gregory, R. A. and Romberger, J. A. (1975) Cambial activity and height of uniseriate vascular rays in conifers. *Bot Gaz* **136:** 246-253.

Helariutta, Y. and Bhalerao, R. (2003) Between xylem and phloem: The genetic control of cambial activity in plants. *Plant Biol* **5:** 465-472.

Hellgren, J. M., Puech, L., Barlow, P., Fink, S., Mellerowicz, E. J. and Sundberg, B. (2003) Auxin and cambial growth rate in poplar. *Silvestria* **268:** 95-105.

Kirschner, H., Sachs, T. and Fahn, A. (1971) Secondary xylem reorientation as a special case of vascular tissue differentiation. *Israel J Bot* **20:** 184-190.

Kitin, P., Funada, R., Sano, Y., Beeckman, H. and Ohtani, J. (1999) Variations in the lengths of fusiform cambial cells and vessel elements in *Kalopanax pictus. Ann Bot* **84:** 621-632.

Kitin, P. B., Fuji, T., Abe, H. and Funada, R. (2004) Anatomy of the vessel network within and between tree rings of *Fraxinus lanuginosa* (Oleaceae). *Am J Bot* **91:** 779-788.

Krabel, D. (2000) Influence of sucrose on cambial activity. In *Cell and Molecular Biology of Wood Formation* (R. A. Savidge, J. R. Barnett and R. Napier, eds.) pp.113-125. BIOS Scientific Publishers, Oxford.

Lachaud, S., Catesson, A.-M. and Bonnemain, J.-L. (1999) Structure and functions of the vascular cambium. *C R Acad Sci, Sér III, Sciences de la Vie* **322:** 633-650.

Langer, K., Ache, P., Geiger, D., Stinzing, A., Arend, M., Wind, C., Regan, S., Fromm, J. and Hedrich, R. (2002) Poplar potassium transporters capable of controlling K$^+$ homeostasis and K$^+$-dependent xylogenesis. *Plant J* **32:** 997-1009.

Larson, P. R. (1976) Development and organization of the secondary vessel system in *Populus grandidentata. Am J Bot* **63:** 369-381.

Larson, P. R. (1994) *The Vascular Cambium: Development and Structure.* Springer-Verlag, Berlin.

Lintilhac, P. M. and Vesecky, T. B. (1984) Stress-induced alignment of division plane in plant tissues grown in vitro. *Nature* **307:** 363-364.

Little, C. H. A. and Savidge, R. A. (1987) The role of plant growth regulators in forest tree cambial growth. *Plant Growth Regul* **6:** 137-169.

Maksymowych, R., Maksymowych, A. B. and Orkwiszewski, J. A. J. (1985) Stem elongation of *Xanthium* plants presented in terms of elemental rates. *Am J Bot* **72:** 914-919.

Mellerowicz, E. J., Baucher, M., Sundberg, B. and Boerjan, W. (2001) Unravelling cell wall formation in the woody dicot stem. *Plant Mol Biol* **47:** 239-274.

Moyle, R., Schrader, J., Stenberg, A., Olsson, O., Saxena, S., Sandberg, G. and Bhalerao, R. P. (2002) Environmental and auxin regulation of wood formation involves members of the *Aux/IAA* gene family in hybrid aspen. *Plant J* **31:** 675-685.

Murakami, Y., Funada, R., Sano, Y. and Ohtani, J. (1999) The differentiation of contact and isolation cells in the xylem ray parenchyma of *Populus maximowiczii. Ann Bot* **84:** 429-435.

Nagl, W. (1990) Differences in genome size and organization in fast and slowly growing trees and other plants. In *Fast Growing Trees and Nitrogen Fixing Trees* (D. Werner and P Müller, eds.) pp. 133-141. Gustav Fischer Verlag, Stuttgart.

Nakielski, J. and Barlow, P.W. (1995) Principal directions of growth and the generation of cell patterns in wild-type and *gib-1* mutant roots of tomato (*Lycopersicon esculentum* Mill.). *Planta* **196:** 30-39.

Newman, I. V. (1956) Pattern in the meristems of vascular plants. I. Cell partition in living apices and in the cambial zone in relation to the concepts of initial cells and apical cells. *Phytomorphol* **6:** 1-19.

Philipson, W. R., Ward, J. M. and Butterfield, B. G. (1971) *The Vascular Cambium. Its Development and Activity.* Chapman and Hall, London.

Price, H. J., Sparrow, A. H. and Nauman, A. F. (1973) Correlations between nuclear volume, cell volume and DNA content in meristematic cells of herbaceous angiosperms. *Experientia* **29:** 1028-1029.

Reiterer, A., Burgert, I., Sinn, G. and Tschegg, S. (2002) The radial reinforcement of the wood structure and its implications on mechanical and fracture mechanical properties: A comparison between two tree species. *J Materials Sci* **37:** 935-940.

Sachs, T. (1981) The controls of the patterned differentiation of vascular tissues. *Adv Bot Res* **9:** 151-262.

Salnikov, V. V., Grimson, M. J., Delmer, D. P. and Haigler, C. H. (2001) Sucrose synthase localizes to cellulose synthesis sites in tracheary elements. *Phytochem* **57:** 823-833.

Sauter, J. J. and Witt, W. (1997) Structure and function of rays: Storage, mobilization, transport. In *Trees – Contributions to Modern Tree Physiology* (H. Rennenberg, W. Eschrich and H. Ziegler, eds.) pp. 177-195. Backhuys Publishers, Leiden.

Savidge, R. A. (1996) Xylogenesis, genetic and environmental regulation: A review. *IAWA J* **17:** 269-310.

Savidge, R. A. (2000) Intrinsic regulation of cambial growth. *J Plant Growth Regul* **20:** 52-77.

Schnepf, E. (1986) Cellular polarity. *Ann Rev Plant Physiol* **37:** 23-47.

Schrader, J., Baba, K., May, S. T., Palme, K., Bennett, M., Bhalerao, R. P. and Sandberg, G. (2003) Polar auxin transport in the wood-forming tissues of hybrid aspen is under simultaneous control of developmental and environmental signals. *Proc Natl Acad Sci U S A* **100:** 10096-10101.

Soh, W. Y. (1990). Origin and development of cambial cells. In *The Vascular Cambium* (M. Iqbal, ed.) pp. 37-62. Research Studies Press, Taunton/John Wiley, New York.

Srivastava, L. M. (1963) Secondary phloem in the Pinaceae. *Univ Calif Publ Bot* **36:** 1-142.

Stewart, C. M. (1966) Excretion and heartwood formation in living trees. *Science* **153:** 1068-1074.

Sundberg, B. and Uggla, C. (1998) Origin and dynamics of indoleacetic acid under polar transport in *Pinus sylvestris*. *Physiol Plant* **104:** 22-29.

Thair, B. W. and Steeves, T. A. (1976) Response of the vascular cambium to reorientation in patch grafts. *Can J Bot* **54:** 361-373.

Thimann, K. V., Reese, K. and Nachmias, V. T. (1992) Actin and the elongation of plant cells. *Protoplasma* **171:** 153-166.

Timell, T. E. (1980) Organization and ultrastructure of the dormant cambial zone in compression wood of *Picea abies*. *Wood Sci Technol* **14:** 161-179.

Tuominen, H., Puech, L., Regan, S., Fink, S., Olsson, O. and Sundberg, B. (2000) Cambial-region-specific expression of the *Agrobacterium iaa* genes in transgenic aspen visualized by a linked *uida* reporter gene. *Plant Physiol* **123:** 531-541.

Uggla, C., Mellerowicz, E.J. and Sundberg, B. (1998) Indole-3-acetic acid controls cambial growth in Scots pine by positional signalling. *Plant Physiol* **117:** 113-121.

van Bel, A. J. E. (1990) Xylem-phloem exchange via the rays: The undervalued route of transport. *J Exp Bot* **41:** 631-644.

Venning, F. D. (1953) The influence of major mineral nutrient deficiencies on growth and tissue differentiation in the hypocotyl of Marglobe tomato, *Lycopersicon esculentum* Mill. *Phytomorphol* **3:** 315-326.

Wareing, P. F. and Roberts, D. L. (1956) Photoperiodic control of cambial activity in *Robinia pseudacacia* L. *New Phytol* **55:** 356-366.

Warren Wilson, J. M. (1978) The position of regenerating cambia: Auxin/sucrose ratio and the gradient induction hypothesis. *Proc Roy Soc London* **B, 203:** 153-176.

Whalley, B. E. (1950) Increase in girth of the cambium in *Thuja occidentalis* L. *Can J Res, Sect C* **28:** 331-340.

Wilson, B. F. (1963) Increase in cell wall surface area during enlargement of cambial derivatives in *Abies concolor*. *Am J Bot* **50:** 95-102.

Wilson, B. F. (1964) A model for cell production by the cambium of conifers. In *The Formation of Wood in Forest Trees* (M. H. Zimmermann, ed.) pp. 19-36. Academic Press, New York.

Włoch, W. and Wawrzyniak, S. (1990) The configuration of events and cell growth activity in the storeyed cambium of the linden (*Tilia cordata* Mill.). *Acta Soc Bot Polon* **59:** 25-43.

Wolkinger, F. (1970) Morphologie und systematische Verbreitung der lebenden Holzfasern bei Sträuchern und Bäumen II. Zur Histologie. *Holzforschung* **24:** 141-147.

Ye, Z.-H. (2002) Vascular tissue differentiation and pattern formation in plants. *Annu Rev Plant Biol* **53:** 183-202.

Zagórska-Marek, B. (1984) Pseudotransverse divisions and intrusive elongation of fusiform initials in the storeyed cambium of *Tilia*. *Can J Bot* **62:** 20-27.

Zasada, J. C. and Zahner, R. (1969) Vessel element development in the earlywood of red oak (*Quercus rubra*). *Can J Bot* **47:** 1965-1971.

Zee, S.-Y. (1968) Ontogeny of cambium and phloem in the epicotyl of *Pisum sativum*. *Aust J Bot* **16:** 419-426.

Zhong, R., Burk, D. H. and Ye, Z.-H. (2001) Fibers. A model for studying cell differentiation, cell elongation, and cell wall biosynthesis. *Plant Physiol* **126:** 477-479.

# 15

# Structure-Function Relationships in Sapwood Water Transport and Storage

*Barbara L. Gartner and Frederick C. Meinzer*

Primary production by plants requires the loss of substantial quantities of water when the stomata are open for carbon assimilation. The delivery of that water to the leaves occurs through the xylem. The structure, condition, and quantity of the xylem control not only the transport efficiency but also the release of water from storage. For example, if there is high resistance to water flow in the stem, then less water is available to the leaves, so less primary production can occur. High resistance can result from wood material with low conductivity, from having only a small amount of conductive wood, or from having very slow release of stored water to the transpiration stream. The subject of this chapter is the efficiency with which different parts of the sapwood transport, store, and release water, and how the structure of the wood affects these processes. Particularly, we describe the radial patterns of axial water transport, their anatomical and physiological causes, the effect that sapwood width and wood structure, especially density, have on water transport, and determinants of sapwood water storage properties.

We define sapwood as the xylem that is conductive to water, although sapwood can also be defined on the basis of color, respiration, and parenchyma vitality. Although our discussion refers to sapwood versus heartwood, many species also have intermediate wood, which is located between the sapwood and heartwood, and is said to have intermediate characteristics (Hillis, 1987). However, some intermediate woods may not have living cells (Hillis, 1987), and more research is needed to delineate which functions are still active in this zone for different species. Likewise, wood has a variety of types, but for simplicity we discuss only three: coniferous, and angiosperm ring-porous and diffuse-porous. These classifications capture much of the interspecific variation in gross wood structure in the temperate and arctic zones, but because most tropical species are diffuse-porous and have extremely variable vessel groupings and/or parenchyma patterns, additional classifications may be more useful for tropical woods.

# Radial Changes in Wood Anatomical Characteristics and Hydraulic Properties

The anatomy of the wood follows systematic patterns of variation that depend on the ring number outward from the pith, the height in the tree, and the radial growth rate. Also, because the xylem transports water for several years, there may be changes in its water transport properties as the cells age. This section describes these basic patterns of wood anatomy, focusing on radial variation, and how they affect the hydraulic conductivity of sapwood.

## Within-Bole Patterns of Wood Anatomy and Properties

In all woody plants, the anatomy of wood changes systematically along a radial (pith-to-bark) direction. The magnitude of the changes depends on the species and the type of wood (e.g., diffuse versus ring-porous). Superimposed on this variability is that caused by environmental factors such as the amount and timing of rainfall and temperature. The most obvious change to the naked eye is the pith-to-bark decrease in growth ring width. At the base of the tree, the first several rings near the pith are usually narrow, produced when the tree was growing slowly while becoming established. Beyond that zone, the rings are wide, then gradually become narrow, associated with slower growth and/or with having a larger cylinder around which to place the new wood produced by a constant-sized crown (Duff and Nolan, 1953). Depending on the species, the percentage of latewood within each ring can increase (common in conifers), decrease (common in ring-porous species), or stay the same with increasing cambial age. Diffuse-porous species are special cases because they have weaker anatomical gradients across the growth ring, so the pith-to-bark changes in growth rate probably have less of an effect on their anatomy.

The second type of radial change in wood structure is referred to as the juvenile/mature wood transition. Juvenile wood (JW) denotes the wood adjacent to the pith. Its properties (anatomical, mechanical, hydraulic, chemical composition) change dramatically from one growth ring to the next (Larson *et al.*, 2001; Senft *et al.*, 1985). In contrast, the properties of mature wood (MW) are relatively constant from ring to ring (if environment remains constant). The extent of the JW zone (in number of growth rings from the pith) depends largely on genetics (Zobel and Sprague,

1998). The juvenile wood zone is usually better correlated with number of growth rings than with distance from the pith (Zobel and Sprague, 1998). In conifers, JW may persist well past 100 years from the pith (Wellwood *et al.*, 1974), but in general the overwhelming majority of the total change will have occurred in the first 10 to 20 years. Most properties either increase or decrease extremely quickly and then reach an asymptote. The radial pattern can be somewhat different with vertical position, but changes with height for the same cambial age are usually small compared to the changes with cambial age at the same height.

In all woody plants studied, the JW shows a gradual increase in the length of the tracheids and/or libriform fibers, usually accompanied by an increase in their diameter with increasing tree age (Zobel and Sprague, 1998). The magnitude of this and other anatomical changes is much higher in gymnosperms than in angiosperms (Panshin and deZeeuw, 1980). For example, in *Pseudotsuga menziesii*, tracheid length increases from 1 to 5 mm from growth rings 2 to 25 (Megraw, 1985) and earlywood tracheid diameter increases from 22 to 40 μm over the same age interval (Spicer and Gartner, 2001). In the ring-porous oak *Quercus garryana*, tracheid length increases from 1.0 to 1.2 mm and earlywood vessel diameter increases from 150 to 250 μm from growth rings 2 to 25 (Lei *et al.*, 1996). Diffuse-porous species have similar increases in cell length and diameter as ring-porous ones.

In diffuse-porous angiosperms, the other radial changes are highly species-specific. As a general rule, their wood has the most constant pith-to-bark properties of any wood type. For example, red alder (*Alnus rubra* Bong) shows no significant change in wood density across the radius (Gartner *et al.*, 1997; Harrington and DeBell, 1980) or at radial growth rates that vary by a factor of 5 (Lei *et al.*, 1997). There may be some difference in the proportion of the growth ring that is vessel, libriform fiber, and parenchyma, and there can be modest increases in vessel diameter (e.g., Lei *et al.*, 1996).

In contrast to diffuse-porous species, ring-porous angiosperms have more significant changes in their wood from pith to bark. In general, these species maintain a similar earlywood width, with the decrease in ring-width at the expense of latewood (Phelps and Workman, 1994; Zhang and Zhong, 1991). Thus, in contrast to the conifers, wood density decreases and proportion of vessel area increases through the JW to the MW.

In conifers, significant changes in wood structure from the JW to MW also include a tendency toward having a higher latewood proportion, thicker cell walls in the earlywood and latewood, denser wood (mostly due to denser latewood) and a change in the cell wall ultrastructure: Microfibrils within the S2 layer decrease from 45 degrees (with respect to axial) in the JW to about 5 degrees in the MW (Larson *et al.*, 2001). The combination of a decrease in microfibril angle and an increase in wood density result in wood that is stronger and stiffer axially in the MW than in JW. The increase in earlywood cell length and diameter and the increase in latewood proportion probably have the largest effect on hydraulics.

Within the mature wood, if the environment stays the same, the wood made in 1 year is similar to the wood made in a previous year. The exception is that in trees that are declining because of environmental degradation, disease, or very old age, the new wood can be different. For example, in *Abies alba*, declining trees had latewood zones of 1 to 2 rows, compared to a broad zone in healthy trees, and the cell wall thickness of the latewood in the declining trees tended to be lower (Schmitt *et al.*, 2003). In *Pinus mugo*, two types of root pathogen caused a decrease in growth ring width compared to uninfected individuals, but the decrease was gradual in trees infected with one type and abrupt for the other. (Cherubini *et al.*, 2002). No data were given on whether earlywood proportion or any of the cell structures were altered as the growth ring widths decreased. In the forest products industry, it is known that very old trees can have much reduced wood density in their outer rings (i.e., Wellwood *et al.*, 1974).

As the wood progresses from being outer to inner sapwood, its properties may change. The outer sapwood of MW may differ from inner sapwood in having a higher percentage of living parenchyma cells (reviewed in Gartner *et al.*, 2000), fewer cells blocked by tyloses (Cochard and Tyree, 1990), and fewer incrusted or aspirated pits (Mark and Crews, 1973; Nobuchi and Harada, 1983). The rates of change with xylem age are species- and property-specific.

Another potential difference is moisture content. There are many individual reports of moisture content variation across the radius. For example, ring-porous species are reported to have saturated earlywood vessels only in the outer growth ring, with the earlywood being embolized in older rings (Cochard and Tyree, 1990; Ellmore and Ewers, 1985; Granier *et al.*, 1994). However, the most comprehensive compendium is for the trees of Bulgaria: Nikolov and Enchev (1967) reported moisture content every centimeter across the north-south transect of the stem xylem for 12 coniferous species and 33 hardwood species. About half of the coniferous species had very constant moisture content across the sapwood, with an abrupt decrease in the heartwood (e.g., *Abies alba, Pinus nigricans,* and some individuals of *Pinus sylvestris*). The other coniferous species had a decline in moisture content, often close to linear, from the outer to the inner sapwood (e.g., *Juniperus excelsa, Taxus baccata, Pseudotsuga menziesii, Larix europea,* and *Pinus strobus*). It was more difficult to discern the sapwood moisture content patterns in the hardwoods because it is not always clear where the sap/heart boundary is from inspection of the moisture contents. Roughly half of the species had relatively constant moisture contents across what appeared to be the sapwood (e.g., *Quercus, Fraxinus, Celtis, Aesculus*). The remaining hardwoods had a variety of moisture content patterns in the sapwood, such as a gradual (*Fagus sylvatica, Populus, Alnus glabrosa*) or steep (*Robinia pseudoacacia*) decrease, or a gradual (*Ailanthus*) or steep (*Gleditsia*) increase from the outer sapwood inward. These patterns do not follow differences in ring porosity versus diffuse porosity. Using both soft x-ray and cryoscanning electron microscopy, Utsumi *et al.* (2003) have shown in several conifer species (*Picea jezoensis, Larix kaempferi,* and *Abies sachalinensis*) that the outer ring has higher moisture content than do the inner growth rings. Similarly, Lu *et al.* (2000) showed that the outer growth ring in the angiosperm *Mangifera indica* had higher moisture content than did the other growth rings.

### Effects of Wood Anatomy and Characteristics on Specific Conductivity

Specific conductivity ($k_s$) is usually higher in the outer than the inner sapwood. As discussed here, $k_s$ refers to the water volume per time that will pass a unit cross-sectional area of sapwood with a given pressure gradient across it. The higher $k_s$ in outer than inner sapwood can be caused by differences in wood structure associated with a decline in radial growth rate or JW/MW changes, and/or from blockages that accumulate in the inner sapwood as it ages. The total sap flow depends on both $k_s$ and sapwood width (see next section).

In MW, the effect of radial growth rate on $k_s$ depends on its effect on latewood proportion. The changes in wood structure that result from the typical decrease in radial growth rate with age may be confounded with the changes occurring in the JW zone. In a study of *Pseudotsuga menziesii*, $k_s$ was positively correlated with height growth, but there was no correlation with diameter growth (Domec and Gartner, 2003).

In diffuse-porous hardwoods, MW should have higher $k_s$ than JW because of the gradual increase in earlywood vessel diameter (Zimmermann, 1978) and the direct relationship between flow and radius to the fourth power. In ring-porous hardwoods, increases in $k_s$ toward the bark are caused by two factors: the increase in earlywood vessel diameter and in the decrease in latewood proportion. In hardwoods, the longer cells in MW than JW should have only a negligible effect on $k_s$ because water moves primarily through the perforations at the ends of vessels (rather than pits), and the perforations contribute less to flow resistance than do pits. In gymnosperms, MW is expected to have higher $k_s$ than JW both because the tracheids are wider and because they are longer. These factors more than compensate for MW's lower earlywood proportion (Mencuccini *et al.*, 1997). Tracheid length is important if a large part of the hydraulic resistance is associated with pit membranes (Petty and Puritch, 1970; Pothier *et al.*, 1989). Water must traverse twice as many pit membranes per meter of stem if the tracheids are half the length.

Because there are no development-caused differences in the anatomy of wood produced from year to year within the MW, outer sapwood should not vary in $k_s$ from year to year. However, there is usually a decline in $k_s$ from outer to inner sapwood, in spite of the lack of changes in the cellular structure. This decrease has been observed in conifers (e.g., Booker and Kininmonth, 1978; Comstock, 1965; Domec and Gartner, 2001, 2002a; Markstrom and Hann, 1972; Spicer and Gartner, 2001) and ring-porous species (Ellmore and Ewers, 1985; see Granier et al., 1994). We do not know of research on diffuse-porous species. One study showed that the pits were "immature and nearly imperforate" in the outer sapwood (Mark and Crews, 1973), from which one would predict lower $k_s$ in the outer sapwood, with an increase inward. To our knowledge, no other research followed up the study by Mark and Crews.

The most likely cause for these declines is a higher incidence of blockages in the inner sapwood. These blockages could be in the form of air (emboli) and/or gums, resins, etc. (extractives). Indirect evidence that inner sapwood may have more embolisms than outer sapwood is found in the studies that show lower moisture content in inner than outer sapwood (above). Decrease in relative water content is associated with a decrease in $k_s$ (Edwards and Jarvis, 1982; Puritch, 1971) and sap flow (Granier et al., 2000). Because extractive content influences heartwood durability, nearly all research on extractives has focused on heartwood rather than sapwood. One article on pit membranes, however, reported an increase in membrane incrustation from outer to inner sapwood (Mark and Crews, 1973).

### Patterns of Sapwood Width

Generally, in both conifers and hardwoods, sapwood width is positively correlated with tree health, vigor, crown class, or radial growth rate (Brix and Mitchell, 1983; de Kort, 1993; Hillis, 1987; Lassen and Okkonen, 1969; Sellin, 1996; Smith et al., 1966). The tree's design criteria for sapwood quantity are unclear because although the pipe model and its modifications often explain sapwood quantity (Margolis et al., 1995; Whitehead et al., 1984), in many cases they do not (e.g., Gartner, 2002). The relationships are stronger between growth rate and sapwood width than between growth rate and number of rings in sapwood. That is, the amount of wood needed for physiological processes is independent of how that wood was developed.

There is little research on the width of sapwood in relation to stem age. In some species, sapwood width of individuals within similar environments is very constant. Examples are the genus *Eucalyptus* (Hillis, 1987) and the species *Pseudotsuga menziesii* (Brix and Mitchell, 1983), although it should be noted that in *P. menziesii*, the sapwood is wider in the bottom meter (Gartner, 2002), and that when the trees are several hundred years old, the sapwood width diminishes (BLG, personal observation). In contrast, in some species, sapwood width increases with tree age. For example, in *Picea*, the sapwood width increases until the cambium attains an age of between 20 and 90 years, depending on species and dominance class of the trees (Hazenberg and Yang, 1991; Sellin, 1996).

## How Species-Specific Characteristics of Sapwood Affect Whole-Tree Water Transport

Many of the radial and developmental changes in sapwood anatomy, structure, and hydraulic properties described in the previous section are largely species-specific and are expected to have an impact on spatial patterns of water movement within the sapwood. For example, radial profiles of axial flow may be associated with species-specific differences in hydraulic architecture that determine the nature of hydraulic connections between leaves and different layers of sapwood. Here we discuss how species-specific variation in fundamental traits such as hydraulic architecture and sapwood quantity, water storage capacity, and density influence water transport within the tree.

### Sapwood Quantity

It is common knowledge that the sapwood cross-sectional area for a given stem diameter varies widely among tree species. For individuals with stem diameters from 0.5 to 1.0 m, radial sapwood depth can range from less than 6 cm in species such as *Eucalyptus regnans* (Wullschleger *et al.*, 1998) and *Pseudotsuga menziesii* (Phillips *et al.*, 2002), to 10 to 20 cm in *Pinus ponderosa* (Domec and Gartner, 2003), and greater than 20 cm in a number of tropical species (James *et al.*, 2003; Meinzer *et al.*, 2001). Despite its implications for rates of whole-tree water use and sap flow per unit sapwood area (sap flux density), the consequences of this wide variation in sapwood area among trees of similar size remain largely unexplored. However, improvements in techniques for measuring sap flow have facilitated determination of the detailed radial profiles of axial flow necessary to analyze how different relationships between sapwood area and tree size among species influence water transport at different scales (e.g., James *et al.*, 2002; Jiménez *et al.*, 2000).

Plant vascular systems, and therefore water use, have been proposed to scale allometrically as power functions of body mass or size (West *et al.*, 1999):

$$Y = Y_0 M^b \tag{1}$$

where $M$ is mass, $b$ is a power exponent, and $Y_0$ is a normalization constant. Although exhaustive tests of the universality of this model have yet to be conducted, it is supported by some empirical data on plant water utilization (Enquist *et al.*, 1998; Meinzer *et al.*, 2001). If the model proves to be generally valid, it would greatly simplify prediction of water utilization over a large range of tree size, and in stands comprising multiple species. Given the large range of sapwood area among trees of similar size, however, it seems clear that multiple allometric functions, rather than a single universal function, are required for predicting hydroactive xylem area as a function of tree size. Nevertheless, differing amounts of sapwood in individuals of similar size do not necessarily preclude similar rates of whole-tree water use as explained later.

Whole-tree water use can be estimated by multiplying the mass flow of sap per unit sapwood area, or sap flux density, by the cross-sectional area of sapwood at the point where flow is measured. Thus, if the allometric scaling model presented previously is valid despite large differences in sapwood area for trees of the same size, it can be hypothesized that a universal proportionality constant describes the relationship between sap flux density and sapwood area. That is, for a given sapwood cross-sectional area, sap flux density would be identical across species regardless of tree diameter. Figure 15.1 shows the relationship between maximum sap flux density in the outer 2 cm of sapwood and sapwood area for five temperate coniferous species and 24 largely tropical angiosperm species. In both groups, sap flux density initially declines steeply, then more gradually with increasing sapwood area. When the relationships are linearized, it is clear that a single function adequately describes the relationship between sap flux density and sapwood area within each group. Sap flux density was consistently greater in the angiosperms than in the gymnosperms. Whether these differences would be maintained when whole-tree water use is estimated from the product of sap flux density and sapwood area depends on the profile of axial flow over the remaining depth of the sapwood. If the differences persist at the whole-tree level, then a single allometric equation cannot be used to scale water use in both angiosperms and gymnosperms. Initial determinations of radial profiles of axial flow for some of the species represented in the figure suggest that for a given stem diameter or basal sapwood area, whole-tree water use of conifers may be substantially lower than that of angiosperms (James *et al.*, 2003; Meinzer, unpublished observations).

It is tempting to infer that the overall greater sap flux density in angiosperms than conifers (Fig. 15.1) is associated with the presence of vessels in the former versus tracheids in the latter. This is probably an oversimplification because sap flux density is not governed by the hydraulic properties of the xylem elements alone. For instance, both the whole-tree leaf area to sapwood area ratio ($A_l$:$A_s$) and the ratio of xylem element lumen area to sapwood area are expected to influence sap flux density. The decline in sap flux density with increasing sapwood area observed in both angiosperms and conifers appears to be consistent with reports that $A_l$:$A_s$

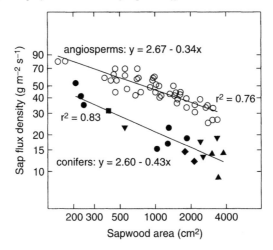

**Figure 15.1** Maximum sap flux density measured with the heat dissipation method (Granier, 1987) in the outermost 2 cm of sapwood in relation to total sapwood area for individuals of five temperate coniferous species (filled symbols) and 24 angiosperm species (open symbols). Coniferous species: (▲) *Pinus ponderosa*, (●) *Pseudotsuga menziesii*, (♦) *Thuja plicata*, and (▼) *Tsuga heterophylla* (unpublished data of Meinzer); (■) *Pinus sylvestris* (Nadezhdina *et al.*, 2002). The angiosperm species are a subset of those listed in Table 1 of Meinzer (2003).

typically decreases with increasing tree height (McDowell *et al.*, 2002). It should be noted, however, that sapwood area is not necessarily a reliable surrogate for tree height because of the species-specific differences in the relationship between sapwood area and tree size described previously. It is interesting that despite the large differences in sap flux density between angiosperms and gymnosperms over the range of sapwood area shown in Fig. 15.1, the Y-intercepts of the two regressions at zero sapwood area are similar. The differences in sap flux density are thus chiefly a consequence of the steeper negative slope for gymnosperms. If the relationships in Fig. 15.1 are governed by $A_l:A_s$, then it could be hypothesized that $A_l:A_s$ may decline more steeply with increasing tree size in gymnosperms than in angiosperms. If this proves to be true, it may indicate that greater hydraulic constraints associated with tracheids than vessels require larger developmental adjustments in hydraulic architecture (e.g., $A_l:A_s$) in gymnosperms in order to maintain sufficiency of water transport to leaves (Becker *et al.*, 1999).

### Radial Profiles of Axial Sap Flow

Numerous studies have reported a range of radial profiles of axial sap flow, but the causes of these patterns are not entirely known. The main sap flow patterns from the outer sapwood inward are even, decreasing, sharply decreasing, peaked (Phillips *et al.*, 1996), and erratic (Fig. 15.2). Even flow

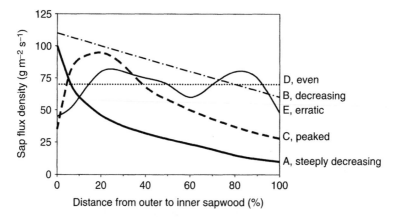

**Figure 15.2**  Schematic diagram of the radial patterns of axial sapflow: (A) steeply decreasing, (B) decreasing, (C) peaked, (D) even, (E) erratic. Bottom axis is the proportional distance across the sapwood, with the cambium on the far left and the sapwood/heartwood interface on the far right.

denotes similar sap flow at all measured radial depths in the sapwood. The decreasing and sharply decreasing patterns denote a gradual and more abrupt decline, respectively, from the values in the outer sapwood to those near the sap/heart border. The peaked pattern (called Gaussian by Phillips *et al.*, 1996) shows highest sap flow at some distance interior to the outer sapwood, usually within 1 to 3 cm of the cambium. Lastly, the random flow pattern shows values that go up and down across the radius in a nonsystematic manner.

To interpret radial variation in axial sap flow, more studies are needed of petiole connections to the stem xylem (Tison, 1902; Elliott, 1937; Maton and Gartner, 2005), the pathways for radial water movement in the stem xylem, and sapwood width. Water is withdrawn only from the points at which the leaf petioles attach to the xylem. In trees with leaves that survive one year or less, all the water is withdrawn through the outer ring of the sapwood. If the stem xylem has high radial resistance to water flow, then these trees should have most of the flow in the outer sapwood with a very steep decline (pattern A, Fig. 15.2). Such a pattern of sharply decreasing flow has been reported for the ring-porous deciduous trees *Quercus robur* (Granier *et al.*, 1994) and *Q. velutina* (Miller *et al.*, 1980), as well as the diffuse-porous deciduous species *Salix fragilis* (Cermak *et al.*, 1984). One might expect decreasing axial sap flow from bark inward (pattern B, Fig. 15.2) in two circumstances: trees with only current year leaves and moderate resistance to radial flow, or trees with leaves that persist for a number of years, with either moderate or strong radial resistance. This pattern of decreasing flow is commonly reported for angiosperms: the diffuse-porous deciduous species *Cordia alliodora* (James *et al.*, 2002), *Liriodendron tulipifera*. (Wullschleger and King, 2000), the diffuse-porous evergreen species *Citrus sinensis* (Cohen *et al.*, 1981), and several species of *Populus* (Edwards and Booker, 1984) and some individuals of the *Quercus* species mentioned previously. This pattern has also been reported in two conifer species: *Pinus taeda* (Phillips *et al.*, 1996) and *Pseudotsuga menziesii* (two of the four individuals, Cohen *et al.*, 1984). Most conifers exhibit peaked flow (below), and it is possible that these conifer individuals have peaked flow but that the sensors missed the peak.

Peaked flow (pattern C, Fig. 15.2) is an interesting case because there are few mechanistic hypotheses to explain its occurrence. In conifers, peaked flow has been reported for *Picea exelsa* (Cermak *et al.*, 1992), *Picea abies* (Cermak *et al.*, 1992), *Picea engelmannii* (Mark and Crews, 1973; Swanson, 1971), a few individuals of *Pseudotsuga menziesii* (Cohen *et al.*, 1984), *Pinus contorta* (Mark and Crews, 1973), and *Pinus sylvestris* (Nadezhdina *et al.*, 2002; Waring and Roberts, 1979). In angiosperms it has been reported for the deciduous species *Prunus serotina* and *Populus canescens* (Nadezhdina *et al.*, 2002) and for the evergreen species *Rhododendron ponticum* (Nadezhdina *et al.*, 2002), *Mangifera indica* (Lu *et al.*, 2000), *Ficus insipida* (James *et al.*, 2002), and *Laurus azorica, Persea indica, Ilex perado, Myrica faya*, and *Erica arborea* (Jiménez *et al.*, 2000). To our knowledge, none of these angiosperms are ring-porous.

Could peaked flow be caused by deeply inserted leaves that had high transpiration rates? Probably not. Deep leaf insertion can never explain peaked flow in deciduous species. In species with long-lived leaves, leaves would probably not stay attached long enough to explain the flow: If one assumes that the peak flow is 1.5 cm from the cambium and that radial growth is 0.3 cm/year, then the foliage attached to the peak radial position would be 5 years old. It is highly unlikely that the cohort of 5-year old foliage would cause the highest flux. Possible explanations are structural, but the only structure that appears able to increase, rather than decrease conductivity once the cells are produced, is the pit membranes. Mark and Crews (1973) reported that the pit membranes were not yet open in the growth rings near the cambium in *Picea engelmannii*, and that their amount of openness correlated with the flow they observed. This observation leads to the hypothesis that the pit membrane pores need to become enlarged beyond the cambial zone, perhaps by erosion during transpiration. Yet another possible explanation is that there are more air blockages in the outer sapwood than in the location where the flow peaks, but this explanation is not consistent with reports that moisture content appears highest in the outer sapwood. In the case of *Mangifera*, the researchers reported that the peaked flow could not be accounted for by the radial patterns of wood density and relative water content (Lu *et al.*, 2000). Jiménez *et al.*

(2002) found that individuals with sparser crowns had wider radial peaks of flow, and that individuals with more dense crowns had more compact flow. More attention is needed to explain the cause of peaked flow.

Even flow (pattern D, Fig. 15.2) could be expected in trees with any pattern of leaf longevity, as long as the wood has low radial resistance. It is logical to suggest that such low radial resistance would be adaptive in species that have wide sapwood. This pattern can also be incorrectly inferred from having sap flow gauges that miss the peaks in patterns A, B, or C. It has been reported in a miscellany of species and circumstances: the diffuse-porous brevideciduous *Anacardium excelsum* (James *et al.*, 2002), the diffuse-porous deciduous species *Liquidambar styraciflua* (Phillips *et al.*, 1996), and two *Populus* species (Edwards and Booker, 1984). Additionally, it was reported for the suppressed individuals of the diffuse-porous evergreen species *Laurus azorica* and *Ilex perado* (Jiménez *et al.*, 2000), and the ring-porous deciduous species *Quercus alba*, but the researchers surmised that they missed the high flow in the outer earlywood (Phillips *et al.*, 1996).

The last pattern, erratic flow (pattern E, Fig. 15.2), could result if the wood had uneven patterns within the stem that gave zones of high sap flow and low sap flow, and then the sensors were placed in these zones, rather than integrating over more "average" wood. Such wood could develop in a ring-porous species that had slow growth rate in one location (thus high earlywood percentage) and fast growth rate in another area (thus low earlywood percentage), or in a tropical species that had broad bands of different tissues. In *Schefflera morototoni*, the erratic flow followed closely the radial pattern of lumen area (James *et al.*, 2002). In *Populus canescens* (Nadezhdina *et al.*, 2002) and in *Eucalyptus globulus* (Zang *et al.*, 1996), the researchers surmised they had measured earlywood in some spots and latewood in others.

## Sapwood Water Storage

Water stored in stems has long been recognized as an important factor in plant-water relations. In large trees, seasonal courses of capacitive discharge and recharge of water stored in sapwood have been measured and modeled (Waring and Running, 1978; Waring *et al.*, 1979). On a shorter timescale, relatively little information exists on the contribution of sapwood water storage to daily water consumption. The dynamics of discharge and recharge of water storage compartments in trees can be characterized using the lag between changes in sap flow in the upper crown (a surrogate for transpiration) and sap flow near the base of the tree (Goldstein *et al.*, 1998; Phillips *et al.*, 1997). Estimates of the contribution of stored water to daily transpiration vary widely, ranging from 10% to 20% (Goldstein *et al.*, 1998; Kobayashi and Tanaka, 2001; Lostau *et al.*, 1996) to as much as 30% to 50% (Holbrook and Sinclair, 1992; Waring *et al.*, 1979). On the other hand, it was suggested that little stored water was withdrawn from the trunk of a *Larix decidua* tree, whereas water stored in branches accounted for 24% of the total daily transpiration (Schulze *et al.*, 1985).

Water released from both elastic compartments, such as parenchymatous tissue, and inelastic compartments, such as apoplastic capillary spaces and vessel and tracheid lumens during cavitation, can contribute to sapwood capacitance (Holbrook, 1995). Water withdrawn from living cells in stems of *Thuja occidentalis* was estimated to contribute about 6% to the total daily transpirational loss (Tyree and Yang, 1990). It has been proposed that daily cycles of cavitation and embolism followed by refilling of xylem elements may have a positive influence on leaf water balance by transiently releasing stored water into the transpiration stream (Lo Gullo and Salleo, 1992). Mounting evidence that embolism repair can occur over very short timescales, even when the surrounding intact xylem is still under tension (see Chapter 18), suggests that lumens of xylem elements constitute a dynamic and reusable water storage compartment.

Capacitance has been formally incorporated into Ohm's law analog models of water transport along the soil/plant/atmosphere continuum (e.g., Cowan, 1972; Phillips *et al.*, 1997). It is typically defined as the ratio of change in water content to change in water potential of a tissue. This relationship specifies the absolute volume of water that can be exchanged with storage tissues over the normal operating range of water potential for those tissues in a given species. Estimates of capacitance obtained from *in situ* measurements on intact trees vary from approximately 0.4 to 2 kg $MPa^{-1}$ for a number of coniferous and angiosperm species (Kobayashi and Tanaka, 2001; Milne, 1989; Tazikawa *et al.*, 1996; Tyree, 1988; Wronski *et al.*, 1985). It is unclear, however, whether this range of values largely reflects intrinsic, species-specific differences in the biophysical properties of the storage tissues involved, or merely differences in the size of the storage compartment, because none of the estimates was normalized for differences in tree size or the total volume of storage tissue. Species-specific values of sapwood capacitance normalized on a sapwood volume basis ranged from 83 to 416 kg $m^{-3}$ $MPa^{-1}$ among four co-occurring tropical forest tree species (Meinzer *et al.*, 2003), confirming that intrinsic differences in sapwood water storage capacity can be profound over the normal physiological operating range of stem water potential.

Despite the central role of sapwood water storage in whole-plant water relations, specific relationships between short-term cycles of capacitive discharge and recharge and the daily dynamics of stomatal behavior, transpiration, and plant water balance have not been studied until relatively recently. Capacitive exchange of water between storage compartments and the transpiration stream leads to daily fluctuations in apparent soil-to-leaf hydraulic conductance, provoking dynamic stomatal responses that maintain the balance between transpiration and hydraulic conductance, thereby limiting daily fluctuations in leaf water potential (Andrade *et al.*, 1998; Meinzer, 2002). Consistent with this, Williams *et al.* (1996) attributed the afternoon decline in $CO_2$ uptake in a mixed *Quercus-Acer* stand to partial stomatal closure in response to depletion of stored water, and Goldstein *et al.* (1998) reported that tropical forest trees with greater water storage capacity maintained maximum rates of transpiration for a substantially longer fraction of the day than trees with smaller water storage capacity.

In a recent study, it was shown that a suite of whole-tree water transport characteristics scaled with species-specific variation in sapwood capacitance (Fig. 15.3). Daily minimum branch water potential and soil-to-branch hydraulic conductance increased linearly with sapwood capacitance. In addition, maximum sap velocity, as estimated from the transit time of deuterated water injected as a tracer, decreased linearly with increasing capacitance. Although strong linear relationships between sapwood capacitance and features such as branch water status, whole-tree hydraulic conductance, and sap velocity do not necessarily imply direct mechanistic linkages, it was notable that based on the four species studied, the scaling of these features with capacitance appeared to be species-independent. Comparative studies of whole-plant water transport would therefore benefit from consideration of the constraints that sapwood biophysical features, such as capacitance, place on physiological functioning.

## Sapwood Density

Wood density ranges from <0.2 g cm$^{-3}$ in species such as *Ochroma pyramidale* (balsa) to >1.0 g cm$^{-3}$ in species such as *Diospyros ebenum* (ebony), and is an important determinant of xylem water transport and storage properties and whole-plant water relations. For example, the xylem tension threshold for 50% loss of specific conductivity by cavitation showed a strong positive correlation with wood density among several species spanning a relatively broad range of wood density (Hacke *et al.*, 2001). Although the relationships were distinct for angiosperms and gymnosperms, there was a common relationship between increasing wood density and increasing resistance to cavitation within each group. Wood density may also prove to be an essential normalizing variable for allometric scaling models that predict growth as a function of tree size. Growth rates of 45 co-occurring tropical tree species scaled with the 3/4 power of mass, but species-specific relationships with identical slopes and different intercepts did not collapse onto a common line until changes in diameter were normalized by the corresponding values of wood density (Enquist *et al.*, 1999). In view of the well-established positive correlation between biomass production and transpiration, these results imply an interaction between water transport and wood density.

Roderick and Berry (2001) developed a model to estimate how wood structure and density should affect water transport. One modeling result

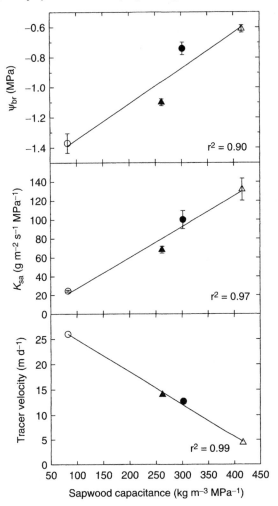

**Figure 15.3** Daily minimum water potential of terminal branches ($\Psi_{br}$), soil to terminal branch hydraulic conductance per unit basal sapwood area ($K_{sa}$), and axial transport rate of a deuterated water tracer signal in relation to sapwood capacitance for four co-occurring tropical forest tree species (○) *Cordia alliodora*, (▲) *Ficus insipida*, (●) *Anacardium excelsum*, and (△) *Schefflera morototoni* (after Meinzer *et al.*, 2003).

was that in softwoods, wood density should be negatively correlated with temperature during the growing season, because if the temperature is lower, then the water is more viscous and a lower density wood is needed to maintain the same flux. This pattern is apparent in many conifers, which have lower wood density at higher elevation or higher latitude. The model further predicted that there would be no such correlation in hardwoods because another term, related to the variability of conduit diameter, swamps the term in which viscosity was important. Again, this prediction is consistent with observations.

The widely reported negative correlation between sapwood saturated water content and density implies that the intrinsic capacitance of sapwood (see previously) diminishes with increasing density (Meinzer *et al.*, 2003). Low capacitance in species with dense wood could reduce the relative contribution of stored water to their daily water budget. However, compensatory behavior, such as stomatal regulation that allows stem water potential to fall to more negative values in species with denser wood, may stabilize the relative contribution of sapwood water storage to daily transpiration in species comprising a broad range of wood density.

There is some evidence that the gain in cavitation resistance associated with increasing wood density is associated with a cost in terms of reduced sapwood hydraulic conductivity. In six co-occurring Hawaiian dry forest species specific hydraulic conductivity of branches increased fivefold as wood density decreased from 0.65 to 0.5 g cm$^{-3}$ (Stratton *et al.*, 2000). Similarly, leaf-specific hydraulic conductivity of upper canopy branches of 20 co-occurring Panamanian forest species showed a sixfold increase as wood density decreased from 0.7 to 0.35 g cm$^{-3}$ (Santiago *et al.*, 2004). Although numerous reports of hydraulic properties of woody stems are available, their relationship with wood density is difficult to document because corresponding values of wood density are rarely reported. Sapwood density and hydraulic conductivity averaged over entire stem cross sections may conceal within-ring variation in density and conductivity that may be associated with within-ring partitioning of water conduction and storage functions (Domec and Gartner, 2002b), or water transport and mechanical functions (Mencuccini *et al.*, 1997).

If both hydraulic conductivity and sapwood water storage capacity are negatively correlated with wood density, it is reasonable to expect that species with greater wood density may experience both larger daily fluctuations in leaf water deficits and more extreme seasonal water deficits. Consistent with this, a strong positive correlation ($R^2 = 0.96$) between the magnitude of daily fluctuations in leaf water potential and wood density was noted among 27 diverse species with wood densities ranging from 0.14 to 0.9 g cm$^{-3}$ (Meinzer, 2003). The daily fluctuation in leaf water potential increased from c. 0.5 to 2.5 MPa over this range of wood density. The 27 species surveyed occupied habitats ranging from the humid tropics to deserts. The leaf water potential corresponding to turgor loss should be associated with the minimum leaf water potential normally experienced by a species, and therefore should provide an integrated measure of prevailing leaf water deficits. A survey of 12 species for which values of both wood density and leaf turgor loss point determined by the pressure-volume method were available revealed a strong negative correlation between the turgor loss point and wood density (Fig. 15.4). The turgor loss point declined from c. $-1.2$ to $-4.5$ MPa over a range of wood density from 0.16 to 0.90 g cm$^{-3}$. The preceding examples suggest that variation in wood density is a strong predictor of variation in a suite of characteristics related to regulation of leaf water deficits and avoidance of turgor loss, and that this feature constrains physiological options related to plant water economy, leading to functional convergence across a broad range of species.

## Prospects for Further Research

A wealth of detailed information is available on developmental and spatial variation in xylem anatomy and structure in trees. Similarly, advances in techniques for measuring sap flow have contributed to a rapidly expanding literature on spatial and temporal variation in water movement through

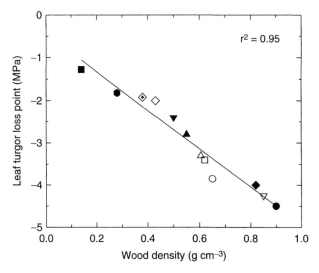

**Figure 15.4** Leaf water potential at the point of turgor loss in relation to wood density. ($\triangledown$) *Acacia greggii* (Nilsen *et al.*, 1984); ($\diamondsuit$) *Liriodendron tulipifera* (Roberts *et al.*, 1980); ($\blacktriangledown$) *Metrosideros polymorpha* (Stratton *et al.*, 2000); ($\triangle$) *Nesoluma polynesicum* (Stratton *et al.*, 2000); ($\bigcirc$) *Nestegis sandwicensis* (Stratton *et al.*, 2000); ($\diamondsuit$) *Populus* spp. (Tyree *et al.*, 1978); ($\square$) *Pouteria sandwicensis* (Stratton *et al.*, 2000); ($\blacklozenge$) *Prosopis glandulosa* (Nilsen *et al.*, 1981); ($\blacksquare$) *Pseudobombax septenatum* (Machado and Tyree, 1994); ($\bullet$) *Rhizophora mangle* (Melcher *et al.*, 2001); ($\bullet$) *Schefflera morototoni* (Tyree *et al.*, 1991); ($\blacktriangle$) *Schinus terebinthifolius* (Stratton *et al.*, 2000). Values of wood density were obtained from Alden (1995), Meinzer (2003), and Stratton *et al.* (2000).

trees. However, improved mechanistic understanding of relationships between xylem water transport and sapwood biophysical properties is hampered by a scarcity of studies in which water movement and sapwood properties have been measured in the same individuals. Therefore, to establish mechanistic linkages between xylem structure and water transport in intact trees, comprehensive characterizations of water movement, sapwood properties, and hydraulic architecture across a broad range of scale and tree size are required. Moreover, many more species must be examined because it is clear that there is more than one evolutionary solution to the problems of water transport.

To assimilate carbon, trees lose water. The amount they lose and how it travels through the sapwood are apparently quite variable. For a given sap flux, plants can co-vary factors such as sapwood permeability (and the anatomical means by which that is achieved), sapwood area, the depth of water transport within the sapwood, the extent to which stored water is accessed (daily and seasonally), and the timing of water transport (daily and seasonally). Are there suites of characteristics that often co-vary? If so, what are their ecological and anatomical patterns? Waring and Roberts (1979) and Dye *et al.* (1991) speculated that the inner sapwood primarily supplied water to the older branches. Does pruning shift sap flux outward in the sapwood? In some but not all coniferous species, petioles break their vascular connections with the previous year's xylem and make new connections annually (Tison, 1902; Maton and Gartner, 2005). Is this pattern unusual or the norm? Does peaked flow occur because of higher driving force on inner growth rings, or highter $k_s$ there? In which ecological settings and with which other characteristics (sapwood depth, permeability, radial and temporal patterns of transport, deciduous/evergreen foliage) is capacitance most pronounced? How do species maintain the moisture content differences across the sapwood, and how do they maintain different moisture contents in the sapwood and the heartwood? An emphasis on the functional tradeoffs among these characteristics may help us see a number of distinct strategies for water transport, which would greatly help our understanding, which is currently on a species- and age-specific basis.

A special case of the previous discussion is why co-occurring tropical trees of a given diameter appear to have the same sapwood depth (Meinzer *et al.*, 2001), but temperate ones vary from one extreme to the other. Likewise, more research is clearly needed on understanding sap flux in species without annual growth rings. Because of the clear role in water transport, much of the emphasis in ecological wood anatomy has been in vessel and tracheid characteristics. However, there is a myriad of patterns of xylem parenchyma, especially in tropical hardwoods, which merit a functional understanding. For example, the longitudinal parenchyma can be dispersed or aggregated. If it is aggregated, it can be near the vessels or apparently distanced from them ("apparently" because although it appears distanced in a cross section, the three-dimensional configuration of parenchyma, including the rays, has not been characterized). It can form tangential or radial bands, or a combination. The radially oriented parenchyma (in the rays) can be abundant or rare, and the cells can be aggregated into very wide multiseriate rays, or uniseriate rays (either of which can be tall or short). Within the rays, there can be procumbent cells and upright ones, and there can be rays within the same individual that are made up of different combinations of these geometric types. What are the roles of these different parenchyma cells and patterns? In what ecological settings are they found, and what is their role in water storage/release and water transport?

Another area that raises many questions is the striking relationship of wood density to the vulnerability of wood to embolism (Hacke *et al.*, 2001). On what structural features do these relationships depend? Given that wood density and the anatomy of the transport cells can vary independently of one another, it would not appear that there should be a relationship. For example, the density of hardwoods can vary simply by having an increase in libriform fiber cells and a decrease in parenchyma cells, with no effect on vessels. This line of research is an excellent example of recent progress in synthesizing data from diverse species and wood strategies. It shows, for example, that many hardwoods fall on one line, and that softwoods fall on another. It is also an example of a line of research that opens many questions about why the observed patterns are there. Many questions remain about structural causes of water transport and the strategies that are used in different ecological settings. These areas will continue to benefit from both the detailed species- and age-level research, and from the multispecies syntheses.

## Acknowledgments

Some of the data presented were obtained with support from National Science Foundation grant IBN 99-05012 to FCM, and with support from the Wind River Canopy Crane Research Facility located within the Wind River Experimental Forest, T. T. Munger Research Natural Area. The facility is a cooperative venture among the University of Washington, the USDA Forest Service Pacific Northwest Research Station and Gifford Pinchot National Forest. BLG wishes to acknowledge the special USDA grant to Oregon State University for wood utilization research.

# References

Alden, H. A. (1995) *Hardwoods of North America. General Technical Report FPL GTR-83.* USDA Forest Service Forest Products Laboratory, Madison, WI.

Andrade J. L., Meinzer F. C., Goldstein G., Holbrook N. M., Cavelier J., Jackson P. and Silvera K. (1998) Regulation of water flux through trunks, branches and leaves in trees of a lowland tropical forest. *Oecologia* 115: 463-471.

Becker, P., Tyree, M. T. and Tsuda, M. (1999) Hydraulic conductances of angiosperms versus conifers: Similar transport sufficiency at the whole-plant level. *Tree Physiol* 19: 445-452.

Booker, R. E. and Kininmonth, J. A. (1978) Variation in longitudinal permeability of green radiata pine wood. *N Z J For Sci* 8: 295-308.

Brix, H. and Mitchell, A. K. (1983) Thinning and nitrogen fertilization effects on sapwood development and relationships of foliage quantity to sapwood area and basal area in Douglas-fir. *Can J For Res* 13: 384-389.

Cermak, J., Cienciala, E., Kucera, J. and Hallgren, J.-E. (1992) Radial velocity profiles of water flow in trunks of Norway spruce and oak and the response of spruce to severing. *Tree Physiol* 10: 367-380.

Cermak, J., Jenik, J., Kucera, J. and Zidek, V. (1984) Xylem water flow in a crack willow tree (*Salix fragilis* L.) in relation to diurnal changes of environment. *Oecologia* 64: 145-151.

Cherubini, P., Fontana, G., Rigling, D., Dobbertin, M., Brang, P. and Innes, J. L. (2002) Tree-life history prior to death: Two fungal root pathogens affect tree-ring growth differently. *J Ecol* 90: 839-850.

Cochard, H. and Tyree, M. T. (1990) Xylem dysfunction in *Quercus*: Vessel sizes, tyloses, cavitation and seasonal changes in embolism. *Tree Physiol* 6: 393-407.

Cohen, Y., Fuchs, M. and Green, G. C. (1981) Improvement of the heat pulse technique method for determining sap flow in trees. *Plant Cell Environ* 4: 391-397.

Cohen, Y., Kellhier, F. M. and Black, T. A. (1984) Determination of sap flow in Douglas-fir trees using the heat pulse technique. *Can J For Res* 15: 422-428.

Comstock, G. L. (1965) Longitudinal permeability of green eastern hemlock. *Forest Prod J* 15: 441-449.

Cowan, I. R. (1972) An electrical analogue of evaporation from, and flow of water in plants. *Planta* 106: 221-226.

de Kort, I. (1993) Relationships between sapwood amount, latewood percentage, moisture content and crown vitality of Douglas fir, *Pseudotsuga menziesii. IAWA J* 14: 413-427.

Domec, J. C. and Gartner, B. L. (2001) Cavitation and water storage capacity in bole xylem segments of mature and young Douglas-fir trees. *Trees* 15: 204-214.

Domec, J. C. and Gartner, B. L. (2002a) Age- and position-related changes in hydraulic versus mechanical dysfunction of xylem: Inferring the design criteria for Douglas-fir wood structure. *Tree Physiol* 22: 91-104.

Domec, J. C. and Gartner, B. L. (2002b) How do water transport and water storage differ in coniferous earlywood and latewood? *J Exp Bot* 53: 2369-2379.

Domec, J.-C. and Gartner, B. L. (2003) Relationship between growth rates and xylem hydraulic characteristics in young, mature, and old-growth ponderosa pine trees. *Plant Cell Environ* 26: 471-483.

Duff, G. H. and Nolan, N. J. (1953) Growth and morphogenesis in the Canadian forest species I. The controls of cambial and apical activity in *Pinus resinosa. Can J Bot* 31: 471-513.

Dye, P. J., Olbrich, B. W. and Poulter, A. G. (1991) The effect of growth rings in *Pinus patula* on heat pulse velocities and sap flow measurement. *J Exp Bot* 42: 867-870.

Edwards, W. R. N. and Booker, R. E. (1984) Radial variation in the axial conductivity of *Populus* and its significance in heat pulse velocity measurement. *J Exp Bot* 35: 551-561.

Edwards, W. R. N. and Jarvis, P. G. (1982) Relations between water content, potential and permeability in stems of conifers. *Plant Cell Environ* **5**: 271-277.

Elliott, J. H. (1937) The development of the vascular system in evergreen leaves more than one year old. *Ann Bot N S* **1**: 107-127.

Ellmore, G. S. and Ewers, F. W. (1985) Hydraulic conductivity in trunk xylem of elm, *Ulmus americana. IAWA Bull N. S.* **6**: 303-307.

Enquist, B. J, Brown, J. H. and West G. B. (1998) Allometric scaling of plant energetics and population density. *Nature* **395**: 163-165.

Enquist B. J., West, G. B., Charnov, E. L. and Brown, J. H. (1999) Allometric scaling of production and life-history variation in vascular plants. *Nature* **401**: 907-911.

Gartner, B. L. (2002) Sapwood and inner bark quantities in relation to leaf area and wood density in Douglas-fir. *IAWA J* **23**: 267-285.

Gartner, B. L., Baker, D. C. and Spicer, R. (2000) Distribution and vitality of xylem rays in relation to tree leaf area in Douglas-fir. *IAWA J* **21**: 389-401.

Gartner, B. L., Lei, H. and Milota, M. R. (1997) Variation in the anatomy and specific gravity of wood within and between trees of red alder (*Alnus rubra* Bong.). *Wood Fiber Sci* **29**: 10-20.

Goldstein, G., Andrade, J. L., Meinzer, F. C., Holbrook, N. M., Cavelier, J., Jackson, P. and Celis, A. (1998) Stem water storage and diurnal patterns of water use in tropical forest canopy trees. *Plant Cell Environ* **21**: 397-406.

Granier, A. (1987) Evaluation of transpiration in a Douglas-fir stand by means of sap flow measurements. *Tree Physiol* **3**: 309-320.

Granier, A., Anfodillo, T., Sabatti, M., Cochard, H., Dreyer, E., Tomasi, M., Valentini, R. and Bréda, N. (1994) Axial and radial water flow in the trunks of oak trees: A quantitative and qualitative analysis. *Tree Physiol* **14**: 1383-1396.

Granier, A., Biron, P. and D. Lemoine (2000) Water balance, transpiration and canopy conductance in two beech stands. *Agr For Met* **100**: 291-308.

Hacke, U. G., Sperry, J. S., Pockman, W. T., Davis, S. D. and McCulloh, K. A. (2001) Trends in wood density and structure are linked to prevention of xylem implosion by negative pressure. *Oecologia* **126**: 457-461.

Harrington, C. A. and DeBell, D. S. (1980) Variation in specific gravity of red alder (*Alnus rubra* Bong.). *Can J For Res* **10**: 293-299.

Hazenberg, G. and Yang, K. C. (1991) The relationship of tree age with sapwood and heartwood width in black spruce, *Picea mariana* (Mill.) B. S. P. *Holzforschung* **45**: 317-320.

Hillis, W. E. (1987). *Heartwood and Tree Exudates.* Springer-Verlag, Berlin.

Holbrook, N. M. (1995) Stem water storage. In *Plant Stems: Physiology and Functional Morphology* (B.L. Gartner, ed.) pp. 151-174. Academic Press, San Diego.

Holbrook, N. M. and Sinclair, T. R. (1992) Water balance in the arborescent palm, *Sabal palmetto.* II. Transpiration and stem water storage. *Plant Cell Environ* **15**: 401-409.

James, S. A., Clearwater, M. J., Meinzer, F. C. and Goldstein, G. (2002) Heat dissipation sensors of variable length for the measurement of sap flow in trees with deep sapwood. *Tree Physiol* **22**: 277-283.

James, S. A., Meinzer, F. C., Goldstein, G., Woodruff, D., Jones, T., Restom, T., Mejia, M., Clearwater, M. and Campanello, P. (2003) Axial and radial water transport and internal water storage in tropical forest canopy trees. *Oecologia* **134**: 37-45.

Jiménez, M. S., Nadezhdina, N., Cermák, J. and Morales, D. (2000) Radial variation in sap flow in five laurel forest tree species in Tenerife, Canary Islands. *Tree Physiol* **20**:1149-1156.

Kobayashi, Y. and Tanaka, T. (2001) Water flow and hydraulic characteristics of Japanese red pine and oak trees. *Hydrol Proc* **15**: 1731-1750.

Larson, P. R., Kretschmann, D. E., Clark, A. I. and Isebrands, J. D. (2001) Formation and properties of juvenile wood in southern pines: A synopsis. FPL-GTR-129. USDA Forest Service, Forest Products Laboratory. Madison, WI.

Lassen, L. E. and Okkonen, E. A. (1969) Sapwood thickness of Douglas-fir and five other western softwoods. USDA Forest Service Research Paper, FPL 124, Madison, WI.

Lei, H., Gartner, B. L. and Milota, M. R. (1997) Effect of growth rate on the anatomy, specific gravity, and bending properties of wood from 7-year-old red alder (*Alnus rubra*). *Can J For Res* **27**: 80-85.

Lei, H., Milota, M. R. and Gartner, B. L. (1996) Between- and within-tree variation in the anatomy and specific gravity of wood in Oregon white oak (*Quercus garryana* Dougl.). *IAWA J* **17**: 445-461.

Lo Gullo, M. A. and Salleo, S. (1992) Water storage in the wood and xylem cavitation in 1-year-old twigs of *Populus deltoides* Bartr. *Plant Cell Environ* **15**: 431-438.

Loustau, D., Berbiger, P., Roumagnac, P., Arruda-Pacheco, C., David, J. S., Ferreira, M. I., Pereira, J. S., and Tavares, R. (1996) Transpiration of a 64-year-old maritime pine stand in Portugal. I. Seasonal course of water flux through maritime pine. *Oecologia* **107**: 33-42.

Lu, P., Muller, W. J. and Chacko, E. K. (2000) Spatial variations in xylem sap flux density in the trunk of orchard-grown, mature mango trees under changing soil water conditions. *Tree Physiol* **20**: 683-692.

Machado, J.-L. and Tyree, M. T. (1994) Patterns of hydraulic architecture and water relations of two tropical canopy trees with contrasting leaf phenologies: *Ochroma pyramidale* and *Pseudobombax septenatum*. *Tree Physiol* **14**: 219-240.

Margolis, H., Oren, R., Whitehead, D. and Kaufmann, M. R. (1995) Leaf area dynamics of conifer forests. In *Ecophysiology of Coniferous Forests* (W. K. Smith and T. M. Hinckley, eds.) pp. 181-223. Academic Press, San Diego.

Mark, W. R. and Crews, D. L. (1973) Heat-pulse velocity and bordered pit condition in living Engelman spruce and lodgepole pine trees. *Forest Sci* **19**: 291-296.

Markstrom, D. C. and Hann, R. A. (1972) Seasonal variation in wood permeability and stem moisture content of three Rocky Mountain softwoods. RM-212. USDA Forest Service, Rocky Mountain Forest and Range Experimental Station, CO.

Maton, C. and Gartner, B. L. (2005) Do gymnosperm needles pull water through the xylem produced in the same year as the needle? *Am J Bot* **92**: 123-131.

McDowell, N., Barnard, H., Bond, B. J., Hinckley, T., Hubbard, R. M., Ishii, H., Kostner, B., Magnani, F., Marshall, J. D., Meinzer, F. C., Phillips, N., Ryan, M. G. and Whitehead, D. (2002) The relationship between tree height and leaf area:sapwood area ratio. *Oecologia* **132**: 12-20.

Megraw, R. A. (1985) *Wood Quality Factors in Loblolly Pine: The Influence of Tree Age, Position in Tree, and Cultural Practice on Wood Specific Gravity, Fiber Length, and Fibril Angle*. Tappi Press, Atlanta, GA.

Meinzer, F. C. (2002) Co-ordination of vapour and liquid phase water transport properties in plants. *Plant Cell Environ* **25**: 265-274.

Meinzer, F. C. (2003) Functional convergence in plant responses to the environment. *Oecologia* **134**: 1-11.

Meinzer, F. C., Goldstein, G. and Andrade, J. L. (2001) Regulation of water flux through tropical forest canopy trees: Do universal rules apply? *Tree Physiol* **21**: 19-26.

Meinzer, F. C., James, S. A., Goldstein, G. and Woodruff, D. (2003) Whole-tree water transport scales with sapwood capacitance in tropical forest canopy trees. *Plant Cell Environ* **26**: 1147-1155.

Melcher, P. J., Goldstein, G., Meinzer, F. C., Yount, D. E., Jones, T. J., Holbrook, N. M. and Huang, C. X. (2001) Water relations of coastal and estuarine *Rhizophora mangle*: Xylem pressure potential and dynamics of embolism formation and repair. *Oecologia* **126**: 182-192.

Mencuccini, M., Grace, J. and Fioravanti, M. (1997) Biomechanical and hydraulic determinants of tree structure in Scots pine: Anatomical characteristics. *Tree Physiol* **17**: 105-113.

Miller, D. R., Vavrina, C. A. and Christensen, T. W. (1980) Measurement of sap flow and transpiration in ring-porous oaks using a heat pulse velocity technique. *Forest Sci* **26**: 485-494.

Milne, R. (1989) Diurnal water storage in the stem of *Picea sitchensis* (Bong.) Carr. *Plant Cell Environ* **12**: 63-72.

Nadezhdina, N., Cermak, J. and Ceulemans, R. (2002) Radial patterns of sap flow in woody stems of dominant and understory species: Scaling errors associated with positioning of sensors. *Tree Physiol* **22**: 907-918.

Nikolov, S. and E. Enchev (1967) *Vlaznost na darvesinata v surovo sastojanie (Moisture Content of Green Wood).* Zemizdat, Sofia, Bulgaria.

Nilsen, E. T., Rundel, P. W. and Sharifi, M. R. (1981) Summer water relations of the desert phreatophyte *Prosopis glandulosa* in the Sonoran Desert of Southern California. *Oecologia* **50**: 271-276.

Nilsen, E. T., Sharifi, M. R. and Rundel, P. W. (1984) Comparative water relations of phreatophytes in the Sonoran Desert of California. *Ecology* **65**: 767-778.

Nobuchi, T. and Harada, H. (1983) Physiological features of the "white zone" of sugi (*Cryptomeria japonica* D. Don)–Cytological structure and moisture content. *Mokuzai Gakkaishi* **29**: 824-832.

Panshin, A. J. and de Zeeuw, C. (1980) *Textbook of Wood Technology: Structure, Identification, Properties, and Uses of the Commercial Woods of the United States.* McGraw-Hill, New York.

Petty, J. A. and Puritch, G. S. (1970) The effect of drying on the structure and permeability of wood of *Abies grandis*. *Wood Sci Technol* **4**: 140-154.

Phelps, J. E. and Workman, Jr., E. C. (1994) Vessel area studies in white oak (*Quercus alba* L.). *Wood Fiber Sci* **26**: 315-322.

Phillips, N., Bond, B. J., McDowell, N. G. and Ryan, M. G. (2002) Canopy and hydraulic conductance in young, mature and old Douglas-fir trees. *Tree Physiol* **22**: 205-212.

Phillips, N., Nagchaudhuri, A., Oren, R. and Katul, G. (1997) Time constant for water transport in loblolly pine trees estimated from time series of evaporative demand and stem sapflow. *Trees* **11**: 412-419.

Phillips, N., Oren, R. and Zimmermann, R. (1996) Radial patterns of xylem sap flow in non-, diffuse- and ring-porous tree species. *Plant Cell Environ* **19**: 983-990.

Pothier, D., Margolis, H. A., Poliquin, J. and Waring, R. H. (1989) Relation between the permeability and the anatomy of jack pine sapwood with stand development. *Can J For Res* **19**: 1564-1570.

Puritch, G. S. (1971) Water permeability of the wood of grand fir (*Abies grandis* (Dougl.) Lindl.) in relation to infestation by the balsam wooly aphid, *Adelges piceae* (Ratz.). *J Exp Bot* **22**: 936-945.

Roberts, S. W., Strain, B. R., and Knoerr, K. R. (1980) Seasonal patterns of leaf water relations in four co-occurring forest tree species: Parameters from pressure-volume curves. *Oecologia* **46**: 330-337.

Roderick, M. L. and Berry, S. L. (2001). Linking wood density with tree growth and environment: A theoretical analysis based on the motion of water. *New Phytol* **149**: 473-485.

Santiago, L. S., Goldstein, G., Meinzer, F. C., Fisher, J. B., Machado, K., Woodruff, D. and Jones, T. (2004) Leaf photosynthetic traits scale with hydraulic conductivity and wood density in Panamanian forest canopy trees. Oecotogia **140**: 543-550.

Sellin, A. (1996) Sapwood amount in *Picea abies* (L.) Karst determined by tree age and radial growth rate. *Holzforschung* **50**: 291-296.

Senft, J. F., Bendtsen, B. A. and Galligan W. L. (1985) Weak wood: Fast-grown trees make problem lumber. *J Forestry* **83**: 476-484.

Schmitt, U., Grunwald, C., Gricar, J., Koch, G. and Cufar, K. (2003) Wall structure of terminal latewood tracheids of healthy and declining silver fir trees in the Dinaric region, Slovenia. *IAWA J* **24**: 41-52.

Schulze, E.-D., Cermak, J., Matyssek, R., Penka, M., Zimmermann, R., Vasicek, F., Gries, W. and Kucera, J. (1985) Canopy transpiration and water fluxes in the trunk of *Larix* and *Picea* trees: A comparison of xylem flow, porometer, and cuvette measurements. *Oecologia* **66**: 475-483.

Smith, J. H. G., Walters, J. and Wellwood, R. W. (1966) Variation in sapwood thickness of Douglas-fir in relation to tree and section characteristics. *For Sci* **1:** 97-103.

Spicer, R. and Gartner, B. L. (2001) The effects of cambial age and position within the stem on specific conductivity in Douglas-fir (*Pseudotsuga menziesii*) sapwood. *Trees* **15:** 222-229.

Stratton, L., Goldstein, G. and Meinzer, F. C. (2000) Stem water storage capacity and efficiency of water transport: Their functional significance in a Hawaiian dry forest. *Plant Cell Environ* **23:** 99-106.

Swanson, R. H. (1971) Velocity distribution patterns in ascending xylem sap during transpiration. Symposium on Flow: Its measurement and control in science and industry. Canadian Forestry Service Paper no. 4/2/171.

Tazikawa, H., Hayami, K., Kubota, J. and Tsukamoto, Y. (1996). Variations of volumetric water content and water potential in tree stems because of transpiration. *J Jpn For Soc* **78:** 66-73.

Tison, A. (1902) Les traces foliares des coniferes dans leur rapport avec l'epaissisement de la tige. *Memoires de la Societe Linneenne de Normandie* **21:** 61-83.

Tyree, M. T. (1988) A dynamic model for water flow in a single tree: Evidence that models must account for hydraulic architecture. *Tree Physiol* **4:** 195-217.

Tyree, M. T., Cheung, Y. N. S., Macgregor, M. E. and Talbot, A. J. B. (1978) The characteristics of seasonal and ontogenetic changes in the tissue-water relations of *Acer, Populus, Tsuga* and *Picea. Can J Bot* **56:** 635-647.

Tyree, M. T. Snyderman, D. A., Wilmot, T. R. and Machado, J. L. (1991) Water relations and hydraulic architecture of a tropical tree (*Schefflera morototoni*): Data, models and a comparison with two temperate species (*Acer saccharum* and *Thuja occidentalis*). *Plant Physiol* **96:** 1105-1113.

Tyree, M. T. and Yang, S. D. (1990) Water-storage capacity of *Thuja, Tsuga* and *Acer* stems measured by dehydration isotherms: The contribution of capillary water and cavitation. *Planta* **182:** 420-426.

Utsumi, Y., Sano, Y., Funada, R., Ohtani, J. and Fujikawa, S. (2003) Seasonal and perennial changes in the distribution of water in the sapwood of conifers in a sub-frigid zone. *Plant Physiol* **131:** 1826-1833.

Waring, R. H. and Roberts J. M., (1979) Estimating water flux through stems of Scots pine with tritiated water and phosphorus-32. *J Exp Bot* **30:** 459-471.

Waring R. H. and Running, S. W. (1978) Sapwood water storage: Its contribution to transpiration and effect upon the water conductance through the stems of old-growth Douglas-fir. *Plant Cell Environ* **1:** 131-140.

Waring, R. H., Whitehead, D. and Jarvis, P. G. (1979) The contribution of stored water to transpiration in Scots pine. *Plant Cell Environ* **2:** 309-317.

Wellwood, R. W., Sastry, C. B. R., Micko, M. M. and Paszner, L. (1974) On some possible specific gravity, holo- and alpha-cellulose, tracheid weight/length and cellulose crystallinity relationships in a 500-year-old Douglas-fir tree. *Holzforschung* **28:** 91-94.

West, G. B., Brown, J. H. and Enquist, B. J. (1999) A general model for the structure and allometry of plant vascular systems. *Nature* **400:** 664-667.

Whitehead, D., Edwards, W. R. N. and Jarvis, P. G. (1984) Conducting sapwood area, foliage, and permeability in mature trees of *Picea sitchensis* and *Pinus contorta. Can J For Res* **14:** 940-947.

Williams, M., Rastetter, E. B., Fernandes, D. N., Goulden, M. L., Wofsy, S. C., Shaver, G. R., Melillo, J. M., Munger J. W., Fan, S.-M. and Nadelhoffer, K. J. (1996) Modelling the soil-plant-atmosphere continuum in a *Quercus-Acer* stand at Harvard Forest: The regulation of stomatal conductance by light, nitrogen and soil/plant hydraulic properties. *Plant Cell Environ* **19:** 911-927.

Wronski, E. B., Holmes, J. W. and Turner, N. C. (1985) Phase and amplitude relations between transpiration, water potential and stem shrinkage. *Plant Cell Environ* **8:** 613-622.

Wullschleger, S. D. and King, A. W. (2000) Radial variation in sap velocity as a function of stem diameter and sapwood thickness in yellow-poplar trees. *Tree Physiol* **20:** 511-518.

Wullschleger, S. D., Meinzer, F. C. and Vertessy, R. A. (1998) A review of whole-plant water use studies in trees. *Tree Physiol* **18:** 499-512.

Zang, D., Beadle, C. L. and White, D. A. (1996) Variation of sapflow velocity in *Eucalyptus globulus* with position in sapwood and use of a correction coefficient. *Tree Physiol* **16:** 697-703.

Zhang, S. Y. and Zhong, Y. (1991) Effect of growth rate on specific gravity of East-Liaoning oak (*Quercus liaotungensis*) wood. *Can J For Res* **21:** 255-260.

Zimmermann, M. H. (1978) Hydraulic architecture of some diffuse-porous trees. *Can J Bot* **56:** 2286-2295.

Zobel, B. J. and Sprague, J. R. (1998) *Juvenile Wood in Forest Trees.* Springer-Verlag, Berlin.

# 16

## Efficiency Versus Safety Tradeoffs for Water Conduction in Angiosperm Vessels Versus Gymnosperm Tracheids

*Uwe G. Hacke, John S. Sperry, and Jarmila Pittermann*

Efficient water transport allows more $CO_2$ uptake per cost of transpired water. The cost of transpiration includes investment in the root-to-leaf xylem pipeline and its protection from failure (Givnish, 1986). Major epochs of land plant history are marked by improved transport efficiency—from the evolution of tracheids and the emergence of vascular plants to the evolution of vessels in derived lineages (Bailey, 1953; Carlquist, 1975; Sperry, 2003). At a finer scale, xylem function is linked to ecology and growth form as shown by the adaptive value of cavitation resistance (Davis *et al.*, 1998; Hacke *et al.*, 2000; Pockman and Sperry, 2000) and the correspondence between conducting capacity and photosynthetic capacity (Brodribb and Feild, 2000; Stiller *et al.*, 2003).

The evolution of greater xylem conductivity cannot compromise safety from transport failure. Transport is threatened by cavitation of the xylem sap (Dixon and Joly, 1895; Pickard, 1981), and collapse of the xylem conduits by implosion (Carlquist, 1975). These problems limit the dimensions of xylem conduits and require strong conduit walls, features that can reduce the conductivity per unit investment in the conduit network. Natural selection should favor structures that can beat or at least improve on these efficiency versus safety tradeoffs. Discovering the rules that define these conflicting functions is important for interpreting the functional and evolutionary significance of diverse xylem anatomies.

In this chapter we attempt to quantify structure-function relations by summarizing results from a detailed biomechanical model of pit and conduit function and related observations (Hacke *et al.*, 2004; Sperry and

Hacke, 2004). Tradeoffs are compared between gymnosperm and angiosperm wood. This comparison tests whether fundamental differences in xylem structure between these groups have an impact on safety versus efficiency relationships. Although gymnosperms are much less diverse than angiosperms, conifers dominate much of the world's forests and include the tallest (*Sequoia sempervirens*, >100 m) and oldest (*Pinus longaeva*, over 4500 years) organisms on earth. Their tracheid-based xylem, though relatively unspecialized, must serve them well in these circumstances. The more specialized angiosperm xylem of vessels and fibers may not be uniformly advantageous, an observation enforced by the possible return to tracheid-based wood in some basal angiosperms (Feild *et al.*, 2002). Interconduit pitting also differs substantially. Many gymnosperms possess circular bordered pits with torus-margo membranes (Bauch *et al.*, 1972), whereas angiosperms show a wider variety of pit shape and mostly have uniformly thin homogenous pit membranes (Fig. 16.1A). Given the crucial importance of pit function, these differences could have a major influence on efficiency versus safety tradeoffs in the two comparison groups.

## Tradeoffs in Interconduit Pit Function

Interconduit pit structure directly influences the hydraulic conductivity of xylem, the resistance to cavitation, and the strength of the wall against implosion. Conductivity versus safety come into direct conflict, because a pit that is more permeable to water flow is likely to be a pit more vulnerable to cavitation by air-seeding and more likely to cause wall failure.

Pit structure and function differs significantly between angiosperms and gymnosperms (Fig. 16.1). The pits of angiosperms block air entry by capillary forces at the relatively small pores in their homogenous pit membranes (Fig. 16.1B). These membranes can aspirate when sealed (Fig. 16.1B), although observations are limited (Thomas, 1972). Air-seeding occurs through pores in the membrane spanning the aperture (Fig. 16.1C) (Crombie *et al.*, 1985; Sperry *et al.*, 1996). The exact mechanism of air-seeding is not known. The pores may already be present in the membrane or may be created by plastic yielding or rupture of the stretching membrane.

The inter-tracheid pits in the early-wood of a majority of conifers and also *Ginkgo biloba* have a torus-margo membrane (Fig. 16.1) (Bauch *et al.*, 1972). Weak capillary forces at the relatively large pores of the margo are sufficient to aspirate the nonporous torus across the pit aperture, which seals off the pit against air (Fig. 16.1B) (Dixon and Joly, 1895; Petty, 1972). Petty's work suggests that there is no capillary-seeding in this type of pit—in early-wood pits the margo aspirates before leaking air. Air-seeding appar-

ANGIOSPERM                    GYMNOSPERM

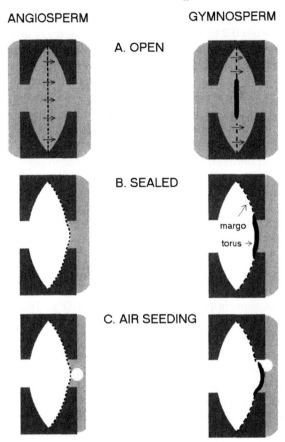

**Figure 16.1** Function of interconduit pits in angiosperms and gymnosperms (mostly conifers) with torus-margo pit structure. Circular bordered pits shown in transverse median section (see also Fig. 16.5) (A) Open configuration allowing flow of water between conduits with minimal deflection of the membrane. (B) Sealed configuration blocking air passage from embolized conduit (*left*) to water-filled ones under negative pressure (*right*). (C) The air-seeding process where the pressure difference across the pit is sufficient to force air into the water-filled conduit and nucleate cavitation. From Hacke *et al.* (2004) *American Journal of Botany* 91:386-400. Copyright Botanical Society of America; adapted with permission.

ently occurs when the torus is pushed through the aperture enough to allow air to leak past its edge (Fig. 16.1C) (Sperry and Tyree, 1990; Domec and Gartner, 2002). It is not clear whether the stretching of the margo is reversible or occurs by plastic yielding or rupture.

There is no evidence that either type of pit is necessarily more resistant to air-seeding than the other. Sperry and Hacke (2004) and Hacke *et al.* (2004) surveyed 17 gymnosperm and 27 angiosperm species of diverse families and habitats for cavitation resistance. The "air-seed pressure" (equal and opposite to the negative pressure causing a 50% drop in hydraulic conductivity by cavitation) ranged from 0.2 to 11.3 MPa in roots and stems of angiosperms versus 0.8 to 11.8 MPa in gymnosperms. In both groups, the more vulnerable xylem was from roots and from species from more mesic habitats. Although the average air-seed pressure was greater in gymnosperms (4.8 MPa) than angiosperms (3.6 MPa), the similar ranges indicate that neither xylem type is necessarily more vulnerable to cavitation by air-seeding than the other.

### Pit Conductivity Versus Air-Seed Pressure

Is there a tradeoff between pit conductivity and air-seed pressure? Do torus-margo pits have a different hydraulic conductivity per pit area than angiosperm pits? Although the air-seed pressure is easy to measure, direct measurements of pit conductivity are much more difficult. To obtain a theoretical answer to these questions, we developed a biomechanical model of pit function that links pit structure to air-seed pressure and conductivity (Figs. 16.2, 16.3) (Sperry and Hacke, 2004) and applied it to the survey of gymnosperm and angiosperm species mentioned previously. Pit membranes were assumed to be composed of multiple sheets of parallel cellulose microfibrils superimposed to create a mesh (Fig. 16.2, lower right). Membrane mechanics and air-seeding were modeled by extending Petty's analysis of pit aspiration (Fig. 16.2). Hydraulic conductivity of the pit membrane mesh was modeled from analytical approximations of numerical solutions for the conductivity of infinitely thin porous plates (Fig. 16.3, right). Conductivity of the pit aperture was calculated from similar equations for flow through pores of finite length (Fig. 16.3, left). The model was used to calculate pit conductivities (hydraulic conductance per membrane area) that corresponded with measured air-seed pressure and pit dimensions of the gymnosperm and angiosperm xylem samples. For the sake of simplicity, the analysis was limited to pits of circular shape.

The model predicts that the torus-margo pits of gymnosperms are much more efficient in optimizing a safety versus efficiency tradeoff. Torus-margo pits achieved between 3 to 60 times the pit conductivity (conductance per membrane area) of angiosperm pits over their similar range of air-seed pressure, with the greater advantage at higher pressure (Fig. 16.4). Angiosperm pits showed a much steeper tradeoff with conductivity declining by 50-fold from low to high air-seed pressure versus only a 3-fold decline for gymnosperm pits (Hacke *et al.*, 2004).

The secret to the torus-margo design is that it uncouples membrane pore diameter from air-seeding. Margo pores in earlywood tracheids can be large without compromising safety because their capillary forces only have to be great enough to deflect the torus as the water drains out of an embolizing tracheid (Sperry and Tyree, 1990; Domec and Gartner, 2002). In contrast, the pores of the homogenous pit membranes of angiosperms must be narrow to provide a capillary seal. Although the margo is only

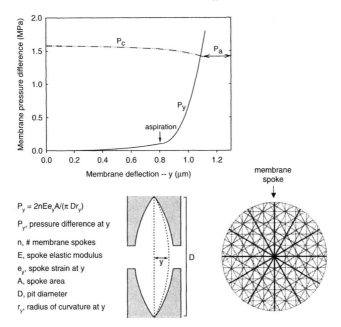

**Figure 16.2** Calculating the air-seed pressure (Sperry and Hacke, 2004). The equation (modified from Petty, 1972) gives the pressure difference ($P_y$) required for a given membrane deflection (y, pit section insert) as a function of the number and properties of the microfibril spokes (*membrane insert*) and of pit geometry. Default spoke elastic modulus was 5 GPa. The solid curve (*graph*) shows the increase in $P_y$ with deflection. At aspiration, the equation was modified for continued deflection through the smaller aperture. For homogenous pits, air-seeding occurred when $P_y$ rose sufficiently to blow air through pores in the membrane ($P_a$, *arrow*). The pore pressure ($P_c$, *dash-dotted line*) was calculated from the capillary equation and accounted for pore expansion in the stretching membrane. For torus-margo pits, the air-seeding occurred when the margo stretched to the aperture lip. This example is for angiosperm pits of average membrane diameter (4.9 μm) and aperture diameter (1.6 μm) (Sperry and Hacke, 2004). From Sperry and Hacke (2004) *American Journal of Botany* 91:369-385. Copyright Botanical Society of America; adapted with permission.

about 75% of the membrane area, conductivity of the membrane is more sensitive to pore diameter (a third power relationship; Fig. 16.3, right) than to pore number. There is a steeper tradeoff in angiosperm pits because their conductivity is membrane-limited and membrane conductivity declines sharply with narrower pores required for higher air-seed pressure. There is a flatter tradeoff in gymnosperm's pits because their conductivity was predicted to be aperture-limited, and aperture conductivity declines only slightly with air-seed pressure (Hacke *et al.*, 2004).

This analysis of pit conductivity awaits testing against actual measurements. Nevertheless, it raises some immediate questions. Does the conductivity versus air-seeding tradeoff at the scale of a single pit propagate to the scale of the entire conduit network? And, if torus-margo pits are so efficient

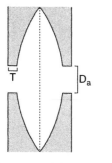
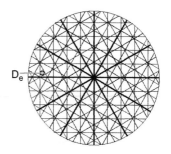

Aperture resistance                    Membrane resistance

$R_a = (128\,T\upsilon)\,/\,(\pi D_a{}^4) + (24\upsilon/\,D_a{}^3)$        $R_m = [24\upsilon/\,(N_p\,D_e{}^3)]\,f(K)$

$\upsilon$, viscosity                          $N_p$, pore number
T, aperture length                     K, (pore area) / (total area)
$D_a$, aperture diameter               $D_e$, equivalent pore diameter

**Figure 16.3**  Calculating pit hydraulic conductivity (Sperry and Hacke, 2004). Left: Hydraulic resistance of a single pit aperture was calculated from aperture diameter ($D_a$) and depth ($T$) using Dagan *et al.*'s (1982) solution for flow through circular pores of finite length. Right: Hydraulic resistance of the membrane was calculated from membrane pore diameter ($D_e$) using Tio and Sadhal's (1994) solution for flow through circular pores in infinitely thin porous meshes. An "equivalent" pore diameter ($D_e$) was used that corresponded to the average conductivity per pore in the heterogeneously porous membrane. The $f(K)$ term is a function correcting for the interaction between closely spaced pores in the mesh (Tio and Sadhal, 1994). Total pit resistance was the sum of membrane and aperture resistances. It was expressed on a pit membrane area basis.

at optimizing this tradeoff, why are they rare in angiosperm xylem? We return to these conundrums after considering the effect of pit structure on wall strength.

### Air-Seed Pressure Versus Implosion Pressure of Pitted Walls

Not only must pitted conduit walls have sufficient resistance to air-seeding, they also need to be strong enough to avoid breaking while holding back the air. Pits are seemingly well designed in this respect (Carlquist, 1988). The structurally weak pit membrane and chamber are located near the neutral axis of the wall with respect to bending stress, and so should not materially reduce strength. The arching borders place strong wall material far from the neutral axis where it counts most in stress reduction. This suggests that the main determinants of wall strength will be aperture size and wall thickness. Widely spaced, narrow apertures and thick walls will increase strength, but tend to reduce conductivity—particularly on a per unit wall investment basis. Such a tradeoff could result in wall strength scaling closely

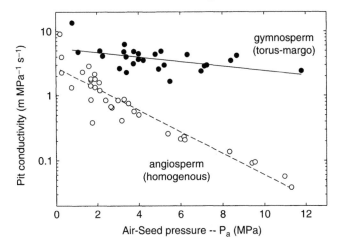

**Figure 16.4**    The tradeoff between pit hydraulic conductivity per membrane area and air-seed pressure ($P_a$). Conductivities were calculated from pit dimensions and $P_a$'s measured for stem and root xylem samples. Torus-margo pit membranes of the gymnosperm species were far superior to homogenous membranes of the angiosperms in achieving a much greater conductivity and showing less of a decline with $P_a$. From Hacke *et al.* (2004) *American Journal of Botany* 91:386-400 and from Sperry & Hacke (2004) *American Journal of Botany* 91:369-385. Copyright Botanical Society of America; adapted with permission.

with air-seed pressure—a stronger wall being unnecessary once air-seeding has eliminated the wall stress caused by the air-water pressure difference.

The strength of pitted walls can be estimated from formulas for bending stress in perforated plates of mechanically isotropic material (Fig. 16.5) (Sperry and Hacke, 2004). These relationships predict that the pressure difference required to bend the wall to rupture (the implosion pressure), is proportional to the square of the "thickness-to-span" ratio (Fig. 16.5, $t / b$). A thicker wall for a given transverse width (span) is a stronger wall. The "ligament efficiency" (Fig. 16.5, $L_e$) may also influence wall strength. This is the distance between aperture edges divided by the distance between aperture centers. At a maximum $L_e$ of 1, there are no apertures and the wall is strongest; as $L_e$ decreases from one, more of the wall is occupied by apertures and it becomes weaker. A third parameter is the "moment ratio," which accounts for the weakening effect of the pit chamber (Fig. 16.5, $I_h / I_s$). This is the second moment of area of the wall ($I$) with the hollowed-out chamber present ($I_h$) divided by the $I$ of the wall with just aperture-sized holes and no chamber ($I_s$). The more the chamber removes structurally significant material from the wall remote from the neutral axis, the lower the moment ratio, and the weaker the wall. The three dimensional ratios were determined for conduits of mean hydraulic diameter for the survey of gymnosperm and angiosperm species (Sperry and Hacke, 2004; Hacke *et al.*,

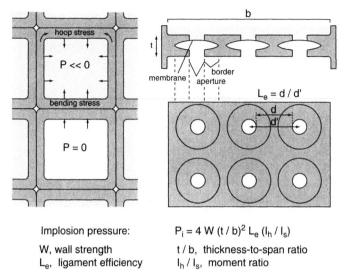

Implosion pressure:

W, wall strength

$L_e$, ligament efficiency

$P_i = 4 W (t / b)^2 L_e (I_h / I_s)$

$t / b$, thickness-to-span ratio

$I_h / I_s$, moment ratio

**Figure 16.5** Calculating implosion pressure (Sperry and Hacke, 2004). Left diagram illustrates hoop and bending stresses in conduit walls from negative xylem pressure. Bending stresses likely exceed hoop stresses. The implosion pressure ($P_i$) is the pressure difference across the bending wall causing failure. It increases with the strength of the solid wall material ($W$ = 80 MPa) (Hacke *et al.*, 2001), thickness-to-span ratio ($t/b$), (*upper right*), ligament efficiency ($L_e$, lower right), and moment ratio ($I_h / I_s$, ratio of second moment of area from central neutral axis with ($I_h$) versus without ($I_s$) a pit chamber present). Equation is modified from Young (1989) and O'Donnell and Langer (1962).

2004). Conduits of the mean hydraulic diameter are approximately the same conduits that will be cavitating at the air-seed pressure.

Angiosperm vessels showed a close correspondence between the implosion pressure of the pitted wall versus its air-seed pressure (Fig. 16.6A) (Sperry and Hacke, 2004). A near 1:1 scaling of implosion and air-seed pressure is maximally efficient because there is presumably no need to reinforce the wall beyond the air-seeding point. The average safety factor (implosion/air-seed pressure) was 1.8. In gymnosperm xylem there was also a close correlation between implosion and air-seeding, but with larger safety factors than for angiosperm vessels (Fig. 16.6A) (Hacke *et al.*, 2004). Safety factors averaged 4.8 in stem tracheids and 2.4 in root tracheids. The positive intercept for stem tracheids indicated additional wall reinforcement that was not related to air-seed pressure. This extra reinforcement and higher safety factor was expected, given the additional need for stem tracheids to support the shoot against gravity and wind. Root tracheids are freed from this requirement because roots are largely supported by soil, although they still must bear some tensile loading. Angiosperm vessels, whether in stems or roots, are presumably relieved of bearing additional stresses because they are embedded in supporting fibers. Of importance, the safety factors we quote are overestimates of the *in situ* value because they do not take into account additional wall stresses from wind, gravity, and other factors. These additional factors should be most important in gymnosperms (especially stems), and true *in situ* safety factors may not vary as much between xylem types as Fig. 16.6A might suggest.

The increase in implosion pressure with air-seed pressure in both angiosperms and gymnosperms was largely the result of increased thickness-to-span of the pitted walls (Hacke *et al.*, 2004; Sperry and Hacke, 2004). The $L_e$ averaged $0.65 \pm 0.1$ in angiosperms and $0.71 \pm 0.01$ in gymnosperms and showed no strong trend with air-seed pressure. These values suggest that the presence of pit apertures weakens the wall 29% to 35% relative to an imperforate wall across both groups. The moment ratio ($I_h/I_s$) was not significantly different from 1 in angiosperms and averaged $0.94 \pm 0.01$ for the larger pit dimensions of gymnosperms. This means that chambers caused at most a 6% weakening of the wall, consistent with them being centered on the neutral axis of the wall.

These estimates of implosion pressure remain to be tested, and there are several uncertainties in the analysis. The analogy between the pitted conduit wall and a flat perforated plate does not take into account that the

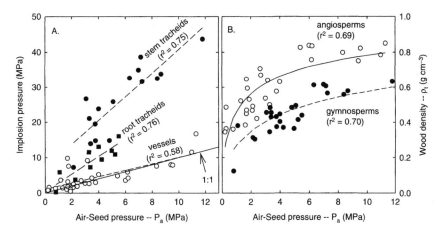

**Figure 16.6** (A) Implosion pressure ($P_i$) versus air-seed pressure ($P_a$) for root and stem xylem samples. Organs were not different in angiosperm species and are pooled. The 1:1 line represents a minimum safety factor from implosion ($P_i/P_a$) of one. Angiosperm vessels had less of a safety factor from implosion than gymnosperm stem tracheids. Root tracheids were intermediate. From Hacke *et al.* (2004) *American Journal of Botany* 91:386-400. Copyright Botanical Society of America; adapted with permission. (B) Wood density ($\rho_t$) versus $P_a$ for same samples as in A. Curves are best fit of Eqn 1 to the pooled data of stems and roots. Gymnosperm xylem is less dense than angiosperm xylem for the same $P_a$.

conduit is embedded in a matrix of tissue, the geometry of which should influence the stress distribution. Nor does it consider the effect of helical thickenings, perforation plate rims, and other local reinforcements of the conduit wall. In view of the apparent importance of wall reinforcement, these features could be adaptive in further economizing the construction of pressure-resistant conduits (Carlquist, 1988). Although a more sophisticated mechanical analysis of conduit wall stress is desirable, the data do show that thickness-to-span ratio is related to cavitation resistance, and the simplistic mechanical analysis suggests the reason is related to support of the wall against negative pressure.

### Safety from Air-Seeding Versus Cost of Conduit Construction

The link between greater air-seed pressure and increasing thickness-to-span ratio suggests an important cost of cavitation resistance by air-seeding. A higher thickness-to-span ratio for a conduit translates into more wall cross-sectional area per conduit area. To maintain a constant safety factor from implosion while increasing air-seed pressure, the plant must invest proportionally more wall material per conduit volume. This cost is independent of the conduit diameter. A conduit 10 or 100 µm in diameter but with the same thickness-to-span will have the same strength against implosion and cost the same per unit volume (Hacke *et al.*, 2001).

Conduits make up a significant fraction of wood volume, especially in gymnosperm xylem. There should be a relationship, at least in gymnosperm wood, between air-seed pressure and the total wood density ($\rho_t$) via the implosion pressure and thickness-to-span ratio of the conduits. Using the expression for implosion pressure in Fig. 16.5 and expressing $(t/b)^2$ in terms of the equivalent density, the $\rho_t$ versus $P_a$ relationship takes the form (Hacke *et al.*, 2001):

$$\rho_t = \rho_w - \rho_w((P_a/k)^{0.5}+1)^{-2} \tag{1}$$

where $\rho_w$ is the density of the solid wall ($\approx 1$ g cm$^{-3}$; Hacke *et al.*, 2001). The $k$ term is a constant (in pressure units) that depends on the implosion safety factor ($P_i/P_a$), ligament efficiency ($L_e$), the strength of wall material ($W$), and the proportionality constant relating the thickness-to-span ratio for conduits of mean hydraulic diameter to the area-averaged ratio of the entire wood sample. Since the $L_e$ was either constant with air-seed pressure or varied little, treating it as a constant was appropriate.

Using $k$ as a curve-fitting parameter, eqn 1 provides a good fit to the curvilinear relationship between wood density and air-seed pressure within gymnosperm and angiosperm woods (Fig. 16.6B). The thickness-to-span ratio, and its equivalent in terms of wood density (Fig. 16.6), are the best anatomical correlates of cavitation resistance that we know of. The two groups did have different $k$ constants ($k = 7.7$ MPa for angiosperms versus

35 MPa for gymnosperms), as expected given their different safety factors, but more importantly their different wood structures. The presence of fibers in angiosperm wood makes for a very different proportionality between $t / b$ of the vessels versus the corresponding area-averaged value for the entire tissue. Scatter in the density versus air-seeding relationship is at least partly due to variation in this proportionality. Gymnosperm wood with relatively more latewood, for example, should have a greater density for the same hydraulic mean $t / b$ and air-seed pressure than wood with more homogenous tracheids.

That there is any relationship at all between wood density and air-seed pressure in angiosperms means that the density of the fiber matrix is proportional to the density of the conduit network, at least in most species (Hacke *et al.*, 2001). This is a somewhat surprising result because fibers are generally viewed as providing mechanical support for the axis—a function that should be independent of the air-seeding requirements of the conduit network. Perhaps the fiber matrix is also involved in mechanical protection of the conduit walls. This needs to be evaluated with a tissue-level analysis of wall stresses from negative pressure. If fibers are required to bear such stresses, then the cost of constructing a vessel network includes fiber reinforcement. Heavily reinforced matrix cells could be particularly important in xylem with wide and long vessels, such as in ring-porous trees.

The analysis shown in Fig. 16.6 is best interpreted as setting a lower boundary on thickness-to-span ratio and density for a given air-seed pressure. The need to avoid implosion means that cavitation-resistant wood could never be light and have low thickness-to-span values. The converse, however, is physically possible. There is nothing to prevent cavitation-susceptible wood from being dense with high thickness-to-span ratio except economy of construction or some independent advantage of these traits. Dense wood does not lead to taller trees, as the relatively light wood of redwoods and sequoias proves. It can also be correlated with reduced water storage capacity and hydraulic conductivity (see Chapter 15) as well as slower growth rates. An advantage of dense wood could be that it promotes more diversity of crown architecture. Dense wood has a higher flexural stiffness for the same cross-sectional area, which may facilitate the support of spreading crowns. In any event, one would expect the accumulation of data of this sort to increase the scatter of data above, but not below, a circumscribed lower boundary set by implosion considerations.

The density versus air-seeding relationship, or at least its lower boundary, represents the construction cost associated with cavitation resistance. The cost curve is relatively steep for low air-seed pressure, becoming more forgiving at higher pressures (Fig. 16.6B). This predicts that plants adapted to mesic and hydric habitats have smaller safety margins from cavitation (because excess safety is much more expensive) than arid-adapted plants. This does appear to be the case, based on relationships between minimum negative pressures and the cavitation pressure required to cause complete hydraulic failure (Pockman and Sperry, 2000).

The wood density result suggests an advantage of gymnosperm wood structure: cheaper cavitation resistance. Gymnosperms as a group achieve the necessary reinforcement against conduit implosion for less wood investment (density) than angiosperms (Fig. 16.6B) despite the fact that tracheids have higher safety-factors from implosion than vessels (Fig. 16.6A). The extra reinforcement of stem tracheids is apparently less costly than the fiber matrix in angiosperms. When conditions favor stress tolerance over rapid resource uptake, this advantage of gymnosperm wood may offset its disadvantage in tending to have lower hydraulic conductivity than many angiosperm woods. The net benefit may contribute to the continued success of gymnosperms in many habitats.

## Conduit Size and the Conductivity Versus Air-Seeding Tradeoff

Thus far, we have identified two safety versus efficiency tradeoffs: a tradeoff between increasing safety from air-seeding and decreasing pit conductivity (Fig. 16.4) and a tradeoff between increased safety from air-seeding and increased construction cost (Fig. 16.6). For both tradeoffs, gymnosperm xylem holds the advantage—more conductive pits and cheaper cavitation resistance. However, neither tradeoff incorporates the effect of greater length and diameter of angiosperm vessels versus gymnosperm tracheids.

The conductivity versus air-seed pressure tradeoff at the pit level can be compensated for by adjustments in conduit length and width. The resistance per unit length (resistivity) of an entire conduit ($R_c$) can be approximated as the sum of the lumen resistivity ($R_l$) and the pit resistance per conduit length (= pit resistivity—$R_p$; Lancashire and Ennos, 2002):

$$R_c = R_l + R_p \tag{2}$$

(The fact that resistances are additive in series justifies the switch from conductance terminology.) Higher pit resistance associated with greater air-seed pressure causes $R_p$ to increase along with the total conduit resistivity (Fig. 16.7, upward $+P_a$ arrow). However, the increase in $R_p$ can be prevented simply by increasing the conduit length (Fig. 16.7, downward $+L$ arrow). If the conduit is long enough, $R_p$ can be rendered negligible regardless of the air-seed pressure, and $R_c \approx R_l$ (Fig. 16.7; 1:1 diagonal). Alternatively an increase in $R_p$ can be balanced by increasing conduit width and causing a drop in $R_l$ (Fig. 16.7, diagonal $+D$ arrow).

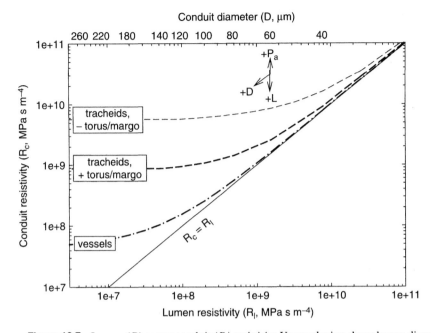

**Figure 16.7** Lumen ($R_l$) versus conduit ($R_c$) resistivity. Upper abscissa shows lumen diameters corresponding to lumen resistivity. If pit resistivity is negligible, conduits will fall on the diagonal $R_c = R_l$ line. Upward arrow indicates the increase in conduit resistivity with air-seed pressure ($+P_a$) because of increased pit resistance. This can be countered by an increase in conduit length ($+L$, downward arrow), which decreases the pit resistivity (resistance per length). Conduit resistivity can also be reduced by increasing lumen diameter ($+D$, diagonal arrow). Dashed curves represent tracheids 3.6 mm long with (heavy) versus without (light) a torus-margo membrane. The air-seed pressure was constant at the average of 4.8 MPa for our gymnosperm species (Hacke *et al.*, 2004). Dash-dotted curve represents vessels with a constant length of 3.6 cm and same air-seed pressure. Curves depart from the $R_c = R_l$ diagonal as lumen diameter increases and pit resistivity becomes significant. The curve becomes flat when $R_c = R_p$, and the y intercept gives the difference in $R_p$ between conduit types.

## Tracheids

Tracheids, being single-celled, are arguably limited in their maximum volume (Lancashire and Ennos, 2002). This places developmental constraints on their length and width, and their minimum conduit resistivity. We predicted their minimum resistivity by combining our model of pitted wall resistance with the conduit resistivity analysis of Lancashire and Ennos (2002). We assumed that 50% of the wall area was occupied by pits, and that the air-seed pressure was the average for our gymnosperm species (4.8 MPa for conifers) (Hacke *et al.*, 2004). We did not determine tracheid lengths. However, Panshin and de Zeeuw (1970) provided a table of average tracheid lengths from stem wood of 43 species of US conifers. The mean of their sample is 3.62 ± 1.1 mm with a maximum of 7.4 mm (*Sequoia sempervirens*).

The dashed curves in Fig. 16.7 correspond to stem tracheids of this average length and show the minimum conduit resistivity as a function of tracheid diameter. Narrow tracheids have a higher lumen resistivity than pit resistivity, and at least potentially the $R_c$ can approximate $R_l$ (Fig. 16.7, $R_c = R_l$ line). Wider tracheids have a lower lumen resistivity than pit resistivity, and $R_c \approx R_p$. Although $R_p$ decreases a small amount with increasing tracheid diameter (because of more wall area for pitting), it is essentially constant, and above a certain diameter there is no significant decrease in conduit resistivity as shown by the flattening trajectory of the dashed curves in Fig. 16.7. Tracheids can have a conduit resistivity anywhere above the dashed curve, but the limitation on their size should prevent them from achieving any resistivity substantially below the curve.

Our analysis for tracheids suggests the $R_p$ may be negligible for narrow diameters, but become increasingly limiting in wider tracheids (Fig. 16.7). This is consistent with the results of Schulte and Gibson (1988) in their survey of tracheids with homogenous membranes. They found the pit resistivity increased from 14% to 84% of the total conduit resistivity with an increase in lumen diameter from 13 to 32 μm. A more extensive survey of pit resistivity would be useful to confirm this trend.

The importance of the torus-margo membrane is illustrated by comparing the upper light dashed curve for tracheids modeled to have angiosperm-like homogenous pit membranes with the lower bold dashed curve for tracheids with torus-margo membranes. Aside from the pit type, these tracheids were the same as those with the native torus-margo membrane ($L = 3.62$ mm, $P_a = 4.8$ MPa, 50% wall pitted). The torus-margo membrane decreased the minimum resistivity by up to nearly an order of magnitude at the wider tracheid diameters (heavy dashed curve below the light dashed curve). It seems doubtful that large conifer trees could exist without the benefit of these specialized membranes that get the most out of the inherently size-limited tracheid. In a similar comparison, we estimated that the conductivity of gymnosperm stems would drop by an average of 40% if torus-margo pits were swapped with angiosperm pits, all else being equal (Hacke *et al.*, 2004).

Maximum tracheid diameters in gymnosperm stems would likely be less than 60 to 80 μm, since there is little benefit of increasing diameter any further because the resistivity is already pit-limited (Fig. 16.7, see upper axis for conduit diameters corresponding with lumen resistivities). Using similar arguments, Lancashire and Ennos (2002) predict a maximum effective tracheid diameter of 98 μm. In fact, the observed maximum diameters for conifer trees of the United States are similar to or less than this range (25-80 μm, average = 52 μm) (Panshin and de Zeeuw, 1970). The validity of these calculations is probably most dependent on the representation of pit resistance, which in turn is a function of the resistance per pit and the number of pits per conduit. There is little data for either of these parameters across a wide range of species.

For the gymnosperms in our survey (Hacke *et al.*, 2004), there was a significant trend for smaller tracheid diameters (and hence lower conductivities) with increasing air-seed pressure (Fig. 16.8, solid symbols). We suspect that this relationship is an indirect effect of the need for high thickness-to-span ratios in tracheids with high air-seed pressure (Fig. 16.6A). Wide tracheids in combination with high thickness-to-span ratios require more wall material per cell to be synthesized. There may be a limit to the wall thickness that can be built by a single cell in terms of developmental timing and assembly capacity. Such a limit would require that high thickness-to-span ratios be associated with narrower tracheids (T. Speck, personal communication). This is consistent with our observation that most of the variation in the thickness-to-span ratio in conifers is in conduit diameter rather than wall thickness.

### Vessels

The importance of the increase in conduit length made possible by the evolution of multicellular vessels is illustrated by comparing the lower tracheid dashed line in Fig. 16.7 (+ torus/margo) with the dash-dotted vessel curve. The vessel curve was modeled for an order of magnitude increase in conduit length, from 3.62 mm to 3.62 cm. This length increase more than offsets the relatively high pit resistance of the homogenous pit membrane of vessels, resulting in another order of magnitude decrease in conduit resistivity. Maximum "profitable" diameters rise to 140 to 180 µm for this length, consistent with the tendency for vessels to reach greater maximum diameters than tracheids (Fig. 16.7, see upper axis for conduit diameter). Of course vessels can be longer than 3.62 cm, meaning their resistivity could drop considerably lower than this arbitrary curve along with achieving even wider maximum diameters. The advantages of the greater size range in vessels would make evolving a more efficient torus-margo pit unnecessary. This may at least in part explain their absence in the vast majority of angiosperms.

There are limits to the vessel strategy. Longer and wider vessels amplify the effects of a single cavitation or damage event on conductivity (Comstock and Sperry, 2000). Wider vessels also become more vulnerable to cavitation by freezing and thawing (Ewers, 1985).

We have hypothesized previously that vessels should be long enough to virtually eliminate the contribution of pit resistivity (Sperry and Hacke, 2004). In support of this, stem-shortening experiments have shown no decrease in stem resistivity as more end walls were removed, suggesting they have negligible resistivity (Chiu and Ewers, 1993), although this experiment bears repeating on more diverse xylem types. However, Hagen-Poiseuille lumen resistivities are generally significantly lower than measured values in vessel as well as tracheid-based xylem (Ewers, 1985), suggesting persistent pit resistivity even in vessel-bearing species. If true, this implies a constraint on vessel length. It may be difficult for plants to control vessel length. Vessel length distributions are strongly short-skewed (Zimmermann and Jeje, 1981). A recent analysis indicates this distribution is consistent with vessel ends being the result of pure chance, hence a low probability of a long longitudinal file of vessel elements without an end (Nijsse, 2004). Increasing the average vessel length to eliminate significant pit resisitivity may require an inordinate number of overly long vessels with their attendant disadvantages. A persistent pit limitation even in vessels may explain why a torus-margo structure has evolved in a few angiosperms (Wheeler, 1983).

In our sample of angiosperms (Sperry and Hacke, 2004), we saw no relationship between vessel diameter and air-seed pressure (Fig. 16.8, open symbols). At best, such comparisons yield statistically significant but weak trends of smaller diameters with larger air-seed pressure (Tyree *et al.*, 1994; Pockman and Sperry, 2000; but see Martinez-Vilalta *et al.*, 2002). This argues against any necessary tradeoff between vessel conductivity and safety from air-seeding. To the extent one exists, it is highly variable. It could be underlain by a less stringent limitation on wall thickness than what we suggest is responsible for the stronger trend in tracheids.

## Conductivity Versus Safety from Cavitation by Freeze-Thaw

From the previous section, it is clear that any tradeoff between conductivity and cavitation by air-seeding is complex and the result of limitations on conduit size that are only indirectly related to the air-seeding process. In contrast, there is evidence for a direct tradeoff between xylem conductivity and resistance to cavitation by freeze-thaw cycles. Wider conduits, which tend to have greater hydraulic conductivity (Fig. 16.7), are also more vulnerable to cavitation by freeze-thaw events (Ewers, 1985; Davis *et al.*, 1999 and see Chapters 19 and 20). This is presumably because freezing in wider conduits traps larger bubbles in the ice, and these larger bubbles are more likely to trigger cavitation during thaw. Of importance, vessels and tracheids of the same narrow diameter appear to be equally safe from cavitation, and become equally vulnerable as their diameter increases (Pittermann and Sperry, 2003). The tradeoff between lumen conductivity and vulnerability to cavitation by freezing seems to be identical for the two conduit types.

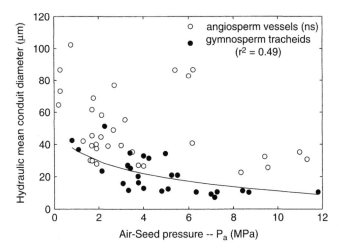

**Figure 16.8** Hydraulic mean conduit diameter versus air-seed pressure ($P_a$) for angiosperm vessels and gymnosperm tracheids from root and stem samples (pooled). There was no significant relationship between vessel diameter and $P_a$ for angiosperms. Gymnosperm tracheids showed a significant decline in lumen diameter with increasing $P_a$. From Hacke *et al.* (2004) *American Journal of Botany* 91:386-400. Copyright Botanical Society of America; adapted with permission.

## Discussion

The gymnosperm versus angiosperm comparison provides some insight into structure-function relationships, which in turn have implications for evolutionary and ecological trends. Probably the most important implication of the analysis of pit conductivity is the marked superiority of the torus-margo pit membrane mechanism over the homogenous membrane in achieving much greater pit conductivity for a given air-seed pressure (Figs. 16.1 and 16.4). The homogenous membrane, being more widespread phylogenetically in tracheids of seedless vascular plants as well as in angiosperms, is undoubtedly the ancestral condition. The specialized torus-margo structure improves substantially the conductivity of length-limited tracheids (Fig. 16.7, dashed curves) and in this regard represents an alternative to the evolution of vessels. The theoretical conductivities of vessels versus torus-margo tracheids for diameters below 40 μm are similar (Fig. 16.7), and measured conductivities of gymnosperm wood overlap considerably with the low end of the angiosperm range (Becker *et al.*, 1999; Brodribb and Feild, 2000). The torus-margo innovation may have been crucial to the continued success of conifers in a world dominated by angiosperms, particularly in temperate regions where large diameter conduits have the disadvantage of greater vulnerability to freeze-thaw events. Only a few conifers are known to lack a torus-margo membrane, and the only gymnosperm division with tracheid-based transport and homogenous membranes are the cycads (Bauch *et al.*, 1972), which are of limited distribution.

If vessels represent an alternative to torus-margo membranes, they should have evolved from the same starting point: tracheids without torus-margo membranes. This may be the case, with vessel-bearing ferns, some gnetophytes, and angiosperms possibly all springing from ancestors with tracheids lacking torus-margo pitting (Carlquist, 1975). In ferns and angiosperms, at least, the tracheids can often be scalariform-pitted. This pit shape is incompatible with a torus-margo mechanism, but does have the probable advantage of packing more pit membrane area per unit tracheid wall and so maximizing the inherently poor conductivity of the homogenous membrane. The next step in this evolutionary trajectory would be the scalariform perforation plate and vessels (Bailey, 1953). It is unknown how scalariform pitting influences the mechanical strength of the membrane and wall. Limitations of this sort could underlie the predominance of circular pitting in many angiosperm vessels despite a possible scalariform ancestry.

The straightforward conductivity versus air-seeding tradeoff at the pit level (Fig. 16.4) does not directly propagate to a similar tradeoff at the whole-conduit or xylem level because of the compensating effects of conduit length and width (Fig. 16.7). The only link may be an indirect one through physiological or developmental limitations on maximum wall thickness. If thickness is limited it becomes impossible to achieve a large thickness-to-span ratio in a large diameter conduit, so the diameter must decrease. Wall thickness in tracheary elements is probably related to the longevity of the protoplast. Whereas earlywood tracheids and vessels rarely live longer than 1 or 2 weeks, latewood tracheids may live longer than one to three months (Schweingruber, 1996). The short life span of earlywood conduits may allow for only limited wall thickening (Schweingruber, 1996).

Potentially independent of the complex conductivity versus air-seeding relationship is the tradeoff between safety from air-seeding and economy of conduit construction (Fig. 16.6). Gymnosperms hold an advantage in this tradeoff, growing cavitation-resistant wood at a lower cost of mechanical strength than angiosperms. Fibers are more expensive per unit volume than tracheids and may also be involved in protecting vessel strength, factors that make angiosperm wood inherently more expensive.

Fibers of course also lead to greater maximum mechanical strength in angiosperm "hardwoods" versus gymnosperm "softwoods." We have emphasized that high density and increased wood strength can be a disadvantage to the extent it is a costly by-product of achieving high air-seed pressure, but high wood strength and its correlate in high Young's elastic modulus may be advantageous in its own right (Wagner *et al.*, 1998). To the extent that there are independent advantages of maximally strong and stiff wood, such as facilitating a spreading crown architecture, angiosperm wood with its fibers can better exploit these advantages than conifer wood.

A greater architectural repertoire may contribute to the wider range of niche specialization in angiosperm versus gymnosperm trees.

Of interest, in both wood types there is a tendency for high wood density to correlate with low hydraulic conductivity (see Chapter 15). This is understandable in conifers where, if wall thickness is limited, dense wood can be achieved only by narrower tracheid lumens. In angiosperms, however, where density and vessel size are at least theoretically uncoupled, a density versus conductivity tradeoff requires a more complex explanation. Sorting out the cause and effect of interactions between wood strength, crown mechanics, and xylem hydraulics across functional and phylogenetic wood types will require more study.

Our conclusions support the common perception that conifers are better adapted to efficient stress tolerance than they are to efficient resource capture—a condition that allows them to compete well in resource-limited habitats where abiotic stress is significant (Woodward, 1995; Willis *et al.*, 1998). Gymnosperm xylem cannot achieve the high conductivities of the most efficient angiosperm woods (Fig. 16.8), and gymnosperms do not achieve the same high water use and gas exchange rates as some angiosperms. However, wood conductivities of angiosperms and gymnosperms do overlap considerably (Becker *et al.*, 1999) thanks to the efficiency of torus-margo pitting (Figs. 16.4 and 16.7) and the greater number of conduits per wood area. In circumstances where large diameter conduits are a liability, gymnosperm and angiosperm conductivities should not be very different (Fig. 16.7). This may contribute to the dominance of conifers in many temperate and boreal forests where evergreen habit and tolerance of freeze-thaw cycles (hence narrow conduits) is advantageous (Woodward, 1995). That gymnosperms can achieve air-seeding resistance at much less cost in wood material than angiosperms should favor conifers in seasonally arid locales and areas subject to winter desiccation where extreme negative xylem pressures are encountered. This may favor the extensive pinyon-juniper woodlands in the intermountain west of the United States, as well as the predominance of conifers at temperate tree lines. Cheaper cavitation resistance in gymnosperms may also explain why stem xylem of some species exhibits relatively large safety margins from cavitation versus angiosperms. Gymnosperms may be able to better afford the luxury of safety against the rare extreme stress event, a factor consistent with the impressive longevity of many conifer species.

## Acknowledgments

Funding to the authors was provided by the National Science Foundation (IBN-0112213) during research and preparation of the chapter.

# References

Bailey, I. W. (1953) Evolution of the tracheary tissue of land plants. *Am J Bot* **40**: 4-8.

Bauch, J. W., Liese, W. and Schultze, R. (1972) The morphological variability of the bordered pit membranes in gymnosperms. *Wood Science and Technology* **6**: 165-184.

Becker, P., Tyree, M. T. and Tsuda, M. (1999) Hydraulic conductances of angiosperm versus conifers: Similar transport sufficiency at the whole-plant level. *Tree Physiol* **19**: 445-452.

Brodribb, T. J. and Feild, T. S. (2000) Stem hydraulic supply is linked to leaf photosynthetic capacity: Evidence from New Caledonian and Tasmanian rainforests. *Plant Cell Environ* **23**: 1381-1388.

Carlquist, S. (1975) *Ecological Strategies of Xylem Evolution.* University of California Press, Berkeley.

Carlquist, S. (1988) *Comparative Wood Anatomy.* Springer-Verlag, Berlin.

Chiu, S. T. and Ewers, F. W. (1993) The effect of segment length on conductance measurements in *Lonicera fragrantissima. J Exp Bot* **44**: 175-181.

Comstock, J. P. and Sperry, J. S. (2000) Tansley review no. 119. Some theoretical considerations of optimal conduit length for water transport in plants. *New Phytol* **148**: 195-218.

Crombie, D. S., Hipkins, M. F. and Milburn, J. A. (1985) Gas penetration of pit membranes in the xylem of *Rhododendron* as the cause of acoustically detectable sap cavitation. *Aus J Plant Physiol* **12**: 445-454.

Dagan, Z., Weibaum, S. and Pfeffer, R. (1982) An infinite-series solution for the creeping motion through an orifice of finite length. *J Fluid Mech* **115**: 505-523.

Davis, S. D., Kolb, K. J. and Barton, K. P. (1998) Ecophysiological processes and demographic patterns in the structuring of California chaparral. In *Landscape Disturbance and Biodiversity in Mediterranean-type Ecosystems* (P. W. Rundel, G. Montenegro, and F. Jaksic, eds.) pp. 297-310. Springer-Verlag, Berlin.

Davis, S. D., Sperry, J. S. and Hacke, U. G. (1999) The relationship between xylem conduit diameter and cavitation caused by freeze-thaw events. *Am J Bot* **86**: 1367-1372.

Dixon, H. and Joly, J. (1895). On the ascent of sap. *Philos. Trans. R. Soc. London B* **186**: 563-576.

Domec, J.-C. and Gartner, B. L. (2002) How do water transport and water storage differ in coniferous earlywood and latewood? *J Exp Bot* **53**: 2369-2379.

Ewers, F. W. (1985) Xylem structure and water conduction in conifer trees, dicot trees, and lianas. *IAWA Bull* **6**: 309-317.

Feild, T. S., Brodribb, T. J. and Holbrook, N. M. (2002) Hardly a relict: Freezing and the evolution of vesselless wood in Winteraceae. *Evolution* **56**: 464-478.

Givnish, T. J. (1986). On the use of optimality arguments. In *On the Economy of Plant Form and Function* (T. J. Givnish, ed.) pp. 3-9. Cambridge University Press, Cambridge.

Hacke, U. G., Sperry, J. S. and Pittermann, J. (2000) Drought experience and cavitation resistance in six desert shrubs of the Great Basin, Utah. *Basic Appl Ecol* **1**: 31-41.

Hacke, U. G., Sperry, J. S. and Pittermann, J. (2004) Analysis of circular bordered pit function. II. Gymnosperm tracheids with torus-margo pit membranes. *Am J Bot* **91**: 386-400.

Hacke, U. G., Sperry, J. S., Pockman, W. P., Davis, S. D. and McCulloh, K. A. (2001) Trends in wood density and structure are linked to prevention of xylem implosion by negative pressure. *Oecologia* **126**: 457-461.

Lancashire, J. R. and Ennos, A. R. (2002) Modelling the hydrodynamic resistance of bordered pits. *J Exp Bot* **53**: 1485-1493.

Martinez-Vilalta, J., Prat, E., Oliveras, I. and Pinol, J. (2002) Xylem hydraulic properties of roots and stems of nine Mediterranean woody species. *Oecologia* **133**: 19-29.

Nijsse, J. (2004) On the mechanism of xylem vessel length regulation. *Plant Physiol* **134**: 32-34.

O'Donnell, W. J. and Langer, B. F. (1962) Design of perforated plates. *ASME Journal of Engineering for Industry, Trans. ASME Series B* **84**: 307-320.

Panshin, A. J. and de Zeeuw, C. (1970) *Textbook of Wood Technology*. McGraw-Hill, New York.

Petty, J. A. (1972) The aspiration of bordered pits in conifer wood. *Proc R Soc Lond B* **181:** 395-406.

Pickard, W. F. (1981) The ascent of sap in plants. *Prog Biophys Mol Biol* **37:** 181-229.

Pittermann, J. and Sperry, J. S. (2003) Tracheid diameter determines the extent of freeze-thaw induced cavitation in conifers. *Tree Physiol* **23:** 907-914.

Pockman, W. T. and Sperry, J. S. (2000) Vulnerability to xylem cavitation and the distribution of Sonoran Desert vegetation. *Am J Bot* **87:** 1287-1299.

Schulte, P. J. and Gibson, A. C. (1988) Hydraulic conductance and tracheid anatomy in six species of extant seed plants. *Can J Bot* **66:** 1073-1079.

Schweingruber, F. (1996). *Tree Rings and Environment. Dendroecology*. Paul Haupt Publishers, Berne, Switzerland.

Sperry, J. S. (2003) Evolution of water transport and xylem structure. *Int J Plant Sci* **164:** S115-S127.

Sperry, J. S. and Hacke, U. G. (2004) Analysis of circular bordered pit function. I. Angiosperm vessels with homogenous pit membranes. *Am J Bot* **91:** 369-385.

Sperry, J. S., Saliendra, N. Z., Pockman, W. T., Cochard, H., Cruiziat, P., Davis, S. D., Ewers, F. W. and Tyree, M. T. (1996) New evidence for large negative xylem pressures and their measurement by the pressure chamber method. *Plant Cell Environ* **19:** 427-436.

Sperry, J. S. and Tyree, M. T. (1990) Water-stress-induced xylem embolism in three species of conifers. *Plant Cell Environ* **13:** 427-436.

Stiller, V., Lafitte, H. R. and Sperry, J. S. (2003) Hydraulic properties of rice (*Oryza sativa* L.) and the response of gas exchange to water stress. *Plant Physiol* **132:** 1698-1706.

Thomas, R. J. (1972) Bordered pit aspiration in angiosperms. *Wood Fiber* **3:** 236-237.

Tio, K. K. and Sadhal, S. S. (1994) Boundary conditions for stokes flows near a porous membrane. *Appl Sci Res* **52:** 1-20.

Tyree, M., Davis, S. and Cochard, H. (1994) Biophysical perspectives of xylem evolution: Is there a tradeoff of hydraulic efficiency for vulnerability to dysfunction? *IAWA J* **15:** 335-360.

Wagner, K. R., Ewers, F. W. and Davis, S. D. (1998) Tradeoffs between hydraulic efficiency and mechanical strength in the stems of four co-occurring species of chaparral shrubs. *Oecologia* **117:** 53-62.

Wheeler, E. A. (1983) Intervascular pit membranes in Ulmus and Celtis native to the USA. *IAWA Bull* **4:** 79-88.

Willis, K. J., Bennett, K. D. and Birks, H. J. B. (1998) The late Quaternary dynamics of pines in Europe. In *Ecology and Biogeography of Pinus* (D. M. Richardson, ed.) pp. 107-121. Cambridge University Press, Cambridge.

Woodward, F. (1995) Ecophysiological controls of conifer distribution. In *Ecophysiology of Coniferous Forests* (W. Smith and T. Hinckley, eds.) pp. 79-94. Academic Press, San Diego.

Young, W. C. (1989). *Roark's Formulas for Stress and Strain*. McGraw Hill, New York.

Zimmermann, M. H. and Jeje, A. A. (1981) Vessel length distribution of some American woody plants. *Can J Bot* **59:** 1882-1892.

# 17

# Vascular Constraints and Long Distance Transport in Dicots

*Colin M. Orians, Benjamin Babst, and Amy E. Zanne*

Plants are often assumed to be capable of freely translocating resources from one part of the plant to all other parts, a condition that we describe as integrated. In reality, many vascular plants are sectorial, meaning that the movement of nutrients, photosynthate, and other substances is restricted to specific vascular, xylem or phloem, pathways (see reviews by Murray *et al.*, 1982; Watson and Casper, 1984; Orians and Jones, 2001). Sectorial plants consist of numerous Integrated Physiological Units (IPUs) whose pathways are relatively independent of each other (Watson and Casper, 1984; Watson, 1986). Integrated and sectorial define two ends of what in reality is a continuum among species. Surprisingly little research has examined the patterns of sectoriality across plant species, the ecological consequences of sectoriality, or the evolutionary pressures selecting for greater sectoriality or integration.

In this chapter, we review why sectorial transport is common and briefly discuss how it interacts with environmental heterogeneity. We end with a description of new techniques for quantifying sectoriality and a discussion of the ecological conditions that might favor evolution of independent sectors or increased integration. Our goal is to stimulate research on (1) the ecological consequences of sectoriality, and (2) the evolutionary forces that may determine the extent of sectoriality. We restrict our discussion to dicots; monocots are generally integrated because of the large number of interconnected vascular traces and the numerous vascular bundles that come together in complex combinations at nodes (Zimmermann and Brown, 1971; Watson and Casper, 1984).

## Vascular Architecture of Xylem and Phloem

Vascular restrictions in long-distance transport exist for both xylem and phloem. In this section, we describe how xylem and phloem anatomy constrains lateral transport of resources and end with a discussion of the conditions that might lead to greater integration in sectored plants.

### Long Distance Transport in the Xylem

*Xylem Anatomy and Transport*    Examination of the ultrastructure of the dicot xylem helps explain xylem sectoriality and the capacity for integration. Xylem vessels arise from elongated cells (often referred to as vessel elements) stacked on top of one another. The end walls between vessel elements are open at one or both ends (= perforation plate) facilitating rapid longitudinal or axial flow. In contrast, the sidewalls are lignified, blocking most lateral transfer between adjacent elements. A limited amount of xylem sap moves between adjacent vessels via small openings, termed *intervessel pits*, allowing some degree of integration in many species (Fig. 17.1) (Carlquist, 2001; Tyree and Zimmermann, 2002). Because of this anatomical structure, resistance to longitudinal flow is much lower than the resistance to lateral flow.

Lateral flow between vessels should be a function of both wood and vessel anatomy and the pressure differential between adjacent xylem cells. Anatomical considerations include the density of vessels, the frequency of contact between adjacent vessels, and the number and size of intervessel pits (Tyree and Zimmerman, 2002; Orians *et al.*, 2004). In total, species having a high proportion of solitary vessels or vessel groups and fewer pits (or smaller pit area) should be more sectored, whereas species having a high proportion of randomly scattered vessels coming in and out of contact and many more pits (or larger pit area) should be more integrated. Because the transport of water through the vascular system is primarily via bulk flow of

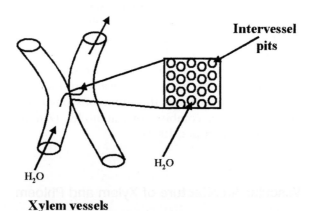

**Figure 17.1**    Lateral water movement between vessels via intervessel pits is possible at contact points as vessels weave in and out of contact with one another.

water, where water flows toward the source of tension (i.e., intercellular menisci in transpiring leaves) via the path of least resistance, differences in transpiration rates between sectors should facilitate lateral flow through existing pit pathways unidirectionally toward the source of greatest tension.

***General Patterns*** Short-term isotope labeling studies have shown that phosphorous transport in the xylem travels from particular lateral roots to specific leaves or branches (reviewed by Watson and Casper, 1984). In *Mentha piperita* (peppermint) orthostichous leaves, directly above the labeled roots, are the tissues that accumulate the isotope (Rinne and Langston, 1960). Long-term consequences of this uneven transport can occur. In *Lycopersicon esculentum*, fertilization of isolated lateral roots resulted in chemical, morphological, and growth responses within the orthostichous sector: larger leaves, lower concentrations of phenolics, and greater growth of side-shoots (Orians *et al.*, 2002).

The previous examples all focused on primary vascular tissue of herbaceous plants. In these species, xylem vessels occur in vascular bundles that come in and out of contact, facilitating the exchange of resources among sectors. Vascular constraints on lateral transport are thought to be especially important in woody plants (Marschner, 1995). Although secondary vessels in woody plants move in and out of contact with one another, vessels vary in their isolation, and the distance around the bole of woody plants can be considerable. Horwath *et al.* (1992) reported fertilizer burn in portions of the crown when nitrogen was injected into the trunk of a *Populus* clone, and Richard Dickson (personal communication) observed differential leaf growth in split-root fertilization experiments with *Populus deltoides* seedlings. Nonetheless differences among species occur in sectoriality. Early studies by Kozlowski and Winget (1963) used dyes to show that patterns of water movement vary among forest tree species. More recently, Orians *et al.* (2004, using stable isotopes, showed that *Betula papyrifera* and *B. lenta* are more integrated than *Acer rubrum* and *A. saccharum*. Examing the ecological consequences of sectoriality in woody plants is more challenging because the potential for spiral wood grain makes it difficult to predict which leaves or branches are connected to specific roots (Kozlowski and Winget, 1963; Tyree and Zimmermann, 2002).

***Modifications of Sectoriality in Xylem Transport*** When damage occurs to the vascular system, transpiration can pull water through adjacent vessels and increase integration. This result is best illustrated by double saw cut experiments (Tyree and Zimmermann, 2002). If two saw cuts are made half way through the trunk from opposite sides, a tree can still survive as long as the two cuts are placed at some distance apart along the length of the trunk. Since the direct pathway is mainly blocked (all but those vessels with pathways twisted enough to circumvent the cuts), the lateral pathway becomes the primary path of water movement. This flexibility of water transport also explains why plants can survive when only a portion of the root system has access to water (Hansen and Dickson, 1979; Fort *et al.*, 1998). Thus, when xylem flow is interrupted, single root sections have the potential to provide both water and mineral resources to the entire plant.

More recently, Zwieniecki *et al.* (2003) found that higher ion concentration in the xylem facilitates the movement of water between sectors. This result led the authors to postulate that plants might manipulate ion concentration to increase lateral transport of resources: nutrients, plant growth hormones, or signal molecules involved in plant defense induction. Although intriguing, further work is required to determine if plants can increase or decrease the level of integration by modifying the ion concentration of xylem sap.

## Long Distance Transport in the Phloem

*Phloem Anatomy and Transport*    Examination of phloem architecture and function is instructive in understanding the sectored nature of phloem. The basic units of phloem tissue are the sieve cells, which are connected end-to-end via sieve plates to form a pipeline that connects one organ to others. The anatomy of sieve cells has evolved so that the path of least resistance is longitudinal, through the sieve plates (van Bel *et al.*, 2002). Plasmodesmata connecting adjacent sieve elements exist, but are not numerous. Thus, relatively little lateral transport would be expected between adjacent sieve tubes. Unlike xylem, which is under negative pressure, phloem transport is driven by positive pressure, powered by loading sugars into the sieve cells at source locations. Because sieve elements are under positive pressure, they are not entirely isolated from surrounding parenchyma tissue. There is a cycle of leakage from and active retrieval by sieve cells (see Chapter 10). This leakage-retrieval cycle may provide one means by which substances might be moved between phloem sectors, as substances leaked by one sieve cell may be "retrieved" by another sieve cell. In herbaceous plants, sieve tubes exist in vascular bundles that come in and out of contact with each other, and in woody plants, the winding nature of the vascular system results in periodic contact between sieve elements. Thus there is the potential for exchange of contents among sieve elements as they come in and out of contact, especially if there is a pressure or concentration gradient among the bundles.

*General Patterns*    Isotope labeling studies with $^{14}C$-labeled $CO_2$ have shown that, within a given shoot, mature source leaves export photosynthate through the phloem to specific leaves, leaf parts, or files of seeds based on leaf orthostichy: the phyllotactic arrangement of leaves (Larson and Dickson, 1973; Watson and Casper, 1984). Defoliation studies have also provided evidence of sectoriality; defoliation within vertical files of cultivated sunflower resulted in unfilled seeds in those sectors of the flower vascularly connected to the defoliated sectors (Prokofyev *et al.*, 1957).

In addition, the phloem is involved in the long distance transport of nitrogen (as amino acids produced in the leaves), solutes such as potassium and phosphorus (Fisher, 2000), proteins and mRNA (see Lucas, 1997), and signaling molecules involved in chemical induction (Narvaez-Vasquez *et al.*, 1994). The importance of phloem in the long distance transport of these resources is best illustrated by examining the research focused on damage-induced chemical signaling. Davis *et al.* (1991) combined wounding and $^{14}CO_2$ labeling to show that import of photosynthate from a damaged leaf depended on sink strength and vascular connectivity and that $^{14}C$ accumulation in distant leaves was positively correlated with expression of wound-induced mRNAs.

Sectoriality in phloem transport also extends from the shoots to the roots (Steiber and Beringer, 1984; Marshall, 1996). As a consequence, defoliation of particular branches can limit the growth and survival of specific roots (Cook and Stoddard, 1960; Murphy and Watson, 1996).

***Modifications of Sectoriality in Phloem Transport*** Manipulation of source-sink relations within a plant may lead to the breakdown of sectoriality. In soybean, localized defoliation of one sector and depodding of the other sector allowed photosynthate to cross from source leaves in one sector to pods (sinks) in the other sector (Noodén *et al.*, 1978). This immediate circumvention hints that barriers to lateral flow between sieve elements are not absolute. We offer two mechanisms to explain the preceding example. First, when pressure in one sector is decreased (i.e., defoliation, which eliminates phloem loading in the defoliated sector), phloem sap may be able to move through plasmodesmata from adjacent sectors that are still at a high pressure. Second, based on the leakage/retrieval system in phloem (see Chapter 10), removal of sinks (i.e., pods) from one sector and removal of sources (i.e., leaves) from another sector may result in a unidirectional leakage from overfilled sieve elements in the de-podded sector, and subsequent capture by sieve cells in the defoliated sector. The ability of plants to circumvent cuts by the formation of new vascular connections as the plant grows further indicates that sectoriality is plastic (see Sachs *et al.*, 1993 for discussion). It is also possible that an active change in transport between sectors occurs, perhaps by specialized transfer cells or by regulatory modification of existing plasmodesmata to allow larger openings between adjacent sieve elements (Pate and Dieter Jeschke, 1995; Fisher, 2000).

## Coupling Environmental Heterogeneity and Sectoriality

For sectoriality to have ecological consequences on plant performance, growth, and resistance to herbivores and reproduction, environmental factors must vary in space or time at scales relevant to individual plants. Although many factors may be important, we focus on three (soil nutrients, light, and damage by herbivores) and provide evidence that they do indeed vary at the scale of individual plants.

### Nutrients

The availability of soil nutrients is highly variable in space and time (Jackson and Caldwell, 1993; Stark, 1994). As a consequence, only a portion of a plant's roots may have access to nutrient-rich patches. Most plants are adept at exploiting these patches. Roots typically proliferate, increase their nutrient uptake rate, and/or live longer in these nutrient rich patches (Caldwell, 1994; Robinson, 1994; van Vuuren *et al.*, 1996). The net effect is greater metabolic activity, nutrient uptake, and plant growth. Moreover, these highly active roots are sinks for photosynthate (Caldwell, 1994). Thus a positive feedback loop may exist within a sector. High nutrient availability leads to greater capacity for photosynthesis and high root activity, increased carbon import, more root growth, and higher nutrient uptake (see Chapter 13).

Patchy water availability, however, may alter the effects of patchy nutrients on aboveground growth, morphology, and chemistry. Specifically, if water and nutrient availability coincide (primarily available to the same roots), we would expect no evidence for sectoriality. Recent experiments with tomato tested this prediction. We imposed three treatments: (1) fertilizer was applied to a lateral pot containing roots from one sector and both pots were watered (treatment $P_nU_w$; simulating a localized nutrient-rich patch and uniform water), (2) nutrients were applied to the main pot with roots from all sectors and both pots were watered (treatment $U_nU_w$; nutrients and water available to all sectors of a plant), and (3) nutrients and water were applied to the lateral pot and minimal water was applied to the main pot (treatment $P_nP_w$ simulating a localized nutrient and water patch). As predicted, when nutrients were patchy and water was readily available to all roots, sectoriality was observed: The orthostichous leaf 7 was larger ($P_nU_w > U_nU_w = P_nP_w$; Fig. 17.2). With water available to the entire root system, a pressure gradient across sectors was minimized, thus maintaining IPUs. When water availability was restricted to the same roots receiving the nutrients ($P_nP_w$), however, the effect of patchy fertilizer on leaf mass was eliminated ($U_nU_w = P_nP_w$). The $P_nP_w$ treatment effect was likely due to a pressure differential being generated across sectors; water (and nutrients) flows from the lateral pot to the entire shoot system.

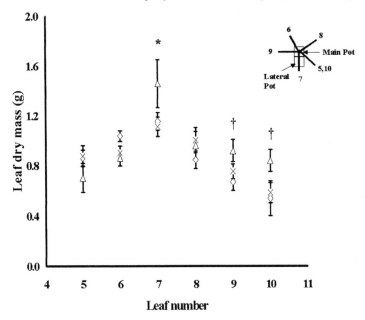

**Figure 17.2**   Effects of patchy (P) or uniform (U) nutrient and water availability on plant differences in leaf mass (g; mean ± 1 standard error) of tomato grown in split-pots. In all treatments, the orthostichous leaf 7 has direct connections to the fertilized lateral roots while leaves 9 and 10 have partial connections and leaves 6 and 8 have minimal connections. $\Delta = P_n U_w$, $\times = U_n U_w$; $\Diamond = P_n P_w$ (see text for description of treatments). Only in the $P_n U_w$ treatment was there evidence for sectorial growth. * denotes significance differences at $P < 0.05$ and † denotes significance differences at $P < 0.1$ among treatments for specific leaves. N = 5 plants per treatment.

## Light

Light availability differs considerably over small and large scales. Different sectors (leaves or branches) of the crown often receive different amounts of light at different times of the day (Chazdon and Fetcher, 1984). For example, plants growing in the forest understory often only receive light for brief periods of time (sunflecks) and plants growing at forest edges might receive light to only one or a few branches. Plants are adept at exploiting this variability (Woodward, 1990). Most of the carbon gained by plants growing in the forest understory comes from the exploitation of the sunflecks (Chazdon and Fetcher, 1984), and when the availability of light is more constant, the branches receiving high light exhibit enhanced growth rates. Again, a positive feedback loop is likely created within a sector. High light availability leads to enhanced carbon gain and delivery to specific shoots and roots leading to increased leaf area, greater transpiration and nutrient delivery, increased growth into the high light patch, and again enhanced carbon gain.

### Herbivore Damage

Extensive evidence exists that the distribution and intensity of damage by herbivores is patchy and varies in intensity (Denno and McClure, 1983). Damage can cause both local and systemic changes in plant characters. The specific changes depend on the distribution of damage (local or widespread), the agent of damage (i.e., the species of herbivore), and the extent of initial damage (Karban and Baldwin, 1997). Since systemic induction depends on the movement of signal molecules through the vascular system, patterns of induction can vary spatially as a function of vascular connectivity (Davis *et al.*, 1991; Jones *et al.*, 1993; Orians *et al.*, 2000; Schittko and Baldwin, 2003). Hence, an initial herbivore attack may result in a patchwork of variability in plant tissue quality, which may affect plant-herbivore interactions over ecological and evolutionary time scales (Karban *et al.*, 1997).

# Techniques and Prospects for Further Research

Sectoriality can affect plant growth and development and differs among species. Both xylem and phloem are subject to vascular constraints; however, much less information is available on the nature of interspecific differences in phloem-to-phloem transfer (see reviews by Pate and Dieter Jeschke, 1995; Fisher, 2000). In this final section, we focus on recent techniques useful for quantifying xylem sectoriality and conditions determining patterns of sectoriality across species: hypothesized anatomical mechanisms and selective pressures.

### Techniques for Quantifying Sectoriality

A number of recently developed techniques are available for measuring sectoriality at the scale of individual xylem vessels to the entire plant. Although they provide different absolute estimates for a given species, each technique has its advantages, and we have preliminary evidence that the general ranks from most to least sectored holds across species (e.g., if a given species is more sectored at the branch-level than a second species, this trend should hold at the whole-plant level) (Table 17.1).

Connections between individual vessels can be investigated by injecting stain via micropipettes (sensu Zweiniecki *et al.*, 2001). At the branch level, hydraulic techniques, in which water or stain is forced laterally through a section of a branch, can be used to quantify the resistance to lateral flow (Zweiniecki *et al.*, 2001). Such a measure integrates the effects of pit morphology with the varying amounts of xylem vessel contact. Isotopes (e.g., $^{13}C$ or $^{15}N$), applied to an isolated leaf or root, have been commonly used as tracers to measure both phloem and xylem sectoriality at the level

**Table 17.1** Typical techniques used to quantify interspecific differences in xylem sap movement. Each provides similar estimates of relative sectoriality

| Species | Conductance (leaf-to-leaf)* | Conductance (branch)† | Staining (root-to-shoot)‡ | Isotopes (root-to-shoot)§ |
|---|---|---|---|---|
| *Betula papyrifera* | 0.90 | 0.22 | 94% | 0.97 |
| *Acer saccharum* | 0.76 | 0.02 | 51% | 0.69 |
| *Quercus rubra* | 0.28 | 0.003 | 22%[b] | – |

*Orians *et al.*, unpublished data: calculated as non-orthostichous conductance/orthostichous conductance.

†Orians *et al.*, unpublished data: calculated as indirect conductance/direct conductance.

‡Orians *et al.*, 2004 and unpublished data: calculated as % of branches receiving stain applied to an isolated lateral root.

§Orians *et al.*, 2004: calculated as correlation between relative current year biomass and relative isotope accumulation following 15N application to an isolated lateral root.

of the whole-plant. For example, if a plant is very integrated, $^{15}$N-NO$_3^-$ applied to a lateral root should accumulate in many different branches, and the relative amount of isotope accumulated should be based on which branch is the largest, and consequently the biggest sink (Fig. 17.3). In contrast, the correlation between relative isotope accumulation and branch size should be poor in sectored plants due to the vascular restrictions in resource transport. Using this approach, we have found striking differences in sectoriality among several temperate tree species (Orians *et al.*, 2004).

These techniques have various advantages and disadvantages. As hydraulic techniques can be done with preserved or fresh woody tissue, they can be useful in rapidly determining xylem sectoriality in many species, such as

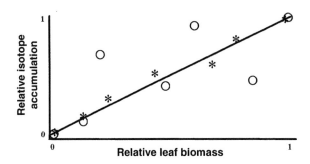

**Figure 17.3** Patterns of relative isotope accumulation in tissue with different relative leaf biomass when the isotopes are applied to tissues (roots or leaves) within a sector. Overall, relative isotope accumulation is positively correlated with relative growth rate for both sectored and integrated plants, but the tightness of the correlation differs between sectored plants (O) and integrated plants (*).

would be done for comparative studies. Hydraulic techniques also allow the manipulation of conditions, such that the extreme possibilities of cross-sector transport can be accurately determined. Isotopes are useful because they can be introduced to roots without damaging the tissue (as occurs in the use of dyes) and accumulation quantified. Isotopes also have the advantage, as they can be applied to living plants, that they are translocated by the plant's internal mechanisms, rather than by an artificial external force, such as pressure. However, hydraulics and traditional stable isotope and radioisotope studies also have a limitation, in that a destructive harvest is required to quantify and describe patterns of sectoriality, giving us a one-time snapshot of isotope distributions for any given individual. New technology allows the rapid measurement and real-time imaging of positron emitting radioactive isotopes (e.g., $^{11}C$ and $^{13}N$) nondestructively (Kiyomiya *et al.,* 2001). Although this new technology is limited to very short-term studies, because of the short half-life of the radioisotopes (e.g., for $^{11}C$ $t_{1/2} = 20.4$ min.), it may be used to determine vascular connectivity within a plant and examine whether certain treatments can alter patterns of sectoriality. In addition, *a priori* knowledge of vascular connections will make feasible experiments that were not possible previously. For example, vascular connections in trees with spiral grain can be mapped and further experiments can then be conducted.

### Potential Selective Pressures Leading to Varying Degrees of Sectoriality

An examination of variation in sectoriality across taxa can be used to refine our understanding of vascular control of resource transport in plants and to suggest selective pressures that might result in the observed variation in the degree of sectoriality. Dicots differ considerably in sectoriality, with variation occurring within both herbaceous and woody species (Shea and Watson, 1989; Watson and Casper, 1984; Zimmerman and Brown, 1971).

***Anatomical Model***    We offer a model for predicting degree of sectoriality based on xylem vessel anatomy across species including characteristics of pit and vessel size and distribution.

$$\text{Degree of Sectoriality} = f(\text{Pit}) + f(\text{Vessel}) \qquad (1)$$

The pit function f(Pit) should include relationships with (1) pit number, (2) individual pit area, and (3) pit membrane porosity. Species with more pits, larger pit areas, and/or with more permeable pit membranes should provide more avenues for interchange between vessels and thus be more integrated. We have found support for some of these relationships (Orians *et al.,* 2004). *Betula papyrifera* had more pits and greater pit area per unit vessel wall than *Acer saccharum* or *Quercus rubra* (Fig. 17.4; see also Color Plate section), and *Betula papyrifera* was more integrated than the two other species (Orians *et al.,* 2004). However, species with numerous or large pits should also be at greater

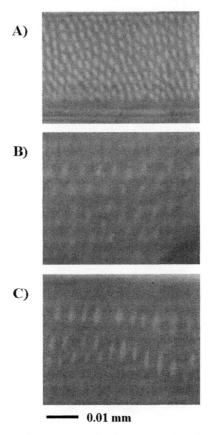

**Figure 17.4**  Xylem pit distributions in (A) *Betula papyrifera*, (B) *Acer saccharum*, and (C) *Quercus rubra*. Note the most dense pitting in *B. papyrifera* and least dense pitting in *Q. rubra*, likely contributing to greater integration in *B. papyrifera* and greater sectoriality in *Q. rubra*. (See also Color Plate section.)

risk of dysfunction, owing to weaker vessel walls or pit membranes, and species with porous pit membranes should be at greater risk of drought-induced embolisms due to air seeding (Hacke and Sperry, 2001).

The vessel function, f(Vessel), should incorporate relationships with vessel (1) number, (2) distribution, and (3) area and length of contact between adjacent vessels. As vessel number increases, the frequency with which vessels contact and share resources should increase. Vessel distribution should also be important. Isolated vessel or vessel groups should lead to greater sectoriality, as they are only infrequently coming into contact with adjacent vessels, whereas randomly scattered vessels should increase the frequency with which vessels contact around the circumference of a stem or branch. As contact length increases between vessels, lateral transfer of resources should increase. The distance of contact depends on the twisting nature of vessels and the amount and type of intervening tissue (such as rays and paratracheal parenchyma cells; Tyree and Zimmermann, 2002). Support for the first two relationships can be seen from cross sections of *B. papyrifera, A. saccharum,* and *Q. rubra* (Fig. 17.5; see also Zimmermann and Brown, 1971). *Betula papyrifera,* the most integrated species, has the greatest number and most randomly scattered vessels, and *Q. rubra,* the most sectored species, has the least number and most isolated vessels (Table 17.1). Finally, vessels that are in close contact with one another over a large distance should increase lateral exchange over those vessels that only briefly contact.

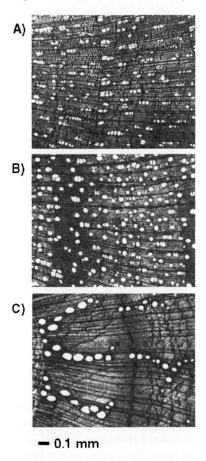

**─ 0.1 mm**

**Figure 17.5**  Xylem cross sections in (A) *Betula papyrifera,* (B) *Acer saccharum,* and (C) *Quercus rubra.* Around the circumference at a given radial distance, note the greater number of randomly scattered vessels in *B. papyrifera* and *A. saccharum* and the fewer and more isolated vessels in *Q. rubra,* likely contributing to greater integration in *B. papyrifera* and *A. saccharum,* and greater sectoriality in *Q. rubra.*

*Selective Pressures on Ring and Diffuse-Porous Species*   We now focus on ring-porous (producing large early spring vessels with vessel diameter decreasing in later wood) and diffuse-porous (producing similar vessel diameters during the year) woody species to highlight potential selective pressures leading to varying degrees of sectoriality. There is increasing evidence that ring-porous trees are more sectored than diffuse-porous trees (Zimmermann and Brown, 1971; Tyree *et al.*, 1994; C. Orians, unpublished data, A. Zanne, unpublished data). Both ring and diffuse porous xylem (Carlquist, 2001; A. Zanne, unpublished data) and degree of sectoriality (Watson and Caper, 1984; Watson, 1986; Orians *et al.*, unpublished data; A. Zanne, unpublished data) are widely scattered across the angiosperm phylogeny, with both the more ancient Eumagnoliids and the more recently evolved Eurosids and Euasterids having ring-porous and highly sectored species. Thus at the level of the major clades, these characters appear to be highly labile; but at the family and genera levels, at least for certain clades, both xylem type and sectoriality are fairly conserved (Zimmermann and Brown, 1971; Carlquist, 2001; Orians *et al.*, 2004; A. Zanne, unpublished data). For instance, analyses on select species native to the northeast United States indicate that *Betula* spp. and *Ostrya* spp. (Betulaceae) are all highly integrated, *Acer* spp. (Sapindaceae) are less integrated, and *Quercus* spp. (Fagaceae) and *Carya* (Juglandaceae) are highly sectored. In the Northeast, Betulaceae and Sapindaceae are diffuse-porous, and Fagaceae and Juglandaceae are ring-porous or semiring-porous. These results suggest that the selective pressures influencing sectoriality and xylem type have been more recent, when families and genera were separating.

   A discussion of the conditions that favor ring-porous species may shed light on the conditions that lead a species to give up the possible advantages of integration in favor of sectoriality. Ring-porous species dominate seasonally xeric upland sites and diffuse-porous species dominate mesic lowland sites (Guthrie, 1989), and sectoriality is frequent in trees from drier habitats (Watson, 1986). In seasonally dry environments, selection is likely to have acted such that species have large vessels available for rapid conduction when rains are available (Tyree *et al.*, 1994). In fact, a negative relationship exists between rate of conductance and amount of precipitation for deciduous species (Maherali *et al.*, 2004); however, large vessels are known to be at higher risk of embolisms due to freeze-thaw cycles (Sperry *et al.*, 1994; Tyree *et al.*, 1994) and drought, at least within a given plant part (Hacke and Sperry, 2001). Since drought-induced embolisms spread through pores in the interconduit pit membranes (Hacke and Sperry, 2001; Tyree and Zimmermann, 2002), isolated vessels are less likely to spread embolisms. Failure of ring-porous species (e.g., *Quercus*) to isolate the large, rapidly conducting vessels could expose plants to the danger of fast spreading embolisms and loss of hydraulic conductance. The many small safe vessels in integrated diffuse-porous species (e.g., *Betula* spp., *Acer* spp.) are less prone to embolisms, especially freezing induced, and thus should not need to be as isolated as ring-porous vessels. Thus, the isolation of embolisms resulting from environmental stresses such as seasonal drought and freeze-thaw cycles could be a driving force selecting for sectoriality.

***Other Selective Pressures*** We now broaden our discussion to consider potential advantages and disadvantages of varying degrees of sectoriality, as well as how multiple pressures act on sectoriality. Our hypothesized list of conditions favoring greater integration or sectoriality is not designed to be exhaustive but merely to provide a framework for further investigation. Plants that are strongly sectored are able to isolate damage or create strong feedback loops within a sector in high resources (e.g., roots in high nutrients connected to leaves in high sunlight). We hypothesize that species susceptible to vascular diseases should be more sectored. Integrated plants, on the other hand, have the ability to translocate resources among parts. Because the success of many clonal plant species depends on the sharing of resources among ramets (de Kroon and van Groenendael, 1997), on average we expect clonal species to be more integrated than nonclonal species. We also suggest that species adapted to environments where damage to the growing shoots is common are likely to be integrated. For example, freezing temperatures at high latitudes (or altitudes) or heavy herbivore pressures should favor species capable of using reserves from more than one sector for regrowth in whichever sector is better positioned for future resource capture. As discussed previously, *Betula* is a genus characteristic of northern latitudes and is relatively integrated, whereas *Quercus* is a speciose genus in xeric Mediterranean habitats and is more sectored. Although these patterns are consistent with the hypothesized outcomes, they do not serve as a test nor predict how environmental conditions interact. As species within genera such as *Betula* and *Quercus* vary in their environmental associations, an examination of sectoriality within these genera should prove particularly fruitful at elucidating key selective pressures.

## Conclusions

We have investigated the impact of sectoriality on long distance transport of resources in dicots by examining evidence from vascular ultrastructure to environmental selective pressures. We suggest that sectoriality is a common plant character, despite the fact that certain conditions might break down sectoriality (e.g., patchy water availability). Insight into the adaptive role that sectoriality plays in species success can be gained by examining the distribution of this character among species while incorporating their evolutionary relations, as well as by examining the environmental selective pressures in the habitats where sectored species are common. Such investigations should be greatly aided by the advent of many new techniques assisting in the determinations of what species are sectored, ranging from determination at the scale of individual cells to whole plants. We suggest that sectoriality is indeed an important plant character, and evidence gathered to date points to the important role of embolisms in influencing sectoriality in long distance transport. As more evidence is gathered, it will be especially intriguing to determine how multiple selective pressures have acted on plant vascular architecture.

## Acknowledgments

We thank Margret van Vuuren, Claire Margerison, Nancy Harris, Adrianna Muir, and Kate Sweeney for assistance in the lab, Maciej Zwieniecki and Lawren Sack for introducing us to the use of hydraulic techniques to measure sectoriality, George Ellmore for assisting us with anatomical techniques to measure vessel and pit sizes, and Zoe Cardon and Arnold Bloom for comments on earlier versions of this chapter. We are grateful to The Andrew Mellon Foundation for providing financial support.

## References

Caldwell, M. M. (1994) Exploiting nutrients in fertile soil microsites. In *Exploitation of Environmental Heterogeneity by Plants: Ecophysiological Processes Above- and Below Ground* (M. M. Caldwell and R. W. Pearcy, eds.) pp. 325-347. Academic Press, San Diego.

Carlquist, S. (2001) *Comparative Wood Anatomy*, 2nd Ed. Springer-Verlag, Berlin.

Chazdon, R. L. and Fetcher, N. (1984) Photosynthetic light environments in a lowland tropical rain forest in Costa Rica. *J Ecol* **72:** 553-564.

Cook, C. W. and Stoddard, L. A. (1960) Physiological responses of big sagebrush to different types of herbage removal. *J Range Manage* **13:** 14-16.

Davis, J. M., Gordon, M. P. and Smit, B. A. (1991) Assimilate movement dictates remote sites of wound-induced gene expression in poplar leaves. *Proc Natl Acad Sci* **88:** 2393-2396.

De Kroon, H. and van Groenendael, J. (1997) *The Ecology and Evolution of Clonal Plants.* Backhuys Publishers, Leiden, Netherlands.

Denno, R. F. and McClure, M. S. (1983) *Variable Plants and Herbivores in Natural and Managed Systems.* Academic Press, New York.

Fisher, D. B. (2000) Long-distance transport. In *Biochemistry and Molecular Biology of Plants* (B. B. Buchanan, W. Gruissem, and R. L. Jones, eds.) pp. 730-784. American Society of Plant Physiologists, Rockville, MD.

Fort, C., Muller, F., Label, P., Granier, A. and Dreyer, E. (1998) Stomatal conductance, growth and root signaling in *Betula pendula* seedlings subjected to partial soil drying. *Tree Physiol* 18: 769-776.

Guthrie, R. L. (1989) Xylem structure and ecological dominance in a forest community. *Am J Bot* 76: 1216-1228.

Hacke, U. G. and Sperry, J. S. (2001) Functional and ecological xylem anatomy. *Perspect Plant Ecol Evol System* 4: 97-115.

Hansen, E. A. and Dickson, R. E. (1979) Water and mineral nutrient transfer between root systems of juvenile *Populus*. *Forest Sci* 25: 247-252.

Horwath, W. W., Paul, E. A. and Pregitzer, K. S. (1992) Injection of nitrogen-15 into trees to study nitrogen cycling in soil. *Soil Sci Soc Am J* 56: 316-319.

Jackson, R. B. and Caldwell, M. M. (1993) Geostatistical patterns of soil heterogeneity around individual perennial plants. *J Ecol* 81: 683-692.

Jones, C. G., Hopper, R. F., Coleman, J. S. and Krischik, V. A. (1993) Control of systemically induced herbivore resistance by plant vascular architecture. *Oecologia* 93: 452-456.

Karban, R., Agrawal, A. A. and Mangel, M. (1997) The benefits of induced defenses against herbivores. Ecology 78: 1351-1355.

Karban, R. and Baldwin, I. T. (1997) *Induced Responses to Herbivory*. University of Chicago Press, Chicago.

Kiyomiya, S., Nakanishi, H., Uchida, H., Tsuji, A., Nishiyama, S., Futatsubashi, M., Tsukada, H., Ishioka, N. S., Watanabe, S., Ito, T., Mizuniwa, C., Osa, A., Matsuhashi, S., Hashimoto, S., Sekine, T. and Mori, S. (2001) Real time visualization of 13N-translocation in rice under different environmental conditions using positron emitting tracer imaging system. *Plant Physiol* 125: 1743-1754.

Kozlowski, T. T. and Winget, C. H. (1963) Patterns of water movement in forest trees. *Botanical Gazette* 124: 301-311.

Larson, P. R. and Dickson, R. E. (1973) Distribution of imported 14C in developing leaves of eastern cottonwood according to phyllotaxy. *Planta* 111: 95-112.

Lucas, W. J. (1997) Application of microinjection techniques to plant nutrition. *Plant and Soil* 196: 175-189.

Maherali, H., Pockman, W. T. and Jackson, R. B. (2004) Adaptive variation in the vulnerability of woody plants to xylem cavitation. *Ecology* 85: 2184-2199.

Marschner, H. (1995) *Mineral Nutrition of Higher Plants*. Academic Press, San Diego.

Marshall, C. (1996) Sectoriality and physiological organisation in herbaceous plants: An overview. *Vegetatio* 127: 9-16.

Murphy, N. and Watson, M. A. (1996) Sectorial root growth in cuttings of *Coleus rehnaltianus* in response to localized aerial defoliation. *Vegetatio* 127: 17-23.

Murray, B. J., Mauk, C. and Noodén, L. D. (1982) Restricted vascular pipelines (and orthostichies) in plants. *What's New in Plant Physiology* 13: 33-36.

Narvaez-Vasquez, J., Orozco-Cardenas, M. L. and Ryan, C. A. (1994) A sulfhydryl reagent modulated systemic signaling for wound-induced and systemin-induced proteinase inhibitor synthesis. *Plant Physiol* 105: 725-730.

Noodén, L. D., Rupp, D. C. and Derman, B. D. (1978) Separation of seed development from monocarpic senescence in soybeans. *Nature* 271: 354-357.

Orians, C. M., Ardón, M. and Mohammad, B. A. (2002) Vascular architecture and patchy nutrient availability generate within-plant heterogeneity in plant traits important to herbivores. *Am J Bot* 89: 270-278.

Orians, C. M. and Jones, C. G. (2001) Plants as resource mosaics: A functional model for predicting patterns of within-plant resource heterogeneity to consumers based on vascular architecture and local environmental variability. *Oikos* 94: 493-504.

Orians, C. M., Pomerleau, J. and Ricco, R. (2000). Vascular architecture generates fine scale variation in the systemic induction of proteinase inhibitors in tomato. *J Chem Ecol* **26:** 471-485.

Orians, C. M., Van Vuuren, M. M. I., Harris, N. L., Babst, B. A. and Ellmore, G. S. (2004) Differential sectoriality in long distance transport in temperate tree species: Evidence from dye flow, $^{15}$N transport, and vessel element pitting. *Trees* **18:** 501-509.

Pate, J. S. and Dieter Jeschke, W. (1995) Role of stems in transport, storage, and circulation of ions and metabolites by the whole plant. In *Plant Stems: Physiology and Functional Morphology* (B. L. Gartner, ed.) pp. 177-204. Academic Press, San Diego.

Prokofyev, A. A., Zhdanova, L. P. and Sobolev, A. M. (1957). Certain regularities in the flow of substances from leaves into reproductive organs. *Sov Plant Physiol* **4:** 402-408.

Rinne, R. W. and Langston, R. G. (1960) Studies on lateral movement of phosphorus 32 in peppermint. *Plant Physiol* **35:** 216-219.

Robinson, D. (1994) The responses of plants to non-uniform supply of nutrients. *New Phytol* **127:** 635-674.

Sachs, T., Novoplansky, A. and Cohen, D. (1993) Plants as competing populations of redundant organs. *Plant Cell Environ* **16:** 765-770.

Schittko, U. and Baldwin, I. T. (2003) Constraints to herbivore-induced systemic responses: Bidirectional signaling along orthostichies in *Nicotiana attenuata*. *J Chem Ecol* **29:** 763-770.

Shea, M. M. and Watson, M. A. (1989) Patterns of leaf and flower removal: Their effect on fruit growth in *Chamaenerion angustifolium* (fireweed). *Am J Bot* **76:** 884-890.

Sperry, J. S., Nichols, K. L., Sullivan, J. E. M. and Eastlack, S. E. (1994) Xylem embolism in ring-porous, diffuse-porous, and coniferous trees of northern Utah and interior Alaska. *Ecology* **75:** 1736-1752.

Stark, J. M. (1994) Causes of soil nutrient heterogeneity at different scales. In *Exploitation of Environmental Heterogeneity by Plants: Ecophysiological Processes Above- and Below Ground* (M. M. Caldwell and R. W. Pearcy, eds.) pp. 255-284. Academic Press, San Diego.

Steiber, J. and Beringer, H. (1984) Dynamic and structural relationships among leaves, roots, and storage tissue in the sugar beet. *Bot Gaz* **145:** 465-473.

Tyree, M. T., Davis, S. D. and Cochard, H. (1994) Biophysical perspectives of xylem evolution: Is there a tradeoff of hydraulic efficiency for vulnerability to dysfunction? *IAWA* **15:** 335-360.

Tyree, M. T. and Zimmermann, M. H. (2002) *Xylem Structure and the Ascent of Sap.* Springer, Berlin.

van Bel, A. J. E., Ehlers, K. and Knoblauch, M. (2002) Sieve elements caught in the act. *Trends Plant Sci* **7:** 126-132.

van Vuuren, M. M. I., Robinson, D. and Griffiths, B. S. (1996) Nutrient inflow and root proliferation during the exploitation of a temporally and spatially discrete source of nitrogen in soil. *Plant Soil* **178:** 185-192.

Watson, M. A. (1986) Integrated physiological units in plants. *Trends Ecol Evol* **1:** 119-123.

Watson, M. A. and Casper, B. B. (1984) Morphogenetic constraints on patterns of carbon distribution in plants. *Annu Rev Ecol Syst* **15:** 233-258.

Woodward, F. I. (1990) From ecosystems to genes: The importance of shade tolerance. *Trends Ecol Evol* **5:** 111-115.

Zimmermann, M. H. and Brown, C. L. (1971) *Trees Structure and Function* Springer-Verlag, New York.

Zwieniecki M. A, Melcher, P. J. and Holbrook, N. M. (2001) Hydrogel control of xylem hydraulic resistance in plants. *Science* **291:** 1059-1062.

Zwieniecki, M. A., Orians, C. M., Melcher, P. J. and Holbrook, N. M. (2003) Ionic control of the lateral exchange of water between vascular bundles in tomato. *J Exp Bot* **54:** 1399-1405.

# Part V

## Limits to Long Distance Transport

# 18

## Embolism Repair and Long Distance Water Transport

*Michael J. Clearwater and Guillermo Goldstein*

"Vessels and tracheids normally contain both air and water, the relative amounts of the two substances varying according to the season and the time of day." Since Haberlandt wrote this statement in 1914, we have slowly come to understand that cavitation and the formation of gas emboli are a common occurrence in the xylem of many plant species (Tyree and Ewers, 1991; Sperry, 1995). The cavitation, or breakage, of a water column may be initiated during water stress by the entry of air through conduit pit membranes, or by the formation of bubbles during the freezing and thawing of xylem sap. Once cavitated, the conduit becomes filled with water vapor and air, resulting in an embolism, or blockage, within the conduit. It is now widely understood that emboli often form during transpiration, but until recently the removal of emboli was not thought to occur except when transpiration ceased and xylem tension was very low. However, a series of steadily more frequent and more convincing reports over the past decade have begun to suggest otherwise—that embolism reversal does occur and that it can happen during transpiration. These reports (see later) are truly surprising. Given the current understanding of long distance water transport as a passive, tension-driven process, refilling of emboli during transpiration should be physically impossible.

According to physical principles, an embolism can be removed only if the pressure in the xylem conduit increases enough to force the gas in the bubble back into solution. The threshold pressure for the embolism to dissolve depends on the gas content of the bubble and the xylem sap, and the radius of the bubble. Using the capillary pressure equation (Yang and Tyree, 1992; Steudle, 2001; Hacke and Sperry, 2003):

$$\psi_{pt} = P - 2T/r \qquad (1)$$

where $\psi_{pt}$ is the threshold xylem pressure potential above which the bubble will collapse, P is the gas pressure, T is the surface tension (0.0728 N m$^{-1}$

at 20° C), and r is the radius of curvature of the gas-water interface (approximately equal to the radius of the conduit). For example, a water vapor bubble in a conduit of 10 μ m radius, P = 0.0023 MPa (water vapor pressure) and $\psi_{pt}$ = − 0.114 MPa relative to atmospheric pressure at sea level. For an air bubble in the same conduit and the xylem sap saturated with air, $\psi_{pt}$ = − 0.015 MPa. Hence emboli will not refill unless the pressure rises to at least 0.1 MPa below atmospheric pressure, and with increasing amounts of air dissolved in the sap and larger xylem conduits, the higher the pressure must be for refilling to occur. There are several examples where refilling in embolized stems has been shown to conform to the capillary equation, when excised stems or intact plants are supplied with water at zero or slightly negative pressure and evaporation is essentially zero (Borghetti *et al.*, 1991; Tyree and Yang, 1992; Sobrado *et al.*, 1992; Lewis *et al.*, 1994; Edwards *et al.*, 1994). During transpiration, however, the xylem pressure of most plants, particularly larger woody plants, is more negative than the "$-2T/r$" threshold and is frequently closer to −1.0 MPa or less. Under these conditions refilling should not be possible unless there is an alternative mechanism that drives bubble dissolution in individual xylem conduits. Here we adopt the terminology of Hacke and Sperry (2003) and refer to refilling at pressures significantly below the "$-2T/r$" threshold as *refilling under tension*. We evaluate the recent evidence and possible explanations for refilling under tension. We also speculate on the overall significance that refilling may have for long distance water transport in plants.

## Evidence for Refilling Under Tension

A review of the recent evidence for refilling under tension provides a colorful account of the difficulty of characterizing what is happening in intact, functioning xylem without perturbing it. Most authors agree that the xylem sap in transpiring plants is under tension and is therefore vulnerable to disruption and the production of artifacts during measurements (Tyree, 1997; Meinzer *et al.*, 2001; Steudle, 2001). A wide variety of techniques, most of them indirect and destructive, have been applied to both generate and measure known xylem tensions, and to document the proportions of gas- and water-filled conduits. Claims have been made for refilling under tension in roots (e.g., McCully *et al.*, 1998; Buchard *et al.*, 1999; Pate and Canny, 1999), stems (Salleo *et al.*, 1996; Zwieniecki and Holbrook, 1998; Melcher *et al.*, 2001), petioles (Canny, 1997; Zwieniecki *et al.*, 2000), and leaves (Canny, 2001b; Lo Gullo *et al.*, 2003), as well as in a range of species that includes herbaceous and woody monocotyledons (McCully, 1999; Pate and Canny 1999), dicotyledons (Berndt *et al.*, 1999; Holbrook *et al.*, 2001), and conifers (Borghetti *et al.*, 1991). Because of the nature of the methods used in these studies, however, results are often prone to experimental error, and many reports of refilling are subject to counterclaims that the results are artifacts of the technique used. Of importance, many accounts also lack a clear record of xylem pressure during refilling at the site of refilling (Salleo and Lo Gullo, 1989; Zwieniecki and Holbrook, 1998; Pate and Canny, 1999; McCully *et al.*, 2000; Facette *et al.*, 2001; Holbrook *et al.*, 2001; Linton and Nobel, 2001; Lo Gullo *et al.*, 2003), thus making it difficult to know whether xylem pressure was below the $-2T/r$ threshold required for refilling under tension (Hacke and Sperry, 2003).

Early discussion concerning the ascent of sap in plants shows that there was general agreement that xylem conduits normally contained both gas and liquid (Strasburger, 1891; Haberlandt, 1914; Priestley, 1935; Greenidge, 1957; Canny, 2001a). Debate over the validity of the cohesion tension theory of long distance water transport centered on the ability of columns of water to withstand significant tension, and whether the xylem contained continuous columns of water. There were many detailed observations of columns of gas and liquid in the xylem (e.g., Haberlandt, 1914; Haines, 1935) and some claims that the xylem was mostly filled with gas at subatmospheric pressure (Preston, 1938). Direct observation of vessel contents in dissected herbaceous and woody stems showed that bubbles are often visible in the xylem, and that they may move, expand, or contract during transpiration (Vesque, 1883; Haines, 1935). In some cases it was concluded that xylem vessels are normally emptied of water and filled with air at low pressure during periods of rapid transpiration, and that they refill at night or during rainy periods (Haberlandt, 1914). However, from these reports alone it is often not possible to determine exactly what the conditions were during refilling. For this reason these studies do not provide unambiguous evidence for fast (daily or hourly) changes in xylem gas content, or a mechanism for refilling of gas-filled conduits while neighboring conduits are under tension. Given our more recent knowledge of the widespread occurrence of xylem cavitation during even mild water stress, it is expected that a significant proportion of xylem conduits in many plants will contain gas bubbles (Sperry, 1995), and that the gas-filled proportion will change during a season (Magnani and Borghetti, 1995). Once a vessel has cavitated, any free water remaining may be drawn to neighboring conduits that are still under tension. When vessels of a range of diameters cavitate, water can be redistributed from larger to smaller conduits in response to capillary forces, thus causing some bubbles to contract while others expand (Crafts, 1939). For the same reason, cavitated vessels may refill during periods of low tension and act as a reservoir of stored water. Direct observation of these changes in dissected plant material can give the impression that an active refilling mechanism is involved (Vesque, 1883;

Crafts, 1939). While the earlier literature does not provide stand-alone evidence for refilling under tension, many of these authors argued strongly that cohesion-tension alone could not explain their observations, and that other, more vital mechanisms must be involved in the transport of sap in the xylem. With suitable caution concerning the generation of artifacts, their methods could be easily adapted to follow changes in xylem contents over shorter timescales, yielding results relevant to the current debate (Canny, 2001a).

Brough *et al.* (1986) used a gamma probe to measure a daily reduction and recovery of stem water content in apple trees. Their study is sometimes cited as an example of refilling of cavitated vessels under tension, but the authors themselves discount a role for cavitation of vessels because xylem tension remained below the threshold required for detection of significant cavitation using acoustic detectors. Recovery of xylem water content and hydraulic conductivity of conifers and some angiosperms over seasonal timescales has also been reported (Waring *et al.*, 1979; Sperry, 1993; Sperry *et al.*, 1994; Magnani and Borghetti, 1995). Root pressure was not observed in these cases, but it is possible that xylem pressures were close to zero during periods of high humidity and rainfall. Refilling in these examples may therefore occur by bubble dissolution at low pressure, capillary rise, or water entry through wet foliage (Grace, 1993). In the same way that direct observation of changes in bubble volume in the xylem is not conclusive evidence for refilling, changes in xylem water content is not evidence for refilling under tension unless significant tension in the xylem is clearly documented while hydroactive conduits are shown to empty and refill.

Refilling under tension has most commonly been inferred from dye uptake experiments and changes in hydraulic conductivity of excised xylem segments (Salleo and Lo Gullo, 1989; Salleo *et al.*, 1996; Zwieniecki and Holbrook, 1998; Tyree *et al.*, 1999; Zwieniecki *et al.*, 2000; Melcher *et al.*, 2001; Hacke and Sperry, 2003). Salleo and Lo Gullo (1989) cut *Vitis* stems under dye solution and recorded some afternoon and overnight recovery in the number of stained conduits of mildly water stressed plants. Recovery during the day was small, and xylem pressure during more significant overnight recovery was not recorded. These authors stressed the risk of underestimating cavitation as a result of capillary refilling of embolized conduits during cutting underwater and therefore recommended cutting stems under reduced pressure (Salleo and Lo Gullo, 1989). Zwieniecki and Holbrook (1998) also observed overnight recovery in the twigs of coniferous and hardwood trees while xylem tensions were low. More convincing hydraulic evidence for afternoon refilling under significant tension in the petioles of transpiring plants has been presented by both Zwieniecki *et al.* (2000) and Bucci *et al.* (2003). In both examples the hydraulic conductivity and dye uptake of petioles recovered in the afternoon after reductions during morning, and xylem tensions were clearly significant during refilling.

Refilling of embolized xylem under tension within minutes or hours has also been clearly demonstrated in *Laurus nobilis*, but only after embolism was artificially induced in excised and dehydrated stems using air injection (Salleo *et al.*, 1996; Tyree *et al.*, 1999). In an earlier study using *L. nobilis*, refilling in soil droughted plants occurred only after 5 days of rewatering (Salleo and Lo Gullo, 1993), leading Hacke and Sperry (2003) to question whether rapid refilling in *L. nobilis* was a unique response to the air-injection method. Magnetic resonance imaging (MRI) has also been used to observe a sudden reduction in xylem water content of stems of potted *Acer rubrum* during air injection, followed by a rapid recovery after injection stopped (M. A. Zwieniecki, personal communication). Refilling after air injection, therefore, may not be representative of a mechanism that operates *in vivo*. However, Hacke and Sperry (2003) did observe partial refilling in *L. nobilis* 24 hours after rewatering of soil droughted plants while stem water potential was clearly significantly below zero. Under identical conditions, little or no refilling was observed in *Acer negundo*, suggesting that refilling under tension does occur in intact *L. nobilis* plants, but more slowly than that induced by the air-injection method (Hacke and Sperry, 2003).

The debate over the existence of a refilling mechanism has been greatly enlivened by the contributions of Martin Canny and co-workers. Based almost entirely on cryoscanning electron microscopy (cryo-SEM), a series of studies have now shown the apparent emptying of xylem vessels early in the day as transpiration begins, and refilling later, paradoxically when xylem tensions are higher. Similar results have been obtained for leaves, petioles, stems, and roots of a range of mostly herbaceous monocotyledons and dicotyledons (Canny, 1997, 1998, 2001b; McCully *et al.*, 1998; Berndt *et al.*, 1999; McCully, 1999; Buchard *et al.*, 1999). Most experiments were on freely transpiring, well-watered plants, and in one case the amount of gas aspirated from *Xanthorrhea* roots seems to confirm the cryo-SEM counts of embolized vessels. In some examples xylem water potential is not measured, the data do not show a clear decline in the proportion of embolized vessels while transpiration continues, or the role of positive root pressure cannot be excluded (Pate and Canny, 1999; McCully *et al.*, 2000; Facette *et al.*, 2001).The most serious criticism, however, has come from evidence showing that a large proportion of the emboli observed using the cryo-SEM are an artifact caused by freezing the xylem under tension (Cochard *et al.*, 2000, 2001). If the tension is relaxed before freezing, the amount of embolism observed is much less and conforms to expectations based on measurements of xylem vulnerability to cavitation. In response, counter-claims were made that the relaxation of tension causes rapid refilling (Canny *et al.*, 2001a, 2001b). The controversy is a reminder that there will almost always be ambiguity and debate over the results of destructive experiments on a system that is thought to be in a meta-stable state at the time of intervention.

Nondestructive techniques for measuring xylem contents are needed to avoid the type of controversy that surrounds the cryo-SEM method. At present, MRI provides the only method for nondestructively imaging the water content of the xylem with sufficient resolution to study the phenomenon of refilling. In general terms, image resolution increases with magnet field strength and decreasing bore and receiver coil diameter, and for MRI microscopy at least half the plant must be threaded or completely inserted into the bore of a magnet and gradient coil 0.02 to 0.1 m in diameter and 1 m or more in length (Gadian, 1995; Ishida *et al.*, 2000). Equipment configuration thus restricts the size and form of plant that can be used, and the cost and location of the equipment can limit the duration and type of treatments that can be applied (e.g., high rates of transpiration or long periods of drought stress may be difficult to achieve). Nevertheless, the nondestructive and noninvasive nature of MRI makes it an exciting tool for studies of long distance transport in plants (Fig. 18.1). Lianas are obvious candidates for MRI studies of xylem refilling because their flexible growth form makes them easier to position in the magnet while still retaining significant leaf area and transpiration outside the magnet, and their large xylem vessels are easier to detect in low resolution micrographs. Holbrook *et al.* (2001) made the first MRI observations of embolism formation and refilling in a Vitis stem, but refilling was observed only after the lights were switched off and measured xylem tension had approached zero. The authors were therefore unable to eliminate the possibility that refilling was the result of positive root pressure. Clearwater and Clark (2003) subjected woody *Ripogonum* (monocotyledon) and *Actinidia* (dicotyledon) vines to a range of treatments while monitoring stem xylem vessels' content using MRI. Cavitation in response to elevated xylem tension was frequently observed, but no refilling was detected at any time, and it was concluded that refilling under tension was unlikely to occur in the species examined (Fig. 18.1). Using MRI we therefore do not yet have unambiguous evidence for embolism repair under tension. However, in the two examples described previously, MRI was applied in a simple way to measure only the presence or absence of water in xylem tissue. More sophisticated MRI experiments have been used to measure water content and bulk flow velocities in plant tissue (Kockenberger *et al.*, 1997; Rokitta *et al.*, 1999; Wistuba *et al.*, 2000; Scheenen *et al.*, 2002), to study the refilling kinetics of resurrection plants (Wagner *et al.*, 2000), and to infer changes in membrane permeability within the component tissues of cucumber stems in response to changes in water status (Scheenen *et al.*, 2002). These examples demonstrate that functional MRI offers promise as a tool to study the phenomenon of embolism repair.

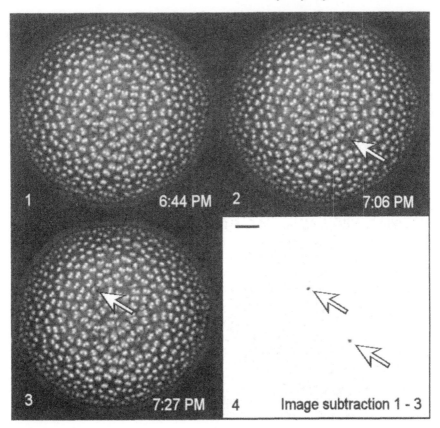

**Figure 18.1** Magnetic resonance imaging of the stem of a monocotyledonous liana, supplejack (*Ripogonum scandens* Forst.) during transpiration. Individual meta-xylem vessels are visible as collateral pairs within each vascular bundle. Sequential images (1-3) show the cavitation of two individual meta-xylem vessels during a period of elevated xylem tension. Image 4 is the result of subtraction of image 1 from image 3, highlighting the location of cavitated vessels. Refilling under tension was not observed with this species (Clearwater and Clark, 2003). Stem diameter is 17 mm, scale bar 2 mm.

To summarize, there are relatively few unambiguous and undisputed accounts that clearly demonstrate refilling of xylem conduits *in vivo* while tension exists in the xylem. However, there is now enough evidence to conclude that not all movement of water within the xylem or between the xylem and other tissues can be accounted for by a cohesion-tension mechanism alone. Some form of refilling mechanism does operate, but it may be restricted to particular species, organs, and even specific locations within the xylem. Next we discuss the possible nature of this mechanism and its significance for long distance transport in plants.

## Potential Refilling Mechanisms

The refilling of xylem conduits under tension represents a paradox for the plant physiologist: How, in a porous plant tissue, can water move into a conduit with enough pressure to remove the embolism while water in neighboring cells remains at much lower pressure? During the last century the concept of the physical cohesion-tension mechanism driving the ascent of sap overtook vitalistic theories of water transport (Milburn, 1979). The xylem is now normally viewed as a passive tissue through which sap is drawn by tension. The idea that refilling under tension may occur has prompted discussion over whether a new paradigm is needed (Holbrook and Zwieniecki, 1999; Tyree *et al.*, 1999), and caused some to reject the cohesion-tension theory outright (Canny, 1995). We divide current hypotheses for refilling under tension into four categories (Table 18.1, Fig. 18.2). The first two invoke solutes within the refilling conduit and simple osmosis to move water into the xylem, but differ in the type of solutes and the location of the semipermeable membrane across which osmosis occurs. The third is based on concepts of local increases in tissue pressure, causing water to be squeezed into the xylem by reverse osmosis. The fourth is a more complex active transport mechanism involving multiple compartments and differences in membrane properties between compartments (asymmetrical membrane). The four hypotheses differ in the location of the driving force for the movement of water and in the predicted solute concentration within the refilling conduit (Table 18.1). Three of the four hypotheses cannot function without a separate mechanism that isolates the refilling conduit, thus preventing refilling water from being lost to the surrounding xylem that is still under tension (Table 18.1). Only one such isolating mechanism has been proposed to date (discussed later with the cell membrane osmosis mechanism). In the pit membrane hypothesis, an isolating mechanism is not required because the walls of the refilling conduit are envisaged as the semipermeable barrier across which osmosis occurs.

## Cell Membrane Osmosis

In theory, refilling can be achieved by the release of salts and/or osmotically active organic molecules from living xylem parenchyma cells surrounding the cavitated vessel (Grace, 1993). If the osmotic potential of sap in the cavitated vessel decreases to a value more negative than the water potential of adjacent conduits or the xylem parenchyma, water will move into the cavitated vessel and the pressure increase above the $-2T/r$ threshold. In its original form, the hypothesis suggests that the semipermeable

**Table 18.1**    Comparison of four hypotheses for embolism repair under tension

| Proposed Mechanism | Location of Driving Force | Solute Content Within Conduit | Isolating Mechanism | References |
|---|---|---|---|---|
| Cell membrane osmosis | Low molecular weight solutes loaded into the refilling conduit | Elevated | Required | Grace, 1993 Salleo *et al.*, 1996 Tyree *et al.*, 1999 |
| Pit membrane osmosis | High molecular weight solutes inside the refilling conduit | Elevated | Not required | Hacke and Sperry, 2003 |
| Tissue pressure | Solute buildup and turgor increase in tissues external to the xylem | Not elevated | Required | Canny, 1995 Canny, 1998 Bucci *et al.*, 2003 |
| Membrane asymmetry | Solute buildup and turgor increase in living xylem cells or living cells outside the xylem | Elevated, but less than first two above | Required | Pickard, 2003a Vesala *et al.*, 2003 |

Solute content is the osmotically active solute concentration within the refilling conduit, compared to adjacent hydroactive xylem. Decreased osmotic potential inside the refilling conduit is the driving force for refilling in the first two mechanisms. In the last mechanism, leakage of solutes accompanies flow into the conduit, but osmotic potential need not be low enough to drive flow. Isolating mechanism refers to the additional requirement for the conduit to be hydraulically isolated from surrounding xylem during refilling.

barrier across which osmosis occurs is the cell membrane (the cell membrane osmosis mechanism, Table 18.1, Fig. 18.2), with the vessels' walls and pit membranes being relatively permeable to water and solutes (Grace, 1993). Salleo *et al.* (1996) hypothesized that solutes are released from the phloem, travel radially through the ray cell walls, and enter the embolized conduits. In their experiments radial transport was enhanced by the use of air injection to generate cavitation. Alternatively, xylem parenchyma may be the main site of solute release from the symplast. Netting (2000) proposed that xylem parenchyma cells respond to xylem tension and facilitate refilling through a mechanism resembling stomatal closure, with a mechanosensitive calcium signaling pathway that activates aquaporins and solute transporters and the release of solutes and water into the embolized conduit. However, measurements of the concentration of solutes in sap during refilling have so far not shown an increase in osmotic potential

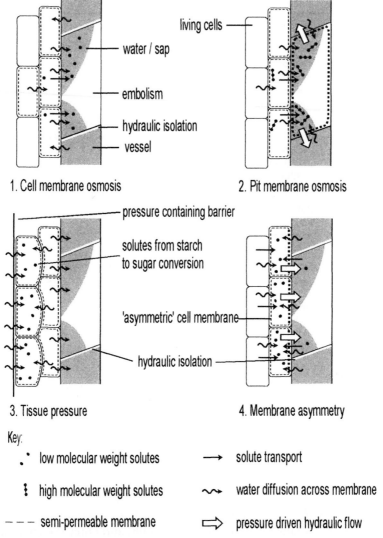

1. Cell membrane osmosis

2. Pit membrane osmosis

3. Tissue pressure

4. Membrane asymmetry

Key:

. ∙ low molecular weight solutes ⟶ solute transport

❀ high molecular weight solutes ∿ water diffusion across membrane

− − − semi-permeable membrane ⟹ pressure driven hydraulic flow

**Figure 18.2** Schematic illustration of the four hypotheses for refilling under tension, listed in Table 18.1 and discussed in more detail in the text. Each diagram shows an embolized and refilling vessel and adjacent living cells. Depending on the hypothesis in question, the living cells involved in refilling may be the xylem parenchyma, or tissues external to the xylem. In 1, 3, and 4, the membranes involved in generating the driving force for refilling are those of living cells. In 2, the key semipermeable membrane is formed by the pit membranes of the vessel primary wall. The scenario illustrated in 3 is the diffusion of water from compressed cells into the xylem (reverse osmosis). An alternative is that hydraulic flow could occur, resulting in water and solutes moving into the xylem. The asymmetrical membrane in 4 indicates that the cell membrane of the central compartment has different properties at its inner and outer boundaries.

sufficient to drive refilling. Tyree *et al.* (1999) measured elevated concentrations of $K^+$ and $Ca^{2+}$ in the sap from laurel stems during refilling, after they had been pretreated with a $K^+$ and indole-3-acetic acid (IAA) solution that enhanced refilling. Bulk sap osmolality was below the level needed to drive refilling, but it is expected that solutes present in refilling cells would be diluted to some degree by sap from uncavitated vessels (Grace, 1993; Tyree *et al.*, 1999; McCully, 1999). Radiographic probe microanalysis has also shown that the solute content of individual vessels that appeared to be refilling when viewed with cryo-SEM is also below the threshold required for an osmotic refilling mechanism (McCully *et al.*, 1998; Tyree *et al.*, 1999; McCully, 1999). Therefore, Tyree *et al.* (1999) rejected the osmotic mechanism and suggested that a new paradigm may be required. However, this conclusion rests on the use of the cryo-SEM method for detecting embolized conduits. If the majority of vessels that appear to be refilling are an artifact of the freezing process (discussed previously), the solute content of sap in these cells should be the same as uncavitated vessels carrying normal xylem sap. Further attempts to measure the osmolality of sap in individual refilling vessels would still be a useful test of the osmotic refilling mechanism.

Holbrook and Zwieniecki (1999) proposed a mechanism by which water in a refilling xylem conduit could remain isolated from neighboring conduits until refilling was complete. Without compartmentalization of the refilling process, water and any low molecular weight solutes released into the conduit would be quickly lost to neighboring unembolized conduits that contain water at much lower potentials. Their proposal, termed the *pit valve hypothesis* by Hacke and Sperry (2003), is that the high contact angle of coalescing water droplets on the hydrophobic inner walls of the refilling conduit, combined with the geometry of the bordered pits, allows positive pressures to develop that are high enough to dissolve gas within the lumen, but not high enough to force the air/water interface across the bordered pit channel. They also hypothesized that the angle of the bordered pit chamber and chemistry of the pit walls also mean that the volume of gas remaining at the time that the meniscus touches the pit membrane is minimized, thus ensuring a stable transition to the full tension of the transpiration stream (Holbrook and Zwieniecki, 1999). In support of their hypothesis, Zwieniecki and Holbrook (2000) were able to calculate the maximum pressure that could be contained by the meniscus in the pit chamber from measurements of pit geometry and droplet contact angle. For six species this pressure varied between 0.07 and 0.3 MPa, indicating that the pits could act as isolating valves during refilling. Direct measurement of the radial conductivity of individual vessels of *Fraxinus americana* also showed a step increase in conductivity at an applied pressure of approximately 0.4 MPa, in agreement with the calculated pressure threshold (Zwieniecki *et al.*, 2001).

Two criticisms have been directed against the pit valve hypothesis. Firstly, vessel secondary walls are composed of a water saturated matrix of cellulose fibers that should be highly permeable to water, thus preventing hydraulic isolation between vessels even when the pit's membranes are blocked by air (Steudle, 2001). However, the secondary cell walls are also impregnated with lignin, a hydrophobic substance that may effectively "waterproof" the cells walls (Holbrook and Zwieniecki, 1999). During measurements it is difficult to separate flow across the pit membrane from flow through the secondary walls. The radial permeability of vessels is high, but these measurements usually include the more permeable pit pathway (e.g., Peterson and Steudle, 1993). Yet when Zwieniecki *et al.* (2001) measured the radial conductance of previously aspirated *Fraxinus* vessels directly, presumably with the pit valves blocked by air, conductance was still relatively high (60% to 90%) compared to the maximum value after the valves were opened by increasing applied pressure. Although there were doubts over whether the aspiration treatment successfully blocked all of the pits, high radial conductance with the pit valves closed may prevent isolation of vessels during transpiration by this mechanism (Zwieniecki *et al.*, 2001). The second difficulty for the pit valve hypothesis is the need for gas remaining in the pit chambers to dissolve in all pits at the same time (Hacke and Sperry, 2003). The geometry of the pit chamber and the droplet contact angle of the pit walls may minimize the volume of gas present at the instant the meniscus contacts the pit membrane (Holbrook and Zwieniecki, 1999), but there must be some variation between pits. The volume of gas remaining in the slowest-filling pits will place an upper limit on the tension that can be sustained after reconnection. More detailed measurements of vessel radial conductivity and the properties of bordered pits are needed before it is possible to speculate further on whether bordered pits do act as pressure-containing valves while vessels refill.

## Pit Membrane Osmosis

Refilling of *L. nobilis* xylem under tension over 24 and 48 hours led Hacke and Sperry (2003) to propose a modification of the osmotic refilling mechanism. Their "pit membrane hypothesis" (Table 18.1, Fig. 18.2) proposes that refilling is achieved by the release of a high molecular weight solute that is impermeable to the pit membrane. In this case, the cell walls and pit membrane act as the semipermeable barrier (Fig. 18.2), obviating the need for an isolating mechanism such as the pit valve hypothesis. Given a sufficiently negative osmotic potential, water could be drawn from all surrounding tissues, including hydroactive conduits and the xylem parenchyma.

The large molecular weight solute should also be trapped within the vessel and therefore not present in extracted sap, and may not be detected using radiographic microprobe analysis (Hacke and Sperry, 2003). The possible identity of the solute or solutes and the mechanism by which it is transported into the refilling conduit are unknown. However, the concept of the xylem conduit wall acting as an imperfect semipermeable membrane has previously been implicated in the refilling and upward movement of water in the xylem in spring (Priestley, 1935). Large amounts of sugars are released from ray contact cells into the vessels of some tree species in spring (Höll, 1975), and osmosis across the cell wall is thought to be involved in the development of stem pressure in maple trees (Johnson and Tyree, 1992; Hacke and Sperry, 2003). Borghetti *et al.* (1991) speculated that gel-like properties of the vessel wall lining might attract water into the conduit, possibly explaining refilling observed in conifer wood below the $-2T/r$ threshold. The presence of high molecular weight mucilage and gel compounds in the xylem has also been suggested as important for maintaining a continuous water transport pathway from the roots to the leaves (Zimmermann *et al.*, 1994, 2002).

Pickard (Chapter 1) questioned the pit hypothesis membrane because it does not account for hydraulic flow through the pores in the pit membrane. Pores even a few nanometers (Choat *et al.*, 2003) in diameter should support hydraulic flow. The rate of refilling by a pit membrane mechanism, therefore, will be the net result of inward diffusion through all water filled pits and outward hydraulic flow through those pits that remain hydraulically connected to the transpiration stream (Fig. 18.2).

## Tissue Pressure

The two hypothesized refilling mechanisms discussed so far are both osmotic in nature, in that they require solute to be released into the xylem conduit. Based on available evidence, these hypotheses are attractive because they are parsimonious and involve water diffusing down a gradient in chemical potential. If solutes are not released into the xylem, an explanation for refilling must account for the movement of water against apparent gradients in chemical potential. Terms such as *pumping* and *extruding* are often resorted to when discussing the puzzle of refilling under tension, although at the cellular level there is no known mechanism for the active transport of water against a gradient in water potential (Zimmermann and Steudle, 1978; Milburn, 1979; but see also Loo *et al.*, 2002). In this context the word *active* refers to work done to directly transport water molecules across a boundary.

As part of his alternative theory of plant water transport, Canny (1995, 1998) proposed a refilling mechanism whereby the expansion of living tissue surrounding the xylem exerts a compensating "tissue" pressure on cells close to the xylem that causes water to move into the refilling conduits (Table 18.1, Fig. 18.2). According to the theory, tissue pressure and the vigor of the refilling process vary with changes in osmotic potential of cells in the swelling tissue. In organs with a prominent sheath of starch cells surrounding the vascular tissue, such as in many leaf petioles, cell osmotic potential and the refilling process may be regulated by hydrolysis of starch into sugars during periods of increased tensions and increased risk of cavitation (Canny, 1998). Canny's explanation for embolism repair has been the subject of extensive debate, including questions over the concept of tissue pressure as distinct from turgor pressure at the cellular level and the possibility that tissue pressure could result in the sustained movement of water into the xylem (Comstock, 1999; Tyree, 1999; Canny, 1999).

Recent observations by Bucci *et al.* (2003) using the petioles of Brazilian savanna trees showed that a midday decrease in petiole specific conductivity was correlated with decreases in water potential, and that an afternoon recovery of conductivity was correlated with decreases in bundle sheath starch and increases in petiole sugar content. Afternoon recovery was enhanced by water applied to the abraded petiole surface, and prevented by cortex removal or longitudinal cuts in the cortex. Based on the requirement for an intact cortex and the inferred changes in bundle sheath osmotic potential, Bucci *et al.* (2003) concluded that refilling may have been the result of local imbalances in tissue pressure that cause the release of water into cavitated vessels. They suggested that increases in tissue volume are partially constrained by the petiole cortex, resulting in a transient pressure imbalance that drives radial water movement in the direction of the embolized vessels, thereby refilling them and restoring water flow. Their proposed mechanism is similar in some aspects to that of Canny's, but it is only seen as operating on a local scale, causing dynamic changes in petiole hydraulic conductivity, rather than supporting the entire water transport system (Bucci *et al.*, 2003). An intriguing feature of the starch sheath observations of Canny (1998) and Bucci *et al.* (2003) is that starch disappears during the day, the opposite of the pattern expected in leaf mesophyll cells, where starch normally accumulates during the day and is hydrolyzed at night (e.g., Fondy and Geiger, 1985). A starch sheath surrounds the vascular tissue in stems and petioles of many species, but it is not known whether diurnal changes in starch content are common, nor whether it is involved in any way in water transport.

The concept put forward by both of these authors suggests that increases in tissue pressure results in elevated turgor within living cells close to the xylem. The resulting increase in water potential within these cells should result in the redistribution of water by diffusion to areas of lower water potential. If the water potential in the compressed cells is elevated above that of water in the refilling conduit, relatively pure water could diffuse into the conduit (reverse osmosis). The solute content of the refilling vessel may not increase (Table 18.1, Fig. 18.2). An alternative explanation could be that the increase in pressure results in hydraulic flow of water and solutes from the compressed cells, via unknown channels, into the conduit (extrusion). This second scenario is similar to the membrane asymmetry hypotheses described later. At present it is not clear what the source of water for the expansion of the starch sheath might be, or how the water from the compressed cells might flow into the refilling conduit, rather than toward the expanding sheath. Bucci *et al.* (2003) found that refilling was enhanced by water applied to the abraded surface of the petioles. They suggest that earlier in the day, water may move from the xylem to cortical tissues, but that during refilling, pressure imbalances cause water to move towards the xylem. It can also be speculated that refilling would be enhanced if the permeability of the membranes to water between the expanding cells and compressed cells was lower than that between the compressed cells and the refilling conduit (Fig. 18.2).

## Membrane Asymmetry

While there is no known mechanism for direct transport of water, it is possible to construct thermodynamically valid scenarios in which water moves against an apparent gradient in water potential. All involve the passive movement of water in response to the active transport of solutes. These active water transport mechanisms require a greater degree of complexity than has so far been demonstrated for the xylem (Zimmermann and Steudle, 1978; Holbrook and Zwieniecki, 1999), but are plausible given our rapidly expanding knowledge of plant cell ultrastructure. Theoretically, active transport of water can occur as an overall property of a system in which two membranes differing in permeability to water and solutes are arranged in series, creating three compartments (Curran and MacIntosh, 1962). Provided a concentration difference is established between the inner and outer compartments by active solute transport across the membranes, water can flow between the outer compartments against its own concentration gradient (Zimmermann and Steudle, 1978). Key features of the system are the requirement for more than two compartments, asymmetry in membrane properties, and the need for metabolic energy to drive solute transport (Table 18.1, Fig. 18.2).

Both Pickard (2003a) and Vesala *et al.* (2003) have recently applied double membrane concepts in numerical models describing exudation into the xylem. Pickard (2003a) produced a model for root pressure, predicting the flux of solutes and water between soil, symplast, and vascular compartments. In a situation analogous to the problem of refilling under tension, the occurrence of root pressure has long been ascribed to the loading of solutes into the xylem and the subsequent osmotic uptake of water. Just as direct measurement of xylem solutes has so far failed to confirm an osmotic embolism repair mechanism, it has frequently been shown that xylem solute concentration in exuding roots is also insufficient to explain the observed behavior (Kramer and Boyer, 1995; Enns *et al.*, 1998; Pickard, 2003a). The model successfully mimics the behavior of exuding roots and is directly applicable to the problem of refilling under tension (Pickard, 2003b) (see Chapter 1). The mechanisms for root pressure and refilling under tension may in fact be similar, raising the possibility that species that develop root pressure are also more likely to exhibit an active refilling mechanism in distal branches and leaves. In Pickard's (2003a) model tissue external to the xylem parenchyma provides a source of solutes and water, and during refilling solutes are loaded by membrane transporters into the xylem parenchyma, causing water uptake and an increase in turgor. If membrane permeability to solutes and water is higher adjacent to the xylem conduits, water and solutes can "leak" into them. The leaked solutes must be scavenged back into the symplast by an active solute transport mechanism. Water thus flows into the xylem even though the osmotic potential of the xylem apoplast remains higher than the external soil solution. It is important to note that the model requires more than one pathway for water to move through membranes between the compartments. Larger-diameter pores, of unknown identity, permit the pressure driven hydraulic flow of water and solutes into the xylem. At the same time, osmotic diffusion along chemical potential gradients occurs between the external compartment and symplast and between the xylem and symplast through more conventional pathways such as plasmalemma-bound aquaporins (Curran and MacIntosh, 1962; Pickard, 2003a).

Vesala *et al.* (2003) also recognized that asymmetry in membrane properties is a key requirement for directional "leaking" of water and solutes into refilling conduits. However, their model does not include the active transport of solutes. Water initially floods from the turgid xylem parenchyma into the embolized conduit after the reflection coefficient of the membrane adjacent to the conduit decreases, possibly as a result of the opening of some form of aquaporin (water channel). The flow then slows as the turgor of the xylem parenchyma declines. Without active solute transport, the rate of refilling and volume of vessel that can be refilled is limited by the ratio of xylem parenchyma volume to conduit volume, and by limits on the quantity of solutes present in the parenchyma and eventually accumulated in the conduit. The Vesala *et al.* (2003) model therefore predicts that refilling is only possible under limited conditions of low xylem tension (pressures higher than < 0.2 MPa) and with relatively small diameter vessels. Their result serves to illustrate that if refilling does occur under significant tension, it must involve active solute transport and significant energy requirements. Although clearly highly speculative, multicompartment models involving complex membrane properties demonstrate that active water transport during refilling is at least theoretically possible and suggest that much can be learnt about the location and nature of the various compartments and transport pathways that may be involved (Pickard, 2003a).

There are notable similarities between the multiple compartment model of Pickard (2003a) and the pressure-based hypotheses of Canny (1998) and Bucci *et al.* (2003, Table 19.1). All propose that refilling is achieved by the pressure driven flow of water into embolized vessels, and is related to increases in the osmotic potential of cells surrounding the vascular tissue. Pickard's model for root pressure proposes that the osmotic potential is raised by solute transport into these cells, but in the petioles considered by Canny and Bucci, the same effect may occur if starch is hydrolyzed to sugar in the starch sheath of petioles. There may be an important role for the phloem in providing a pathway for the transport of solutes such as mineral ions or carbohydrates that are required for changes in turgor pressure, and possibly as one of the tissue compartments directly involved in the refilling mechanism. The various explanations differ in the scale and range of tissues and cells over which changes in turgor and tissue pressure result in redistribution of water, but all recognize that tissues external to the xylem must be important, and at least imply some form of asymmetry in the properties of the cells and cell membranes involved.

## Vascular Anatomy and Refilling

Although a variety of explanations have been put forward to explain embolism repair, there is a common denominator among them. Phloem activity and living cells in contact with the xylem are a prerequisite for refilling under tension. Phloem may be involved directly, by releasing solutes and water close to or into embolized vessels or indirectly by providing a source of sugars and other solutes to other tissues, driving solute transport and osmotic activity. Salleo *et al.* (1996) found that refilling of *L. nobilis* was enhanced by KCl and IAA solutions applied to the exposed cortex before air injection, and proposed that IAA promoted refilling by stimulating phloem loading and the release of solutes into the apoplast. Stems girdled proximally to the site of measurement did not refill, suggesting that a messenger transported in the phloem stimulated refilling. Zwieniecki and

Holbrook (1998) showed that girdling reduced overnight refilling in ash, but had little effect on maple and red spruce. They hypothesized that refilling was an energy demanding process that required phloem activity and an adequate supply of carbohydrates. Later, $HgCl_2$ and girdling were also found to inhibit afternoon recovery in two other species, further suggesting the involvement of living cells in the refilling process (Zwieniecki *et al.*, 2000). In addition to supplying carbohydrates and solutes, the phloem may also be a source of water for refilling. Milburn (1996) pointed out that the release of water from unloading sieve tubes in the phloem (Münch water) provides a continuous source of water, possibly at elevated pressure, that may have a role in the refilling or prevention of emboli. The flow of Münch water varies diurnally and is maximal at midday, the period when refilling is predicted to be most active (Milburn, 1996).

As well as the often-close association between the phloem and xylem, almost all plants contain living cells within the xylem (Holbrook, 1995). While ray and xylem parenchyma have long been recognized as important sites for the storage of carbohydrates and other reserves (Carlquist, 2001), their close association with the xylem conduits also suggests a role for these cells in axial water transport and the phenomenon of refilling under tension. Depending on the actual mechanism driving refilling, the xylem parenchyma may be the site of synthesis or release of solutes into the refilling conduit, or the point from which water is released into the conduit by diffusion or pressure driven hydraulic flow (Fig. 18.2). Xylem parenchyma typically show high levels of expression of aquaporin (water channel) proteins, a feature of tissues involved in bulk water and solute flux (Yamada *et al.*, 1995; Kirch *et al.*, 2000). They also show structural and biochemical features in keeping with the high metabolic activity required for active loading and unloading of solutes, and there is evidence for specialized development of the membrane and cytoplasm and concentration of aquaporins in regions adjacent to pits between the parenchyma cell and vessel (De Boer and Volkov, 2003). Little is so far known about spatial variation in cellular ultrastructure between and within living cells of the xylem, but it is likely that further work will reveal at least some of the polarity in hydraulic conductance and transporter behavior predicted for an active refilling mechanism to operate (Tyerman *et al.*, 2002). It is also interesting that axial parenchyma cells are often rich in starch (Carlquist, 2001), a trait already implicated in the refilling of xylem in leaf petioles (Canny, 1998; Bucci *et al.*, 2003). Further studies are needed to assess whether the starch content of the xylem parenchyma fluctuates diurnally and to determine whether the mechanism suggested for embolism repair in petioles can also be applied to larger stems.

Vascular anatomy varies radially and axially within a plant (see Chapter 15). For the reason that living cells and significant metabolic activity must be required for an active refilling mechanism, embolism repair may be more likely in organs where there is closer contact between the phloem, xylem, and other living cells. Leaves, petioles, and small terminal branches are consequently the best candidate sites for active refilling. Smaller, distal organs are also the sites where xylem tension and vulnerability to cavitation are highest (Zimmermann, 1983). Refilling under tension may be less likely or not occur at all in main stems and larger branches, particularly in those woody species with diffuse porous anatomy and relatively large sapwood cross-sectional areas where a large proportion of functioning xylem vessels are more distant from the phloem.

Xylem conduit diameter should also be an important factor in refilling because of the volume of water and time needed to refill a large diameter vessel (Vesala *et al.*, 2003). Narrower conduits should have a higher pit surface area and xylem parenchyma contact area-to-volume ratio, and may therefore be more likely to refill. Within the primary xylem of leaves, herbs, and monocotyledons, protoxylem vessels are usually narrower than metaxylem vessels (Raven *et al.*, 1986), and should refill faster. Canny (2001b) concluded that refilling was more active in the lacunae and protoxylem vessels of lateral and finer veins of maize leaves because of their smaller volume, more permeable walls, and closer association with parenchyma cells. Leaves may also be better material than woody stems for studying an active refilling mechanism because metabolic processes and changes in membrane permeability should be easier to monitor (Lo Gullo *et al.*, 2003).

It is also interesting to speculate on the significance of differences in xylem anatomy between the various taxonomic groups for refilling. Spring refilling of winter embolism in conifers has been documented several times (Sperry, 1993; Sperry *et al.*, 1994; Mayr *et al.*, 2002), and in some cases refilling may have occurred below the $-2T/r$ threshold (Borghetti *et al.*, 1991). Conifers do not exhibit root pressure and differ in xylem and conduit anatomy from angiosperms. Refilling may occur in conifers, but by a different mechanism than those proposed for angiosperms. Similarly, the anatomy of monocotyledons suggests that there may be closer contact between individual xylem conduits and living parenchyma and phloem cells than in the dicotyledons. The lack of secondary growth in most monocotyledons means that an active refilling mechanism may be more important for maintaining the integrity of the xylem. There is already extensive, cryo-SEM evidence for refilling in maize (McCully *et al.*, 1998; Shane and McCully, 1999; McCully, 1999; Canny, 2001b), but when using MRI, no evidence was found for refilling in the stems of a woody monocotyledonous liana (Clearwater and Clark, 2003). However, newer results suggest that refilling under tension does occur in rice leaves and that it is an active (energy requiring) process (V. Stiller and J. Sperry, unpublished manuscript).

## Conclusions: How Important Is Embolism Repair?

Evidence is fast accumulating that some form of refilling mechanism does operate in the xylem of some species. If it is a common phenomenon, refilling under tension has important implications for our understanding of long distance transport in plants. The involvement of living cells suggests that some plants can exert a degree of active control over the hydraulic properties of their xylem. Instead of being an irreversible dysfunction, variation in conductance caused by cavitation may be part of important regulatory mechanisms. The hydraulic conductance of the xylem in species exhibiting refilling may be a dynamic balance between embolism formation and repair that varies throughout the day in response to changes in xylem tension and the vigor of the refilling mechanism (Bucci *et al.*, 2003). In many species stomatal conductance is closely coordinated with the hydraulic conductance of the soil to leaf pathway (Meinzer *et al.*, 2001; Hubbard *et al.*, 2001). Transpiration is usually regulated in a way that allows maximum xylem tension to approach the threshold at which significant cavitation occurs (Jones and Sutherland, 1991; Bond and Kavanagh, 1999). This fine balance between normal xylem tension and the loss of conductivity suggests a functional role for cavitation as part of a feedback mechanism linking stomatal regulation to hydraulic conductance and plant water status (Bond and Kavanagh, 1999; Salleo *et al.*, 2000). If cavitation is reversible, it is possible that a daily cycle of embolism followed by refilling in distal stems and leaves contributes to the regulation of leaf gas exchange (Lo Gullo *et al.*, 2003). In evolutionary terms, an active refilling mechanism is likely to be an energetically expensive and complex process that must confer some form of adaptive benefit. A sensitive signaling mechanism linked directly to the onset of hydraulic failure in the xylem would integrate aspects of both the supply and demand for water at the leaf surface, allowing the plant to maintain maximum rates of photosynthesis and respond quickly to short-term changes in evaporative demand.

Hypotheses to explain refilling are almost as abundant as the actual reports of refilling. Four hypotheses have been described in this chapter. Each requires more detailed experimental work before we can begin to eliminate some of these competing ideas. The hypotheses vary from the simple release of osmolytes into the vessel lumen to detailed models involving multiple tissues or compartments and complex membrane properties. Most also require a mechanism that isolates the refilling conduit from negative pressures in the surrounding xylem. Controversy has arisen because we lack repeatable and nondestructive methods for measuring xylem contents. We also need to develop a consistent model system or species for the study of embolism repair. Some of the confusion to date over the existence of a refilling mechanism may well be related to differences between species and variation within individual plants (Hacke and Sperry, 2003). More information is needed to document the prevalence of refilling, including comparisons between herbaceous plants, lianas and trees, hardwoods and conifers, and tree taxa of differing xylem anatomy and ecology (Meinzer *et al.*, 2001). For example, does refilling under tension occur in conifers, or is it restricted to angiosperms? Is refilling more important in monocotyledons? Is refilling more active in the leaves of C4 than C3 plants because of a higher water content and reduced proportion of intercellular air space (Canny, 2001b)? As more information is gathered, an exciting picture may emerge of xylem water transport as a more dynamic process than is currently recognized, one that is closely integrated with other vital processes within the plant.

## Acknowledgments

We thank John Sperry, Rick Meinzer, Tim Brodribb, Maciej Zwieniecki, and Missy Holbrook for helpful comments on earlier versions of the manuscript; and Martin Canny for helpful discussion and translations of some of the older literature.

## References

Berndt, M. L., McCully, M. E. and Canny, M. J. (1999) Is xylem embolism and refilling involved in the rapid wilting and recovery of plants following root cooling and rewarming? A cryo-microscope investigation. *Plant Biol* **1:** 506-515.

Bond, B. J. and Kavanagh, K. L. (1999) Stomatal behavior of four woody species in relation to leaf-specific hydraulic conductance and threshold water potential. *Tree Physiol* **19:** 503-510.

Borghetti, M., Edwards, W. R. N., Grace, J., Jarvis, P. G. and Raschi, A. (1991) The refilling of embolized xylem in *Pinus sylvestris* L. *Plant Cell Environ* **14:** 357-369.

Brough, D. W., Jones, H. G. and Grace, J. (1986) Diurnal changes in water content of the stems of apple trees, as influenced by irrigation. *Plant Cell Environ* **9:** 1-7.

Bucci S. J., Scholz F. G., Goldstein G. and Meinzer F. C. (2003) Dynamic changes in hydraulic conductivity in petioles of two savanna tree species: Factors and mechanisms contributing to the refilling of embolized vessels. *Plant Cell Environ* **26:** 1633-1645.

Buchard, C., McCully, M. E. and Canny, M. (1999) Daily embolism and refilling of root xylem vessels in three dicotyledonous crop plants. *Agronomie* **19:** 97-106.

Canny, M. J. (1995) A new theory for the ascent of sap-cohesion supported by tissue pressure. *Ann Bot* **75:** 343-357.

Canny, M. J. (1997) Vessel contents during transpiration: Embolisms and refilling. *Am J Bot* **84:** 1223-1230.

Canny, M. J. (1998) Applications of the compensating pressure theory of water transport. *Am J Bot* **85:** 897-909.

Canny, M. J. (1999) The forgotten component of plant water potential. *Plant Biol* **1:** 595-597.

Canny, M. J. (2001a) Contributions to the debate on water transport. *Am J Bot* **88:** 43-46.

Canny, M. J. (2001b) Embolisms and refilling in the maize leaf lamina, and the role of the protoxylem lacuna. *Am J Bot* **88:** 47-51.

Canny, M. J., Huang C. X. and McCully M. E. (2001a) The cohesion theory debate continues. *Trends Plant Sci* **6:** 454-455.

Canny, M. J., McCully M. E. and Huang C. X. (2001b) Cryo-scanning electron microscopy observations of vessel content during transpiration in walnut petioles. Facts or artefacts? *Plant Physiol Biochem* **39:** 555-563.

Carlquist, S. J. (2001). *Comparative Wood Anatomy: Systematic, Ecological and Evolutionary Aspects of Dicotyledon Wood,* 2nd Ed. Springer-Verlag, New York.

Choat, B., Ball, M., Luly, J. and Holtum, J. (2003) Pit membrane porosity and water stress-induced cavitation in four co-existing dry rainforest tree species. *Plant Physiol* **131:** 41-48.

Clearwater, M. J. and Clark, C. J. (2003) In vivo magnetic resonance imaging of xylem vessel contents in woody lianas. *Plant Cell Environ* **26:** 1205-1214.

Cochard, H., Ameglio, T. and Cruiziat, P. (2001) Vessel content debate revisited. *Trends Plant Sci* **6:** 13.

Cochard, H., Bodet, C., Ameglio, T. and Cruiziat, P. (2000) Cryo-scanning electron microscopy observations of vessel content during transpiration in walnut petioles. Facts or artifacts? *Plant Physiol* **124:** 1191-1202.

Comstock, J. P. (1999) Why Canny's theory doesn't hold water. *Am J Bot* **86:** 1077-1081.

Crafts, A. S. (1939) Solute transport in plants. *Science* **90:** 337-338.

Curran, P. F. and MacIntosh, J. R. (1962) A model system for biological water transport. *Nature* **193:** 347-348.

De Boer, A. H. and Volkov, V. (2003) Logistics of water and salt transport through the plant: Structure and functioning of the xylem. *Plant Cell Environ* **26:** 87-101.

Edwards, W. R. N., Jarvis, P. G., Grace, J. and Moncrieff, J. B. (1994) Reversing cavitation in tracheids of *Pinus sylvestris* L. under negative water potentials. *Plant Cell Environ* **17:** 389-397.

Enns, L. C., McCully, M. E. and Canny, M. J. (1998) Solute concentrations in xylem sap along vessels of maize primary roots at high root pressure. *J Exp Bot* **49:** 1539-1544.

Facette, M. R., McCully, M. E., Shane, M. W. and Canny, M. J. (2001) Measurements of the time to refill embolized vessels. *Plant Physiol Biochem* **39:** 59-66.

Fondy, B. R. and Geiger, D. R. (1985) Diurnal changes in allocation of newly fixed carbon in exporting sugar beet leaves. *Plant Physiol* **78:** 753-757.

Gadian, D. G. (1995) *NMR and Its Application to Living Systems,* 2nd Ed. Oxford University Press, Oxford.

Grace, J. (1993) Refilling of embolized xylem. In *Water Transport in Plants Under Climate Stress* (M. Borghetti, J. Grace and A. Raschi, eds.) pp. 51-62. Cambridge University Press, Cambridge.

Greenidge, K. N. H. (1957) Ascent of sap. *Annu Rev Plant Physiol* **8:** 237-256.

Haberlandt, G. (1914) *Physiological Plant Anatomy.* Translated by M. Drummond. MacMillan, London.

Hacke, U. G. and Sperry, J. S. (2003) Limits to xylem refilling under negative pressure in *Laurus nobilis* and *Acer negundo. Plant Cell Environ* **26:** 303-311.

Haines, F. M. (1935) Observations on the occurrence of air in conducting tracts. *Ann Bot* **49:** 367-379.

Holbrook, N. M. (1995) Stem water storage. In *Plant Stems: Physiological and Functional Morphology* (B. L. Gartner, ed.) pp. 151-174. Academic Press, San Diego.

Holbrook, N. M., Ahrens, E. T., Burns, M. J. and Zwieniecki, M. A. (2001) In vivo observation of cavitation and embolism repair using magnetic resonance imaging. *Plant Physiol* **126:** 27-31.

Holbrook, N. M. and Zwieniecki, M. A. (1999) Embolism repair and xylem tension: Do we need a miracle? *Plant Physiol* **120**: 7-10.

Höll, W. (1975) Radial transport in rays. In *Transport in Plants I. Phloem Transport. Encyclopedia of Plant Physiology, New Series*, Vol. 1 (M. H. Zimmermann and J. A. Milburn, eds.) pp. 432-450. Springer-Verlag, Berlin.

Hubbard, R. M., Ryan, M. G., Stiller, V. and Sperry, J. S. (2001) Stomatal conductance and photosynthesis vary linearly with plant hydraulic conductance in ponderosa pine. *Plant Cell Environ* **24**: 113-121.

Ishida, N., Koizumi, M. and Kano, H. (2000) The NMR microscope: A unique and promising tool for plant science. *Ann Bot* **86**: 259-278.

Johnson, R. W. and Tyree, M. T. (1992) Effect of stem water-content on sap flow from dormant maple and butternut stems: Induction of sap flow in butternut. *Plant Physiol* **100**: 853-858.

Jones, H. G. and Sutherland, R. A. (1991) Stomatal control of xylem embolism. *Plant Cell Environ* **14**: 607-612.

Kirch, H. H., Vera-Estrella, R., Golldack, D., Quigley, F., Michalowski, C. B., Barkla, B. J. and Bohnert, H. J. (2000) Expression of water channel proteins in *Mesembryanthemum crystallinum. Plant Physiol* **123**: 111-124.

Kockenberger, W., Pope, J. M., Xia, Y., Jeffrey, K. R., Komor, E. and Callaghan, P. T. (1997) A non-invasive measurement of phloem and xylem water flow in castor bean seedlings by nuclear magnetic resonance microimaging. *Planta* **201**: 53-63.

Kramer, P. J. and Boyer, J. S. (1995) *Water Relations of Plants and Soils*. Academic Press, San Diego.

Lewis, A. M., Harnden, V. D. and Tyree, M. T. (1994) Collapse of water-stress emboli in the tracheids of *Thuja occidentalis* L. *Plant Physiol* **106**: 1639-1646.

Linton, M. J. and Nobel, P. S. (2001) Hydraulic conductivity, xylem cavitation, and water potential for succulent leaves of *Agave deserti* and *Agave tequilana. Int J Plant Sci* **162**: 747-754.

Lo Gullo, M. A., Nardini, A., Trifilo, P. and Salleo, S. (2003) Changes in leaf hydraulics and stomatal conductance following drought stress and irrigation in *Ceratonia siliqua* (Carob tree). *Physiol Plantarum* **117**: 186-194.

Loo, D. D. F., Wright, E. M. and Zeuthen, T. (2002) Water pumps. *J Physiol-London* **542**: 53-60.

Magnani, F. and Borghetti, M. (1995) Interpretation of seasonal-changes of xylem embolism and plant hydraulic resistance in *Fagus sylvatica. Plant Cell Environ* **18**: 689-696.

Mayr, S., Wolfschwenger, M. and Bauer, H. (2002) Winter-drought induced embolism in Norway spruce (*Picea abies*) at the Alpine timberline. *Physiol Plantarum* **115**: 74-80.

McCully, M. E. (1999) Root xylem embolisms and refilling. Relation to water potentials of soil, roots, and leaves, and osmotic potentials of root xylem sap. *Plant Physiol* **119**: 1001-1008.

McCully, M. E., Baker, A. N., Shane, M. W., Huang, C. X., Ling, L. E. C. and Canny, M. J. (2000) The reliability of cryoSEM for the observation and quantification of xylem embolisms and quantitative analysis of xylem sap in situ. *J Microsc* **198**: 24-33.

McCully, M. E., Huang, C. X. and Ling, L. E. C. (1998) Daily embolism and refilling of xylem vessels in the roots of field-grown maize. *New Phytol* **138**: 327-342.

Meinzer, F. C., Clearwater, M. J. and Goldstein, G. (2001) Water transport in trees: Current perspectives, new insights and some controversies. *Environ Exp Bot* **45**: 239-262.

Melcher, P. J., Goldstein, G., Meinzer, F. C., Yount, D. E., Jones, T. J., Holbrook, N. M. and Huang, C. X. (2001) Water relations of coastal and estuarine Rhizophora mangle: Xylem pressure potential and dynamics of embolism formation and repair. *Oecologia* **126**: 182-192.

Milburn, J. A. (1979) *Water Flow in Plants*. Longman, London.

Milburn, J. A. (1996) Sap ascent in vascular plants: Challengers to the cohesion theory ignore the significance of immature xylem and the recycling of Munch water. *Ann Bot* **78**: 399-407.

Netting, A. G. (2000). pH, abscisic acid and the integration of metabolism in plants under stressed and non-stressed conditions: Cellular responses to stress and their implication for plant water relations. *J Exp Bot* **51**: 147-158.

Pate, J. S. and Canny, M. J. (1999) Quantification of vessel embolisms by direct observation: A comparison of two methods. *New Phytol* **141**: 33-43.

Peterson, C. A. and Steudle, E. (1993) Lateral hydraulic conductivity of early metaxylem vessels in *Zea-mays* l roots. *Planta* **189**: 288-297.

Pickard, W. F. (2003a). The riddle of root pressure. I. Putting Maxwell's demon to rest. *Functional Plant Biology* **30**: 121-134.

Pickard, W. F. (2003b) The riddle of root pressure. II. Root exudation at extreme osmolalities. *Funct Plant Biol* **30**: 135-141.

Preston, R. D. (1938) The contents of the vessels of *Fraxinus americana* L., with respect to the ascent of sap. *Ann Bot* **2**: 1-21.

Priestley, J. H. (1935) Sap ascent in the tree. *Sci Prog* **117**: 42-56.

Raven, P. H., Evert, R. F. and Eichhorn, S. E. (1986) *Biology of Plants*, 4th Ed. Worth, New York.

Rokitta, M., Peuke, A. D., Zimmermann, U. and Haase, A. (1999) Dynamic studies of phloem and xylem flow in fully differentiated plants by fast nuclear-magnetic-resonance microimaging. *Protoplasma* **209**: 126-131.

Salleo, S. and Lo Gullo, M. A. (1989) Xylem cavitation in nodes and internodes of *Vitis vinifera* L. plants subjected to water stress: Limits of restoration of water conduction in cavitated xylem conduits. In *Structural and Functional Responses to Environmental Stresses: Water Shortage* (K. H. Kreeb, H. Richter and T. M. Hinckley, eds.) pp. 33-42. SPB Academic Publishing, The Hague.

Salleo, S. and Lo Gullo, M. A. (1993) Drought resistance strategies and vulnerability to cavitation of some Mediterranean sclerophyllous trees. In *Water Transport in Plants Under Climate Stress* (M. Borghetti, J. Grace and A. Raschi, eds.) pp. 99-113. Cambridge University Press, Cambridge.

Salleo, S., Lo Gullo, M. A., DePaoli, D. and Zippo, M. (1996) Xylem recovery from cavitation-induced embolism in young plants of *Laurus nobilis*: A possible mechanism. *New Phytol* **132**: 47-56.

Salleo, S., Nardini, A., Pitt, F. and Lo Gullo, M. A. (2000) Xylem cavitation and hydraulic control of stomatal conductance in Laurel (*Laurus nobilis* L.). *Plant Cell Environ* **23**: 71-79.

Scheenen, T., Heemskerk, A., de Jager, A., Vergeldt, F. and Van As, H. (2002) Functional imaging of plants: A nuclear magnetic resonance study of a cucumber plant. *Biophys J* **82**: 481-492.

Shane, M. W. and McCully, M. E. (1999) Root xylem embolisms: Implications for water flow to the shoot in single-rooted maize plants. *Aust J Plant Physiol* **26**: 107-114.

Sobrado, M. A., Grace, J. and Jarvis, P. G. (1992) The limits to xylem embolism recovery in *Pinus sylvestris* L. *J Exp Bot* **43**: 831-836.

Sperry, J. S. (1993) Winter xylem embolism and spring recovery in *Betula cordifolia, Fagus grandifolia, Abies balsamea* and *Picea rubens*. In *Water Transport in Plants Under Climate Stress*. (M. Borghetti, J. Grace and A. Raschi, eds.) pp. 86-98. Cambridge University Press, Cambridge.

Sperry, J. S. (1995) Limitations on stem water transport and their consequences. In *Plant Stems: Physiological and Functional Morphology* (B. L. Gartner, ed.) pp. 105-124. Academic Press, San Diego.

Sperry, J. S., Nichols, K. L., Sullivan, J. E. M. and Eastlack, S. E. (1994) Xylem embolism in ring-porous, diffuse-porous, and coniferous trees of northern Utah and interior Alaska. *Ecology* **75**: 1736-1752.

Steudle, E. (2001) The cohesion-tension mechanism and the acquisition of water by plant roots. *Annu Rev Plant Physiol Plant Mol Biol* **52**: 847-875.

Strasburger, E. (1891) *Über den bau und die Verrichtungen der Lietungsbahnen in den Pflanzen.* Gustav Fischer, Jena, Germany.

Tyerman, S. D., Niemietz, C. M. and Bramley, H. (2002) Plant aquaporins: Multifunctional water and solute channels with expanding roles. *Plant Cell Environ* **25**: 173-194.

Tyree, M. T. (1997) The cohesion-tension theory of sap ascent: Current controversies. *J Exp Bot* **48**: 1753-1765.

Tyree, M. T. (1999) The forgotten component of plant water potential: A reply—tissue pressures are not additive in the way M. J. Canny suggests. *Plant Biol* **1**: 598-601.

Tyree, M. T. and Ewers, F. W. (1991) The hydraulic architecture of trees and other woody-plants. *New Phytol* **119**: 345-360.

Tyree, M. T., Salleo, S., Nardini, A., Lo Gullo, M. A. and Mosca, R. (1999) Refilling of embolized vessels in young stems of laurel. Do we need a new paradigm? *Plant Physiol* **120**: 11-21.

Tyree, M. T. and Yang, S. D. (1992) Hydraulic conductivity recovery versus water-pressure in xylem of *Acer saccharum*. *Plant Physiol* **100**: 669-676.

Vesala, T., Holtta, T., Peramaki, M. and Nikinmaa, E. (2003) Refilling of a hydraulically isolated embolized xylem vessel: Model calculations. *Ann Bot* **91**: 419-428.

Vesque, J. (1883) Observations directe du mouvement de l'eau dans les vaisseaux. *Annales des Sciences Naturelles VI Botanique* **15**: 5-15.

Wagner, H. J., Schneider, H., Mimietz, S., Wistuba, N., Rokitta, M., Krohne, G., Haase, A. and Zimmermann, U. (2000) Xylem conduits of a resurrection plant contain a unique lipid lining and refill following a distinct pattern after desiccation. *New Phytol* **148**: 239-255.

Waring, R. H., Whitehead, D. and Jarvis, P. G. (1979) The contribution of stored water to transpiration in Scots pine. *Plant Cell Environ* **2**: 309-318.

Wistuba, N., Reich, R., Wagner, H. J., Zhu, J. J., Schneider, H., Bentrup, F. W., Haase, A. and Zimmermann, U. (2000) Xylem flow and its driving forces in a tropical liana: Concomitant flow-sensitive NMR imaging and pressure probe measurements. *Plant Biol* **2**: 579-582.

Yamada, S., Katsuhara, M., Kelly, W. B., Michalowski, C. B. and Bohnert, H. J. (1995) A family of transcripts encoding water channel proteins: Tissue-specific expression in the common ice plant. *Plant Cell* **7**: 1129-1142.

Yang, S. and Tyree, M. T. (1992) A theoretical-model of hydraulic conductivity recovery from embolism with comparison to experimental-data on *Acer saccharum*. *Plant Cell Environ* **15**: 633-643.

Zimmermann, M. H. (1983) *Xylem Structure and the Ascent of Sap*. Springer-Verlag, New York.

Zimmermann, U., Schneider, H., Wegner, L. H., Wagner, H. J., Szimtenings, M., Haase, A. and Bentrup, F. W. (2002) What are the driving forces for water lifting in the xylem conduit? *Physiol Plantarum* **114**: 327-335.

Zimmermann, U. and Steudle, E. (1978) Physical aspects of water relations of plant cells. *Adv Bot Res* **6**: 45-117.

Zimmermann, U., Zhu, J. J., Meinzer F. C., Goldstein, G., Schneider, H., Zimmermann, G., Benkert, R., Thurmer, F., Melcher, P., Webb, D. and Haase, A. (1994) High-molecular-weight organic-compounds in the xylem sap of mangroves: Implications for long-distance water transport. *Botan Acta* **107**: 218-229.

Zwieniecki, M. A. and Holbrook, N. M. (1998) Diurnal variation in xylem hydraulic conductivity in white ash (*Fraxinus americana* L.), red maple (*Acer rubrum* L.) and red spruce (*Picea rubens* Sarg.). *Plant Cell Environ* **21**: 1173-1180.

Zwieniecki, M. A. and Holbrook, N. M. (2000) Bordered pit structure and vessel wall surface properties: Implications for embolism repair. *Plant Physiol* **123**: 1015-1020.

Zwieniecki, M. A., Hutyra, L., Thompson, M. V. and Holbrook, N. M. (2000) Dynamic changes in petiole specific conductivity in red maple (*Acer rubrum* L.), tulip tree (*Liriodendron tulipifera* L.) and northern fox grape (*Vitis labrusca* L.). *Plant Cell Environ* **23**: 407-414.

Zwieniecki, M. A., Melcher, P. J. and Holbrook, N. M. (2001) Hydraulic properties of individual xylem vessels of *Fraxinus americana*. *J Exp Bot* **52**: 257-264.

# 19

## Impacts of Freezing on Long Distance Transport in Woody Plants

*Jeannine Cavender-Bares*

Freezing temperatures can cause lethal injuries in living plant tissues and are a major factor limiting the long distance transport of water in the xylem and phloem. The ability of different species to avoid or tolerate freezing stress through various mechanisms can go a long way in explaining species-geographic distributions (Parker, 1963; George *et al.*, 1974; Burke *et al.*, 1976). Plant species that live in freezing climates have three options for survival: (1) die, leaving a protected seed bank behind, (2) remain active, or (3) become dormant. While the first strategy is used by annual herbaceous species, long-lived woody species that leave their above-ground biomass in place require mechanisms that allow persistence during winter.

Plants that remain active during winter are subject to freezing of living and nonliving tissues, including the vascular system. In living tissues, freezing can cause intracellular ice formation, which can kill the cells, or extracellular ice formation, which may protect the cells from damage (Kuroda *et al.*, 1997). If extracellular freezing occurs, cellular dehydration becomes the main problem, although mechanical damage to cell membranes may also occur (Fujikawa and Kuroda, 2000). Within the nonliving conduits of the xylem, freezing temperatures are likely to cause cavitation resulting in losses of xylem function, depending on the size of the conduits as well as the freezing temperature and tension in the xylem. Phloem transport is also limited by cold temperatures as sap becomes more viscous and as sink strength and rates of phloem unloading decline, slowing the bulk flow of phloem sap. To remain active in winter, therefore, plants must prevent intracellular freezing in their living tissues, including in the phloem system, and maintain some xylem function, even if the soil is frozen. Woody plants that become dormant during winter discard their most vulnerable living tissues, but still need to prevent intracellular freezing in stems and roots, and must be able to recover xylem and phloem function in the spring.

The potentially lethal stresses caused by freezing and low temperatures and the strategies woody plants use to survive these stresses are the subject of this chapter. In the first part of the chapter, I discuss the primary mechanisms plants use to control freezing dynamics and reduce freezing injury in living tissues, as well as the physiological transformations that take place during cold acclimation. In the second part of the chapter, I focus on the limitations to long distance transport in the xylem and phloem and how the vascular systems of woody plants have adapted.

## Survival of Living Tissues at Low Temperatures

In plants, widespread death of living cells caused by intracellular freezing can be prevented by structural features and chemical properties that lower the freezing point of water and direct where freezing occurs in the plant. The freezing point of water can be lowered by several different factors, including increased osmotic concentration (Fig. 19.1A), the absence of ice nucleators, the presence of hydrophilic surfaces and macromolecules that lower the potential energy of water, and by increased viscosity, which delays ice formation (Wolfe *et al.*, 2002). Water often occurs in finely divided volumes in plants so that the freezing temperature of water can vary substantially across different tissues and even adjacent cells, giving plants some control over freezing dynamics. Resistance to freezing in plants is based either on tolerance to extracellular ice formation, which causes cell dehydration, or on avoidance of freezing, particularly through supercooling.

### Extracellular Freezing and Dehydration Tolerance
In most cold-tolerant woody species, extracellular freezing is an important mechanism for preventing intracellular freezing (Kuroda *et al.*, 1997; Fujikawa and Kuroda, 2000). As temperatures drop, extracellular ice is formed on the surface of the cell wall, in lumens of nonliving fibers or vessels or in the extracellular spaces (Guy, 1990; Ameglio *et al.*, 2001b). The formation of ice crystals lowers the water potential because ice has a lower chemical potential than liquid water corresponding to approximately $-1$ MPa per °C (Fig. 19.1B) (Hansen and Beck, 1988). Water molecules from inside the cell are drawn to the ice crystal surface outside the cell and freeze. As liquid water moves out of the cell, the concentration of dissolved solutes inside the cell increases. The increase in osmotic concentration lowers the entropy of the water molecules and depresses the freezing point by 1.86° C per mole of solute dissolved per kg of water (Fig. 19.1B) (Chang, 1991). The redistribution of water from inside cells to extracellular ice masses causes cells to shrink in volume and has large effects on protoplas-

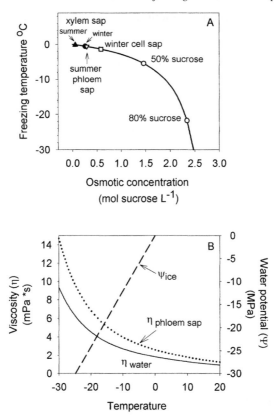

**Figure 19.1** (A) The freezing point depression resulting from increased solute concentration, calculated as a 1.86° C drop in freezing temperature per mol solute per kg water. Molal concentrations were converted to molar concentrations based on the molecular weight of sucrose. Approximate concentrations of summer (*filled triangle*) and winter (*filled circle*) xylem sap in sugar maple (Kozlowski and Pallardy, 1997), and summer phloem sap from castor bean, *Ricinus communis* (*open triangle*) (Taiz and Zeiger, 1998) are shown, as well as winter cell sap from cold-hardy ivy, *Hedera helix*, grown under 5° C days and −1° C nights (Hansen and Beck, 1988). Note that the freezing depression in the xylem is very small. Two sucrose solutions (% = g sucrose $g^{-1}$ solution) are also shown that have osmotic concentrations equivalent to those that might be expected during winter in partially dehydrated living cells, including the phloem, in cold-adapted species as a result of extracellular freezing. (B) Relationship between viscosity and water potential of ice with temperature. Solid curve shows the exponential increase in viscosity with decreasing temperature for pure water (based on Lide, 1993); dotted curve is for a 10% sucrose solution (10 mg $mL^{-1}$) (based on Mathlouthi and Génotelle, 1995) of the same approximate osmotic concentration as measured phloem sap (Taiz and Zeiger, 1998). The linear decline in the water potential of ice is calculated as $\Psi_{ice} = RT/V_w^* \ln(P_{ice}/P_{liquid})$, where $\Psi$ is water potential, R is the universal gas constant, T is temperature, $V_w$ is the molar volume of water, and $P_{ice}$ and $P_{liquid}$ represent water vapor pressure of ice and supercooled water, respectively. This calculation agrees well with measured psychometric values (Hansen and Beck, 1988).

mic concentration and on interactions between the plasmalemma and cell walls (Guy, 2003).

Extracellular freezing protects cells from intracellular freezing but causes dehydration inside the cell proportional to the temperature below freezing (Kubler, 1983). For woody plants occurring in temperate regions, cellular dehydration is thought to be the primary cause of freezing injury (Tranquillini, 1982; Ashworth *et al.*, 1993; Webb and Steponkus, 1993; Fujikawa, 1997). The sensitivity to dehydration of the living cells in wood, particularly the xylem ray parenchyma cells that are more sensitive than other woody tissues, may ultimately determine freezing tolerance of woody species. Whereas cortical or cambial cells in most hardwood species respond to freezing in nature by extracellular freezing, cells of the xylem ray tissues tend to avoid the associated dehydration through supercooling (Fujikawa, 1997).

## Supercooling

Deep supercooling in plant tissues prevents intracellular freezing while limiting the degree of cellular dehydration (Fujikawa *et al.*, 1996). Supercooling refers to the cooling of a liquid below the freezing temperature that is expected based on the solute concentration. It can occur readily in very small volumes of water, where surface properties influence the free energy of water, particularly in the absence of nucleation particles or agents that initiate ice-crystal formation. During supercooling, cells are not subjected to the effects of freezing, and cell functions can be maintained, albeit at a reduced rate. Supercooling is a common phenomenon in woody plants, both in leaves and the living cells of the xylem, including the xylem ray parenchyma cells (Guy, 2003). Limited supercooling to $-10°$ C, referred to as shallow supercooling, has even been demonstrated experimentally in the xylem parenchyma of trees native to nonfrost tropical and subtropical zones, including *Ficus elastica, F. microcarpa, Mangifera indica, Hibiscus rosa-sinensis,* and *Schefflera arboricola,* indicating that it may be an inherent property of the anatomical structure of wood (Kuroda *et al.*, 1997).

Shallow supercooling may occur without extracellular freezing but can quickly lead to rapid intracellular freezing, killing cells, if nucleation occurs. Deep supercooling generally occurs in combination with extracellular freezing and increasing solute concentrations inside cells. Both viscosity and surface properties of membranes and macromolecules influence the supercooling process. As temperatures decrease, the viscosity of liquid water increases exponentially (Fig. 19.1B) (Cho *et al.*, 1999). As solute concentration increases during extracellular freezing, water becomes even more viscous, particularly at low temperatures (Fig. 19.1B) (Mathlouthi and Génotelle, 1995). As a consequence, water molecules are slower to diffuse and to rotate so that ice nuclei are less likely to form. With high viscosity, attraction to hydrophilic surfaces also becomes increasingly important, particularly for water in small volumes, and water is less likely to diffuse from the region near a hydrophilic surface to a region where ice has already started to form. Increasing viscosity and hydrophilic membranes thus promote supercooling and decrease dehydration in cells (Wolfe *et al.*, 2002).

As subzero temperatures decline, the likelihood of ice nucleation increases (even in the absence of nucleating particles) until −40° C, when homogeneous nucleation of water occurs. This causes a breakdown of supercooling leading to intracellular freezing in the living cells although the −40° C limit can be lowered in proportion to the osmotic concentration of the cell sap or raised by the presence of heterogeneous nucleators (e.g., caused by bacteria or various injuries) (Guy, 2003). Some boreal hardwood species, including *Salix sachalinensis*, *Populus sieboldii*, *Betula platyphylla*, and *B. pubescens*, that grow in areas where the minimum air temperature reaches −50° C or below have xylem parenchyma cells that undergo deep supercooling in concert with extracellular freezing (Gusta *et al.*, 1983; Kuroda *et al.*, 2003). In these species, high osmotic concentration in the xylem ray parenchyma cells extends the limit of supercooling below −50° C, and these cells become only partially dehydrated. Other tree species that exist in extremely cold environments where the annual minimal temperature is significantly less than −40° C, including red osier dogwood (*Cornus sericea*) (Kuroda *et al.*, 2003), are generally thought to rely only on extracellular freezing, as this mechanism has the potential to allow survival even at the temperature of liquid nitrogen (−196° C) (Sakai and Larcher, 1987). Species that undergo extracellular freezing at such low temperatures must have mechanisms for dealing with the resulting severe dehydration stress.

## Cold Acclimation

Sakai (1970) observed that under natural conditions, the freezing resistance of trees increases when the daily minimum temperature falls to subzero for a week and that the degree of freezing resistance depends on the air temperature at which the plants are wintering. In a series of experiments with different willow species, he showed that whether species actually tolerate freezing depends on the climate they experience and whether they have been cold acclimated. The acclimation process, or cold hardening, is often triggered in response to a decrease in the photoperiod (sensed by phytochrome) and exposure to cold temperatures. After growth ceases in response to declining photoperiod, some degree of freezing resistance often develops even without exposure to low temperature (Napp-Zinn, 1984). Subsequent changes occur in response to cold temperatures. What actually occurs during cold acclimation of plants? In woody species, this process involves a suite of complex changes in anatomy and functioning of living and nonliving tissues, although not in areas of active growth. Growing shoots are cold sensitive and are not able to increase their freezing resistance even when acclimated to cold temperatures. In nongrowing tissues, the following changes have been observed in species that become cold acclimated:

- alterations in cell wall properties
- alterations in the plasma membrane, including lipid composition
- development of an amorphous layer of xylem parenchyma cells
- increased abscisic acid (ABA) concentration in different tissues and organs
- changes in gene expression and upregulation of dehydrins as well as enzymes involved in sugar metabolism
- accumulation of sugars in living and nonliving tissues

### Cell Wall Structure

Composition and properties of cell walls change during cold acclimation, including increases in cell-wall thickness and tensile strength and decreases in pore size, changes in the amounts of lignin and suberin, and increases in cell-wall-associated proteins. These changes are important, as the walls are at the interface between extracellular ice and the cell protoplasm. Temperature-dependent changes in the freezing behavior of xylem parenchyma cells in hardwood species are thought to be controlled by changes in cell wall properties (Fujikawa and Kuroda, 2000; Fujikawa, 2002).

The porosity of the cell wall is thought to play a role in the flow of cellular water to extracellular ice. During cold acclimation in some woody species, deposition of pectin in the cell wall reduces the size of the microcapillaries in the walls, thus increasing dehydration resistance and preventing the growth of ice crystals into the cell. In the spring, pectin is enzymatically removed again, and supercooling to low temperatures is no longer possible (Wisniewski *et al.*, 1991). The rigidity of the cell wall prevents cell contraction and collapse during freezing that could otherwise occur owing to the growth of extracellular ice crystals and protects against membrane damage (Burke *et al.*, 1976; Rajashekar and Lafta, 1996). Rigid cell walls may also be important in preventing heterogeneous nucleation of ice formation, thus allowing deep supercooling (Wisniewski and Ashworth, 1985; Ashworth *et al.*, 1993). Phenolic cross-linking between cell wall polymers, deposition of lipids on the cell wall, and deposition of extensin on the cell wall during cold acclimation may increase rigidity.

### Plasma Membrane

The plasma membrane has been isolated as the primary site of freezing injury in many cold-sensitive plants (Stepkonkus *et al.*, 1983; Steponkus,

1984). The interaction between the plasma membrane and the cell wall differs between cold acclimated and nonacclimated specimens of red osier dogwood, a highly cold tolerant woody species (Ristic and Ashworth, 1995). In subarctic trees, which tolerate temperatures below −40° C, low temperatures cause the synthesis of membrane lipids with less saturated fatty acids allowing membranes to remain flexible at low temperatures. Along with the cell wall, the plasma membrane is likely to be important as a barrier against the seeding of ice into the cells.

### Amorphous Layer of Xylem Parenchyma

The xylem parenchyma cells that border vessels in angiosperms, called contact cells (see section on xylem refilling), are characterized by having a wall layer deposited between the plasma membrane of the parenchyma cell and the adjacent vessel-parenchyma pit membrane, called an amorphous layer or protective layer. This layer is likely to be important in the freezing resistance of plants as it may increase dehydration resistance. Seasonal alterations in the structure of this layer may affect its permeability and thus regulate the response of xylem parenchyma to freezing temperatures (Schaffer and Wisniewski, 1989).

### Increased Abscisic Acid Levels and Altered Gene Expression

ABA is likely to play an important role in cold acclimation processes in many kinds of plants. ABA has a direct role in the response to cell desiccation that occurs during freezing, and it is also likely to be involved in the control of gene expression during cold acclimation. It has been hypothesized that low temperatures induce increased synthesis of ABA in plants, which in turn triggers the expression of genes involved in freezing tolerance. In a number of plants, including winter wheat, bromegrass, and *Arabidopsis*, exogenous application of ABA induces cold-acclimation (e.g., Churchill *et al.*, 1998). In sugar maple, the ABA concentration of xylem sap was found to increase 10-fold in late autumn, reaching a maximum before maximum cold hardiness in buds and roots (Bertrand *et al.*, 1997). The rise in ABA could be induced by freezing dehydration (Hartung and Davies, 1991). The increased apoplastic ABA concentration could then influence gene expression during cold acclimation (Guy, 1990).

During the development of freezing tolerance in cold-hardened buds and roots of different species, upregulation of proteins in the dehydrin family have been observed after increases in ABA levels (Wisniewski *et al.*, 1996; Bertrand *et al.*, 1997; Sarnighausen *et al.*, 2002). Dehydrins are hydrophilic and heat stable and are thought to play a role in membrane stabilization and in the stabilization of the linkages between the cell wall and the plasma membrane during cold stress (Wisniewski *et al.*, 1999). Some of these proteins have been classified as antifreeze proteins (Wisniewski *et al.*, 1999), which are known to aid in low temperature resistance of animals (e.g., Ewart *et al.*, 1999). In plants, they may prevent or restrict the growth of ice crystals and control the sites of ice formation (Griffith *et al.*, 1997), although they are unlikely to actually prevent plants from freezing, owing to their minimal impact on freezing depression (0.3° C) (Pearce, 2001).

### Accumulation of Sugars

Proteins involved in starch-to-sugar conversion and in sucrose synthesis and degradation that control the sugar cycle in trees also appear to be upregulated during cold acclimation (Schrader and Sauter, 2002). Sugar accumulation in living and nonliving tissues during winter in cold-hardy trees has long been observed and is thought to be important in cold acclimation because it increases viscosity, which reduces ice crystal formation, helps stabilize membranes by binding to the free phosphate groups of membrane lipids, maintains respiration in living cells, and allows cell metabolism to recover after freezing (Schrader and Sauter, 2002; Guy, 2003; Wong *et al.*. 2003). It also lowers the freezing temperature of sap, although this has little effect until osmotic concentrations are quite high (Fig. 19.1A), as can occur in living cells during extracellular freezing. Upregulation of sucrose-phosphate synthase and sucrose synthase (which actually degrades sucrose), observed in *Populus,* may allow tight coupling of the regulation of sucrose biosynthesis and breakdown with temperature during cold acclimation in trees (Schrader and Sauter, 2002).

In sugar maple, starch is accumulated in the xylem ray tissues in late summer and early fall. As temperatures drop during the cold season, starch is hydrolyzed and soluble sugars accumulate. A sharp decline in the level of starch in the wood tissues corresponds to the accumulation of a large pool of sugars, including sucrose, fructose, glucose, stachyose, raffinose, and xylose. At the end of dormancy, the levels of soluble sugars decline and starch levels increase. Sugar concentrations increase again before leaf out, as carbon is mobilized for primary growth activities, including flowering and shoot and root growth (Wong *et al.*, 2003). This coincides with the timing of vessel repair and replacement (discussed later).

## Impacts of Freezing on Water-Conducting Conduits of the Xylem

### Xylem Tension and Conduit Size

In woody plants, freezing occurs in the apoplastic space of xylem conduits at much higher temperatures than inside the living cells of wood (Fujikawa *et al.*, 1996). In nonhardened plants, freezing exotherms of stems occur between $-1°$ C and $-5°$ C, presumably indicating freezing of xylem sap. Because air is 1000 times less soluble in ice than in water, when sap freezes in xylem vessels and tracheids, air is released from solution and forms bubbles that become trapped in the ice (Scholander *et al.*, 1953). Upon thawing, the bubbles may dissolve back into the sap or expand causing conduits to embolize. Whether embolism occurs is thought to depend on the balance between the pressures exerted on the air-water meniscus.

A number of studies indicate that across diverse taxa, larger xylem conduits are more vulnerable to embolism by freezing than smaller conduits (Ewers, 1985; Sperry and Sullivan, 1992; Lo Gullo and Salleo, 1993; Davis *et al.*, 1999; Feild and Brodribb, 2001; Pittermann and Sperry, 2003). The main explanation is that larger bubbles form in larger xylem conduits. Larger bubbles are more difficult to dissolve upon thawing and dissolution time increases approximately with the square of the initial bubble diameter (Ewers, 1985; Yang and Tyree, 1992). The tension during thawing and the timing of the onset of tension upon thawing influences whether dissolution or expansion occurs (Tyree and Zimmermann, 2002). If the xylem pressure potential ($P_x$) is more negative than the surface tension forces acting to compress the bubble, then the bubble expands. Surface tension forces are given by Laplace's law, and are equal to $2T/r$, where $T$ is the surface tension of water (0.0728 Pa m at $20°$ C) and $r$ is the bubble radius. If these forces overcome the tension in the xylem, the bubble contracts and dissolves. This can be seen from the following equation:

$$P_b = 2T/r + P_x \qquad (P_b \geq \text{pure vacuum})$$

(Yang and Tyree, 1992), where $P_b$ is the internal pressure of the bubble. The larger the bubble and the lower the xylem pressure, the lower the bubble pressure, increasing the likelihood of expansion. In contrast, the higher the xylem pressure and the smaller the bubble, the more likely the bubble will dissolve. Assumptions and caveats of these theoretical expectations have been discussed in detail elsewhere (Sperry and Robson, 2001).

When data from similar diameter stems from various authors, representing a wide array of species, are plotted together, this broad relationship between vessel diameter and vulnerability to freezing is apparent, indicating a continuum of freezing sensitivity ranging from the southern hemisphere conifer, *Diselma archeri,* with a mean vessel diameter of 11.4 µm and no loss of conductivity after freezing, to the ring-porous deciduous oak, *Quercus gambelii,* with a mean conduit diameter of 71 µm (hydraulically weighted diameter of 99 µm) and almost complete loss of conductivity after freezing (Fig. 19.2). The residual variation in the dependence of freezing-induced embolism on mean conduit diameter (Fig. 19.2A and B) is high (either for a linear or a sigmoidal function), however, possibly as a result of differences in freezing and thawing conditions, which could change dynamics of bubble formation. Alternatively, the loss of conductivity after freezing may be more directly related to the distribution of vessel diameters in a stem or to the hydraulically weighted vessel diameter (Fig. 19.2B). If most of the stem conductance is through a small number of large diameter vessels that have a high probability of embolizing after a freeze-thaw event (ring-porous species), these species show higher losses of conductivity after freezing even if their average vessel diameter is relatively low, as flow rate in a conduit is estimated to be proportional to the 4th power of the radius according to the Hagen-Poiseuille relationship. The correlation between conduit diameter and loss of conductivity also depends on having a large range of mean conduit diameters. Among 17 species of co-occurring Florida oaks, which share important similarities in wood and anatomical properties and for which diameter variation falls within a relatively restricted range, loss of conductivity after freezing is not dependent on vessel diameter (Cavender-Bares and Holbrook, 2001). Nevertheless, these values fall within the expected range of the larger sample of woody species (Fig. 19.2A and B).

Other aspects of xylem anatomy may also be important. Conduit lengths may be important depending on the physics of bubble formation (i.e., whether many bubbles form throughout the conduit that can coalesce

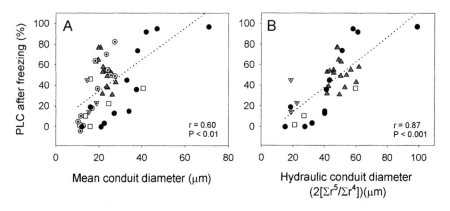

●   Davis et al., 1999, -12°C to -15°C (freeze/thaw: 0.25/0.45°C min⁻¹), -0.5 MPa (centrifugation)
▲   Cavender-Bares and Holbrook, 2001, -15°C (freeze/thaw: 0.4/0.6°C min⁻¹), -0.02MPa (suction)
◉   Feild and Brodribb, 2001, -10°C (freeze/thaw est.: 1.0/0.6°C min⁻¹), -0.5 MPa (dehydration)
▽   Cordero and Nilsen, 2002, -10°C (freeze/thaw 0.1/0.1°C min⁻¹), -0.02 MPa (suction)
□   Pitterman and Sperry, 2003, -15°C (freeze/thaw 0.17/0.2°C min⁻¹) -0.5 MPa (centrifugation)

**Figure 19.2**   Relationship of percent loss of hydraulic conductivity in woody plant stems to mean conduit diameters from multiple authors (A) and hydraulically weighted conduit diameter, calculated as $2^*$ $(\Sigma r^5/\Sigma r^4)$, where r is conduit radius, for available data from the same authors (B). Freezing temperature, xylem tension, and freeze/thaw rates are indicated with authors for each symbol. Linear correlation coefficients are given in each plot. A sigmoidal function would be preferred over a linear fit if there is a conduit diameter corresponding to a cavitation threshold, although nonlinear functions fit poorly to the combined data.

to form large bubbles or whether they form in longitudinal files that do not coalesce) (Pittermann and Sperry, 2003). In Norway spruce, Mayr *et al.* (2003b) found that leader shoots have significantly lower freezing-induced embolism, despite larger diameter tracheids, than twigs. They attributed the lack of freezing embolism in leader shoots to smaller pit dimensions and lack of compression wood.

### Temperature Dependence of Xylem Cavitation

In addition to conduit diameter, xylem sensitivity to freezing appears to be dependent on the minimum temperature experienced (Lo Gullo and Salleo, 1993; Kuroda *et al.*, 1997; Pockman and Sperry, 1997; Pittermann and Sperry, 2003; Ball, personal communication; J. Cavender-Bares, P. Cortes, R. Joffre, S. Rambal, unpublished), as well as on the rate of freezing (Kikuta and Richter, 2003) and thawing (Sperry, 1995; Cordero and Nilsen, 2002) (see Chapter 20), and the number of freeze-thaw cycles the plant has experienced previously (Lemoine *et al.*, 1999; Mayr *et al.*, 2003a). Various authors have shown that loss of conductivity to freezing can vary threefold with a temperature decline from −5° C to −15° C, depending on the species (Fig. 19.3). The redistribution of water as extracellular freezing occurs into other parts of the wood may increase tension in the xylem that would increase the likelihood of bubble expansion on thawing. Kikuta and Richter (2003) showed that ultrasonic acoustic emissions (typically thought

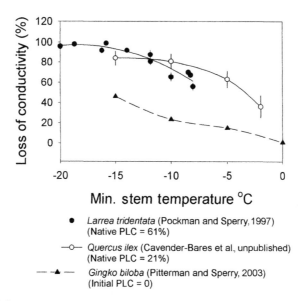

**Figure 19.3** PLC in different species with decreasing minimum freezing temperature.

to correlate with cavitation events) become much more frequent in rachides of non–cold-hardened black walnut below −12° C, well after the prominent xylem exotherm occurred, signaling potentially important changes in xylem that could contribute to reduced function after thawing (Fig. 19.4). One interpretation of these results is that numerous freezing events, not resulting in measurable exotherms, occurred in very small diameter vessels or fibers as a result of the breakdown of supercooling with decreasing temperatures. Alternatively, these signals could have resulted from disruption of living cells of the xylem, such as the xylem ray parenchyma. The number of acoustic emissions was much lower in saturated stems than in unsaturated stems ($\Psi$ = ~ −0.7 MPa), indicating that tension in the xylem increased the number of acoustic emissions.

LoGullo and Salleo (1993) showed that within a given stem, larger vessels were more likely to embolize than smaller vessels as a result of freezing, and the number of embolized vessels increased with decreasing temperature. They identified critical vessel diameters for a given temperature that were susceptible to xylem embolism. They argued that narrow conduits are likely to stay functional at colder temperatures than larger vessels because narrow vessels contain liquid water at lower free energy, and therefore have a lower freezing point and hence a greater capacity for supercooling. LoGullo and Salleo (1993) also found higher recovery of xylem embolism after freezing in smaller vessels, which they speculated to be the result of redissolution of air in sap after high root water absorption relative to transpiration. The per-

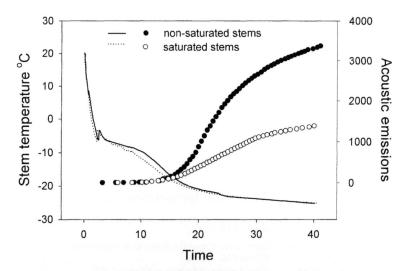

**Figure 19.4** Rise in ultrasonic acoustic emissions with decreasing stem temperatures below the freezing point in stems of *Juglans nigra* for both a saturated and an unsaturated stem. (Replotted from Kikuta and Richter, 2003.)

centage of xylem conduits that recovered was negatively correlated with conduit diameter, presumably because it is more difficult to expel larger bubbles from wider xylem conduits (Lo Gullo and Salleo, 1993).

The interaction of stem diameter (surface area-to-volume ratio), vessel diameter and volume, xylem tension, freezing temperature, and anatomical features that influence the way water is redistributed in the wood during freezing are likely to influence the degree of susceptibility of xylem to freezing under natural conditions. In *Fagus sylvatica*, apical shoots freeze more quickly than older shoots with larger vessels (Lemoine *et al.*, 1999), most likely because of higher surface-to-volume ratios of apical versus second- or third-year shoots. Apical shoots have narrower vessels than those of older shoots, but show higher embolism after freeze-thaw events. Lemoine *et al.* (1999) observed that water exudation as a consequence of volume expansion and redistribution of water during freezing lowered the water content of the smaller shoots. They hypothesized that this induced higher xylem tension on thawing and triggered embolism formation in the apex. Although bubbles were likely to be smaller in these terminal shoots, the capillary pressure they developed was not high enough to compensate for the decrease in xylem pressure. They concluded that embolism formation depends more on dynamics of sap freezing than on xylem characteristics.

### Links Between Xylem Properties, Phenology, and Climate

Xylem vulnerability to freezing stress has been shown to vary considerably among species, and xylem properties appear to be correlated with phenology (Lechowicz, 1984; Wang *et al.*, 1992; Cavender-Bares and Holbrook, 2001a) and climatic distribution (Noshiro and Baas, 2000; Cordero and Nilsen, 2002). Hydraulic properties and patterns of vulnerability to freezing show biogeographical patterns, suggesting that freezing-induced embolism is likely to be an important factor limiting species ranges. Xylem conduit length and diameter are significantly negatively correlated to latitude among species within the genus *Cornus* with longer vessels (and fibers) and larger-diameter vessels occurring in warmer climates (Noshiro and Baas, 2000). Among congeners of *Rhododendron*, species from habitats with a higher frequency of freeze-thaw events have smaller vessel sizes (Cordero and Nilsen, 2002). However, the *Rhododendron* species with the smallest mean (or hydraulically weighted) vessel diameter, *R. catawbiense*, is the most sensitive to freezing. Refilling following freeze-thaw events is apparently the main mechanism for overcoming freezing stress for this plant and is more important than preventing embolism (Cordero and Nilsen, 2002).

### Mechanisms of Maintaining or Recovering Xylem Function

To survive freezing, plants must minimize embolism to allow continued function or resume xylem function following embolism. Conifers tend to minimize xylem dysfunction by having very narrow conduits. After a freeze-thaw event, only a few tracheids with big air bubbles cavitate (Sucoff, 1969). As these bubbles expand, tension is released, which allows bubbles in the surrounding tracheids to dissolve. In distal shoots, embolism may protect the older parts of the branch from freezing damage. Still, to maintain winter activity in very cold climates, some transport must continue even while xylem sap is frozen. Continued water transport in frozen wood may be possible through cell wall capillaries at temperatures below freezing (Sparks *et al.*, 2000). Also, the process of freezing and thawing of stems can physically move water within the xylem (Zimmermann, 1983). During freezing, the increase in volume forces water to move centripetally, either from smaller to larger stems or into the heartwood (Robson and Petty, 1993). In several northern species, including lodgepole pine and western larch, more than 25% of the water in wood was liquid even at −15° C. This unfrozen water is mainly found within the cell wall and may be available for transport (Sparks *et al.*, 2000).

Deciduous broadleaved trees, which are dormant in winter, tend to be more vulnerable to winter cavitation and may contain less liquid water at temperatures below freezing than wood of conifers. Rather than maintaining xylem function during winter, they recover it after the cold season either by the replacement of embolized vessels with new functional vessels or by refilling of embolized vessels. All plants with secondary cambium can replace xylem, but its efficiency during bud break depends on the timing of radial growth resumption (Ameglio, 2002). Some species, but not all, use the strategy of refilling embolized vessels through positive pressures in the xylem. For pressures to develop in the xylem vessels, an osmotic pressure difference must develop between the vessels and the neighboring compartment, the symplast (contact cells and xylem parenchyma), separated by a semipermeable cell membrane (Yang and Tyree, 1992).

Walnut and maple exhibit positive pressures in the xylem sap during winter. These are associated with high xylem sap sugar concentration (Ameglio and Cruiziat, 1992; Ameglio *et al.*, 2001b). Maple is unusual in that its xylem takes up water upon freezing, and xylem tension is reduced or even positive upon thawing, which reduces embolism formation (Sperry *et al.*, 1988; Tyree, 1983). In temperate, ring-porous oaks and in peach, however, positive xylem pressures have never been observed (Ameglio *et al.*, 2002). Freezing injury to roots that prevents development of positive pressure in the xylem after spring thawing can impair the ability of some species to refill vessels and recover from freezing-induced xylem embolism (Zhu *et al.*, 2002).

In many species, including most hardwoods, sugar concentration in the xylem sap is regulated by the xylem axial and ray parenchyma, which are living cells with large pits connecting them to vessels. These cells are the sites of sugar secretion in the xylem sap in spring and are involved in carbohydrate metabolism and the translocation, storage, and mobilization of nutrients (Schaffer and Wisniewski, 1989). Hence, xylem parenchyma are likely to govern the ability of trees to refill embolized xylem vessels after freeze-thaw events. Temperature may determine whether xylem parenchyma secrete or absorb sugars, with secretion dominating at low temperatures. At low, nonfreezing temperatures, the starch that is stored in parenchyma cells in the xylem is converted to sugars, particularly sucrose (Marvin *et al.*, 1967) and is then secreted by contact cells into the xylem sap (Ameglio and Cruiziat, 1992). In many temperate trees, positive xylem pressures are associated with high sugar concentrations in the xylem sap in the early spring, but not in the winter (Essiamah, 1980). Such trees have a period in the early spring when starch is hydrolyzed to sugars and released in the xylem sap.

Defoliation of walnut trees reduced winter osmotic concentrations and xylem pressures (Ameglio *et al.*, 2001b). The role of osmotic concentration in reversing winter embolism may explain why defoliated trees have a reduced ability to survive winter. Any factor, such as insect attack, pathogens, or severe summer drought, that results in a reduction of stored carbohydrates can reduce winter xylem pressures, which would reduce the ability of the plant to reverse embolism (Ameglio *et al.*, 2001b).

Ameglio *et al.* (2001a) found that the osmotic concentration of the xylem sap doubled in response to a freeze thaw event and the symplastic stem water content decreased as a result of sucrose secretion by contact cells. The increase in osmotic concentration was followed by xylem pressure increases as xylem sap drew water from the living cells of the xylem, and possibly from other unfrozen parts of the stem and roots. Surprisingly, these pressures increased almost sixfold over the value predicted by enhanced osmotic concentration alone (Ameglio *et al.*, 2001a). They hypothesized that some of the unexplained additional pressure might also have been associated with the expansion of water during the phase change from liquid to solid. The ice within vessels could have exerted pressure on the surrounding fluid until the ice was completely melted. Alternatively, changes in gas pressures with temperature in the xylem fibers could have caused water from the fibers to be expulsed.

## Impacts of Cold Temperatures and Freezing on the Phloem

### Physical Effects of Low Temperatures

Transport in the living phloem of woody plants is driven by mass flow from high pressure to low pressure regions throughout the phloem network (Muench, 1930; van Bel, 2003). As negative pressures do not occur in the phloem, freezing-induced embolism is not a risk as it is in the xylem. Yet low temperatures pose other problems for phloem transport. Phloem sap is significantly more viscous than xylem sap, and phloem viscosity can increase by an order of magnitude with decreasing temperatures (Fig. 19.1A), posing significant limitations to bulk flow. Intercellular diffusion rates also decrease with colder temperatures, and functioning of organelles in companion cells that regulate phloem transport may be impaired. Plasmodesmata are thought to narrow in response to cold, reducing the efficiency of symplastic transport (Gamelei *et al.*, 1994). Decreasing sink strength with reduced respiration and growth activity in winter should also reduce the pressure gradient along the phloem pathway and limit phloem transport (Hoffmann-Thoma *et al.*, 2001). Consequently, in deciduous plants, as growth ceases in winter, the phloem is likely to become inactive. In temperate woody dicots, new phloem (inner bark) is put down by the vascular cambium in the spring and previous phloem collapses as xylem expands. Phloem differentiation begins in the buds and new shoots and extends basipetally downward. It can precede or follow xylem development by several weeks in the spring depending on the species. Annual phloem growth rings may often be observed in the bark of temperate trees. Renewal of phloem tissue and collapse of older sieve tubes is thought to occur in a similar way in tropical evergreen species (Zimmermann, 1961). In most conifers, phloem activity and differentiation continue through the winter (Priestley, 1930). An interesting question is the extent to which previous year phloem can be reused in different species. In long-lived monocots, phloem sieve tubes can survive multiple decades owing to the maintenance and support of the companion cells (van Bel, 2003).

### Extracellular Freezing and Bark Shrinkage

At freezing temperatures, bark shrinks as a result of water movement out of sieve tube elements into the apoplast (Zweifel and Hasler, 2000; Ameglio *et al.*, 2001a). Cortical cells of phloem tissue, in contrast to xylem parenchyma, contract in response to freezing as extracellular ice forms and cells become dehydrated with the outflow of cellular water (Ashworth *et al.*, 1993). The lignified xylem parenchyma cells have thicker cell walls than that of bark tissue and are also attached to adjacent cells in all directions, which prevents contraction. In Norway spruce, a subalpine conifer, the loss of water in the bark (including in the cambium, phloem, and parenchyma) is dependent on contact with the xylem (Zweifel and Hasler, 2000). Exposed twigs in the periphery of the tree are likely to freeze first when the temperature drops below the freezing point. Once freezing begins, nucleation (seeding) occurs and freezing spreads to other parts of the tree. Partial freezing in the xylem can cause a strong water potential gradient between the phloem and the xylem, causing water to move into the xylem from the phloem and allowing supercooling to occur in the phloem, protecting the live phloem cells from freezing damage. Increased osmotic concentration in the phloem allows for rapid rehydration after thawing. Zweifel and Hasler (2000) hypothesized that Norway spruce can remain active in winter because on sunny days, the crown gathers enough warmth to rehydrate the bark in the upper tree, even as the lower stem remains frozen. In the reactivated part of the crown, small amounts of photosynthesis and transpiration can occur and water is withdrawn from internal reserves, rather than from the soil. As the winter progresses, the crown and the bark of the upper stem become more dehydrated than the stem parts near the ground.

In contrast to Norway spruce, bark contraction in black walnut is not dependent on contact with the xylem (Ameglio *et al.*, 2001a). Extracellular water in the bark freezes first since it has a lower solute concentration than intracellular vacuolar and cytoplasmic water. The extracellular ice crystals then draw water out of the living bark cells through the plasma membrane. The porous bark allows the ice to expand even as the water loss in the living cells causes bark contraction. Contraction increases as plants become cold hardened, presumably because extracellular freezing increases relative to intracellular freezing.

### Symplastic Versus Apoplastic Phloem Loaders and Climatic Distribution

The hypothesis that cold stress may have been important in driving the evolution of an apoplastic loading strategy arose from the observation that plant species with high plasmodesmatal frequency (thought to be correlated with symplastic loading) evolved in tropical climates while plants with minor vein anatomy consistent with apoplastic loading evolved more recently in temperate climates (Gamelei, 1989). The corollary to this hypothesis is that apoplastic loaders may have greater transport efficiency at low temperatures (Gamelei *et al.*, 1994) and that symplastic loaders should be more vulnerable to cold temperatures than apoplastic loaders, which would be evidenced by reduced export efficiency from leaves and starch buildup in winter chloroplasts of mesophyll and bundle sheath cells. One of the underlying assumptions is that the greater the number of plasmodesmata at a given cellular interface, the greater is the potential for symplastic transport through that interface. However, species that are classified as symplastic loaders on account of plasmodesmatal connectivity may turn out to load differently (see Chapter 3; Turgeon and Medville, 1998).

There are several reasons why symplastic species and species with greater numbers of plasmodesmata may be more vulnerable to cold stress. There is some evidence that translocation through plasmodesmata begins to be inhibited at 10° C and blockage occurs, with the temperature minimum of symplastic transport thought to be between 2 and 4° C (Geiger and Sovonick, 1975). It has also been suggested that apoplastic loading, which requires a proton pump, could be more active at lower temperatures than symplastic loading, which depends on diffusion (Gamelei, 1991). Many symplastic loaders transport raffinose and stachyose and other raffinose family oligosaccharides (RFO) in addition to sucrose. These are synthesized from sucrose and galactose in intermediary cells and form a polymer trap that allows sucrose to diffuse into intermediary cells and into sieve elements while preventing large RFOs to diffuse back into the mesophyll. Raffinose shows much lower solubility at low temperatures than sucrose and should be poorly transported at cold temperatures (Turgeon, 1995), although other RFOs apparently have higher solubility at low temperatures (Lambers *et al.*, 1998).

However, recent experiments show no difference in cold sensitivity between symplastic and apoplastic loaders (Hoffmann-Thoma *et al.*, 2001). In a study of three broadleaved evergreen species that were shown to be symplastic loaders, reduced phloem loading was not observed in winter, and sugar export was maintained in cold-acclimated leaves. Phloem loading type was based on transport sugars and limitation by p-chloromercuribenzenesulfonic acid, which impairs sucrose carriers involved in apoplastic loading. During periods of cold temperatures, sugar exudation actually increased (Fig. 19.5). In addition, no damage to intermediary cell vacuoles or other structures were found after exposure to natural freezing episodes (−9 to −11° C). Hoffmann-Thoma *et al.* (2001) hypothesized that the additional sugar export in winter was necessary for the energy requirements of cold acclimation, including the production of solutes, osmotic adjustment, protein synthesis, and reorganization of membrane lipids. Starch did not accumulate in winter chloroplasts of mesophyll and bundle sheath cells, and much higher amounts were found in these cells in summer.

Low temperatures may impose greater limits on the symplastic pathway, but cold hardening is able to overcome these limitations. In plants from eight families, including four apoplastic loaders and four symplastic loaders, no significant differences were observed in response to chilling at 10° C, and there was no evidence that phloem loading was impaired in symplastic species (Schrier *et al.*, 2000). Phloem loading was fully operative at 10° C, and there was no difference in exudation of sugars between symplastic and apoplastic species. For all species, however, export of sugars in response to long-term exposure to 10° C was slower, as carbohydrates were lost more slowly and phloem loading was slightly lower. Starch content in leaves increased after long-term exposure to low temperatures in all plants. In all 10° C cold-shocked plants, cold sensitivity was apparent as indicated by a reduced export of sugars and total nonstructural carbohydrate. This sensitivity was explained by a possible cold-induced narrowing of the plasmodesmata in the mesophyll to sieve element trajectory (Schrier *et al.*, 2000). Thus, low temperatures

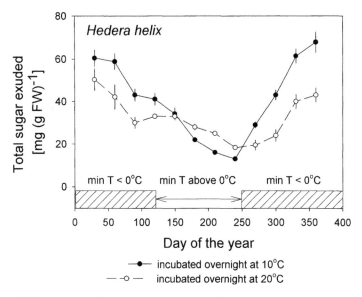

**Figure 19.5** Sugar exudation levels of leaves in different seasons for the evergreen symplastic phloem loading species, *Hedera helix*, growing outdoors but incubated overnight at either 10° or 20° C. For each experiment, total sugar quantity of overnight exudates was analyzed by high-performance anion exchange for at least four leaves of each species for each chamber temperature. Bar at bottom indicates timing of when daily minimum temperatures were below zero. Redrawn from Hoffman-Thoma *et al.* (2001).

clearly impact phloem transport, but symplastic phloem loading is not more vulnerable to chilling or freezing stress than apoplastic loading.

## Conclusions

Freezing and cold temperatures pose clear problems to living tissues and long distance transport systems in plants. These physical limitations are partially overcome by cold acclimation in cold-hardy plants, which involves a suite of biochemical and physiological transformations triggered by photoperiod and cold temperatures. The impacts of freezing on xylem transport depend in large part on xylem vessel diameters. Yet other anatomical features of xylem that influence their vulnerability to freezing embolism are only beginning to be understood. The mechanisms by which declining temperatures below freezing continue to impact xylem function are not well understood. Much is left to be learned about the influence of the redistribution of water during freezing on water content and tension in xylem, and how this affects dynamics of cavitation and sap flow. For example, how does freezing tolerance of living xylem parenchyma influence xylem function? Although the impacts of freezing on living cells and on xylem in woody plants have been intensively studied, much less is known about the effects of subzero temperatures on phloem and its role, if any, in regulating sugar mobilization and xylem repair in the spring. How active is the phloem in winter across different plant taxonomic groups, and what role does the phloem play across diverse taxa in cold acclimation including in the transport of ABA and synthesized proteins? Studies that can address the impacts of freezing on cellular processes and on long-distance transport in both the xylem and the phloem will help provide an integrated understanding of plant function in response to freezing stress.

## Acknowledgments

I thank Maciej Zwieniecki, N. Michele Holbrook, Lawren Sack, Kent Cavender-Bares, and an anonymous reviewer for helpful comments. I am also grateful to Silvia Kikuta and Aart van Bel for sending data that were used in Figs. 19.4 and 19.5.

## References

Ameglio, T., Bodet, C., Lacointe, A. and Cochard, H. (2002) Winter embolism, mechanisms of xylem hydraulic conductivity recovery and springtime growth patterns in walnut and peach trees. *Tree Physiology* **22:** 1211-1220.

Ameglio, T., Cochard, H. and Ewers, F. W. (2001a) Stem diameter variations and cold hardiness in walnut trees. *J Exp Bot* **52:** 2135-2142.

Ameglio, T. and Cruiziat, P. (1992) Tension pressure alternation in walnut xylem sap during winter: The role of winter temperature. *Comptes Rendus De L Academie Des Sciences Serie Iii-Sciences De La Vie-Life Sciences* **315:** 29-435.

Ameglio, T., Ewers, F. W., Cochard, H., Martignac, M., Vandame, M., Bodet, C. and Cruiziat, P. (2001b) Winter stem xylem pressure in walnut trees: Effects of carbohydrates, cooling and freezing. *Tree Physiol* **21:** 387-394.

Ashworth, E. N., Malone, S. R. and Ristic, Z. (1993) Response of Woody plant cells to dehydrative stress. *Int J Plant Sci* **154:** 90-99.

Bertrand, A., Robitaille, G., Castonguay, Y., Nadeau, P. and Boutin, R. (1997) Changes in ABA and gene expression in cold-acclimated sugar maple. *Tree Physiol* **17:** 31-37.

Burke, M., Gusta, L., Quamme, H., Weiser, C. and Li, P. (1976) Freezing injury in plants. *Annu Rev Plant Physiol* **27:** 507-528.

Cavender-Bares, J. and Holbrook, N. M. (2001) Hydraulic properties and freezing-induced xylem cavitation in evergreen and deciduous oaks with contrasting habitats. *Plant Cell Environ* **24:** 1243-1256.

Chang, R. (1991) *Chemistry.* New York, McGraw-Hill.

Cho, C. H., Urquidi, J. and Robinson, G. W. (1999) Molecular-level description of temperature and pressure effects on the viscosity of water. *J Chem Phys* **111:** 10171-10176.

Churchill, G. C., Reaney, M. J. T., Abrams, S. R. and Gusta, L. V. (1998) Effects of abscisic acid and abscisic acid analogs on the induction of freezing tolerance of winter rye (*Secale cereale* L.) seedlings. *J Plant Reg* **25:** 35-45.

Cordero, R. A. and Nilsen, E. T. (2002) Effects of summer drought and winter freezing on stem hydraulic conductivity of Rhododendron species from contrasting climates. *Tree Physiol* **22:** 919-928.

Davis, S. D., Sperry, J. S. and Hacke, U. G. (1999) The relationship between xylem conduit diameter and cavitation caused by freezing. *Am J Bot* **86:** 1367-1372.

Essiamah, S. K. (1980) Spring sap of trees. *Ber Deutsch Bot Ges* **93:** 257-267.

Ewart, K. V., Lin, Q. and Hew, C. L. (1999) Structure, function and evolution of antifreeze proteins. *Cell Mol Life Sci* **55:** 271-283.

Ewers, F. (1985) Xylem structure and water conduction in conifer trees, dicot trees, and lianas. *IAWA Bull* **6:** 309-317.

Feild, T. S. and Brodribb, T. (2001) Stem water transport and freeze-thaw xylem embolism in conifers and angiosperms in a Tasmanian treeline heath. *Oecologia* **127:** 314-320.

Fujikawa, S. (1997) Seasonal changes in dehydration tolerance of xylem ray parenchyma cells of *Stylax obassia* twigs that survive freezing temperatures by deep supercooling. *Protoplasma* **197:** 34-44.

Fujikawa, S. (2002) Structural characteristics of xylem parenchyma cell walls of trees in relation to the freezing adaptation. *Mokuzai Gakkaishi* **48:** 323-331.

Fujikawa, S. and Kuroda, K. (2000) Cryo-scanning electron microscopic study on freezing behavior of xylem ray parenchyma cells in hardwood species. *Micron* **31:** 669-686.

Fujikawa, S., Kuroda, K. and Ohtani, J. (1996) Seasonal changes in the low-temperature behavior of xylem ray parenchyma cells in Red Osier Dogwood (*Cornus sericea* L.) with respect to extracellular freezing and supercooling. *Micron* **27:** 181-191.

Gamelei, Y. (1989) Structure and function of leaf minor veins in trees and herbs. *Trees* **3:** 96-110.

Gamelei, Y. (1991) Phloem loading and its development related to plant evolution from trees to herbs. *Trees* **5:** 50-64.

Gamelei, Y., van Bel, A. J. E., Pakhomova, M. and Sjutkina, A. (1994) Effects of temperature on the conformation of the endoplasmic reticulum and on starch accumulation in leaves with the symplasmic minor-vein configuration. *Planta* **194:** 443-453.

Geiger, D. R. and Sovonick, S. A. (1975) Effects of temperature, anoxia and other metabolic inhibiters. In Transport in plants. I. Phloem transport. *Encyclopedia of Plant Physiology, New Series,* Vol 1, (M. H. Zimmerman and J. A. Milburn, eds.) pp. 256-288. Springer, Berlin.

George, M., Pellett, H. and Johnson, A. (1974) Low temperature exotherms and woody distribution. *Hort Sci* **9:** 519-522.

Griffith, M., Antikainen, M., Hon, W. C., PihakaskiMaunsbach, K., Yu, X. M., Chun, J. U. and Yang, D. S. C. (1997) Antifreeze proteins in winter rye. *Physiol Plantarum* **100:** 327-332.

Gusta, L., Tyler, N. and Chen, T. (1983) Deep undercooling in woody taxa growing north of the −40° C isotherm. *Plant Physiol* **62:** 899-901.

Guy, C. L. (1990) Cold-acclimation and freezing stress tolerance: The role of protein-metabolism. *Annu Rev Plant Physiol Plant Mol Biol* **41:** 187-223.

Guy, C. L. (2003) Freezing tolerance of plants: Current understanding and selected emerging concepts. *Can J Bot-Revue Canadienne De Botanique* **81:** 1216-1223.

Hansen, J. and Beck, E. (1988) Evidence for ideal and non-ideal equilibrium freezing of leaf water in frosthardy ivy (*Hedera helix*) and winter barley (*Hordeum vulgare*). *Botan Acta* **101:** 76-82.

Hartung, W. and Davies, W. (1991) Drought-induced changes in physiology and ABA. In *Abscisic Acid Physiology and Biochemistry* (W. Davies and H. Jones, eds.) pp. 63-77. Bios Scientific Publishers, Oxford, U.K.

Hoffmann-Thoma, G., van Bel, A. J. E. and Ehlers, K. (2001) Ultrastructure of minor-vein phloem and assimilate export in summer and winter leaves of the symplasmically loading evergreens *Ajuga reptans* L., *Aucuba japonica* Thunb., and *Hedera helix* L. *Planta* **212:** 231-242.

Kikuta, S. and Richter, H. (2003) Ultrasound acoustic emissions from freezing xylem. *Plant Cell Environ* **26:** 383-388.

Kozlowski, T. and Pallardy, S. (1997) *Physiology of Woody Plants.* Academic Press, New York.

Kubler, H. (1983) Mechanism of frost crack formation in trees: A review and synthesis. *For Sci* **29:** 559-568.

Kuroda, K., Kasuga, J., Arakawa, K. and Fujikawa, S. (2003) Xylem ray parenchyma cells in boreal hardwood species respond to subfreezing temperatures by deep supercooling that is accompanied by incomplete desiccation. *Plant Physiol* **131:** 736-744.

Kuroda, K., Ohtani, J. and Fujikawa, S. (1997) Supercooling of xylem ray parenchyma cells in tropical and subtropical hardwood species. *Trees Structure Function* **12:** 97-106.

Lambers, H., Chapin, F. S., III and Pons, T. L. (1998) *Plant Physiological Ecology.* Springer, Berlin.

Lechowicz, M. J. (1984) Why do temperate deciduous trees leaf out at different times? Adaptation and ecology of forest communities. *Am Naturalist* **124:** 821-842.

Lemoine, D., Granier, A. and Cochard, H. (1999) Mechanism of freeze-induced embolism in *Fagus sylvatica* L. *Trees Structure Function* **13:** 206-210.

Lide, D. R. (1993) *Handbook of Chemistry and Physics.* CRC Press, London.

Lo Gullo, M. A. and Salleo, S. (1993) Different vulnerabilities of *Quercus ilex* L. to freeze- and summer drought-induced xylem embolism: An ecological interpretation. *Plant Cell Environ* **16:** 511-519.

Marvin, J. W., Morselli, M. and Laing, F. M. (1967) A correlation between sugar concentration and volume yields in sugar maple: An 18-year study. *Forest Sci* **13:** 346-351.

Mathlouthi, M. and Génotelle, J. (1995) Rheological properties of sucrose solutions and suspensions. In *Sucrose: Properties and Applications* (M. Mathlouthi and P. Reiser, eds.) pp. 126-154. Blackie Academic & Professional, London.

Mayr, S., Gruber, A. and Bauer, H. (2003a) Repeated freeze-thaw cycles induce embolism in drought stressed conifers (Norway spruce, stone pine). *Planta* **217:** 436-441.

Mayr, S., Rothart, B. and Damon, B. (2003b) Hydraulic efficiency and safety of leader shoots and twigs in Norway spruce growing at the alpine timberline. *J Exp Bot* **54:** 2563-2568.

Muench, E. (1930) *Die Stoffbewegungen in der Pflanze.* Gustav Fischer, Jena, Germany.

Napp-Zinn, K. (1984) Light and vernalization. In *Light and the Flowering Process* (D. Vince-Prue, B. Thomas and K. Cockshull, eds.) pp. 75-88. Academic Press, London.

Noshiro, S. and Baas, P. (2000) Latitudinal trends in wood anatomy within species and genera: Case study in *Cornus* S.L. (Cornaceae). *Am J Bot* **87:** 1495-1506.

Parker, J. (1963) Cold resistance in woody plants. *Bot Rev* **29:** 123-201.

Pearce, R. S. (2001) Plant freezing and damage. *Ann Bot* **87:** 417-424.

Pittermann, J. and Sperry, J. (2003) Tracheid diameter is the key trait determining the extent of freezing-induced embolism in conifers. *Tree Physiol* **23:** 907-914.

Pockman, W. T. and Sperry, J. S. (1997) Freezing-induced xylem cavitation and the northern limit of *Larrea tridentata*. *Oecologia* **109:** 19-27.

Priestley, J. H. (1930) Studies in the physiology of cambial activity. III. The seasonal activity of the cambium. *New Phytol* **29:** 316-354.

Rajashekar, C. and Lafta, A. (1996) Cell-wall changes and cell tension in response to cold acclimation and exogenous abscisic acid in leaves and cell cultures. *Plant Physiol* **111:** 605-612.

Ristic, Z. and Ashworth, E. N. (1995) Response of xylem ray parenchyma cells of supercooling wood tissues to freezing stress: Microscopic study. *Int J Plant Sci* **156:** 784-792.

Robson, D. and Petty, J. A. (1993) Freezing and thawing in conifer xylem. In *Water Transport in Plants Under Climate Stress* (M. Borghetti, J. Grace, and A. Raschi, eds.) pp. 75-85. Cambridge University Press, New York.

Sakai, A. (1970) Freezing resistance in willows from different climates. *Ecology* **51:** 485-491.

Sakai, A. and Larcher, W. (1987) *Frost Survival of Plants: Responses and Adaptations to Freezing Stress.* Springer-Verlag, Berlin.

Sarnighausen, E., Karlson, D. and Ashworth, E. (2002) Seasonal regulation of a 24-kDa protein from red-osier dogwood (*Cornus sericea*) xylem. *Tree Physiol* **22:** 423-430.

Schaffer, K. and Wisniewski, M. (1989) Development of the amorphous layer (protective layer) in xylem parenchyma of cv. Golden Delicious Apple, cv. Loring Peach, and Willow. *Am J Bot* **76:** 1569-1582.

Scholander, P., Flagg, W., Hock, R. and Irving, L. (1953) Studies on the physiology of frozen plants and animals in the arctic. *J Cell Comp Physiol* **42 (Suppl 1):** 1-56.

Schrader, S. and Sauter, J. J. (2002) Seasonal changes of sucrose-phosphate synthase and sucrose synthase activities in poplar wood (*Populus x canadensis Moench < robusta >*) and their possible role in carbohydrate metabolism. *J Plant Physiol* **159:** 833-843.

Schrier, A. A., Hoffmann-Thoma, G. and van Bel, A. J. E. (2000) Temperature effects on symplasmic and apoplasmic phloem loading and loading-associated carbohydrate processing. *Aust J Plant Physiol* **27:** 769-778.

Sparks, J. P., Campbell, G. S. and Black, R. A. (2000) Liquid water content of wood tissue at temperatures below 0° C. *Can J Forest Re-Rev Canadienne De Recherche Forestiere* **30:** 624-630.

Sperry, J. S. (1995) Limitations on stem water transport and their consequences. In *Plant Stems: Physiology and Functional Morphology* (B. L. Gartner, ed.) pp. 105-124. Academic Press, San Diego, California.

Sperry, J. S., Donnelly, J. R. and Tyree, M. T. (1988) Seasonal occurrence of xylem embolism in sugar maple (*Acer saccharum*). *Am J Bot* **75:** 1212-1218.

Sperry, J. S. and Robson, D. (2001) Xylem cavitation and freezing in conifers. In *Conifer Cold Hardiness* (F. Bigras and S. Colombo, eds.) pp. 121-136. Kluwer Academic Publishers, Netherlands.

Sperry, J. S. and Sullivan, J. E. M. (1992) Xylem embolism in response to freeze-thaw cycles and water-stress in ring-porous, diffuse-porous, and conifer species. *Plant Physiol* **100:** 605-613.

Stepkonkus, P., Dowgert, M. and Gordon-Kamm, W. (1983) Destabilization of the plasma membrane of isolated plant protoplasts during a freeze-thaw cycle: The influence of cold acclimation. *Cryobiology* **20:** 448-465.

Steponkus, P. L. (1984) Role of the plasma membrane in freezing injury and cold acclimation. *Annu Rev Plant Physiol* **35:** 543-584.

Sucoff, E. (1969) Freezing in conifer xylem sap and the cohesiontension theory. *Physiol Plantarum* **22:** 424-431.

Taiz, L. and Zeiger, E. (1998) *Plant Physiology.* Sinauer Associates, Sunderland, Massachusetts.

Tranquillini, W. (1982) Frost-drought and its ecological significance. In *Physiological Plant Ecology II: Water Relations and Carbon Assimilation. Encyclopedia of Plant Physiology* (O. Lange, P. Nobel, C. Osmond and H. Ziegler, eds.) pp. 379-400. Springer-Verlag, Berlin.

Turgeon, R. (1995) The selection of raffinose family oligosaccharides as translocates in higher plants. In *Carbon Partitioning and Source-Sink Interactions in Plants* (M. Madore and W. Lucas, eds.) pp. 195-203. American Society of Plant Physiologists, Rockville.

Turgeon, R. and Medville, R. (1998) The absence of phloem loading in willow leaves. *Proc Natl Acad Sci* **95:** 12055-12060.

Tyree, M. T. (1983) Maple sap uptake, exudation and pressure changes correlated with freezing exotherms and thawing endotherms. *Plant Physiol* **73:** 277-285.

Tyree, M. T. and Zimmermann, M. H. (2002) *Xylem Structure and the Ascent of Sap.* Springer, Berlin.

van Bel, A. J. E. (2003) The phloem, a miracle of ingenuity. *Plant Cell Environ* **26:** 125-149.

Wang, J., Ives, N. E. and Lechowicz, M. J. (1992) The relation of foliar phenology to xylem embolism in trees. *Funct Ecol* **6**: 469-475.

Webb, M. S. and Steponkus, P. L. (1993) Freeze-induced membrane ultrastructural alterations in rye (*Secale cereale*) leaves. *Plant Physiol* **101**: 955-963.

Wisniewski, M., Close, T., Artlip, T. and Arora, R. (1996) Seasonal patterns of dehydrins and 70-kDa heat-shock proteins in bark tissues in eight species of woody plants. *Physiol Plantarum* **96**: 496-505.

Wisniewski, M., Davis, G. and Schaffer, K. (1991) Mediation of deep supercooling of peach and dogwood by enzymatic modifications in cell-wall structure. *Planta* **184**: 254-260.

Wisniewski, M., Webb, R., Balsamo, R., Close, T. J., Yu, X. M. and Griffith, M. (1999) Purification, immunolocalization, cryoprotective, and antifreeze activity of PCA60: A dehydrin from peach (*Prunus persica*). *Physiol Plantarum* **105**: 600-608.

Wisniewski, M. E. and Ashworth, E. N. (1985) Changes in the ultrastructure of xylem parenchyma cells of peach (*Prunus persica*) and red oak (*Quercus rubra*) in response to a freezing stress. *Am J Bot* **72**: 1364-1376.

Wolfe, J., Bryant, G. and Koster, K. L. (2002) What is 'unfreezable water', how unfreezable is it, and how much is there? *CryoLetters* **23**: 157-166.

Wong, B. L., Baggett, K. L. and Rye, A. H. (2003) Seasonal patterns of reserve and soluble carbohydrates in mature sugar maple (*Acer saccharum*). *Can J Bot-Rev Canadienne De Botanique* **81**: 780-788.

Yang, S. and Tyree, M. T. (1992) A theoretical-model of hydraulic conductivity recovery from embolism with comparison to experimental-data on *Acer saccharum*. *Plant Cell Environ* **15**: 633-643.

Zhu, X. B., Cox, R. M., Bourque, C. P. A. and Arp, P. A. (2002) Thaw effects on cold-hardiness parameters in yellow birch. *Can J Bot-Rev Canadienne De Botanique* **80**: 390-398.

Zimmermann, M. H. (1961) Movement of organic substances in trees. *Science* **133**: 73-79.

Zimmerman, M. H. (1983) *Xylem Structure and the Ascent of Sap*. Springer-Verlag, New York.

Zweifel, R. and Hasler, R. (2000) Frost-induced reversible shrinkage of bark of mature subalpine conifers. *Agric Forest Meteorol* **102**: 213-222.

# 20

# Interactive Effects of Freezing and Drought on Long Distance Transport: A Case Study of Chaparral Shrubs of California

*Stephen D. Davis, Frank W. Ewers, R. Brandon Pratt, Pamela L. Brown, and Timothy J. Bowen*

A primary objective of plant ecology is to explain the origin and mainte-nance of different vegetation types and species associations. In the case of evergreen chaparral shrubs of southern California, responses to freezing and drought appear to play a significant role in shaping the distribution patterns exhibited by different species. The mechanisms by which these two environmental factors influence performance and survival involve irre-versible disruption of xylem water transport from soil to leaves as a result of freeze-thaw-induced or water stress-induced embolism (Langan *et al.*, 1997; Davis *et al.*, 2002).

## Evolutionary History of Chaparral in Relation to Freezing and Drought

Chaparral is defined as a plant community consisting of evergreen, sclero-phyllous shrubs adapted to a Mediterranean-type climate (Cooper, 1922). Among the five Mediterranean-type climate regions of the world, California chaparral experiences the most consistent and persistent peri-ods of summer drought (Cowling *et al.*, 2004). Descriptions of the evolu-tionary origin of chaparral frequently cite preadaptation to drought tolerance, especially the loss of summer rainfall, as a major cause of chap-arral emergence and recent expansion (Minnich, 1985; Axelrod, 1989; Keeley, 1999). Adaptation to freezing and the interactions of freezing with water stress may be particularly important because the geographical origin of California vegetation consisted of an ecotone between cold-elements of the North from the Arcto-Tertiary flora (one half of species) and temperate elements of the South in the Madro-Tertiary flora (one third of species), with the remaining flora primarily originating in desert ecosystems (Raven, 1977; Raven and Axelrod, 1978). Furthermore, glacial and interglacial periods during the Pleistocene caused repeated mixing of cold and temperate plant assemblages, across latitudes and elevations (Axelrod, 1981; Keeley, 1999). Chaparral expanded during recent global xerotherms (8000-1000 Before Present; Raven and Axelrod, 1978; Stine, 1994), during which time, the evolutionary derived, nonsprouting species of *Ceanothus* increased (Wells, 1969; Zedler, 1994; Davis *et al.*, 1999). These nonsprouting, post-fire obligate seeders are especially dependent on the persistent function of xylem long distance transport during extended drought (Davis *et al.*, 2002), freezing temperatures (Langan *et al.*, 1997), or, on rare occasions, a combination of drought in late fall with an early seasonal freeze (Ewers *et al.*, 2003). This dependence on persistent xylem function occurs because nonsprouters do not form dormant buds or carbon reserves in a woody lignotuber like sprouting species and are thus unable to survive via vegetative regrowth after shoot removal by wildfire, freezing, or extreme drought (James, 1984). Thus nonsprouters should theoretically be more edaphically and climatically specialized in their microenvironments (Wells, 1969; Zedler, 1994), in contrast to species of *Ceanothus*, *Rhus*, and *Malosma*, which have been observed to vigorously resprout after shoot removal by wildfire and freezing (DeSouza *et al.*, 1986; Frazer and Davis, 1988; Thomas and Davis, 1989; Stoddard and Davis, 1990; Langan *et al.*, 1997).

## Distribution of *Ceanothus, Rhus,* and *Malosma* Species Along a Freezing Gradient

A profound shift in chaparral species composition occurs from coastal to inland sites of the Santa Monica Mountains of southern California (Fig. 20.1A). Coastal sites rarely experience air temperatures below 0° C, whereas just 5 to 6 km inland, cold valleys experience temperatures as low as −12° C (Fig. 20.1B). On average, seasonal rainfall is about 400 mm and primarily restricted to winter months (December to March). Seasonal drought periods can last 6 to 8 months and may extend, on rare occasions, into the month of December, coincidental with the onset of winter freeze (Ewers *et al.*, 2003). *Ceanothus megacarpus*, *C. spinosus*, and *Malosma laurina* dominate the landscape in coastal exposures, whereas *Ceanothus crassifolius* and *Rhus ovata* dominate inland, cold air drainages. Because *C. spinosus* and *M. laurina* form lignotubers, they can persist immediately upslope of cold valleys through vegetative resprouting after periodic freeze-induced death of shoots (Langan *et al.*, 1997; Ewers *et al.*, 2003; Pratt *et al.*, 2005).

## Distribution of *Ceanothus*, *Rhus*, and *Malosma* Species Corresponds to Freezing Tolerance of Leaves

We have used three different methods to estimate the tolerance of chaparral leaves to freezing temperatures after winter hardening: cellular vital

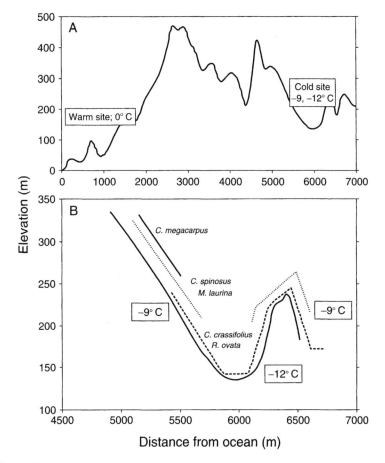

**Figure 20.1**   Distribution of *Ceanothus, Rhus,* and *Malosma* species along a freezing gradient (A) from a warm site near the Pacific Ocean to a cold site 5 to 6 km inland of the Santa Monica Mountains of southern California and (B) enlarged view of cold site showing species distribution relative to minimum air temperature. From Ewers *et al.* (2003) *Oecologia* 136:213-219. *Freeze thaw stress in Ceanothus of southern California chaparral.* Copyright Springer; figures 1b and 1c adapted with permission.

stain, dark-adapted chlorophyll fluorescence (Fv/Fm), and leaf color index (Boorse *et al.*, 1998a, 1998b; Ewers *et al.*, 2003). We have found that the temperature for 50% cell death or 50% loss in leaf activity (LT50) corresponds to the distribution patterns of chaparral species along a coastal to inland freezing gradient (Fig. 20.1). Leaves of *C. megacarpus, C. spinosus,* and *M. laurina* all had LT50 values greater than the minimum temperature of −12° C recorded at inland valleys where *C. crassifolius* and *R. ovata* dominate (Fig. 20.2A). In contrast, both *C. crassifolius* and *R. ovata* had LT50 values 4 to 6° C more negative than the minimum temperature of −12° C (Fig. 20.2A). Although *M. laurina* had the most sensitive leaves to freezing injury (LT50 = −6° C), adults of *M. laurina* can survive microsites that experience freezing events more negative than −6° C by resprouting from lignotubers (Fig. 20.1B) (Langan *et al.*, 1997). These resprouting events occur as frequently as every 3 to 4 years at some sites, resulting in plants that are stunted in height and populations that are low in density (Langan *et al.*, 1997).

## With the Exception of *R. ovata*, Distribution of Chaparral Shrubs Corresponds to Susceptibility to Freeze-Thaw-Induced Embolism

Using freezing chambers to simulate radiation freezes on large branches of *C. megacarpus*, Langan *et al.* (1997) found that freeze-thaw-induced embolism of stem xylem interacts with the degree of water stress, consistent with the theory of gas bubble dissolution in xylem sap at the time of thaw (Yang and Tyree, 1992). That is, the more negative the water potential at the time of thaw, the greater the loss of hydraulic conductivity in stem xylem of *C. megacarpus* (Langan *et al.*, 1997). A water potential of −1 MPa during a freeze-thaw cycle resulted in an increase in xylem embolism over that due solely to water stress from 10% (control) to 35% (freeze-thaw). In contrast, a freeze at a water potential of −5 MPa caused > 90% embolism. We used a freezing centrifuge technique modified after Davis *et al.* (1999) to separate the effects of water stress-caused embolism from freeze-thaw-induced embolism within the same stem samples (unpublished data) (Fig. 20.3). Between −1 and −4 MPa water stress-induced embolism remained below 20%, whereas embolism caused by freezing alone (i.e., above that attributable to water stress) increased from 30% to 80%, with overall embolism exceeding 90% at −4 MPa, for both *C. spinosus* and *C. megacarpus* (Fig. 20.3). These species were selected because they had low predawn water potentials at the time of first winter freezing events (Ewers *et al.*, 2003). Also, *C. crassifolius* was relatively resistant to freeze-thaw-induced embolism at −5 MPa, whereas *C. spinosus* and *C. megacarpus* were susceptible (Fig. 20.2B).

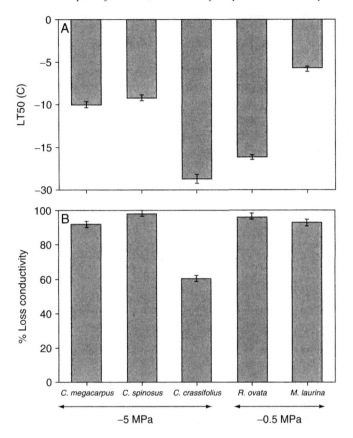

**Figure 20.2**   (A) Leaf temperature at 50% cell death of leaf palisade parenchyma using a vital stain (fluorescein diacetate) in conjunction with epifluorescent microscopy (Nikon, Microphot-FX). Error bars on symbols represent ± 1 SE, n = 6. (B) Effect of freeze-thaw treatments on excised, >2 m long, branches placed in a freezing chamber either at a dehydration level of about –5 MPa (*C. megacarpus, C. spinosus, C. crassifolius*) or fully hydrated at about –0.5 MPa (*R. ovata, M. laurina*). Error bars on symbols represent ± 1SE, n = 6 to 12.

Ewers *et al.* (2003) surveyed the interactive effects of water stress (–5 MPa) and freeze-thaw events on xylem embolism formation in the three *Ceanothus* species shown in Fig. 20.1. They chose –5 MPa because they measured predawn water potentials of –5 MPa in the field for *Ceanothus* species at the time of a first winter freezing event. Freeze-thaw-induced embolism of *M. laurina* and *R. ovata* were done at water potentials of –0.5 MPa because they are deep rooted and maintain high water potentials at predawn when freezing occurs (Poole and Miller, 1975; DeSouza *et al.*, 1986). Ewers and colleagues found that *C. megacarpus* and *C. spinosus* experienced > 90%

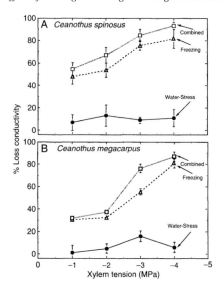

**Figure 20.3** Percent loss in hydraulic conductivity of stem xylem resulting from water stress at −1, −2, −3, and −4 MPa followed by a freeze-thaw event using a freezing centrifuge. The subtraction of percent loss in conductivity resulting from water stress from the total percent loss in conductivity (combined) allowed the calculation of embolism resulting from freeze-thaw alone. Separate stems were used for each xylem tension for six individuals of (A) *C. spinosus* and (B) *C. megacarpus*. Error bars on symbols represent ± 1 SE, n = 6.

embolism, whereas *C. crassifolius* was relatively resistant to freeze-caused embolism at −5 MPa, experiencing only 62% embolism (Fig. 20.2B). Maximum vessel diameters for all species examined were below a critical hydraulic mean value of 43 to 44 μm, which has been found to be relatively safe from embolism caused by freeze-thaw events under modest water potentials of −0.5 MPa (Davis *et al.*, 1999; Pittermann and Sperry, 2003). In contrast to these results, both *M. laurina* and *R. ovata* had maximum vessel diameters of about 60 μm, and, as predicted, freeze-thaw events under well-hydrated conditions (−0.5 MPa) resulted in > 90% embolism for both species (Fig. 20.2B).

Feild and Brodribb (2001) did not find a threshold increase but a linear increase in freeze-thaw-induced embolism, with conduit diameter starting at a mean conduit diameter of about 15 μm. However, they did not measure natural thaw rates in the field, nor did they control or measure thaw rates in the laboratory. When Langan *et al.* (1997) followed a similar method to Feild and Brodribb for *C. megacarpus* (e.g., removed bagged branches from a freezing chamber and allowed to rapidly thaw to room temperature in an air conditioned laboratory), thaw rates increase 10-fold over those measured under field conditions, from 0.08° C/min measured *in situ* to 0.8° C/min measured in the laboratory, which caused xylem embolism to increase from 35% to 74% (Langan *et al.*, 1997). It was concluded that freeze-thaw-induced embolism of stem xylem was a function of thaw rate. Thus in all subsequent experiments, thaw rates were rigorously controlled at 0.08° C/min, as determined under natural field conditions (Langan *et al.*, 1997; Davis *et al.*, 1999; Ewers *et al.*, 2003). If one takes the mean vessel diameter measured for *C. megacarpus* of 28.8 µm and uses the linear regression given by Feild and Brodribb (2001) (Fig. 20.2), predicted embolism is about 75%, close to the 74% embolism observed at an artificially high thaw rate of 0.8° C/min in the laboratory (See figure 5 in Langan *et al.*, 1997).

The obvious question that arises is, how can *R. ovata* persist at the coldest inland valleys when it experiences > 90% embolism as a result of freeze-thaw events? Because there is no evidence of embolism reversal in mature chaparral species (Kolb and Davis, 1994; Williams *et al.*, 1997), one possibility is that *R. ovata* leaves are unusually resistant to water loss and maintain hydration levels long enough for the development of new xylem during spring growth. Pratt *et al.* (2005) has found evidence to support this hypothesis. Figure 20.4 shows that in comparison to *M. laurina*, excised leaves of *R. ovata* take nearly four times as long to lose the same water mass per unit area and reduce their relative water content to minimum physiological levels (relative water content of 0.5). Also, *R. ovata* required more than five times as long to lose dark-adapted chlorophyll fluorescence (Fig. 20.4C). Evidently, *R. ovata* persists in cold inland valleys by having leaves that are resistant to freezing air temperatures (LT50 = −16° C) and extremely low leaf conductance so as to avoid desiccation due to > 90% freeze-caused embolism of stem xylem. New leaf emergence and stem elongation are presumably delayed until late spring after new xylem has been laid down by the vascular cambium (Pratt *et al.*, 2005).

Even though *R. ovata* had > 90% embolism of its stem xylem after multiple freeze-thaw events in December and a relatively low leaf specific conductivity of only 45 mg $MPa^{-1}m^{-1}s^{-1}$ measured in January, this conductivity was sufficient to supply water to leaves at a maximum daily transpiration rate of 11.4 mg $m^{-1}$ $s^{-1}$ measured at a height of about 1m above the soil surface with a pressure gradient > 1 MPa ($\Psi_{predawn} - \Psi_{midday} = 1.6$ MPa). Measurements of transpiration were made on 2 February 2004 using a steady-state porometer. Midday water potential on 2 February 2004 was −3.4 MPa ± 0.63 (SE; n = 10). This was an unusually low water potential for *R. ovata* on what is normally the wettest month of the year (e.g., *M. laurina* at the same study site and date had a midday water potential of −1.7 MPa), but the relatively large gradient in water potential for *R. ovata* increased the driving force for water transport from roots to leaves when leaf specific conductivity of the stem xylem was extremely low. Clearly, leaves on highly embolized stems of *R. ovata* remained partially hydrated and evergreen, allowing them to contribute to new growth

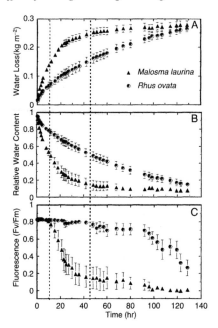

**Figure 20.4** (A) Water loss per unit leaf area, (B) relative water content from pressure-volume curves, and (C) dark-adapted chlorophyll fluorescence (Fv/Fm) for *M. laurina* and *R. ovata*. Data for water loss and fluorescence were collected on excised leaves in which petioles were sealed with epoxy and allowed to dehydrate in an air-conditioned laboratory at 21.5° C and a VPD of 0.95 kPa for 6 days (144 hr) where VPD is vapor pressure difference between the leaf and air. Within 1 day, *M. laurina* leaves had 50% loss in chlorophyll fluorescence (Fv/Fm < 0.4), whereas *R. ovata* retained Fv/Fm > 0.4 for about 5 days. The dotted, vertical lines estimate irreversible damage to leaves for *M. laurina* and *R. ovata*, respectively (relative water content < 0.5; *cf.* Burghardt and Riederer, 2003).

after the onset of new xylem production and full hydration in spring (mean predawn water potentials of –0.6 MPa on 20 April 2004 and new shoot growth detected on 24 May 2004).

## Conclusions

Chaparral species are evergreen and retain metabolically functional leaves throughout the year, requiring uninterrupted supplies of water to remain hydrated (Davis and Mooney, 1986a, 1986b; Thomas and Davis, 1989; Saruwatari and Davis, 1989). Water stress-induced or freeze-thaw-induced embolism of xylem water conduits represents a permanent loss of hydraulic capacity, as there is no evidence of reversal mechanism in adult shrubs (Kolb and Davis, 1994; Williams *et al.*, 1997). Assuming that resistance to freezing- or drought-induced injury incurs some cost, one might surmise that leaves should be about equal in susceptibility to injury by water stress and freezing as xylem conduits are to water stress and freezing-caused embolism. In *C. crassifolius*, stem xylem is more susceptible to water stress than are leaves (whole branchlet dieback with no recovery, Davis *et al.*, 2002), whereas in the chaparral shrub *Quercus berberidifolia* leaves are more susceptible than stem xylem (entire leaf canopy dies in response to severe drought and remains on a shrub until new leaf emergence on stems with winter rains, Pratt, unpublished data). In *R. ovata*, stem xylem is more susceptible to freeze-caused dysfunction than leaves; however, in a closely related species, *M. laurina*, leaves and stems are equally susceptible to freezing injury. Thus the assumption of equal susceptibility to water stress and freezing between leaves and stem xylem does not always hold. Other survival traits may compensate for imbalances in the susceptibility of leaves and stem xylem, such as the high resistance to water loss in leaves of *R. ovata* that allows leaves to persist in a partially hydrated state until the vascular cambium produces new xylem.

The combined action of drought and freezing temperatures are likely to have their greatest impact on nonsprouting species. For example, the stem xylem of *C. megacarpus* is relatively resistant to freeze-caused embolism under well-hydrated conditions, reflecting the presence of small-diameter xylem vessels in this species. However, under water stress conditions, the stem xylem of *C. megacarpus* is extremely susceptible to freezing injury. Thus in this species, stem xylem appears to be more susceptible than leaves to freezing injury when there is an interaction with drought. However, in a closely related species, *C. crassifolius*, both leaves and stems are resistant to freezing, even under water stress. The ability to form sprouts probably allows *C. spinosus* to extend into colder sites than the nonsprouter *C. megacarpus*, but interactions with water stress and freezing exclude *C. spinosus* from the coldest sites where *C. crassifolius* and *R. ovata* dominate. It appears that there is a suite of traits that determines survival of chaparral shrubs in the face of water stress and freezing and that these factors play a significant role in determining plant distribution across the landscape, especially along a freezing gradient. The evolutionary history of chaparral derived from ecotones of cold, warm, and dry climates have selected for an interplay of adaptations that facilitate persistence and diversity in a heterogeneous landscape.

## Acknowledgments

Funding provided by the National Science Foundation (IBN-0130870 and DBI-0243788) during research and preparation of the chapter.

# References

Axelrod, D. I. (1981) Holocene climate changes in relation to vegetation disjunction and speciation. *Am Naturalist* **117**: 847-870.

Axelrod, D. I. (1989). Age and origin of chaparral. In *The California Chaparral: Paradigms Reexamined* (S. C. Keeley, ed.) pp. 7-19. Natural History Museum of Los Angeles County, Los Angeles.

Boorse, G. C., Bosma, T. L, Meyer, A.-C., Ewers, F. W. and Davis, S. D. (1998a) Comparative methods of estimating freezing temperatures and freezing injury in leaves of chaparral shrubs. *Int J Plant Sci* **159**: 513-521.

Boorse, G. C., Ewers, F. W. and Davis, S. D. (1998b) Response of chaparral shrubs to below-freezing temperatures: Acclimation, ecotypes, seedlings vs. adults. *Am J Bot* **85**: 1224-1230.

Burghardt, M. and Riederer, M. (2003) Ecophysiological relevance of cuticular transpiration of deciduous and evergreen plants in relation to stomatal closure and leaf water potential. *J Exp Bot* **54**: 1941-1949.

Cooper, W. S. (1922) *The Broad-Sclerophyll Vegetation of California. An Ecological Study of the Chaparral and Its Related Communities.* Carnegie Institution of Washington Publication 319. Washington, DC.

Cowling, R. M., Ojeda, F., Lamont, B. B. and Rundel, P. W. (2004) Climate stability in Mediterranean-type ecosystems: Implications for the evolution and conservation of biodiversity. In *Ecology, Conservation and Management of Mediterranean Climate Ecosystems. Proceedings of the 10th International Conference of Mediterranean Climate Ecosystems* (M. Arianoutsou and V. P. Papanastasis, eds.) pp. 101(094pdf). Millpress, Rotterdam, Netherlands.

Davis, S. D., Ewers, F. W., Sperry, J. S., Portwood, K. A., Crocker, M. C. and Adams, G. C. (2002) Shoot dieback during prolonged drought in *Ceanothus* chaparral of California: A possible case of hydraulic failure. *Am J Bot* **89**: 820-828.

Davis, S. D., Sperry, J. S. and. Hacke, U. G. (1999) The relationship between xylem conduit diameter and cavitation caused by freeze-thaw events. *Am J Bot* **86**: 1367-1372.

Davis, S. D. and Mooney, H. A. (1986a) Water use patterns of four co-occurring chaparral shrubs. *Oecologia (Berlin)* **70**: 172-177.

Davis, S. D., and Mooney, H. A. (1986b) Tissue water relations of four co-occurring chaparral shrubs. *Oecologia (Berlin)* **70**: 527-535.

DeSouza, J., Silka, P. and Davis, S. D. (1986) Comparative physiology of burned and unburned *Rhus laurina* after chaparral wildfire. *Oecologia (Berlin)* **71**: 63-68.

Ewers, F. W., Lawson, M. C., Bowen, T. J. and Davis, S. D. (2003) Freeze/thaw stress in *Ceanothus* of southern California chaparral. *Oecologia (Berlin)* **136**: 213-219.

Feild, T. S. and Brodribb, T. (2001) Stem water transport and freeze-thaw xylem embolism in conifers and angiosperms in a Tasmanian treeline heath. *Oecologia (Berlin)* **127**: 314-320.

Frazer, J. M. and Davis, S. D. (1988) Differential survival of chaparral seedlings during the first summer drought after wildfire. *Oecologia (Berlin)* **76**: 215-221.

James, S. (1984) Lignotubers and burls—their structure, function and ecological significance in Mediterranean ecosystems. *Bot Rev* **50**: 225-266.

Keeley, J. E. (2000) Chaparral. In *North American Terrestrial Vegetation* (M. G. Barbour and W. D. Billings, eds.) pp. 203-254. Cambridge University Press, Cambridge.

Kolb, K. J. and Davis, S. D. (1994) Drought-induced xylem embolism in co-occurring species of coastal sage and chaparral of California. *Ecology* **75**: 648-659.

Langan, S. J., Ewers, F. W. and Davis, S. D. (1997) Differential susceptibility to xylem embolism caused by freezing and water stress in two species of chaparral shrubs. *Plant Cell Environ* **20**: 425-437.

Minnich, R. A. (1985) Evolutionary convergence of phenotypic plasticity? Responses to summer rain by California chaparral. *Phys Geogr* **6**: 272-287.

Pittermann, J. and Sperry, J. (2003) Tracheid diameter is the key trait determining the extent of freezing-induced embolism in conifers. *Tree Physiol* **23**: 907-914.

Poole, D. K. and Miller, P. C. (1975) Water relations of selected species of chaparral and coastal sage communities. *Ecology* **56**: 1118-1128.

Pratt, R. B., Ewers, F. W., Lawson, M. C., Jacobsen, A. L., Brediger, M. M. and Davis, S. D. (2005) Mechanisms for tolerating freeze-thaw stress of two evergreen chaparral species: *Rhus ovata* and *Malosma laurina* (Anacardiaceae) *Am J of Bot.* In press.

Raven, P. H. (1977) The California flora. In *Terrestrial Vegetation of California* (M. B. Barbour and J. Major, eds.) pp. 109-138. Wiley, New York.

Raven, P. H. and Axelrod, D. I. (1978) Origin and relationships of the California flora. *University of California Publications in Botany* **73**: 1-134.

Saruwatari, M. W. and Davis, S. D. (1989) Tissue water relations of three chaparral shrub species after wildfire. *Oecologia (Berlin)* **80**: 303-308.

Stine, S. (1994) Extreme and persistent drought in California and Patagonia during mediaeval time. *Science* **264**: 546-548.

Stoddard, R. J. and Davis, S. D. (1990) Photosynthesis, water relations, and nutrient status of burned, unburned, and clipped *Rhus laurina* after chaparral wildfire. *Bull South Calif Acad Sci* **89**: 26-38.

Thomas, C. M. and Davis, S. D. (1989) Recovery patterns of three chaparral shrub species after wildfire. *Oecologia (Berlin)* **80**: 309-320.

Wells, P. V. (1969) The relation between mode of reproduction and extent of speciation in woody genera of California chaparral. *Evolution* **23**: 264-267.

Williams, J. E., Davis, S. D. and Portwood, K. A. (1997) Xylem embolism in seedlings and resprouts of *Adenostoma fasciculatum* after fire. *Aust J Bot* **45**: 291-300.

Yang, S. and Tyree, M. T. (1992) A theoretical model of hydraulic conductivity recovery from embolism with comparison to experimental data on *Acer saccharum*. *Plant Cell Environ* **15**: 633-643.

Zedler, P. H. (1994) Plant life history and dynamic specialization in the chaparral/coastal sage shrub flora of southern California. In *Ecology and Biogeography of Mediterranean Ecosystems in Chile, California and Australia* (M. T. K. Arroyo, P. H. Zedler and M. D. Fox, eds.) pp. 89-115. Springer-Verlag, New York.

# 21

# Transport Challenges in Tall Trees

*George W. Koch and Arthur L. Fredeen*

The evolutionary trend of increasing stature in terrestrial plants (Niklas, 1994) has meant an increase in xylem and phloem transport distances, which at their extreme may pose special challenges to the design and function of vascular systems and may constrain the physiological performance and growth of plants. Perhaps the longest transport paths are in tropical lianas, stems of which may extend 240 m (Richards, 1996). Among self-supporting terrestrial plants, heights of nearly 130 m have been claimed for both gymnosperms and angiosperms, although these records are questionable and the trees are no longer standing (Carder, 1995). Currently, individuals of five conifer species of western North America and five angiosperms (four *Eucalyptus* species of Australia and the legume *Koopassia excelsa* of Borneo) exceed 85 m in height, the tallest living trees being coast redwoods (*Sequoia sempervirens*) of northern California, which are up to 113 m tall (Koch *et al.*, 2004). Extensive root systems (Stone and Kalisz, 1991) and complex crown architecture (Sillett and Bailey, 2003) add considerably to transport distances approximated by ground-to-top height.

Older physiological studies with tall trees tested predictions of the cohesion-tension model (Scholander *et al.*, 1965; Tobiessen *et al.*, 1971). In recent years, provocative proposals about the limits to tree height and size growth have sparked debate about the hydraulic and physiological constraints underlying vascular system design and function (Friend, 1993; Ryan and Yoder, 1997; Becker *et al.*, 2000). Improved canopy access via cranes (e.g., McDowell *et al.*, 2002a; Meinzer *et al.*, 1999) and rope-based techniques (Jepson, 2000; Koch *et al.*, 2004) have extended the height range of trees in temperate and tropical forests suitable for well-replicated experiments. Studies of xylem transport in moderately tall trees now abound, but those of individuals close to historic maximum heights for a species, and of long distance transport in the phloem, are scarce. Drawing on the relevant literature, our objectives in this chapter are to examine the constraints on transport imposed by extremes of height and path length, and to synthesize information on the mechanisms that may compensate for these constraints. We emphasize issues of xylem transport, the scant relevant literature making our treatment of phloem transport necessarily brief and speculative. Because stomatal regulation both affects, and is affected by, xylem transport and interacts with phloem transport via its influence on water potential and the production of photosynthate, we devote a final section to a new supply-loss model that links stomatal regulation to hydraulic conductivity of the xylem.

## The Physical Setting and Its Problems

As trees grow taller, two fundamental changes are unavoidable. First, gravity exerts a greater influence on total water potential and increases the energy required to move water from soil to leaves. Second, vascular pathways increase in length, and this may increase the total resistance to water movement from soils to leaves and for phloem transport from sources to sinks. Here we examine the implications of these changes for xylem transport and briefly touch on their potential influence on phloem function.

### Gravity

Because of gravity, the hydrostatic pressure of tracheary elements decreases by c. 0.01 MPa m$^{-1}$. Xylem pressure gradients close to the hydrostatic have been demonstrated in tall gymnosperms and angiosperms (Scholander *et al.*, 1965; Tobiessen *et al.*, 1971; Connor *et al.*, 1977; Koch *et al.*, 2004). Deviations from the expected hydrostatic gradient have been explained as a failure to account for the influence on xylem pressure because of the lateral position of a sampled twig on a transpiring branch (Zimmermann, 1983), and may also arise from uptake of water from foliage or branches wetted by rain or fog (see later). The necessary reduction of maximum xylem pressure because of gravity has several important and related implications for transport processes. First, it counteracts the driving force for xylem transport—that portion of the total root-to-leaf water potential gradient not due to gravity. Second, it may both increase the risk of xylem cavitation and decrease or prevent embolism repair in trees of moderate to great heights. Third, the hydrostatic gradient may underlie height trends in leaf turgor, foliar structure, and stomatal conductance, and these changes can indirectly affect components of xylem and phloem transport.

***Reduced Driving Force for Xylem Transport***    The driving force for xylem transport ($\Delta\Psi_{trans}$) is the total water potential difference between the source and destination ($\Delta\Psi_{tot}$) minus the difference in gravitational potential, $\Delta\Psi_g$. In short plants, $\Delta\Psi_g$ is negligible and $\Delta\Psi_{trans} \approx \Delta\Psi_{tot}$. In tall trees, $\Delta\Psi_g$ may be a large and even dominant fraction of $\Delta\Psi_{tot}$ (McDowell *et al.*, 2002a; Koch *et al.*, 2004). Because of gravity, $\Delta\Psi_{trans}$ should decrease with tree height unless minimum leaf water potential ($\Delta\Psi_{L\,min}$) decreases by at least 0.01 MPa m$^{-1}$ (see Mechanisms Compensating for Height Constraints). A decrease in $\Delta\Psi_{trans}$ causes a proportional reduction in Poiseulle flow velocity, and therefore requires an increase in sapwood area and/or sapwood permeability to maintain total volume flow rate as height increases.

*Height and Xylem Dysfunction*   It is evident that trees grow tall only where water is abundant, and in this sense problems of xylem dysfunction might seem unimportant in tall trees. Because of gravity, however, maximum xylem pressures must decrease with height, a trend amplified by the higher transpiration rates of upper crown versus lower crown foliage. Thus, simply to sustain hydraulic continuity in a static column 100-m high, the upper crown xylem must tolerate pressures of $\leq -1$ MPa, a level sufficient to cause significant cavitation in some species of small trees of riparian habitats (e.g., Tyree *et al.*, 1994b; Pockman and Sperry, 2000).

Given uniform wood hydraulic properties within the crown, the risk of xylem cavitation should increase with height. Domec and Gartner (2002) showed, however, that native embolism was very low in upper branches of mature 45-m tall Douglas-fir, and it was lower there than at the trunk base because the juvenile wood at the treetop was more cavitation resistant. Because the mechanical safety factor exceeded the hydraulic safety factor in wood from all heights, they concluded that selection for wood properties is more likely mediated by hydraulic than mechanical pressures. Koch *et al.* (2004) also showed height-related variation in wood hydraulic properties: xylem pressure for 50% loss of hydraulic conductivity ($\Psi_{P50}$) in redwood was significantly lower for branches high in the crown than low in the crown ($-6.6 \pm 0.5$ MPa at 109 m versus $-5.6 \pm 0.7$ MPa at 57 m). That the $\Psi_{P50}$ was much lower than the lowest xylem tension recorded in the dry season (c. $-1.8$ MPa) suggests that native embolism is generally low at the tops of tall trees.

Why should there be such a large apparent safety margin in redwood? Vulnerability characteristics might reflect water relations' challenges during the seedling phase when soil drought is more important. It may also be the case that cavitation avoidance is critical for extreme height growth, with even a small loss of conducting capacity restricting photosynthesis and leaf development sufficiently to halt growth.

The influence of gravity should also be manifested as a decreasing probability of embolism reversal with height. Although there is accumulating evidence that embolism repair can occur under tension (see Chapters 1 and 18), this has not been demonstrated in conifers and other tall trees (i.e., > 30m). Early models of embolism reversal required that xylem pressure exceed about −0.1 MPa to cause bubbles to redissolve and conduits to refill, the exact pressure depending on conduit radius (Yang and Tyree, 1992). While possible by root pressure in short plants, dissipation of root pressure by gravity would prevent this at heights of a few tens of meters. If, in fact, embolism repair is increasingly limited by height, then this may impose design requirements on xylem structure and stomatal regulation that serve to avoid widespread cavitation in tall trees.

Given that the tallest trees grow where branch and leaf surfaces are frequently wetted by rain and fog, it is interesting to speculate whether surface uptake of water might contribute to refilling of cavitated conduits, either by increasing xylem pressure sufficiently or by allowing refilling under tension by some variation of a pressurization mechanism (e.g., Pickard, 1989; Tyree *et al.*, 1999). Foliar or branch uptake has been documented for many plants (Breazeale *et al.*, 1950), including trees that grow tall (Katz *et al.*, 1989; Boucher *et al.*, 1995; Yates and Hutley, 1995; Burgess and Dawson, 2004). Connor *et al.* (1977) reported water potential gradients that were less steep than the hydrostatic gradient when foliage was wet, and they proposed that water taken up by a wet canopy may be an important aspect of the water relations of the leaves of tall trees. In redwood, Jennings (2003) also observed a predawn gradient less than the hydrostatic when the canopy was wet. Foliar uptake during actual or simulated fog events can drive reverse xylem flow in redwood (Burgess and Dawson, 2004) and can increase water potential of small branches by up to 0.5 MPa (Koch, Sillett, and Jennings, unpublished data). It would be surprising if such uptake did not impact the maintenance of hydraulic sufficiency, especially where heavy fogs are common during the otherwise dry (rainless) season (Dawson, 1998). Clearly, more information is needed on the dynamics of native embolism and hydraulic conductivity in tall trees to guide tests of mechanisms of embolism repair within the constraints on xylem tension imposed by gravity.

***Height and Turgor Maintenance***    As total water potential declines owing to gravity, either the turgor pressure of living tissues must decline or solute concentration must increase to maintain turgor. Friend (1993) speculated that low turgor might contribute to height limitation in tall trees and shift the period of leaf expansion at the treetop to the night when water potential and turgor would reach daily maxima. Supporting this prediction, studies in tall conifers (Koch *et al.*, 2004; Woodruff *et al.*, 2004) demonstrated that leaf turgor declines with height and may be functionally linked to height gradients in foliar morphology, which in turn may impact gas exchange and carbon balance. These studies, both of conifers, determined that osmotic adjustment was insufficient to prevent a reduction in turgor with height. Connor *et al.* (1977) observed a reduction in osmotic potential ($\Psi_S$) with height in *Eucalyptus regnans* of 0.01 MPa m$^{-1}$, equal to the hydrostatic gradient, but foliar turgor was not reported. Additional studies of turgor, osmotic regulation, and tissue elastic properties are needed to reveal how gravity may influence leaf development and gas exchange in tall trees.

*Stomatal Regulation and Photosynthesis* Because of gravity, a leaf at a height of 100 m that uses only water transported from the soil cannot experience a water potential above about −1 MPa, a value that causes at least some reduction in stomatal conductance in many tree species (Kramer and Boyer, 1995). Does the hydrostatic gradient narrow the range of $\Psi_{soil}$ over which stomata can be open? The data are far from conclusive, and comparisons are confounded by results derived by different methods and involving multiple trees of different heights (e.g., Hubbard *et al.*, 1999; Ryan *et al.*, 2000; Schäfer *et al.*, 2000) or indirect measures that are complicated by the influence of varied microclimate within crowns of individual trees (e.g., Koch *et al.*, 2004). On balance, however, it appears that stomatal conductance does decline with height, particularly in very tall trees, and that some of this reduction can be attributed to the influence of gravity on $\Psi_L$.

## Path Length

Because of gravity, great height necessarily affects the energy state of water in the plant. In addition, the long path necessary to deliver water to great height, and to the ends of long branches, may also impose constraints on water supply.

*Transport Time and Root-Leaf Signaling* Perhaps the most obvious implication of greater path length is an increased transport time for xylem and phloem transport. For a constant driving force and geometry of conduits, transport time is directly proportional to path length. Conduit diameter should decrease with height and vascular segmentation (West *et al.*, 1999), and gravity reduces $\Delta\Psi_{trans}$ with height unless minimum leaf water potential decreases in parallel (see later). A simple simulation (Fig. 21.1) using the Hagen-Poiseulle equation demonstrates how transport time from the top to the bottom of a 100 m tree varies across a range of $\Delta\Psi_{trans}$ values and conduit diameters measured for coast redwood (Koch *et al.*, 2004; Pitterman and Sperry, 2003; Panshin and De Zeeuw, 1980). For example, a $\Delta\Psi_{trans}$ of −0.7 MPa/100 m yields a transport time of about 2 to more than 30 days for tracheids of 60 μm to 15 μm diameter, respectively. An average sap velocity of 0.4 m h$^{-1}$ measured for large redwoods by the heat ratio method (S. Burgess, unpublished data) gives a 100 m transport time of

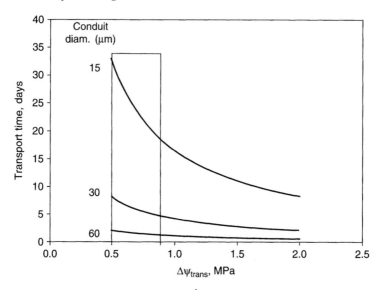

**Figure 21.1** Simulations of the time required for transport of water by Poiseulle flow a distance of 100 m through conduits of different diameters as a function of the pressure gradient for transport, $\Delta\Psi_{trans}$. The rectangle spans the approximate range of independent variables published for redwood, *Sequoia sempervirens* (Pitterman and Sperry, 2003; Koch *et al.*, 2004).

about 10 days, consistent with an average tracheid diameter of somewhat less than 30μm (Pittermann and Sperry, 2003). Although this would seem to uncouple leaves from chemical signaling (e.g., abscisic acid) from roots (Saliendra *et al.*, 1995), soil moisture also changes slowly in environments that support tall trees, and a long transport time need not preclude the importance of root signals. Nevertheless, the near instantaneous sensing of soil/root moisture via hydraulic signaling (i.e., xylem $\Psi$) has obvious temporal and energetic advantages and is more relevant to diel timescales and short-term fluctuations of light and atmospheric moisture (Meinzer, 2002).

***Whole-Tree Resistance*** If water flowed through plants as through a single ideal capillary of fixed diameter, the Hagen-Poiseuille relationship would predict that total path resistance would increase linearly with path length (Nobel, 1999) and decrease in proportion to the fourth power of the radius. However, other factors affect actual resistance: (1) xylem conduits are not ideal tubes and differ in effective diameters, taper, shape, and lengths; (2) conduits are interconnected via pits of various sizes and resistances; (3) the conducting system is a massively parallel network of sapwood of variable cross-sectional area; and (4) the path includes branch junctions with higher resistance than the rest of the flow path.

Furthermore, all of these characteristics may exhibit interspecific, seasonal, and/or ontogenetic variation, making the relationship between path length and resistance complex.

Several studies, using different approaches, have demonstrated that increased path length itself, apart from the associated influence of gravity, significantly increases whole-tree resistance (Mencuccini and Grace, 1996; Hubbard *et al.*, 1999; Schäfer *et al.*, 2000). These results are at odds with models indicating that total resistance should be relatively insensitive to total length (West *et al.*, 1999; McCulloh *et al.*, 2003). Reconciling these differences may require closer examination of the assumptions of theoretical approaches, for example, that of Murray's Law, that vascular tissue serves only a transport and not a support role, which is clearly not the case in conifers (McCulloh *et al.*, 2003). Other uncertainties may also hinder the bridge between theory and reality. For example, how much of total-path resistance is attributable to pits versus lumen resistance and to what extent do branch junctions and branch length contribute to total resistance in very tall trees? That the slopes of the predawn and midday $\Psi_L$-height relationship of foliage quite near to the bole of 110-m tall redwoods were similar and close to the hydrostatic gradient (Koch *et al.*, 2004) implies that the added resistance, if any, may reside in long branches of giant trees. The question of how hydraulic resistance varies along the stem and branch axes and with tree height and size has been central to the debate about the nature of limits to tree height and size growth (e.g., Ryan and Yoder, 1997; Becker *et al.*, 2000; Bond and Ryan, 2000) and underlies the discussion of compensatory mechanisms that is developed more fully later.

### Phloem Transport

Changes in xylem water relations with height and path length should also affect phloem transport because the two systems are coupled hydraulically (see Chapter 2). As xylem $\Psi$ decreases with height, phloem solute concentration should increase to maintain osmotic water uptake from the apoplast and generate the turgor pressure in phloem cells necessary for phloem transport. Longer flow paths may also require a higher $\Psi p$ at the source end to overcome added resistance, and/or changes in sieve element anatomy, which reduce specific resistance. With increased height, the period of most active phloem loading might be expected to shift to the night when xylem water potential is least negative and water can be taken up most readily by sieve elements.

The maximum lengths of phloem transport pathways—that is, the distance from a source of photosynthate to the most distant sink it may serve—are likely well over 50 m and possibly considerably longer. Steep vertical gradients in foliar $\delta^{13}C$ within the crowns of the tallest redwoods (e.g., $-22\%$ at 110 m versus $-30\%$ for the lowest foliage at 50 m) (Koch *et al.*, 2004) argue for much branch-level C autonomy, at least within, say, a 10 m vertical distance. Root $\delta^{13}C$ in 100-m tall redwoods resembles that of the mid crown (Koch, unpublished data), suggesting a transport distance from source leaves to the roots that may exceed 70 m. Both older (Tyree *et al.*, 1974) and newer (Thompson and Holbrook, 2003) models indicate that a Münch-type pressure-flow system can operate over long distances (e.g., > 50 m), although these models differ regarding the design of a system that can maintain adequate delivery rates at such a distant sink. Measurements (Milburn and Kallarackal, 1989) and models (Tyree *et al.*, 1974; Thompson and Holbrook, 2003) indicate average phloem transport velocities of c. 0.2 to 0.8 m hr$^{-1}$, which produce transit times similar to those for the xylem, namely, several days or more to travel 70 m. Consistent with this, a study of temporal variation in stable carbon isotope composition of soil-respired $CO_2$ showed a lag of 1 to 4 days between a change in canopy photosynthesis and the consequent change in root respiration in a 20 to 25 m boreal forest stand (Ekblad and Högberg, 2001). Thompson and Holbrook (2003) used a model to show that for a given transport distance, the transit time is proportional to the inverse of the square of the sieve tube length. Thus, transit time would be reduced by having more sieve tubes of shorter length, and this then would require a relay system as envisioned by Lang (1979).

Long phloem pathways need not imply sluggish long distance communication via the phloem. Thompson and Holbrook (2004) proposed that information about the physiological status of source and sink tissues might be communicated rapidly over long distances by changes in phloem pressure that signal altered rates of loading or unloading, or changes in apoplastic water potential.

A final intriguing idea regarding phloem/xylem integration concerns the possible involvement of other solutes in osmoregulation (see Chapter 11). Because ions such as $K^+$ affect viscosity little while boosting osmotic potential, they might be particularly efficient at enhancing Münch flow. Highly regulated, energy-dependent $K^+$ recirculation between the xylem and phloem may be a key to integrating maintenance of pressure gradients in the face of variations in sucrose loading. Moreover, phloem-to-xylem recirculation could serve the additional function of returning $K^+$ to those regions high in trees where its potential to regulate xylem permeability via hydrogel control is most valuable (Zwieniecki *et al.*, 2001 and see Chapter 11).

## Mechanisms Compensating for Height Constraints

The hydraulic limitation hypothesis of Ryan and Yoder (1997) proposed that, as trees grow taller, total path length resistance increases, causing stomatal conductance to decline and thereby limiting carbon gain and further height growth. This provocative and biophysically reasonable hypothesis sparked a vigorous debate (Becker *et al.*, 2000; Mencuccini and Magnani, 2000; Bond and Ryan, 2000) and stimulated new research on the mechanisms by which trees might compensate for transport constraints imposed by increased height. Here we briefly illustrate some of these compensating mechanisms with selected examples from the literature. As background for the interrelationships among structural and physiological variables involved with hydraulic compensation, the reader is referred to the whole-tree hydraulic model developed by Whitehead and Jarvis (1981) and Whitehead *et al.* (1984), as used by McDowell *et al.* (2002a, 2002b).

### Sapwood Area/Leaf Area Ratio

The functional scaling of stemwood area (or sapwood area, $A_S$) supporting a given leaf area ($A_L$) dates to the pioneering studies of Huber (1928) who measured and compared the stem area/leaf mass ratio (Huber value, Zimmermann, 1983). Because $A_S/A_L$ provides an index of plant water supply relative to demand, it has been proposed that an increase in $A_S/A_L$ might compensate for increasing path length to maintain canopy conductance as tree height increases (Becker *et al.*, 2000). McDowell *et al.* (2002a) analyzed published data for a broad range of species and concluded that $A_S/A_L$ generally does increase with tree height for both tracheid-bearing and vessel-bearing species. However, this adjustment does not always eliminate reductions in canopy conductance with increased height (Schäfer *et al.*, 2000; McDowell *et al.*, 2002b). Adequate explanations are pending for cases where $A_S/A_L$ decreased with height (Köstner *et al.*, unpublished data in McDowell *et al.*, 2002a; Coyea and Margolis, 1992). As for most height-related phenomena, studies of $A_S/A_L$ have been limited to trees of half to two thirds of the maximum known height for a species. New studies of taller individuals should reveal more clearly the significance of $A_S/A_L$ adjustments in mitigating height-related constraints on xylem transport.

### Minimum Leaf Water Potential

To maintain the same $\Delta\Psi_{trans}$ *as height increases*, minimum leaf water potential, $\Psi_{L\,min}$, would have to decrease by 0.01 MPa m$^{-1}$ of vertical height. The $\Psi_{L\,min}$ was lower in taller compared to shorter individuals of Douglas-fir, *Pseudotsuga menziesii* (McDowell *et al.*, 2002b), *Eucalyptus saligna* Sm. (Barnard, 2000), *Pinus contorta* (Yoder *et al.*, 1994), and *E. regnans* (Connor *et al.*, 1977), but not in *Pinus sylvestris* (Magnani *et al.*, 2000) or *P. ponderosa* (Yoder *et al.*, 1994; Hubbard *et al.*, 1999; Ryan *et al.*, 2000). For Douglas-fir, McDowell *et al.* (2002b) found that $\Psi_{L\,min}$ was 0.5 MPa lower in 60-m trees than in 32-m and 15-m trees. They concluded that this contributed more to maintenance of leaf specific hydraulic conductance than did the observed increase in $A_S/A_L$. The ability to maintain stomatal opening at a lower $\Psi_{L\,min}$ in taller than shorter trees may allow an extended period of daily and seasonal gas exchange compared to when $\Psi_{L\,min}$ does not vary with height. Given the strong relationship of $\Psi_L$ to stomatal conductance, photosynthesis, cavitation, and tissue turgor and osmotic relations, understanding the mechanistic basis of $\Psi_{L\,min}$ acclimation is a fundamental challenge in tall tree ecophysiology.

## Sapwood Conductivity

An increase with height in the specific conductivity of sapwood should act to maintain whole-tree water transport as path length increases (Becker *et al.*, 2000). Pothier *et al.* (1989) found an increase in fully hydrated specific conductivity with tree size (stand development), while McDowell *et al.* (2002b) found no significant differences in fully hydrated specific conductivity of bole sapwood among 15-m, 32-m, and 60-m tall Douglas-fir, nor between shoots of 15-m and 60-m trees. Cavitation resistance of bole or branch sapwood increased with height in Douglas-fir (Domec and Gartner, 2001, 2002) and redwood (Koch *et al.*, 2004). The apparent tradeoff between hydraulic efficiency and cavitation resistance (Tyree *et al.*, 1994a) may argue against increases in specific conductivity in tall trees, the extreme height growth of which may best be served by cavitation avoidance (Koch *et al.*, 2004).

Perhaps the most fascinating recent discovery in xylem function was the demonstration that xylem sap inorganic ions could regulate sapwood permeability by causing the swelling of hydrogels in intervessel pit membranes (Zwieniecki *et al.*, 2001). Although this mechanism has not yet been tested in tall trees, and is of minor significance in the tracheid-bearing species examined thus far, this would seem an efficient mechanism for short-term regulation of sapwood permeability. Moreover, it raises the question of whether phloem-to-xylem $K^+$ recirculation might integrate dual roles for $K^+$ in phloem osmotic regulation and xylem sap flow (see Chapter 11). Studies of xylem and phloem ion concentrations across tree height gradients, and experimental tests of their functional significance, may well reveal new aspects of how tall trees deal with the transport constraints associated with great size.

## Viscosity and Stem Temperature

Xylem sap viscosity is essentially that of water, and is likely not influenced much by the range of xylem solute concentrations in most plants. Viscosity decreases by roughly 20% per 10° C increase, and this could significantly influence whole-tree water relations in different thermal environments. We propose that the greater energy inputs into the tops of large trees may in fact facilitate transport where and when it is most needed. Air temperature gradients of 5° C were noted in mid morning over a 20-m distance within the canopy of Norway spruce (Zweifel *et al.*, 2002), and air temperature may vary by 10° C or more within the crown of a 110-m tall coast redwood (Sillett, unpublished data). That temperature-driven changes in viscosity impact leaf gas exchange was first clearly demonstrated by Fredeen and Sage, (1999), and their data are discussed more fully later in the chapter. To our knowledge, the implications of this temperature compensation, essentially an indirect consequence of forest structure, for whole-tree transport have not been examined in tall trees.

### Water Storage

Water stored in plant tissues contributes to transpiration over periods of hours to months (see Holbrook, 1995) and could compensate for the reduction in soil water supply owing to gravity and increased transport resistance as trees grow taller. It is becoming increasingly clear that capacitance, specific conductivity, and cavitation resistance are interdependent features. Gartner and Meinzer (see Chapter 15) examine structural properties that influence sapwood capacitance, and discuss tradeoffs between sapwood capacitance and other sapwood functions. Here we highlight studies that bear on the potential role of foliar, branch, and stem capacitance in relation to the water relations of tall trees.

Leaves and small branches can harbor much of a tree's total capacitance (Tyree, 1988; Zweifel *et al.*, 2001), but whether their importance varies with tree size or species is unknown. Foliar capacitance makes water readily available at the sites of evaporation and might be particularly important in tall trees, which should have longer lag times for withdrawal and transport from distant stems and branches. In Norway spruce saplings, crown and stem reserves were depleted each day and supplied 10% to 65% of daily transpiration (Zweifel *et al.*, 2001), a range similar to that estimated from a model for western red cedar (Tyree, 1988). Both studies concluded that two thirds or more of the stored water was supplied by leaves.

Among trees of different heights, Phillips *et al.* (2003) found the importance of daily storage/discharge cycles in the main bole increased with tree size and height. Over the course of the growing season, bole-stored water contributed 20% to 25% of total daily water use in 60-m tall Douglas-fir trees compared to 7% in 15-m trees. From the same study, comparable figures for ponderosa pine (*P. ponderosa*) were 4% to 20% of total daily water use provided from storage in 36-m trees versus 2% to 4% in 12-m trees, and for Oregon white oak (*Quercus garryana*), 10% to 23% in 25-m trees versus 9% to 13% in 10-m tall trees. In all three species, the contribution of stored water to daily total water use increased as soil moisture declined from early to late in the growing season, and this increase was greater in taller trees. That sapwood volume per unit leaf area increased with height in all species, whereas sapwood volume per unit bole volume decreased with height, was interpreted as evidence that increased storage is an adaptive response to hydraulic constraints in tall trees and not simply a consequence of size. In other studies, the importance of storage did not vary with tree size across a strong climatic gradient (Maherali and DeLucia, 2001). As in temperate ecosystems, greater tree size may improve buffering against dry season reduction in soil moisture in the tropics (Goldstein *et al.*, 1998; Meinzer *et al.*, 1999).

There is evidence that a number of structural and functional features including sapwood/leaf area ratio, minimum leaf water potential, specific conductivity of sapwood, and water storage capacity change with height in ways that tend to compensate for increased path length and a greater influence of gravity in taller individuals of some species and environments. A more complete resolution of their importance will benefit from studies in taller trees and over time periods and locations that may favor particular compensatory mechanisms.

## Interaction Between Stomatal Regulation and Xylem Transport

The hydraulic conductivity and water supply to foliage and the subsequent requirement of that water for growth and losses through gas exchange via stomatal conductivity are inextricably linked. Growth of leaf tissue is turgor driven, yet as a result in large part to gravity, water potential and turgor decrease with tree height and may ultimately limit leaf expansion and further height growth (Koch *et al.*, 2004; Woodruff *et al.*, 2004). Stomatal conductance is also turgor mediated and has long been known to be sensitive to soil and atmospheric drought. In fact, stomatal conductance has been observed to decline with tree height (e.g., Schäfer *et al.*, 2000). However, our description, and to a greater extent our understanding, of the actual linkages between transport and stomata and the mechanism(s) involved are relatively limited. These linkages, if any, would be expected to be most critical in the tallest trees where path lengths and distance between soil resources and sinks are greatest.

That stomata control transpiration in forests, and especially conifer forests, is well supported (McNaughton and Jarvis, 1983). That stomatal conductance is sensitive to xylem conductance (the supply of water) is both self-evident and supported by the bulk of the literature (see Jones, 1998). Yet, it is the precision of the control mechanism that is of interest. In both angiosperm and gymnosperm trees, stomata commonly enable transpiration to operate near or at the threshold of xylem dysfunction (Tyree and Sperry, 1988; Sperry and Sullivan, 1992; Tyree *et al.*, 1994). Despite this seemingly precarious balance between supply of water to leaf tissue (xylem conductance) and loss of water from leaves (transpiration), mature tree mortality from catastrophic xylem embolism is rare (e.g., Kavanagh and Zaerr, 1997). That onset of stem xylem embolisms appears to be correlated with reversible tracheid collapse in needles of pine directly before decreases in stomatal conductance provides a tantalizing new clue as to the nature of the mechanistic linkage between xylem and stomatal conductances (Cochard *et al.*, 2004). However, the prevalence and importance of embolisms and reversible tracheid collapse as causal agents in the moisture status and signal-transduction pathway of tall trees remain to be seen.

The other half of "the equation" governing water balance of foliage is rate of loss. Loss of water from foliage, or transpiration, is the product of leaf conductance (i.e., stomatal conductance in many instances) and the vapor pressure deficit (VPD). In most trees, stomatal conductance increases initially with increase in VPD (the driving gradient for transpiration), then remains constant or decreases above a threshold that is commonly at or near 1 kPa (Anthoni *et al.*, 1999; Bauerle *et al.*, 1999; Ryan *et al.*, 2000). Leaf-level studies suggest that stomata respond to transpiration rather than VPD or relative humidity directly, perhaps through rates of peristomatal (Lange *et al.*, 1971) or mesophyll (Schulze, 1994) water loss. In any event, the literature strongly suggests that stomata are responding on one hand to supply of water from the vascular tissue and, on the other, to transpiration.

Thus, an obvious explanation for the often precarious balance between supply (soil/xylem) and loss (stomatal conductance and VPD) of water from plants (and particularly where long distances separate points of water supply and loss) is that the guard cells (or associated cells) are controllers as well as sensors of water balance. This type of control has been suggested by a number of groups since 1965 (see Maier-Maercker, 1998; Tyree and Sperry, 1988; and Schulze, 1994). Additional discussion of stomatal-xylem interrelationships can be found in Chapter 4 and, at least heuristically, in the model of Buckley *et al.* (2003). Here we present a predominantly conceptual model explaining a way in which supply and loss of water from leaves are coordinated with particular emphasis on issues germane to tall trees.

## Supply:Loss Stomatal Conductance Model

Fredeen and Sage (1999) described a previously unpublished phenomenon by which branchlet-level temperature has a direct effect on transpiration in white spruce (*Picea glauca*) that can in part (~50%) be explained by changes in viscosity of water. This model is expanded here to provide an empirical/mechanistic basis to describe how stomata permit transpiration (water loss function) at $\Psi_{xylem}$ very near, but typically above, the point of catastrophic xylem embolism where loss of stem hydraulic conductivity occurs (water supply function). Such precise regulation may be particularly important in tall trees, in which cavitation avoidance may be a prerequisite for extreme height growth (see previous discussion).

In the model, stomatal closure occurs at the intersection between the function describing water supply to guard cells and that of the rates of evaporative loss of water through the stomata via transpiration. Simply put, and ignoring interactions between guard and epidermal cells, stomata close when loss of water exceeds supply and guard cells lose turgor. This form of stomatal closure may be of special significance for tall trees with long path lengths and foliage that may be ostensibly chemically decoupled from soil moisture levels by many days (Fig. 21.1).

The supply function in our model is given by Darcy's Law, which is proportional to the gradient in water potential between xylem and guard cell, and inversely proportional to the viscosity of water. The loss function is described by a linear relationship between transpiration and VPD (i.e., for a given species, age, location, height, etc.) as was observed for white spruce (Fredeen and Sage, 1999) (Fig. 21.2), although for predictive purposes, a dynamic model of stomata and transpiration involving feedback loops (e.g., Franks and Farquhar, 1999) would be required. Nevertheless, we propose that at least conceptually, the intersection of the supply and loss functions sets the point of stomatal closure (Fig. 21.3). In concept, this model is similar to the graphical representation of plant cell enlargement presented by Taiz and Zeiger (2002), although in the case of guard cells, the enlargement is reversible. Our supply-loss mechanism would automatically provide a temperature-corrected set-point for stomatal closure to restrict transpiration to a rate that could be met by xylem transport. Stomatal closure would occur passively, as hydraulic supply to guard cells falls below their transpiration rate, resulting in loss of guard cell turgor. With stomatal closing, leaf and guard cell transpiration is reduced, slowing the reduction in xylem tension and ultimately bringing supply and loss functions into balance. This entire mechanism would largely circumvent runaway xylem dysfunction, unless of course stress was particularly prolonged or severe, or if nonstomatal loss rates (e.g., transcuticular) became important.

Increased temperature enhances the supply of water to leaves by reducing viscosity (Fredeen and Sage, 1999) and increases the potential loss from leaves via increased leaf-to-air VPD. In this way, positive temperature gradients with tree height should tend to match supply to periods of greatest demand, partially mitigating path length and gravity constraints, as long as the influence of reduced viscosity exceeded that of any increases in VPD.

## Conclusions and Directions for Future Research

A variety of transport challenges face the tallest trees, many of which appear to push against the biophysical limits of their climatic range. A record

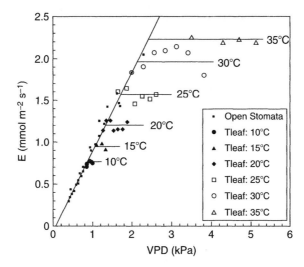

**Figure 21.2** Effect of measurement temperature on the relationship between transpiration (E) and VPD (kPa) in white spruce seedlings under well-watered conditions (modified from Fredeen and Sage, 1999). The initial slope of E versus VPD is shown (E = VPD * 0.95 − 0.07: $r^2$ = 0.98).

redwood tree in a temperate rainforest may face many of the transport challenges that are faced by shorter trees in less hospitable environments. There is a clear need for more transport work on the tallest of our trees in which gradients of important ecophysiological variables are larger and more easily resolved than in shorter trees. This is particularly true when examining the influence of gravity because the hydrostatic gradient appears to account for an increasing fraction of total water potential as trees grow taller. We do not know whether the low maximum water potential and long path length in tall trees preclude certain phenomena such as xylem embolism repair (refilling) and the rapid root-to-shoot signaling that is possible in short-statured plants.

Our understanding of long distance phloem transport, particularly in conifers, and its integration with xylem transport is particularly lacking. Models such as that of Thompson and Holbrook (2003, 2004) have renewed the theoretical exploration of the problems and design solutions of phloem transport over great distances. We hope that the development of new tools (e.g., for *in situ* measurement of phloem pressure and solute concentrations) will soon enable experimental studies to test theory and build an understanding of xylem/phloem integration in tall trees. Finally, testing of the supply-loss model presented here will help determine whether stomata are indeed key sensors and regulators of water potential in tall trees.

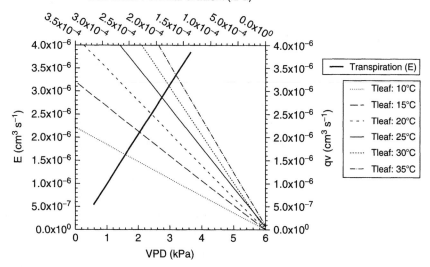

**Figure 21.3** An empirical-mechanistic supply-loss model to explain the linkage between hydraulic supply of water to leaves (qv is the volume flow given by Darcy's Law at various temperatures versus the $\Delta\Psi$ (kPa) between xylem and leaf cell) and the volume loss of water through transpiration (E) versus VPD (kPa). Tracheid radius was assumed to be 23 μm (Ewers, 1985). The effect of temperature on the supply of water and, therefore, the transpiration set-points are the intersection points. It is assumed that the full temperature effect on the water supply function is double the effect of viscosity alone as demonstrated for roots (Kramer, 1940) and foliage (Fredeen and Sage, 1999).

# References

Anthoni, P. M., Law, B. E. and Unsworth, M. H. (1999) Carbon and water vapor exchange of an open-canopied ponderosa pine ecosystem. *Agric For Meteorol* **95**: 151-168.

Barnard, H. R. (2000) Testing the hydraulic limitation hypothesis in fast-growing *Eucalyptus saligna*. M.S. thesis. Colorado State University, Fort Collins, Colorado.

Bauerle, W. L., Hinckley, T. M., Cermak, J., Kucera, J. and Bible, K. (1999) The canopy water relations of old-growth Douglas-fir trees. *Trees Structure Function* **13**: 211-217.

Becker, P., Meinzer, F. C. and Wullschleger, S. D. (2000) Hydraulic limitation of tree height, a critique. *Funct Ecol* **14**: 4-11.

Bond, B. J. and Ryan, M. G. (2000) Comment on 'Hydraulic limitation of tree height: A critique' by Becker, Meinzer and Wullschleger. *Funct Ecol* **14**: 37-40.

Boucher, J. F., Munson, A. D. and Bernier, P. Y. (1995) Foliar absorption of dew influences shoot water potential and root growth in *Pinus strobus* seedlings. *Tree Physiol* **15**: 819-823.

Breazeale, E. L., McGeorge, W. T., and Breazeale, J. F. (1950) Moisture absorption by plants from an atmosphere of high humidity. *Plant Physiol* **25**: 413-419.

Buckley, T. N., Mott, K. A. and Farquhar, G. D. (2003) A hydromechanical and biochemical model of stomatal conductance. *Plant Cell Environ* **26**: 1767-1785.

Burgess, S. S. O. and Dawson, T. E. (2004) The contribution of fog to the water relations of *Sequoia sempervirens* (D.Don)., foliar uptake and prevention of dehydration. *Plant, Cell, and Environ* **27**: 1023-1034.

Carder, A. (1995) *Forest Giants of the World*. Fitzhenry & Whiteside, Ltd., Markham, Ontario, CA.

Cochard H., Froux, F., Mayr, S. and Coutand, C. (2004) Xylem wall collapse in water-stressed pine needles. *Plant Physiol* **134**: 401-408.

Connor, D. J., Legge, N. J. and Turner, N. C. (1977) Water relations of mountain ash (*Eucalyptus regnans* F Muell.) forests. *Aust J Plant Physiol* **4**: 735-762.

Coyea, M. R. and Margolis, H. A. (1992) Factors affecting the relationship between sapwood area and leaf area of balsam fir. *Can J Forest Res* **22**: 1684-1693.

Dawson, T. E. (1998) Fog in the California redwood forest: ecosystem inputs and use by plants. *Oecologia* **117**: 476-485.

Domec, J.-C. and Gartner, B. L. (2001) Cavitation and water storage capacity in bole xylem segments of mature and young Douglas-fir trees. *Trees* **15**: 204-214.

Domec, J.-C. and Gartner, B. L. (2002) Age- and position-related changes in hydraulic versus mechanical dysfunction of xylem, inferring the design criteria for Douglas-fir wood structure. *Tree Physiol* **22**: 91-104.

Ekblad, A. and Högberg, P. (2001) Natural abundance of $^{13}$C in $CO_2$ respired from forest soils reveals speed of link between tree photosynthesis and root respiration. *Oecologia* **127**: 205-311.

Ewers, F. W. (1985) Xylem structure and water conduction in conifer trees, dicot trees, and lianas. *Int Assoc Wood Anat Bull N S* **6**: 309-317.

Franks, P. J. and Farquhar, G. D. (1999) A relationship between humidity response, growth form and photosynthetic operating point in C-3 plants. *Plant Cell Environ* **22**: 1337-1349.

Fredeen, A. L. and Sage, R. F. (1999) Temperature and humidity effects on branchlet gas-exchange in white spruce, an explanation for the increase in transpiration with branchlet temperature. *Trees Structure Funct* **14**: 161-168.

Friend, A. D. (1993) The prediction and physiological significance of tree height. In *Vegetation Dynamics and Global Change* (A. M. Solomon and H. H. Shugart, eds.) pp. 101-115. Chapman and Hall, New York.

Goldstein, G., Andrade, J. L., Meinzer, F. C., Holbrook, N. M., Cavelier, J., Jackson, P. and Celis, A. (1998) Stem water storage and diurnal patterns of water use in tropical forest canopy trees. *Plant Cell Environ* **21**: 397-406.

Holbrook, N. M. (1995) Stem water storage. In *Plant Stems: Physiological and Functional Morphology* (B. L. Gartner, ed.) pp. 151-174. Academic Press, San Diego.

Hubbard, R. M., Bond, B. J. and Ryan, M. G. (1999) Evidence that hydraulic conductance limits photosynthesis in old *Pinus ponderosa* trees. *Tree Physiol*. **19**: 165-172.

Huber, B. (1928) Weitere quantitative Untersuchungen über das Wasserleitungssystem der Pflanzen. *Jb Wiss Bot* **67**: 877-959.

Jennings, G. M. (2003) M.S. thesis. Humboldt State University, Arcata, California.

Jepson, J. (2000) *The Tree Climber's Companion*. Beaver Tree Publishing, Longville, Minnesota.

Jones, H. G. (1998) Stomatal control of photosynthesis and transpiration. *J Exp Bot* **49**: 387-398.

Katz, C., Oren, R., Schulze, E.-D. and Milburn, J. A. (1989) Uptake of water and solutes through twigs of *Picea abies* (L.). Karst. *Trees* **3**: 33-37.

Kavanagh, K. L. and Zaerr, J. B. (1997) Xylem cavitation and loss of hydraulic conductance in western hemlock following planting. *Tree Physiol* **17**: 59-63.

Koch, G. W., Sillett, S. C., Jennings, G. M. and Davis, S. D. (2004) The limits to tree height. *Nature* **428**: 851-854.

Kramer, P. J. (1940) Root resistance as a cause of decreased water absorption by plants at low temperatures. *Plant Physiol* **15**: 63-79.

Kramer, P. J. and Boyer, J. S. (1995) *Water Relations of Plants and Soils.* Academic Press, San Diego.

Lang, A. (1979) A relay mechanism for phloem translocation. *Ann Bot* **44:** 141-145.

Lange, O. L., Losch, R., Schulze, E. D. and Kappen, L. (1971) Response of stomata to changes in humidity. *Planta* **100:** 76-86.

Magnani, F., Mencuccini, M. and Grace, J. (2000). Age-related decline in stand productivity: The role of structural acclimation under hydraulic constraints. *Plant Cell Environ* **23:** 251-264

Maherali, H. and DeLucia, E. H. (2001) Influence of climate-driven shifts in biomass allocation on water transport and storage in ponderosa pine. *Oecologia* **129:** 481-491.

Maier-Maercker, U. (1998) Dynamics of change in stomatal response and water status of *Picea abies* during a persistent drought period, a contribution to the traditional view of plant water relations. *Tree Physiol* **18:** 211-222.

McCulloh, K. A., Sperry, J. S. and Adler, F. R. (2003) Water transport in plants obeys Murray's Law. *Nature* **421:** 939-942.

McDowell, N., Barnard, H., Bond, B. J., Hinckley, T., Hubbard, R. M., Ishii, H., Köstner, B., Magnani, F., Marshall, J. D., Meinzer, F. C., Phillips, N., Ryan, M. G. and Whitehead, D. (2002a) The relationship between tree height and leaf area:sapwood area ratio. *Oecologia* **132:** 12-20.

McDowell, N. G., Phillips, N., Lunch, C., Bond, B. J. and Ryan, M. G. (2002b) An investigation of hydraulic limitation and compensation in large, old Douglas-fir trees. *Tree Physiol* **22:** 763-774.

McNaughton, K. G. and Jarvis, P. G. (1983) Predicting effects of vegetation changes on transpiration and evaporation. In *Water Deficits and Plant Growth* (T. T. Kozlowski, ed.) pp.1-47. Academic Press, New York.

Meinzer, F. C. (2002) Co-ordination of vapour and liquid phase water transport properties in plants. *Plant Cell Environ* **25:** 265-274.

Meinzer, F. C., Andrade, J. L., Goldstein, G., Holbrook, N. M., Cavelie, J. and Wright, S. J. (1999) Partitioning of soil water among canopy trees in a seasonally dry tropical forest. *Oecologia* **121:** 293-301.

Mencuccini, M. and Grace, J. (1996) Hydraulic conductance, light interception, and needle nutrient concentration in Scots pine stands and their relation with net primary productivity. *Tree Physiol* **16:** 459-468.

Mencuccini, M. and Magnani, F. (2000) Comment on Hydraulic limitation of tree height: A critique by Becker, Meinzer and Wullschleger. *Funct Ecol.* **14:** 135-136

Milburn, J. A. and Kallarackal, J. (1989) Physiological aspects of phloem translocation. In *Transport of Photoassimilates* (D. A. Baker and J. A. Milburn, eds.) pp. 264-305. John Wiley & Sons, New York.

Niklas, K. J. (1994) *Plant Allometry: The Scaling of Form and Process.* University of Chicago Press, Chicago.

Nobel, P. S. (1999) *Physicochemical and Environmental Plant Physiology.* 2nd Ed. Academic Press, San Diego.

Panshin, A. J. and de Zeeuw, C. (1980) *Textbook of Wood Technology, Structure, Identification, Properties, and Uses of the Commercial Woods of the United States and Canada.* 4th Ed. McGraw-Hill College, New York.

Phillips, N. G., Ryan, M. G., Bond, B. J., McDowell, N. G., Hinckley, T. M. and Cermak, J. (2003) Reliance on stored water increases with tree size in three species in the Pacific Northwest. *Tree Physiol* **23:** 237-245.

Pickard, W. F. (1989). How might a tracheary element which is embolized by day be healed by night? *J Theor Biol* **141:** 259-279.

Pittermann, J. and Sperry, J. (2003) Tracheid diameter is the key trait determining the extent of freezing-induced embolism in conifers. *Tree Physiol* **23:** 907-914.

Pockman, W. T. and Sperry, J. S. (2000) Vulnerability to cavitation and the distribution of Sonoran desert vegetation. *Am J Bot* **87:** 1287-1299.

Pothier, D., Margolis, H. A. and Waring, R. H. (1989) Patterns of change of saturated sapwood permeability and sapwood conductance with stand development. *Can J For Res* **19**: 432-439.

Richards, P. W. (1996) *The Tropical Rain Forest: An Ecological Study.* 2nd Ed. Cambridge University Press, Cambridge, U.K.

Ryan, M. G. Bond, B. J., Law, B. E., Hubbard, R. M., Woodruff, D., Cienciala, E. and Kucera, J. (2000) Transpiration and whole-tree conductance in ponderosa pine trees of different heights. *Oecologia* **124**: 553-560.

Ryan, M. G. and Yoder, B. J. (1997) Hydraulic limits to tree height and growth. *BioScience* **47**: 235-242.

Saliendra, N. Z., Sperry, J. S. and Comstock, J. P. (1995) Influence of leaf water status on stomatal response to humidity, hydraulic conductance, and soil drought in *Betula occidentalis*. *Planta* **196**: 357-366.

Schäfer, K. V. R., Oren, R. and Tenhunen, J. D. (2000) The effect of tree height on crown level stomatal conductance. *Plant Cell Environ* **23**: 365-375.

Scholander, P. F., Hammel, H. T., Bradstreet, E. D. and Hemmingsen, E. A. (1965) Sap pressure in vascular plants. *Science* **148**: 339-346.

Schulze, E.-D. (1994) The regulation of plant transpiration: Interactions of feedforward, feedback, and futile cycles. In *Flux Control in Biological Systems* (E.-D. Schulze, ed.) pp.203-235. Academic Press, London.

Sillett, S. C. and Bailey, M. G. (2003) Effects of tree crown structure on biomass of the epiphytic fern *Polypodium scouleri* (Polypodiaceae) in redwood forests. *Am J Bot* **90**: 255-261.

Sperry, J. S. and Sullivan, J. E. M. (1992) Xylem embolism in response to freeze-thaw cycles and water stress in ring-porous, diffuse-porous, and conifer species. *Plant Physiol* **100**: 605-613.

Stone, E. L. and Kalisz, P. J. (1991) On the maximum extent of tree roots. *For Ecol Mgt* **46**: 59-102.

Taiz, L. and Zeiger, E. (2002) *Plant Physiology.* 3rd ed. Sinauer Associates, Sunderland, MA.

Thompson, M. V. and Holbrook, N. M. (2003) Application of a single-solute non-steady-state phloem model to the study of long-distance assimilate transport. *J Theor Biol* **220**: 419-455.

Thompson, M. V. and Holbrook, N. M. (2004) Scaling phloem transport: Information transmission. *Plant Cell Environ* **27**: 509-519.

Tobiessen, P., Rundel, P. W. and Stecker, R. E. W. (1971) Water potential gradient in a tall *Sequoiadendron*. *Plant Physiol* **48**: 303-304.

Tyree, M. T. (1988) A dynamic model for water flow in a single tree: Evidence that models must account for hydraulic architecture. *Tree Physiol* **4**: 195-217.

Tyree, M. T., Christy, A. L. and Ferrier, J. M. (1974) A simpler interative steady state solution of Münch pressure-flow systems applied to long and short translocation paths. *Plant Physiol* **54**: 589-600.

Tyree, M. T., Davis, S. D. and Cochard, H. (1994a) Biophysical perspectives of xylem evolution: Is there a tradeoff of hydraulic efficiency for vulnerability to dysfunction? *IAWA Bull* **15**: 335-360.

Tyree, M. T., Kolb, K. J., Rood, S. B. and Patino, S. (1994b) Vulnerability to drought-induced cavitation of riparian cottonwoods in Alberta: A possible role in the decline of the ecosystem? *Tree Physiol* **14**: 455-466.

Tyree, M. T., Salleo, S., Nardini, A., Lo Gullo, M. A. and Mosca, R. (1999) Refilling of embolized vessels in young stems of laurel. Do we need a new paradigm? *Plant Physiol* **120**: 11-21.

Tyree, M. T. and Sperry, J. S. (1988) Do woody plants operate near the point of catastrophic xylem dysfunction caused by dynamic water stresses? Answers from a model. *Plant Physiol* **88**: 574-580.

West, G. B., Brown, J. H. and Enquist, B. J. (1999) A general model for the structure and allometry of plant vascular systems. *Nature* **400**: 664-667.

Whitehead, D., Edwards W. R. N. and Jarvis, P. G. (1984) Conducting sapwood area, foliage area, and permeability in mature trees of *Picea sitchensis* and *Pinus contorta*. *Can J For Res* **14**: 940-947.

Whitehead, D. and Jarvis, P. G. (1981) Coniferous forests and plantations. In *Water Deficits and Plant Growth* (T. T. Kozlowski, ed.) pp. 99-152. Academic Press, New York.

Woodruff, D. R., Bond, B. J. and Meinzer, F. C. (2004) Does turgor limit growth in tall trees? *Plant Cell Environ* **27:** 229-236.

Yang, S. and Tyree, M. T. (1992) A theoretical model of hydraulic conductivity recovery from embolism with comparison to experimental data on *Acer saccharum. Plant Cell Environ* **15:** 633-643.

Yates, D. J. and Hutley, L. B. (1995) Foliar uptake of water by wet leaves of *Sloanea woollsii,* an Australian subtropical rainforest tree. *Aust J Bot* **43:** 157-167.

Yoder, B. J., Ryan, M. G., Waring, R. H., Schoettle, A. W. and Kaufmann, M. R. (1994) Evidence of reduced photosynthetic rates in old trees. *For Sci* **49:** 513-527.

Zimmermann, M. H. (1983) *Xylem Structure and the Ascent of Sap.* Springer-Verlag, Berlin.

Zweifel, R., Böhm, P. and Häsler, R. (2002) Midday stomatal closure in Norway spruce: Reactions in the upper and lower crown. *Tree Physiol* **22:** 1125-1136.

Zweifel, R., Item, H. and Häsler, R. (2001) The link between diurnal stem radius change and tree water relations. *Tree Physiol* **21:** 869-877.

Zweiniecki, M. A., Melcher, P. and Holbrook, N. M. (2001) Hydrogel control of xylem hydraulic resistance in plants. *Science* **291:** 1059-1062.

# 22

# Senescence in Secondary Xylem: Heartwood Formation as an Active Developmental Program

*Rachel Spicer*

Trees are exceptionally long-lived organisms and must be able to respond to the environment throughout their development. This plasticity is particularly remarkable in large woody stems, where vascular development maintains a dynamic balance between the demands for mechanical support and the transport of water and assimilates. Because of the cumulative nature of growth in woody stems, vascular development includes both the creation and senescence of tissue. Rates of cell division and the production of functionally different cell types by the vascular cambium determine the relative amounts of secondary xylem and phloem, and allow for tissue specificity in meeting the demands for support, carbohydrate storage, and efficiency of transport. The amounts of functional xylem and phloem are also continually adjusted through senescence, although the fates of old phloem and xylem are inherently different given the radial nature of cell production by the cambium. Phloem is ultimately partitioned to the outside of the stem by successive phellogens and sloughed off as bark, while nonfunctional xylem remains within the stem and is compartmentalized as heartwood.

Functional secondary xylem must provide mechanical support, meet water transport needs, serve as storage for both water and carbohydrates, and respond to stem wounding through the production of new tissue and the compartmentalization of decay. Although the bulk of secondary xylem is nonliving, it contains an intricate network of living cells oriented both axially and radially that link xylem and phloem (Fig. 22.1; see also Color Plate section), allowing constant exchange between the transpiration stream and adjacent living parenchyma. Sapwood is defined by the presence of these living cells, which make up about 5% to 35% of the tissue volume. As a respiring tissue, sapwood represents a large carbohydrate sink, and it is therefore important that a tree be able to regulate the volume of sapwood to a level suitable for a particular environment. Heartwood formation describes the process of sapwood senescence, in which parenchyma cells die, all physiological activity ceases, and a core of nonfunctional xylem forms in the center of the stem. Retention of this tissue throughout the life of a plant presents new challenges to development, namely, the protection of dead plant material from microorganisms. Many of the events that characterize heartwood formation are part of this process of compartmentalization. All evidence suggests that heartwood formation is an actively regulated stage of woody plant development.

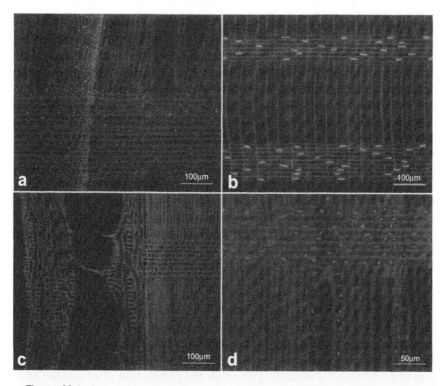

**Figure 22.1**  Living ray and axial parenchyma shown by fluorescent staining (DAPI) of nuclei in radial sections of secondary xylem. (A) Ray and marginal (axial) parenchyma in *Liriodendron tulipifera.* Marginal parenchyma refers to axial parenchyma formed at the growth ring boundary and is characteristic of this species. (B) Two rays in the outermost sapwood of *Tsuga canadensis.* The absence of axial parenchyma is typical of many conifers. (C) Ray parenchyma and a network of axial parenchyma and tracheids surrounding an earlywood vessel in *Quercus rubra.* (D) Ray and axial parenchyma surrounding a latewood vessel in *Fraxinus americana.* Marginal (axial) parenchyma is also shown at the growth ring boundary on the far right. (See also Color Plate section.)

# The Role of Parenchyma Cell Death in Heartwood Formation

## Parenchyma Activity Preceding Cell Death

The hallmark of heartwood formation is the death of both axial and ray parenchyma cells within the narrow transition zone between sapwood and heartwood. Events immediately preceding cell death are responsible for the best known characteristics of heartwood and are themselves suggestive of an active cell metabolism before death. These include rapid synthesis of secondary metabolites, formation of blockages (e.g., tyloses and gels) within conducting elements, and changes in the structure of pits connecting conducting elements with parenchyma and with each other. The overwhelming majority of research in this area has been descriptive work documenting the chemical and structural changes that occur during heartwood formation, rather than exploring the underlying physiology of the cells responsible for these changes. The application of molecular biological tools to xylem development should improve our understanding of the role of parenchyma cells in xylem senescence.

In many conifers and angiosperms, parenchyma in the transition zone synthesize secondary metabolites (often termed *extractives* in the forest products and wood anatomy literature). These compounds are deposited in the walls of neighboring cells and often fill parenchyma lumen, imparting properties of decay resistance, color, and reduced permeability to the heartwood. They are predominantly phenolic compounds found in a range of polymerization states, including lignans, stilbenes, flavonoids, and quinones, with the latter two responsible for the color change associated with heartwood formation (Hillis, 1987). Phenolics are synthesized in the transition zone *in situ* from precursors (Magel *et al.*, 1991; Beritognolo *et al.*, 2002), but likely require the import of carbon skeletons from the phloem and outer sapwood (Magel *et al.*, 1994). Phenolics may accumulate in the vacuole (Wink, 1997; Dehon *et al.*, 2002) and polymerize after rupture of the tonoplast. Enzymes involved in phenolic synthesis show increased activity in the transition zone and are regulated both at the level of transcription and via post-translational modification (Magel, 2000; Beritognolo *et al.*, 2002). Further polymerization may occur via nonenzymatic oxidation, resulting in changes in chemical composition and coloration of the heartwood after parenchyma cell death. Once synthesized, phenolics appear to fill ray and axial parenchyma in the innermost transition zone, and may infiltrate (presumably polymerizing with cell wall components) and coat the walls of adjacent conducting elements. The heartwood of some conifers (members of *Larix* and *Pinus*) also contains water-soluble carbohydrates (Hillis, 1987), but little is known about their origin and distribution. An exhaustive review of heartwood extractive chemistry can be found in Hillis (1987).

Further reductions in permeability occur when axial conduits are blocked through the secretion of gels (gums) and formation of tyloses in angiosperm vessels, and the formation of tylosoids in conifer resin canals. These blockages or occlusions may also form within sapwood, both normally and in response to injury (see later), but they are a defining feature of heartwood formation. Excellent surveys of the occurrence of tyloses and gels in temperate species can be found in Gerry (1914) and Saitoh *et al.* (1993). Whether vessels are occluded by gels or tyloses is correlated with the size of the half-bordered pit (connecting parenchyma and vessel) aperture, with tyloses forming only in vessels with larger pits (Chattaway, 1949; Bonsen and Kucera, 1990; Saitoh *et al.*, 1993), although it is not known whether this observation is of any mechanistic significance. Gels and tyloses originate from both ray and axial parenchyma, but the latter form primarily from rays, even in vessels surrounded by a sheath of axial parenchyma (Chattaway, 1949; Saitoh *et al.*, 1993). In either case, the transfer of material into the vessel occurs through the protective layer, a loose polysaccharide network rich in pectin that surrounds the plasma membrane of vessel-associated parenchyma and is particularly thick at the vessel-parenchyma pits (Chafe, 1974; Chafe and Chauret, 1974). This layer was originally thought to protect (hence the name) living parenchyma from the changing hydrostatic pressure of the transpiration stream, as well as the hydrolytic enzymes associated with cell wall degradation during vessel development. Described as porous and loosely fibrillar, it is now more appropriately credited with serving to increase the contact area between the apoplast and symplast (Chafe, 1974; Barnett *et al.*, 1993).

Gels are pectic polysaccharides synthesized by parenchyma cells and secreted into vessels through half-bordered pits in the transition zone of many angiosperm species. Rioux *et al.* (1998) make a good case for use of the term *gels* rather than the more loosely used *gums*, which vary widely in chemical composition (e.g., compare true gums, latex, kino, and manna) (Hillis, 1987) and are synthesized by epithelial cells and actively secreted into gum canals, the angiosperm equivalent to coniferous resin canals. Although gels make up the bulk of the occluding mass within conduits, they may also occur with a layer rich in lignin and other phenolics (Rioux *et al.*, 1988).

Tyloses are produced by an outgrowth of a ray or axial parenchyma cell that expands into a neighboring vessel. Tylosis growth during heartwood formation can occur over one to several growing seasons, when the protective layer between the parenchyma cell wall and plasma membrane advances into a vessel through a half-bordered pit. Although its role in parenchyma physiology and tylosis formation is not well understood, the protective layer is involved in the transfer (i.e., material appears to move across the protective layer) of a loose cellulosic network similar to primary wall material that expands into a neighboring conduit as the parenchyma protoplast enters through a pit (Chafe, 1974; Rioux *et al.*, 1988). Multiple tyloses typically block a single vessel, and there is much variation in the thickness and lamellar structure of the tylosis wall, the presence or absence of pectin and lignin rich layers, and even pitting between appressed tyloses (Kórán and Côté, 1964; Foster, 1967). In this sense tyloses should be viewed as transdifferentiated parenchyma; their cytoplasm, including the nucleus and functional organelles, migrate through the pit into the vessel behind a moving front of developing cell wall. Tylosoids are similar to tyloses, except that they form in gum and resin canals as outgrowths of the thin-walled epithelial cells lining the canal.

Permeability is further reduced before parenchyma death through changes in bordered pit structure. Conifer pits become aspirated in the transition zone when the impermeable torus is displaced to block flow through tracheids (Hart and Thomas, 1967). Aspirated pits in conifers and angiosperm pits (which lack tori in all but a few cases) are then further sealed through the deposition of secondary metabolites over the pit membrane (Krahmer and Côté, 1963; Yamamoto, 1982).

### Sapwood Longevity and Patterns of Parenchyma Cell Death

There is great variation among species in the length of time that sapwood is maintained, ranging from as few as 2 years in *Quercus rubra*, a ring-porous angiosperm, to over 150 years in *Pinus ponderosa*, a conifer from xeric habitats (R. Spicer, personal observation). Determining the location of the sapwood/heartwood boundary (and therefore the longevity of sapwood) is generally straightforward because by definition all parenchyma cells are dead in the heartwood. However, some cell death may occur within the sapwood, and species vary in the proportion of cells that remain alive in the innermost, oldest sapwood. Most studies suggest a gradual loss of living parenchyma from the cambium to the sapwood/heartwood boundary, but this work is almost exclusively in conifers (Necesaný, 1968; Nobuchi and Harada, 1983; Yang and Tyree, 1992; Gartner *et al.*, 2000). In contrast, parenchyma cells may remain alive up to an abrupt boundary in many angiosperms (Fig. 22.2). It is not known to what extent these radial patterns of cell death are characteristic of a species, or how they vary within a tree (i.e., stem versus branch) or change with the environment. A better understanding of the process of cell death within the sapwood, and how it compares to cell death associated with heartwood formation, will be important in studies of sapwood longevity.

The extent to which tylosis and gel formation occurs in the sapwood may determine whether this apparent difference between angiosperms and conifers in the relative abruptness of parenchyma cell death holds across a broad taxonomic sampling. Tylosis formation culminates in parenchyma cell death and is known to occur in the outer sapwood of a number of

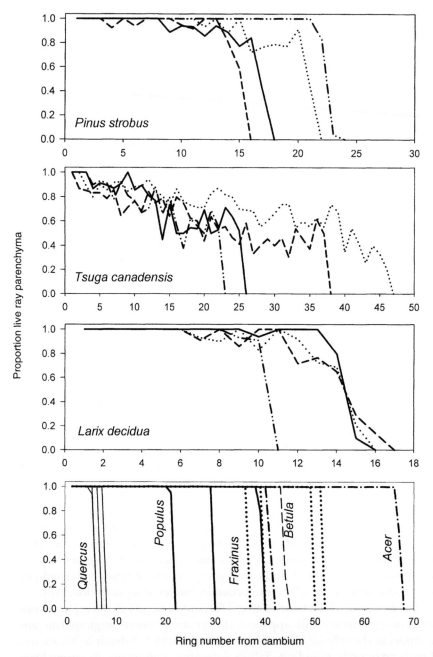

**Figure 22.2**   Proportion of rays that remain alive across a radial transect through the sapwood of conifers and angiosperms. Conifers (*Picea abies* and *Juniperus virginiana* not shown)

angiosperms (Gerry, 1914; Saitoh *et al.*, 1993). Because the development of tyloses in sapwood may span several years, parenchyma death associated with tylosis formation could occur at the sapwood/heartwood boundary, but this is not known. Likewise, it is not clear to what extent gel formation is associated with parenchyma death in the sapwood. Gel occlusions form in the transition zone between sapwood and heartwood, but are best studied as a wound response in infected xylem. Both cases end in cell death, but the time lags involved and role of gel occlusions in the sapwood are not well studied. It may be that the proportion of parenchyma that die in the sapwood of angiosperms during gel and tylosis formation is quite small relative to the proportion of parenchyma that die in the sapwood of conifers.

### Possible Nondevelopmental Causes of Parenchyma Cell Death

Theories for the actual cause of cell death in secondary xylem are numerous but have received surprisingly little attention given the importance of this process in driving heartwood formation. Early proposals suggested the buildup of secondary metabolites (e.g., polyphenolics) created a toxic environment for cells, but this was based on the idea that these compounds were waste products transported radially to the transition zone (Stewart, 1966). We now know that phenolics are synthesized (or at least polymerized) *in situ*, with precursors safely compartmentalized in the vacuole (Necesaný, 1973; Wink, 1997). Toxicity also seems an unlikely agent of cell death because some trees produce little or no secondary metabolites. Some species have been erroneously described as not producing "true" heartwood because of the lack of color change, but they show the same pattern of parenchyma death (e.g., *Populus grandidentata* in Fig. 22.2) as other species.

Gas contents within tree stems differ significantly (reduced $O_2$ and elevated $CO_2$) from that of the ambient atmosphere (Chase, 1934; Carrodus and Triffett, 1975; Eklund, 1993, 2000; Eklund and Klintborg, 2000; Spicer and Holbrook, 2005), and it has been proposed that cells in the transition zone die from anoxic conditions (Panshin and de Zeeuw, 1980; Eklund, 2000). This hypothesis is especially attractive in light of evidence that the

---

show a gradual loss of living parenchyma across the sapwood, whereas 100% of angiosperm ray tissue is alive until an abrupt sapwood/heartwood boundary. The nuclei disintegrate at the boundary, with no trace of nuclear material in the heartwood (nor in the dead ray cells of conifer sapwood). Vital staining indicates that cells remain alive while the nucleus is present. Although not quantified directly, axial parenchyma cells in angiosperms also retained nuclei until the sapwood/heartwood boundary. Each line represents an individual (e.g., four individuals are shown for each conifer). Angiosperms shown are *Quercus rubra, Populus grandidentata, Fraxinus americana, Betula papyrifera,* and *Acer rubrum* (each with a designated line pattern, for example, dotted for *Fraxinus,* dash-dotted for *Acer*).

transpiration stream is the primary source of $O_2$ in stems (del Hierro *et al.*, 2002; Gansert, 2003; Mancuso and Marras, 2003), and that cessation of water flow typically precedes parenchyma cell death (see below). Although there is some evidence for extremely low $O_2$ content in the outer sapwood of conifer wood (e.g., 1% to 5%; Eklund, 2000), other work suggests that $O_2$ levels in the transition zone are not consistent with cell death due to anoxia (Fig. 22.3; Spicer and Holbrook, 2005; see also Pruyn *et al.*, 2002b). In contrast, it is possible that elevated $CO_2$ levels, which may be as high as 10% in conifer stems, could significantly inhibit parenchyma respiration and even lead to cell death (McDowell *et al.*, 1999; Pruyn *et al.*, 2002b). More work is needed to document typical $O_2$ and $CO_2$ contents at different radial depths in stems and to determine the effects of local gas composition on parenchyma physiology.

### Parenchyma Cell Death as Programmed Cell Death

Given that parenchyma cell death terminates a process of tissue senescence and occurs, at least in some species, in a fairly abrupt and coordinated manner, it might logically be considered a form of programmed cell death (PCD). In fact, many aspects of parenchyma cell death in the transition zone fit with our current model of plant PCD. One of the hallmarks of PCD in both plants and animals is the cleavage of nuclear DNA by endonucleases (Gavrieli *et al.*, 1992; Pennell and Lamb, 1997). Although there is no direct evidence for this cleavage in dying parenchyma cells, the nuclei do become irregularly lobed and disintegrate upon death (Yang, 1993), both characteristic of PCD. Evidence that the heterochromatin condenses in the nuclei of dying parenchyma is also suggestive of PCD (Nobuchi and Harada, 1968), as are observations of proliferation of the vacuole (Nobuchi and Harada, 1985). Finally, ethylene, which is implicated in a number of cases of plant PCD (Lam *et al.*, 1999; Gunawardena *et al.*, 2001), has been observed to increase in the transition zone (Eklund and Klintborg, 2000) and suggests that heartwood formation is under internal hormonal regulation.

The significance of recognizing that parenchyma cell death is a form of PCD is considerable. Demonstrating that this process is PCD would underscore the active nature of heartwood formation, define it as a form of tissue senescence, and pave the way for a biological model of this final stage in vascular development. Because sapwood represents a large volume of respiring tissue that requires a supply of carbohydrate, the quantity of sapwood maintained must be regulated and therefore be under developmental control. Also, by providing a new example of plant PCD that is widespread throughout seed plants, such studies would shed light on the sequence of events and common motifs leading to cell death in a diverse group of organisms. Investigations into the signal transduction pathway, ultimately linking hormonal cues to transcriptional and post-transcriptional regulation, will be of both fundamental and commercial importance.

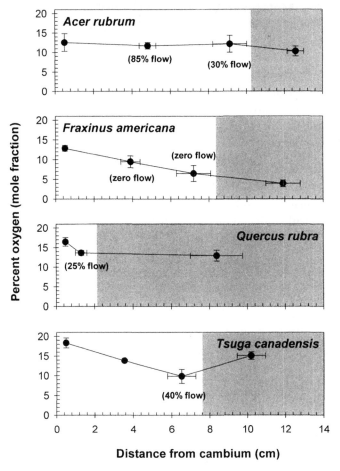

**Figure 22.3** Oxygen contents (percent represents mole fraction, with ambient $O_2$ measuring 21%) at different radial depths in mature tree stems. Percent flow indicates the xylem sap flux density (e.g., Fig. 22.5) at that radial position expressed as percent of flow in the outermost 1 cm of sapwood (no data were available for the middle position in *Tsuga canadensis*). Oxygen was measured by inserting a fiber optic needle-tipped $O_2$ probe through a silicon septum housed at the inner end of permanently installed tubes. Means (and standard error; n = 6) are shown for measurements made in July 2002. Shaded regions indicate heartwood. These values suggest that parenchyma cell death via oxygen deprivation is unlikely. Sap flow data (e.g., Fig. 22.5) at the same radial depths further suggest that $O_2$ can remain high in the absence of water flow through an area (e.g., in the middle and inner sapwood of *Fraxinus*). Similarly, *Acer rubrum* showed no change in $O_2$ content despite a fourfold decrease in sap flow rates in the inner relative to outer sapwood. From Spicer and Holbrook (2005) *Plant Cell Environment.* Copyright Blackwell Publishing; figure 2 adapted with permission.

# Metabolic Activity and Carbohydrate Storage in Aging Sapwood

## Parenchyma Respiration

Rates of respiration (measured either as $O_2$ consumption or $CO_2$ evolution) in sapwood typically decrease toward the sapwood/heartwood boundary (Goodwin and Goddard, 1940; Pruyn *et al.*, 2002a, 2002b), and this has been taken as evidence for a decline in cellular metabolism with age. However, these measurements could be confounded by the physical position of samples within the tree (i.e., proximity to live foliage or cambium), the proportion of living cells in the sapwood, and the time of year. In conifers, respiration increases with height for sapwood of the same cambial age (e.g., the outermost, youngest sapwood; Pruyn *et al.*, 2002b), and does not always differ with radial position, particularly between middle and inner sapwood (Pruyn *et al.*, 2002a). Rather than a steady decline in respiration with tissue age, it may be that the outermost sapwood has a very high rate of respiration, particularly during the growing season. Also, when respiration is expressed on a live tissue basis rather than on a sapwood volume basis (i.e., by quantifying the volume of living parenchyma per unit tissue), there may be no difference between respiration rates in the youngest and oldest sapwood (Fig. 22.4). Similarly, parenchyma from the innermost sapwood of several angiosperm species respired at the same rate despite a difference in age of more than 30 years (Fig. 22.4). Finally, transient increases in metabolic activity in the transition zone have been observed during the dormant season (Höll and Lendzian, 1973; Shain and MacKay, 1973) and likely correspond to parenchyma activity before death. Hence there is little evidence for an inherent decline in metabolic activity as parenchyma cells age.

Histochemical studies show a shift in metabolism with sapwood age such that enzyme levels of the oxidative pentose phosphate pathway are elevated in the transition zone relative to the middle and inner sapwood (Higuchi *et al.*, 1967; Shain and MacKay, 1973). A key intermediate in this pathway is required for the production of phenolics via the shikimic acid pathway, and the timing of this increase in enzyme activity corresponds well with a transient increase in respiration (Shain and MacKay, 1973). Knowledge of the seasonal timing of parenchyma activity during heartwood formation will facilitate future studies on respiration in aging sapwood.

## Parenchyma Contents

When viewed as a form of tissue senescence, it seems logical that both carbohydrate and mineral nutrient concentrations would decrease during the conversion of sapwood into heartwood. The concentrations of starch, lipids, and many soluble sugars (glucose, fructose, and sucrose) are in fact

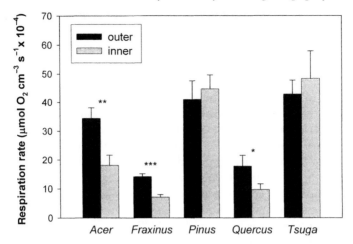

**Figure 22.4** Rates of respiration measured on xylem tissue from inner and outer sapwood. Oxygen consumption was determined by measuring the $O_2$ content in the headspace of sealed vials with a fiber optic $O_2$ sensing system over 36 hours, following equilibration to uniform $O_2$ contents. The live volume fraction of each sample was estimated using image analysis (to determine parenchyma volume) and DAPI staining (as an index of vitality, to determine live parenchyma volume). Average age of tissue (outer/inner) was as follows: *Tsuga* (4/23 yrs), *Acer* (5/38 yrs), *Quercus* (2/3 yrs), *Fraxinus* (5/34 yrs), *Pinus* (3/13 yrs). Inner sapwood of the three angiosperms respired at the same rate despite a difference in age of over 30 years. Respiration of inner sapwood of conifers was as high or higher than that of outer sapwood. Asterisks indicate significance levels for paired t-tests within each species (*P*-values indicated are $\leq 0.05$ (*), $\leq 0.01$ (**), $\leq 0.001$ (***); n = 6).

negligible in the heartwood (Fisher and Höll, 1992; Höll, 2000). In the case of starch, concentrations may decrease gradually through the sapwood, or remain constant and then decline dramatically at the sapwood/heartwood boundary (Höll, 2000; Dehon *et al.*, 2002). The extent to which these reserves are being translocated rather than consumed *in situ* during the synthesis of secondary metabolites or tylosis formation is unclear. In *Robinia pseudoacacia* there is evidence for both increased activity of sucrose synthase and neutral invertase in the transition zone (Hauch and Magel, 1998) and for radial import of carbohydrates (Magel, 2000). Similarly, although nutrient concentrations (N, P, and K) are much lower in the heartwood than in the sapwood, there is little direct evidence for resorption or translocation of these elements from heartwood (Colin-Belgrand *et al.*, 1996). However, the observation that concentration ratios between sapwood and heartwood are very mineral specific, and that trees with low nutrient contents in the sapwood have disproportionately lower contents in the heartwood, have both been considered suggestive of selective recycling (Meerts, 2002).

## Loss of Conductive Function in Secondary Xylem and Phloem

### Cessation of Water Transport in Secondary Xylem

It is worth distinguishing between conducting sapwood and true sapwood, with the former referring to the region of sapwood that functions in water transport, and the latter defined as the region that contains living parenchyma. Many species maintain a much wider band of sapwood (i.e., true sapwood) than is needed to meet hydraulic requirements. This observation has two important implications: first, that the amount of sapwood maintained by a tree is *not* determined solely on the basis of water relations, and second, that water flow through a localized region of xylem is not necessary to support parenchyma metabolism. Living ray tissue at the sapwood/heartwood boundary in *Fraxinus americana* may be as far as 10 cm away from any active flow of water (Fig. 22.3).

Although studies rarely (unfortunately) couple observed spatial patterns of living parenchyma with those of water flow, cessation of water transport precedes cell death, with a time lag between the two events ranging from months to over 10 years (Fig. 22.5) (Gerry, 1914; James *et al.*, 2002). In conifers, there is a narrow (one to two annual rings) and well-defined *white zone* just outside the heartwood, so termed because it appears white, presumably because of its lower moisture content. Pits are aspirated in the white zone (Nobuchi and Harada, 1983), thus precluding water transport, and it is within this zone that the parenchyma cells die. Excepting this narrow white zone, conifer xylem remains conductive throughout the sapwood. In contrast, there is great variation in the length of time (and width) that the sapwood of angiosperms remains conductive, and it is clear that the cessation of water flow may occur many years (30+) before the death of parenchyma cells (Fig. 22.5). Radial patterns of sap flow are reviewed extensively by Gartner and Meinzer (Chapter 15) in this volume.

What actually causes the cessation of water flow is not clear. Conduits in the sapwood may be physically blocked by gel exudation or tylosis formation in angiosperm vessels or by pit aspiration in conifer tracheids, but these events are likely the result of cavitation rather than the cause (i.e., they do not block active flow). Cavitation in older conduits could result from excessive tensions in the xylem generated by the increasingly tortuous path to actively transpiring foliage. Tensions would only have to be excessive relative to the ability of the pit membrane to prevent passage of the air-water interface. There is evidence that bordered pit membranes (which are not true membranes but instead composed of primary cell wall material) in angiosperms are weakened with age and repeated cycles of embolism and refilling, both reversibly and irreversibly (Sperry *et al.*, 1991; Hacke *et al.*, 2001; Stiller and Sperry, 2002). Reversible increases in vulnerability to cavitation are thought to be the result of membrane stretching and altered

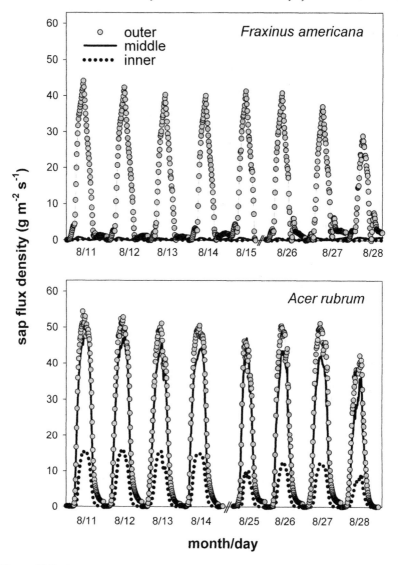

**Figure 22.5**  Sap flux density measured at different radial depths in two angiosperms with wide sapwood (11 cm [35 yrs] and 9 cm [30 yrs] for *Acer* and *Fraxinus*, respectively) using modified Granier sensors (James *et al.*, 2002). The *inner* depth in each species represents sapwood within 1 cm of the sapwood/heartwood boundary (defined by presence of living parenchyma and determined by vital staining with triphenyltetrazolium chloride); *outer* represents the outermost 1 cm of sapwood, with *middle* placed equidistant between the two. Note that although the sap flux pattern of *Fraxinus americana* fits with that expected of a ring-porous species (i.e., only conducting in one or two annual increments), it maintains living parenchyma for more than 30 years. From Spicer and Holbrook (2005) *Plant Cell Environment.* Copyright Blackwell Publishing; figure 3 adapted with permission.

bonding between cellulose microfibrils (Hacke *et al.*, 2001; Stiller and Sperry, 2002), whereas irreversible increases could result from physical rupture of the membrane (Sperry *et al.*, 1991; Hacke *et al.*, 2001). In conifers, although reduced permeability in the innermost sapwood may be attributable to accumulated embolisms (Spicer and Gartner, 2001), direct evidence for increased vulnerability to cavitation in inner relative to outer sapwood is weak (Domec and Gartner, 2001). Finally, it is possible that cavitated vessels might accumulate in older sapwood because of a decreased capacity for refilling. Clearly more work is needed on the effects of xylem aging on pit membrane structure and the ability to refill embolisms, particularly if parenchyma cells are required for the latter.

### Analogous Cessation of Assimilate Transport in Secondary Phloem

Secondary phloem maintains living parenchyma cells for a number of years after conductive elements have ceased to function, much like secondary xylem. Extensive callose deposition (sometimes termed *definitive callose*) in sieve elements marks the end of their functional lifespan. In woody plants this can range from one (e.g., in *Pyrus*; Evert, 1963a, 1963b) to several (e.g., *Tilia* and *Vitis*; Esau, 1948, 1950) years and is accompanied by death of companion cells (in angiosperms) and albuminous cells (in conifers), as well as the death of some parenchyma after the breakdown of starch. However, most parenchyma stay alive for several years and continue to store starch, proteins, and polyphenols (Schneider, 1955; Evert, 1963b); some parenchyma may live for 20 or more years (Grillos and Smith, 1959). These cells remain alive until the development of a new phellogen (Esau, 1965), and ultimately are sloughed off as outer bark.

## Wound Compartmentalization Versus Heartwood Formation

There are references in the literature to *wound wood, wound heartwood,* and *artificial heartwood*. These refer to discolored wood produced in secondary xylem after mechanical injury or microbial infection that provides a mechanism for rapidly compartmentalizing decay and restricting the movement of microorganisms within the stem (Smith, 1997). Wound wood shares many of the characteristics of true heartwood and should serve as a useful companion system for the study of parenchyma physiology. One of the primary functions of parenchyma in the sapwood is to respond to injury, which they do by synthesizing secondary metabolites including suberin and complex phenolics (the composition of which may differ markedly from those produced in normal heartwood of the same tree), blocking conduits through the production of gels and tyloses, and ultimately dying (Schmidt and Liese, 1990, 1993; Smith, 1997; Beckman, 2000). This creates a localized region of wood with reduced permeability, little to no stored carbohydrates, and large deposits of plant defense compounds. Although the cues are quite different—wound wood is clearly triggered by external events, whereas heartwood is initiated in response to internal factors—the process is very similar, and both are likely related forms of PCD. The response to wounding and infection in herbaceous plants is an exceptionally active field of research, and includes work on the hormonal signals and transduction pathway of the PCD-inducing hypersensitive reaction to pathogens (e.g., Alvarez, 2000; Mittler and Rizhsky, 2000; Hauck *et al.*, 2003). As woody plants are developed as model systems in molecular biology, these studies should provide a valuable point of comparison.

## Conclusions and Directions for Future Research

Senescence of vascular tissue is critical in shaping woody stem structure and function. Parenchyma cells drive this process through a dramatic shift in function near the end of their life. In the past, heartwood formation was viewed as an aging process in which a gradual loss of metabolic activity led to cell death, but it is now clear that it defines an active program of tissue senescence. Carbohydrate stores are consumed, a suite of plant defense compounds are produced, and occlusions in the conducting elements render the tissue impermeable. A tree can thus regulate the amount of sapwood by continually, and safely, compartmentalizing nonfunctional xylem toward the center of the stem.

Future work in this area should seek to link parenchyma physiology with the better-studied water transport properties of xylem, and consider the role of both axial and ray parenchyma in the exchange of material with the transpiration stream. One important area is the decoupling of water transport and carbohydrate storage functions in sapwood, and the extent to which parenchyma cell death is linked to cavitation. A better understanding is needed of how the loss of conductive function in xylem occurs with age, and how this process differs across species. Conifers and diffuse-porous species may prove particularly interesting because their sapwood remains conductive for many years, and it is not known if the reduced flow in the inner sapwood is due to cumulative and diffuse dysfunction (e.g., the occasional embolized vessel or tracheid) or simply reduced pressure gradients driving flow. Similarly, work on the basic physiology of ray and axial parenchyma, at all stages of development, will enhance our understanding of this radial route of carbohydrate transport and the way in which cellular metabolism changes with age. This work, in particular, will benefit from the application of molecular biological and imaging techniques, especially when combined with a physiological approach. Finally, given the variation in patterns of loss of physiological function in the xylem, a comparative approach, using a wide range of species with different xylem tissue characteristics, will be especially important in studies of vascular tissue senescence.

## Acknowledgments

I gratefully acknowledge support and facilities provided by the Harvard Forest and the Smithsonian Center for Environmental Research, as well as funding from the United States Environmental Protection Agency and National Science Foundation. I thank Carol Peterson, Peter Barlow, and Frank Ewers for helpful comments on an earlier version of the chapter.

## References

Alvarez, M. A. (2000) Salicylic acid in the machinery of hypersensitive cell death and disease resistance. *Plant Mol Biol* **44:** 429-442.

Barnett, J. R., Cooper, P. and Bonner, L. J. (1993) The protective layer as an extension of the apoplast. *IAWA Bull* **14:** 163-171.

Beckman, C. H. (2000) Phenolic-storing cells: Keys to programmed cell death and periderm formation in wilt disease resistance and in general defence responses in plants. *Physiol Mol Plant Pathol* **57:** 101-110.

Beritognolo, I., Magel, E., Abdel-Latif, A., Charpentier, J.-P., Jay-Allemand, C. and Breton, C. (2002) Expression of genes encoding chalcone synthase, flavanone 3-hydroxylase and dihydroflavonol 4-reductase correlates with flavanol accumulation during heartwood formation in *Juglans nigra. Tree Physiol* **22:** 291-300.

Bonsen, K. J. M. and Kucera, L. J. (1990) Vessel occlusions in plants: Morphological, functional and evolutionary aspects. *IAWA Bull* **11:** 393-399.

Carrodus, B. B. and Triffett, A. C. K. (1975) Analysis of composition of respiratory gases in woody stems by mass spectrometry. *New Phytol* **71:** 713-718.

Chafe, S. C. (1974) Cell wall formation and "protective layer" development in the xylem parenchyma of trembling aspen. *Protoplasm* **80:** 335-354.

Chafe, S. C. and Chauret, G. (1974) Cell wall structure in xylem parenchyma of trembling aspen. *Protoplasm.* **80:** 129-147.

Chase, W. W. (1934) The composition, quantity, and physiological significance of gases in tree stems. *University of Minnesota Agricultural Experimental Station Technical Bulletin* **99:** 1-51.

Chattaway, M. M. (1949) The development of tyloses and secretion of gum in heartwood formation. *Aust J Sci Res Ser B, Biol Sci* **2:** 227-240.

Colin-Belgrand, M., Ranger, J. and Bouchon, J. (1996) Internal nutrient translocation in chestnut tree stemwood. III. Dynamics across an age series of *Castanea sativa* (Miller). *Ann Bot* **78:** 729-740.

Dehon, L., Macheix, J. J. and Durand, M. (2002) Involvement of peroxidases in the formation of the brown coloration of heartwood in *Juglans nigra*. *J Exp Bot* **53**: 303-311.

del Hierro, A. M., Kronberger, W., Hietz, P., Offenthaler, I. and Richter, H. (2002) A new method to determine the oxygen content inside the sapwood of trees. *J Exp Bot* **53**: 559-563.

Domec, J.-C. and Gartner, B. L. (2001) Cavitation and water storage capacity in bole xylem segments of mature and young Douglas-fir trees. *Trees* **15**: 204-214.

Eklund, L. (1993) Seasonal variations of $O_2$, $CO_2$, and ethylene in oak and maple stems. *Can J For Res* **23**: 2608-2610.

Eklund, L. (2000) Internal oxygen levels decrease during the growing season and with increasing stem height. *Trees* **14**: 177-180.

Eklund, L. and Klintborg, A. (2000) Ethylene, oxygen and carbon dioxide in woody stems during growth and dormancy. In *Cell & Molecular Biology of Wood Formation*. (R. Savidge, J. Barnett and R. Napier, eds.) pp. 43-56. BIOS Scientific Publishers, Ltd., Oxford.

Esau, K. (1948) Phloem structure in grapevine, and its seasonal changes. *Hilgardia* **18**: 217-296.

Esau, K. (1950) Development and structure of the phloem tissue. II. *Bot Rev* **16**: 67-114.

Esau, K. (1965) *Plant Anatomy*. John Wiley & Sons, Inc., New York.

Evert, R. F. (1963a) The cambium and seasonal development of the phloem in *Pyrus malus*. *Am J Bot* **50**: 149-159.

Evert, R. F. (1963b) Ontogeny and structure of the secondary phloem in *Pyrus malus*. *Am J Bot* **50**: 8-37.

Fisher, C. and Höll, W. (1992) Food reserves of scots pine (*Pinus sylvestris* L.) II. Seasonal changes and radial distribution of carbohydrate and fall reserves in pine wood. *Trees* **6**: 147-155.

Foster, R. C. (1967) Fine structure of tyloses in three species of the Myrtaceae. *Aust J Bot* **15**: 25-34.

Gansert, D. (2003) Xylem sap flow as a major pathway for oxygen supply to the sapwood of birch (*Betula pubescens* Ehr.). *Plant Cell Environ* **26**: 1803-1814.

Gartner, B. L., Baker, D. C. and Spicer, R. (2000) Distribution and vitality of xylem rays in relation to tree leaf area in Douglas-fir. *IAWA Bull* **21**: 389-401.

Gavrieli, Y., Sherman, Y. and Ben-Sasson, S. A. (1992) Identification of programmed cell death in situ via specific labeling of nuclear DNA fragmentation. *J Cell Biol* **119**: 493-501.

Gerry, E. (1914) Tyloses: Their occurrence and practical significance in some American woods. *J Agric Res* **1**: 445-485.

Goodwin, R. H. and Goddard, D. R. (1940) The oxygen consumption of isolated woody tissues. *Am J Bot* **27**: 234-237.

Grillos, S. J. and Smith, F. H. (1959) The secondary phloem of Douglas-fir. *For Sci* **5**: 377-388.

Gunawardena, A. H., Pearce, D. M., Jackson, M. B., Hawes, C. R. and Evans, D. E. (2001) Characterisation of programmed cell death during aerenchyma formation induced by ethylene or hypoxia in roots of maize (*Zea mays* L.). *Planta*. **212**: 205-214.

Hacke, U. G., Stiller, V., Sperry, J. S., Pitterman, J. and McCulloh, K. A. (2001) Cavitation fatigue. Embolism and refilling cycles can weaken the cavitation resistance of xylem. *Plant Physiol* **125**: 779-786.

Hart, C. A. and Thomas, R. J. (1967) Mechanism of bordered pit aspiration as caused by capillarity. *For Prod J* **17**: 61-68.

Hauch, S. and Magel, E. (1998) Extractable activities and protein content of sucrose-phosphate synthase, sucrose synthase and neutral invertase in trunk wood tissues of *Robinia pseudoacacia* L. are related to cambial wood production and heartwood formation. *Planta* **207**: 266-274.

Hauck, P., Thilmony, R. and He, S. Y. (2003) A *Pseudomonas syringae* type III effector suppresses cell wall-based extracellular defense in susceptible *Arabidopsis* plants. *Proc Natl Acad Sci U S A* **100**: 8577-8582.

Higuchi, T., Shimada, M. and Watanabe, K. (1967). Studies on the mechanism of heartwood formation. V. Change in the pattern of glucose metabolism in heartwood formation. *Mokuzai Gakkaishi* **13**: 269-273.

Hillis, W. E. (1987) *Heartwood and Tree Exudates.* Springer-Verlag, Berlin.

Höll, W. (2000) Distribution, fluctuation and metabolism of food reserves in the wood of trees. In *Cell & Molecular Biology of Wood Formation.* (R. Savidge, J. Barnett and R. Napier, eds.) pp 347-362. BIOS Scientific Publishers, Ltd., Oxford.

Höll, W. and Lendzian, K. (1973) Respiration in the sapwood and heartwood of *Robinia pseudoacacia. Phytochemistry* **12:** 975-977.

James, S. A., Clearwater, M. J., Meinzer, F. C. and Goldstein, G. (2002) Heat dissipation sensors of variable length for the measurement of sap flow in trees with deep sapwood. *Tree Physiol* **22:** 277-283.

Kórán, Z. and Côté, W. A., Jr. (1964) Ultrastructure of tyloses and a theory of their growth mechanism. *IAWA News Bull* **2:** 3-15.

Krahmer, R. L. and Côté, W. A., Jr. (1963) Changes in coniferous wood cells associated with heartwood formation. *Tappi* **46:** 42-49.

Lam, E., Pontier, D. and del Pozo, O. (1999) Die and let live: Programmed cell death in plants. *Curr Opin Plant Biol* **2:** 502-507.

Magel, E. A. (2000) Biochemistry and physiology of heartwood formation. In *Cell & Molecular Biology of Wood Formation.* (R. Savidge, J. Barnett and R. Napier, eds.) pp. 363-376. BIOS Scientific Publishers, Ltd., Oxford.

Magel, E. A., Drouet, A., Claudot, A. C. and Ziegler, H. (1991) Formation of heartwood substances in the stem of *Robinia pseudoacacia* L. I. Distribution of phenylalanine ammonium lyase and chalcone synthase across the trunk. *Trees* **5:** 203-207.

Magel, E. A., Jay-Allemand, C. and Ziegler, H. (1994) Formation of heartwood substances in stemwood of *Robinia pseudoacacia* L. II. Distribution of nonstructural carbohydrates and wood extractives across the trunk. *Trees* **8:** 165-171.

Mancuso, S. and Marras, A. M. (2003) Different pathways of the oxygen supply in the sapwood of young *Olea europaea* trees. *Planta* **216:** 1028-1033.

McDowell, N. G., Marshall, J. D., Qi, J. and Mattson, K. (1999) Direct inhibition of maintenance respiration in western hemlock roots exposed to ambient soil carbon dioxide concentrations. *Tree Physiol* **19:** 599-605.

Meerts, P. (2002) Mineral nutrient concentrations in sapwood and heartwood: A literature review. *Ann For Sci* **59:** 713-722.

Mittler, R. and Rizhsky, L. (2000) Transgene-induced lesion mimic. *Plant MolBiol* **44:** 335-344.

Necesaný, V. (1968) The biophysical characteristics of two types of heartwood formation in *Quercus cerris* L. *Holzforschung* **20:** 49-52.

Necesaný, V. (1973) Kinetics of secondary changes in living xylem I. Time dependent formation of tyloses and polyphenolic substances. *Holzforschung* **27:** 73-76.

Nobuchi, T. and Harada, H. (1968) Electron microscopy of the cytological structure of the ray parenchyma cells associated with heartwood formation of Sugi (*Cryptomeria japonica* D. Don). *Mokuzai Gakkaishi* **14:** 197-202.

Nobuchi, T. and Harada, H. (1983) Physiological features of the "white zone" of Sugi (*Cryptomeria japonica* D. Don): Cytological structure and moisture content. *Mokuzai Gakkaishi* **29:** 824-832.

Nobuchi, T. and Harada, H. (1985) Ultrastructural changes in parenchyma cells of Sugi (*Cryptomeria japonica* D. Don) associated with heartwood formation. *Mokuzai Gakkaishi* **31:** 965-973.

Panshin, A. J. and de Zeeuw, C. (1980) *Textbook of Wood Technology.* McGraw-Hill, New York.

Pennell, R. I. and Lamb, C. (1997) Programmed cell death in plants. *The Plant Cell* **9:** 1157-1168.

Pruyn, M., Gartner, B. L. and Harmon, M. (2002a) Respiratory potential in sapwood of old versus young ponderosa pine trees in the Pacific Northwest. *Tree Physiol* **22:** 105-116.

Pruyn, M. L., Gartner, B. L. and Harmon, M. E. (2002b) Within-stem variation of respiration in *Pseudotsuga menziesii* (Douglas-fir) trees. *New Phytol* **154:** 359-372.

Rioux, D., Nicole, M., Simard, M. and Ouellette, G. B. (1988) Immunocytochemical evidence that secretion of pectin occurs during gel (gum) and tylosis formation in trees. *Phytopathology* **88:** 494-505.

Saitoh, T., Ohtani, J. and Fukazawa, K. (1993) The occurrence and morphology of tyloses and gums in the vessels of Japanese hardwoods. *IAWA Bull* **14:** 359-372.

Schmidtt, U. and Liese, W. (1990) Wound reaction of the parenchyma in *Betula*. *IAWA Bull* **11:** 413-420.

Schmidtt, U. and Liese, W. (1993) Response of xylem parenchyma by suberization in some hardwoods after mechanical injury. *Trees* **8:** 23-30.

Schneider, H. (1955) Ontogeny of lemon tree bark. *Am J Bot* **42:** 893-905.

Shain, L. and MacKay, J. F. G. (1973) Seasonal fluctuation in respiration of aging xylem in relation to heartwood formation in *Pinus radiata*. *Can J Bot* **51:** 737-741.

Smith, K. (1997) Phenolics and compartmentalization in the sapwood of broad-leaved trees. In *Methods in Plant Biochemistry and Molecular Biology*. (W. V. Dashak, ed.) pp.189-198. CRC Press, Boca Raton, Florida.

Sperry, J. S., Perry, A. H. and Sullivan, J. E. M. (1991) Pit membrane degradation and air-embolism formation in ageing xylem vessels of *Populus tremuloides* Michx. *J Exp Bot* **42:** 1399-1406.

Spicer, R. and Gartner, B. L. (2001) The effects of cambial age and position within the stem on specific conductivity in Douglas-fir (*Pseudotsuga menziesii*) sapwood. *Trees* **15:** 222-229.

Spicer, R. and Holbrook, N. M. (2005) Within-stem oxygen concentration and sap flow in four temperate tree species: does long-lived xylem parenchyma experience hypoxia? *Plant Cell Environ* **28**: 192-201.

Stewart, C. M. (1966) Excretion and heartwood formation in living trees. *Science* **153:** 1068-1074.

Stiller, V. and Sperry, J. S. (2002) Cavitation fatigue and its reversal in sunflower (*Helianthus annuus* L.). *J Exp Bot* **53:** 1155-1161.

Wink, M. (1997) Compartmentation of secondary metabolites and xenobiotics in plant vacuoles. In *The Plant Vacuole*. (R. A. Leigh and D. Sanders, eds.) pp. 141-169. Academic Press Limited, London.

Yamamoto, K. (1982) Yearly and seasonal process of maturation of ray parenchyma cells in *Pinus* species. *Res Bull College Experimental Forest, Hokkaido University* **39:** 245-296.

Yang, K. C. (1993) Survival rate and nuclear irregularity index of sapwood ray parenchyma cells in four tree species. *Can J For Res* **23:** 673-679.

Yang, S. and Tyree, M. T. (1992) A theoretical model of hydraulic conductivity recovery from embolism with comparison to experimental data on *Acer saccharum*. *Plant Cell Environ* **15:** 633-643.

# Part VI

## Evolution of Transport Tissues

# 23

# The Evolutionary History of Roots and Leaves

*C. Kevin Boyce*

The long distance transport system of plants links the two primary sites of assimilation from the environment, root and leaf. The evolutionary history underlying that simple statement is extremely complex when the full diversity of vascular plants over the last 400 million years is taken into account. In the larger context of this volume, stems may be considered primarily as the link between root and leaf, but roots and leaves have each evolved independently in a number of plant lineages, and it is only the stem connecting those termini that can be deemed homologous across the vascular plants.

Though it is understandable that the angiosperms that dominate the modern world have been the primary focus of physiological investigation, it is important to keep in mind that the period of angiosperm dominance represents only the last quarter of vascular plant history (Wing *et al.*, 1993; Knoll and Niklas, 1987). A comparative, evolutionary context can allow assessment of the degree to which results from angiosperm exemplars can be extended to other groups and vascular plants as a whole. The fossil record can also point toward living taxa that provide opportunities for physiological comparisons of independently derived but functionally similar structures, such as with roots and leaves.

The first vascular plants consisted of small, unadorned axes, which were responsible both for photosynthesis and assimilation of water and nutrients. Roots have evolved at least twice (Kenrick, 2002a; Gensel *et al.*, 2001; Raven and Edwards, 2001). The roots found in the lycopod and euphyllophyte lineages (Fig. 23.1) have evolved independently and those of free-sporing euphyllophytes differ from the seed plants in key respects. Leaves have had an even more complicated history. The lycopods, again, have independently evolved leaves, simple linear structures. Members of the euphyllophyte clade have evolved laminate leaves at least four times (Boyce and Knoll, 2002), and patterns of morphological evolution have been complex in two of these lineages, the ferns and seed plants.

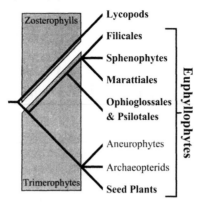

**Figure 23.1** Evolutionary relationships of the vascular plants discussed in this chapter. Lineages with extant members are in boldface. The zosterophylls s. l. are a paraphyletic group of fossils in which the lycopod clade is nested, and the trimerophytes are a paraphyletic group of fossils in which the extant euphyllophyte clades are nested. The sphenophytes include the extant Equisetales, as well as the extinct Sphenophyllales and Pseudoborniales mentioned in the text. The evolutionary diversification represented here happened very rapidly; the basal-most branch between lycophytes and euphyllophytes would have occurred shortly before the beginning of the Devonian (410 MA) and all lineages within the euphyllophytes diverged by the end of the Devonian (355 MA), although much diversification within these lineages occurred over later time. Phylogeny of extant taxa follows Pryer and colleagues (2001).

The significance of this evolutionary history for understanding the physiology of long distance transport is at least threefold. First, transport within the termini of root and leaf is an important part of the transport system as a whole (see Chapters 5, 7, and 8). Second, the fact that stems in early land plants were initially responsible for all functions now associated with root and leaf emphasizes that specialization for long distance transport is itself a secondary function convergently evolved in a number of groups. Third, the initial evolution of lateral organs and subsequent changes in their geometry has likely had a strong, direct influence on stem vascular anatomy through hormonal inputs during development (Stein, 1993). This chapter focuses on the evolutionary relationships of the roots and leaves of extant vascular plants as documented with the fossil record, as well as some of the physiological and developmental implications of these relationships for the transport system.

The role that fossils can play in this discussion differs for the organ types. Roots are rarely preserved and, when found, they may be difficult to assign to any particular clade (Raven and Edwards, 2001). As a result, the fossil record can provide trends in rooting depth through time (Driese and Mora, 2001; Algeo and Scheckler, 1998) and general information about plant body plans (Rothwell, 1995), but discussion of root morphology or anatomy must rely more on information from living taxa placed in a phylogenetic context. Leaf fossils are abundant in many depositional settings, but are typically found detached from the parent axis (Chaloner, 1986). As a result, knowledge of leaf morphologies through time is good for the environments in which fossilization can occur, but it is some times difficult to determine the phylogenetic affinities of fossil leaf taxa, and it is usually not possible to estimate the size and architecture of the parent plant.

It previously had been expected that fossils would play a vital role in phylogeny reconstruction by providing character state combinations not found among living taxa (Donoghue *et al.*, 1989). However, phylogenetic hypotheses inferred using fossil plant morphologies (Nixon *et al.*, 1994; Rothwell and Serbet,1994; Doyle and Donoghue, 1992; Crane, 1985) have been robustly refuted by more recent molecular phylogenies (Magallón and Sanderson, 2002; Bowe *et al.*, 2000; Chaw *et al.*, 2000; Winter *et al.*, 1999). This is a testament to the frequently high degree of evolutionary convergence between distantly related plant lineages. However, the fact that there has been too much convergence for morphology to provide a reliable source of information for phylogeny reconstruction demonstrates that fossils are essential for understanding morphological evolution because inferences of ancestral morphologies and evolutionary patterns from phylogenies will frequently be incorrect. For example, a single origin for roots and for leaves would be the most parsimonious conclusion in light of how common these structures are across the phylogeny of extant vascular plants (Schneider *et al.*, 2002). However, the fossil record indicates that the evolution of plant morphology has been far from parsimonious. In addition to the obscuring of phylogenetic signal by the noise of frequent evolutionary convergence, the morphologies of living plants can also be directly misleading. As the angiosperms have undergone an immense radiation (as did the seed plants before that), other groups that formerly had much greater ecophysiological ranges have been marginalized and undergone a biased loss of morphological diversity. For example, the tree habit and secondary growth have evolved in several groups other than the seed plants, but this would not be guessed from exclusive observation of the modern world.

## Roots

### Evolution of Plant Body Plans and Rooting Structure Function

The first vascular plants did not have roots. A similar situation is found today in *Psilotum*, although apparently a secondary derivation of this state (Fig. 23.1) (Pryer *et al.*, 2001; Bierhorst, 1971). Although of obvious importance, the evolution of roots is only one of several major evolutionary innovations that have affected both how plants interact with their substrate and the possible architectures of aboveground structure. These innovations, which also have included shifts away from sporophyte dependence on the gametophyte and the evolution of bipolar growth, have greatly affected both assimilation and transport processes. Adequate hydraulic supply can be accomplished in bryophytes with only diffusion and wicking of water along the surface of aerial axes (Hébant, 1977). The largest trees require bipolar growth and secondary vascular production in root and stem to supply aboveground tissues. A diversity of morphologies between these end members is documented in the fossil record of vascular plants.

The most morphologically simple Silurian and Early Devonian fossil plants (e.g., *Cooksonia*-like fossils, first appearing approximately 425 million years ago [hereafter abbreviated as MA]) are determinate, dichotomously branching sporophytes a few centimeters tall (Edwards and Fanning, 1985) that have been reconstructed as gametophyte-dependent (Rothwell, 1995). Functions such as nutrient acquisition, substrate anchorage, and symbiotic interactions that are attributed to the sporophyte roots in living vascular plants would likely have been performed exclusively by the gametophyte, although only the sporophyte is known for these particular fossils. A similar arrangement is found in the extant bryophytes, in which the gametophytes have rhizoids for absorption and anchorage. Bryophyte gametophytes also support mycorrhizal symbioses (Read *et al.*, 2000). It has been hypothesized that a fungal partnership was essential for terrestrial colonization by land plants (Pirozynski and Malloch, 1975), and spores from the Ordovician have been interpreted as those of glomalean fungi (Redecker *et al.*, 2000). During the early history of this symbiosis, fungal interaction was likely restricted to the plant gametophyte.

In more morphologically complex Early Devonian taxa, robust, vascular gametophyte fossils with erect axes bearing terminal cups of gametangia are associated with axial vascular sporophytes that would have likely been physiologically independent after a gametophyte-dependent establishment phase early in ontogeny (Remy *et al.*, 1993; Kenrick *et al.*, 1991). The rhizomatous habit of these plants has been suggested as a shared characteristic present in the common ancestor of all living vascular plants (Kenrick and Crane, 1997). After a gametophyte-dependent stage of early ontogeny, the rhizomatous stems of these fossils would have carried out all functions now associated with roots, which evolved only later (Gensel *et al.*, 2001; Raven and Edwards, 2001). Rhizoids have been found on the stems of several fossils of this organization (Gensel *et al.*, 2001). Endomycorrhizal fungi have been found in the stem cortex of such fossils as well (Remy *et al.*, 1994).

The aerial portions of rootless Early Devonian fossils are not known to have been more than about 20-cm tall. With the evolution of roots, there was a dramatic increase in the capacity for hydraulic support of biomass not in contact with the substrate, and by the Late Devonian (350 MA) trees 10- to 20-m tall had evolved in a number of lineages (e.g., the lycopod *Cyclostigma*, the sphenophyte *Pseudobornia*, and the progymnosperm *Archaeopteris*). By the mid Carboniferous (300 MA), the lycopods, which are now exclusively small herbs, had attained heights of 40 to 50 m (for review, see Taylor and Taylor, 1993). Many of these early examples of arborescence differ markedly from what is typically found in the modern world. The tree habit in living seed plants (with some notable angiosperm exceptions) is dependent on possession of secondary tissues and of bipolar growth with a primary root system initiated with the radicle at the same time as the stem is initiated at the beginning of sporophyte ontogeny (Fig. 23.2C). The lycopod, sphenopsid, and progymnosperm/seed plant lineages each independently evolved secondary xylem, but only the last of these lineages evolved bipolar growth. In the first lycophyte and euphyllophyte plants with roots, sporophyte growth was unipolar (Fig.23.2A, B) (examples in Gensel *et al.*, 2001; Rothwell, 1995) with multiple initiations of root systems along a rhizome throughout ontogeny, such root growth being termed homorhizic (Groff and Kaplan, 1988).

Bipolar growth is uniquely associated with the seed plant lineage, but it is not clear when this trait first appeared. Because no definitive evidence of bipolar growth is known before a mid-Carboniferous conifer (Rothwell, 1995; Mapes and Rothwell, 1988), it is at least possible that growth was still unipolar in the earliest seed plants. Alternatively, bipolar growth may already have evolved in the progymnosperm ancestors of the seed plants. A fossil suggested to be a young archaeopterid sporeling appears to have bipolar growth, although with an unusual anatomy (*Eddya*) (Beck, 1967), and bipolar growth has also been suggested for *Archaeopteris* itself (Scheckler communication in Gensel *et al.*, 2001). However, the earliest seed plants more closely resemble a second group of progymnosperms, the aneurophytes (Rothwell, 1987), for which growth characteristics are unknown. Unless eventually documented to be present in aneurophytes and early seed plants, the characteristic of bipolar growth may ultimately join the eustele (Fig. 23.2C), pycnoxylic wood with little included parenchyma, and laminate leaves as independent convergences between the archaeopterids and later seed plants.

During the evolution of nonseed plant examples of the tree habit, several alternatives to bipolar growth have arisen that allow for a sufficient root stock beneath a large upright shoot system. The arborescent lycopods achieved an analogous form of bipolar growth through an initial stem dichotomy, of which one arm becomes the aerial trunk and the other becomes a robust, dichotomizing rooting structure, an arrangement that persists in the rhizomorph of the living Isoetales (Rothwell and Erwin, 1985). The calamitalean sphenophytes had rhizomes of sufficient size for

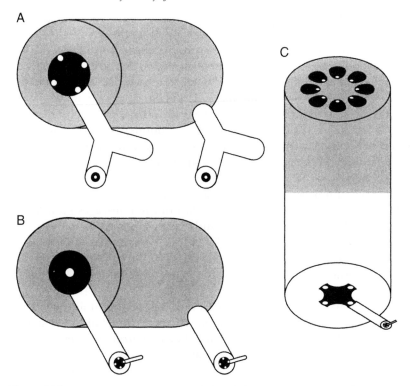

**Figure 23.2**    Vascular anatomy of stem and root. Only xylem and cortex are figured and no secondary growth is shown. Gray cortex indicates a stem, white cortex indicates a root. Black indicates metaxylem and white circles within black indicate location of protoxylem. *Protoxylem* tracheids mature earlier in ontogeny and tend to be small and have simple wall anatomy relative to the *metaxylem* tracheids. *Exarch* xylem maturation—protoxylem is peripheral to the metaxylem—is figured in the stem of A and the roots of B and C. *Endarch* xylem maturation—metaxylem peripheral to the protoxylem—is figured in the root of A and stems of B and C. A *protostele*—a single, central xylem bundle—is figured in stem and root of A and B, as well as the root of C. A *eustele*–multiple discrete vascular bundles surrounding a central pith—is figured in the stem of C. *Unipolar* growth—there is no primary root and there are multiple *homorhizic* initiations of root systems along the stem throughout sporophyte ontogeny—is figured in A and B. *Bipolar* growth—there is a single root system derived only from the radicle at the beginning of sporophyte ontogeny—is figured in C. A has *endogenous* root *initiation* from the stem, but root *branching* due to *exogenous* dichotomies of the root apex. Endogenous root branching is figured in B and C. Plants with these suites of characteristics can be found among (A) lycopods, (B) ferns, and (C) seed plants, but morphologies within these lineages are diverse and the primary intent with these greatly simplified diagrams is to illustrate terminology that may be unfamiliar.

the homorhizic support of arborescent lateral branches, analogous to modern bamboos. Aborescent ferns essentially consist of a unipolar rhizome growing perpendicular to the ground that is supported hydraulically and structurally by a homorhizic root mantle, which grows down to the ground as the rhizome grows up away from it.

**Evidence for Homology and Convergence in Rooting Structures**

Land plant macrofossils first appear in the Silurian, but the first appearance of significant root traces is not until the Early Devonian (Driese and Mora, 2001). By this point, the zosterophylls and trimerophytes were distinct lineages (Kenrick and Crane, 1997), which suggests evolutionary independence for the roots of their living relatives, the lycopods and the ferns, horsetails, and seed plants, respectively (Fig. 23.1). Furthermore, although there may be one or two exceptions that are difficult to interpret, , nearly all accounts of rootlike structures in fossils of this age are in the zosterophyll/lycopod lineage (reviewed in Gensel *et al.*, 2001). Among members of the trimerophyte/euphyllophyte lineage, roots are not found until later in the Devonian (Raven and Edwards, 2001; Fairon-Demaret and Li, 1993). It could be argued that roots were present but not preserved in these plants, but that would not explain the presence of roots in lycophyte fossils and their absence from other equally well-preserved plants from the same localities (Raven and Edwards, 2001). As a result, it is reasonably well accepted that roots evolved independently in the zosterophylls and trimerophytes. The larger question among paleontologists is whether there might be more than two origins: fern and seed plant lineages may have independently evolved roots after divergence from a rootless trimerophyte-grade common ancestor (Gensel *et al.*, 2001; Raven and Edwards, 2001).

Though roots as a whole cannot be considered homologous, a variety of component characteristics of roots must be considered individually. Root hairs, at least, are likely homologous across all plants. Whether termed root hairs or rhizoids, nonvascular emergences from epidermal cells that interact with the substrate are found in the gametophytes of all bryophyte groups (Watson, 1971) and of pteridophytes (Gifford and Foster, 1989) and the sporophytes of vascular plants, including their early fossil ancestors (Kidston and Lang, 1917, 1920a).

Other general aspects of root form may reflect parallel evolution more than true convergence. Though there are many obvious differences between roots and stems, roots are still apically growing axes and therefore may have co-opted aspects of the preexisting developmental programs present in stems. This has been borne out by studies indicating that the stem and root in *Arabidopsis* share several developmental regulatory genes (Dolan and Scheres, 1998). Though demonstrated only in an angiosperm, similar evolutionary processes may have been involved in each of the multiple origins of roots because the tracheophyte common ancestor likely possessed an axial sporophyte. It is unclear whether the endodermis must be independently derived in different lineages. An endodermis with casparian strip is common in the shoots of living pteridophytes (Lersten, 1997), where the functional significance of an endodermis for a subterranean stem of a rhizomatous plant bearing multiple root systems may be the same as for roots proper (for discussions of the significance of the endodermis to root physiology, see Chapter 7). Given that the stem fulfilled the functions of roots before their evolution, the root endodermis might reflect a parallel extension in each evolution of roots of a preexisting characteristic of stems. The endodermis may thereby be homologous across the vascular plants and have been present in the stem of the tracheophyte common ancestor. Fossils contribute little to answering this particular question, as the presence or absence of an endodermis is difficult to assess. Notably, *Lycopodium* has an endodermis in its stem but not its root (Damus *et al.*, 1997), but this may be a derived loss.

Other characteristics must have evolved independently. The presence of a root cap in all lineages must represent convergence: Structures resembling root hairs are found in plants without roots—and can therefore be homologous even if the roots on which they are now borne are not—but nothing resembling a root cap is available to be similarly co-opted for each evolution of roots. Despite the extreme similarities of root caps in each lineage, their convergent evolution seems reasonable because of the functional importance of protecting the apical meristem as it is pushed through the substrate by expansion of more proximal tissues. The stem-derived rooting structures of the arborescent lycopods also evolved an apical structure interpreted as a plug of tissue protecting the apical meristem (Rothwell, 1984). By the same logic applied to the origin of root caps, convergent evolution of roots in lycopods and euphyllophytes requires convergent evolution of endogenous initiation of roots, as this characteristic is unique to roots. In addition, the first fossil occurrences among the zosterophylls of axial systems committed to a rooting function appear to be exogenous derivatives of the stem apex (Kenrick, 2002a), providing further indication that the endogenous initiation of roots has been independently derived more than once.

As with the initiation of root systems on stems, the branching of roots within a root system can be accomplished endogenously (see Fig. 23.2 for distinction between root initiation and branching). This branching is common in all groups except the lycopods where roots branch strictly through dichotomy of the root apex (Schneider *et al.*, 2002; Gifford and Foster, 1989; Eames, 1936). That statement is controversial; some general descriptions of lycopods mention endogenous branching of roots. If not attributable to the lumping of root branching with root initiation, it is here presumed that such descriptions ultimately result from the ambiguity over whether the root-bearing rhizophores of *Selaginella* are stems or roots. Even in this case, however, endogenous refers to within a few cells of the surface of the apical meristem (Lu and Jernstedt, 1996), which is substantially different from the endogenous initiation of a root from the endodermis of a mature axis. Perhaps this subsurface origin of the root apical cell away from the epidermis may be a part of initiating the periclinal divisions establishing the root cap.

## The Evolution of Root Development and Vascular Architecture

Despite independent origins, the structure of roots in different lineages can be remarkably similar. Furthermore, despite the great diversification of shoot morphologies since the Devonian within each major vascular plant lineage, roots have remained relatively uniform. Though there can be a diversity of cortical cell types (see Chapters 7 and 8 and Damus *et al.*, 1997; Schneider, 1996), roots typically possess a simple protostele (Fig. 23.2). The xylem maturation is usually exarch (Fig. 23.2B, C; Gifford and Foster, 1989) except in lycopod roots, which are usually endarch (Fig. 23.2A; Schneider *et al.*, 2002; Eames, 1936). All extant groups include endogenous origin of roots from the endodermis (both lycopods and pteridophytic euphyllophytes) or pericycle (seed plants) of the parent axis (Ogura, 1972). This relative homogeneity of root vascular anatomy across distantly related lineages has been thought to reflect the relative stability of the soil environment in combination either with the early achievement of a structure that optimizes functional efficiency or simply with the maintenance of a primitive anatomy to which the first stems also adhered (Esau, 1953). These explanations essentially imply an early attained physiological efficacy for a function that has remained largely unaltered ever since. Such arguments of functional optimization may well be valid, but little evolutionary change may also reflect nonadaptive constraints on the range of forms that can be taken. In particular, limitations imposed by root development may play a greater role than has been recognized.

Across the vascular plants, vascular patterning in stems has been shown to be dependent on the pattern of auxin input from the exogenous primordia produced at the stem apex (Ma and Steeves, 1992; Sachs, 1991), and this has been applied successfully to the modeling of stelar patterning based on branching geometry in several Devonian fossils (Stein, 1993). Therefore, the dramatic increase in shoot stelar complexity in the Later Devonian and Carboniferous (e.g., Galtier and Phillips, 1996) may be a direct result of the contemporaneous increase in complexity of shoot lateral organs (Boyce and Knoll, 2002; Galtier and Phillips, 1996). Because no such comparable scenario is possible in roots, the simplicity and conservatism of root vascular anatomy may simply be a requirement of root development: Rather than producing exogenous apical laterals that can influence patterning of the parent axis as in stems, the root apex is simple and endogenous laterals are produced only after stelar maturation is complete. This interpretation of the differences between stem and root anatomical disparity seems generally applicable, but there are exceptions, such as the multiple concentric vascular cylinders in the roots of Marattialean ferns (Ogura, 1972), for which it is difficult to account with this hypothesis.

The above is based on the role that has been documented for auxin transport in the vascular patterning of stems. However, it also must be considered that auxin transport through the vasculature is proximally away from the apex in stems and distally toward the apex in roots, as documented at least in seed plants (Sachs, 1991) and lycopods (Wochok and Sussex, 1974). The source of auxin to the differentiating zone of the root being the proximal parent axis, rather than any geometrically complex distal array of apical primordia, may further limit the potential for complex stelar anatomies in roots. Alternatively, this may allow more proximal sites of lateral root production to influence patterning in more distal growing zones. There is often a correspondence between the radial arms of the xylem of the root protostele and the sites of endogenous lateral root production (Bell, 1993).

This correspondence between xylem structure and the sites of endogenous branching may also be relevant to the important distinctions between lycopod and euphyllophyte roots. Unlike euphyllophytes, lycopod roots are typically endarch, rather than exarch, and lack endogenous branching. In stems, the location of protoxylem has been linked to the likely sites of greatest auxin flux derived from transport from apical primordia (Stein, 1993). Auxin also is involved in organization of the root apical meristem and endogenous root initiation (Berleth and Sachs, 2001; Sabatini *et al.*, 1999). If maintenance of meristematic potential in the endodermal zone of roots is coupled with high local auxin concentrations, as seen in the vascular cambium (Uggla *et al.*, 1996), or if an increase in auxin production is otherwise associated with endogenous root initiation, then peripheral protoxylem may reflect the hormonal gradients associated with endogenous branching in the roots of euphyllophytes and more medial protoxylem may reflect the lack of endogenous branching in lycopods. The presence or absence of increased auxin concentrations on the periphery of the vascular cylinder may determine whether protoxylem occurs in a peripheral or more medial position within the xylem. (Lycopods do have exarch stems, again unlike most euphyllophytes. It has been speculated that this results from the hormonal influence of the densely packed microphyll primordia on the stem apex [Kenrick and Crane, 1997], a hypothesis that should be testable through observation of the nonfertile regions of the leafless, but exarch zosterophylls.) This linking of root exarchy with the maintenance of the meristematic capability of endogenous branching perhaps may be testable by careful survey of anatomical changes along the length of lycopod roots (Pixley, 1968) and by survey of euphyllophyte root anatomy in a variety of epiphytic taxa that have apparently lost the capability of endogenous root branching.

The endogenous initiation of roots also may relate to the opposing hormonal environments of root and stem. Vascular connection after the surficial initiation of a lateral organ on a mature axis requires induction of a strand through the differentiated cortical tissues of the parent axis. Such a vascular connection does occur during growth of the surficial axillary buds on stems, but here auxin produced by the growing bud is transported proximally toward the vasculature of the parent axis, inducing the differentiation of a vascular strand (Sachs, 1991). As a result of the apical direction of auxin transport in roots, vascular connection to the parent axis could not be achieved in the same way if root initiation was surficial. Endogenous initiation of roots may simply be necessary for hydraulic continuity with the shoot system.

## Leaves

### Evidence for Homology and Convergence in Leaves

That the leaves found in living plants have evolved independently a number of times may be surprising considering how common leaves are across the vascular plants. However, the fossil record not only demonstrates this fact, but makes it all the more striking by also providing evidence of a number of additional independent evolutions of leaves among extinct groups. Laminate multiveined leaves are found only in the ferns and seed plants among living plants, but they are also found in fossils of the archaeopterid progymnosperms and the sphenophyll lineage of the sphenophytes (and perhaps independently in other extinct sphenophytes as well, *Pseudobornia*, and some Equisetales). Independent evolution is indicated in each of these lineages by the fact that the earliest fossil members of each group did not possess laminate leaves (Boyce and Knoll, 2002). In the leafless Early Devonian plants ancestral to these lineages, photosynthesis was conducted by the stems, which bore stomata and contained cortical airspaces analogous to the spongy mesophyll of leaves (Edwards, 1993). Recent molecular phylogenies (Pryer *et al.*, 2001) suggest there may actually have been additional evolutions of laminate leaves in the Ophioglossales and perhaps also the Marattiales, depending on the correct resolution of the evolutionary relationships between the Marattiales, Filicales, and sphenophytes (Fig. 23.1).

It has been argued that some aspects of the megaphyll typology (Gifford and Foster, 1989) are homologous among all extant euphyllophytes, even if laminate leaves are not (Schneider *et al.*, 2002), but this is highly unlikely. Even homology of a basic nonlaminate frond system is not supported because most of the earliest, superficially frondlike fossils have stemlike, radially symmetrical vascular anatomy rather than frond or leaflike dorsiventral anatomy. Therefore the most that could be homologous between the euphyllophyte lineages is the capacity for distinct horizontal and vertical branches and perhaps for monopodial branching (i.e., lateral organs initiated as small surficial primordia on the apex of the parent axis, as opposed to all branches being derived from dichotomy of the stem apex).

It has frequently been argued that the megaphyllous leaves of the ferns and seed plants are united by the anatomical trait of leaf gaps, local parenchymatous interruptions in the stem xylem immediately distal to the departure of a leaf trace. Even this limited homology is not possible because the concept of a leaf gap requires a pith and vasculature organized in either a eustele (Fig. 23.2) or a siphonostele (a hollow cylinder of xylem) in order for the parenchymatous "gap" to occur, but the earliest fossil members of the living euphyllophyte groups all had protosteles. Furthermore, anatomical details differ between the ferns and seed plants (Taylor and Taylor, 1993). However, the structures termed leaf gaps in different lineages may perhaps be a shared consequence of homologous mechanisms of auxin transport that could not be manifested until more complex stem anatomies evolved in each lineage. Similarly, it has recently been argued that the Fibonacci-based phyllotactic spiral found in many plants results from patterns of epidermal auxin transport in the stem apex (Reinhardt *et al.*, 2003), and this discovery in *Arabidopsis* may well apply to phyllotaxis in the lycopods because details of auxin transport are likely homologous across the vascular plants even if leaves are not.

Regardless of any argument concerning the euphyllophyte leaves, it is generally accepted that the microphylls of the lycopods are not homologous to the leaves of euphyllophytes (although the hypothesis that the microphylls of lycopods and horsetails were homologous had been suggested based on now outdated phylogenies [Gifford and Foster, 1989]). The earliest zosterophylls and trimerophytes were appendageless and the leaves of their descendants bear little similarity, indicating independent origins. It seems widely accepted that the leaves found in euphyllophytes are ultimately derived from modification of stems, although details differ according to different scenarios (Kenrick, 2002b; Boyce and Knoll, 2002; Zimmerman, 1952). A similar hypothesis was proposed for the evolution of microphylls through the reduction of lateral branch systems (Zimmerman, 1952). Although branched microphylls do exist in later Early and Middle Devonian lycopods (*Leclercqia*) (Grierson and Bonamo, 1979), this is after the origin of the structure and the reduction hypothesisea's is no longer favored. Two more plausible theories are the enation and sterilization hypotheses, both of which are consistent with the fossil record (Kenrick and Crane, 1997). By the enation hypothesis (Bower, 1908), microphylls evolve by transition from completely nonvascular epidermal appendages (e.g. the enations of many zosterophylls); (Taylor and Taylor, 1993), through appendages with a trace that travels from the stele through the cortex ending at the appendage base (e.g. Early Devonian *Asteroxylon;* Kidston and Lang, 1920b), to vascularized microphylls. By the sterilization hypothesis, the origin of microphylls is suggested to be from the sterilization of the laterally borne sporangia found in the zosterophyll/lycopod lineage

(Kenrick, 2002b; Kenrick and Crane, 1997). Unlike in the enation hypothesis, the unvascularized appendages found in many zosterophylls (and occasionally even borne on sporangia [*Discalis*] [Hao, 1989]) are thought to be unrelated to the evolution of microphylls in the sterilization hypothesis.

### Diversification of Leaf Morphology and Development

Despite their independent origins, the early leaves of each euphyllophyte lineage are remarkably similar. Fossil plants in Carboniferous coal measures had been studied for decades before it was noticed that most of the leaves from this Age of Ferns did not have leaf-borne sporangia as ferns do, and we now know that many, perhaps most, of these fossils were actually of seed plants (reviewed in Scott, 1909). Furthermore, a basic trend of morphological change through time was replicated in several groups. Each of the euphyllophyte lineages that evolved laminate leaves in the Paleozoic followed a similar sequence of morphologies: dichotomizing structures made of single veined segments, followed by laminar leaves with dichotomous veins that run divergent paths to the distal margin, then leaves with convergent and later reticulate veins that still only terminate along the margins (Fig. 23.3). This repeated morphological pattern is thought to reflect the parallel evolution of marginal meristematic laminar growth in each lineage (Boyce and Knoll, 2002). The association of this venation type with marginal growth was based on developmental studies in ferns (Zurakowski and Gifford, 1988; Pray, 1960, 1962) and has since been tested and confirmed in seed plant leaves and petals with dichotomous, marginally ending veins (*Zamia* and *Ranunculus*) (Boyce, unpublished data).

The ferns and seed plants were the only lineages to complete the morphological sequence described previously without its truncation by extinction of the group. During the late Paleozoic, ferns and seed plants shared the complete range of leaf shapes and venation patterns found among plants presumed to have marginal leaf growth. This range has been maintained by the ferns through to the present, but was largely lost by the seed plants during the Mesozoic. The seed plants came to be dominated by lineages with leaves, typified today by *Ginkgo* and the cycads, with a single order of dense dichotomous veins that run relatively straight, parallel courses exclusively to the distal margin of the lamina, a morphology that is largely absent among ferns (Boyce, unpublished data). This divergence of fern and seed plant leaf morphologies may reflect that leaves have undergone functional shifts similar to what has been discussed for stems and roots. Fern fronds fulfill both photosynthetic and reproductive functions and this state was also found in many Paleozoic seed plant lineages, which bore seeds and pollen organs on foliage leaves (examples in Taylor and Taylor, 1993). These dual functions may have influenced the early history of leaf morphological evolution and contributed to the great morphological

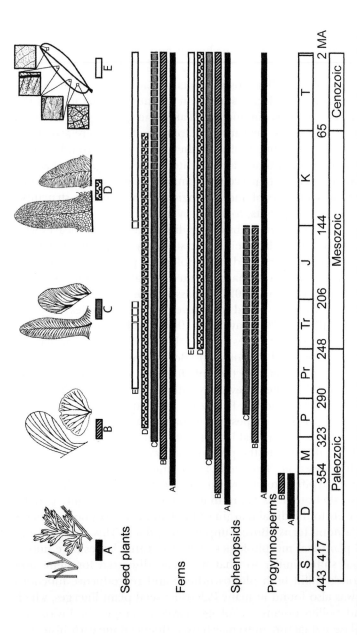

**Figure 23.3**   Stratigraphic distribution of different leaf venation types. Broken lines indicate a morphology is present but rare during a particular time interval. The patterns depicted for each lineage represent the lumped patterns of subgroups within each lineage. For example, the Jurassic gap in more angiospermlike leaves seen in the seed plants reflects the evolution and extinction of gigantopterids and peltasperms during the Permian and Triassic followed by appearance of the angiosperms and Gnetales in the Cretaceous. Basic leaf type groupings are (A) linear leaves with marginal vein endings, (B) laminar leaves with open dichotomous veins that do not converge toward their marginal endings, (C) laminar leaves

*Continued*

diversity found in ferns and Paleozoic seed plants. The loss of much of this morphological range among later (nonangiosperm) seed plants would then reflect the eventual segregation of fertile and vegetative functions into different structures (Boyce, 2005).

The evolution of more angiosperm-like leaves with many orders of reticulate veins with internal vein endings dispersed across the leaf is thought to result from a shift from growth by means of a marginal meristem to diffuse growth throughout the leaf (Boyce and Knoll, 2002), based on the documentation of such growth in several angiosperms (Dolan and Poethig, 1998; Hagemann and Gleissberg, 1996; Poethig and Sussex, 1985; Pray, 1955). As with marginal growth, leaves with morphologies suggesting diffuse leaf growth have evolved independently in a number of groups. Among seed plants, there have been the extant angiosperms and Gnetales, as well as the Permian gigantopterids and Triassic peltasperms. There is *Ophioglossum* and *Christensenia* in the eusporangiate Ophioglossales and Marattiales. Among the leptosporangiate ferns, there are the dipterids with a fossil record extending back into the Triassic, as well as at least four clades among the polypod and dryopterid ferns (Boyce, 2005). The link between diffuse lamina growth and complex venation patterns has so far been documented only in angiosperms, but is also consistent with qualitative observation of appropriate fern groups (Wagner, 1979).

Recognizing the great differences between the leaf types, it has been suggested that angiosperm leaves may represent a complete reinvention of the leaf after passing through a nonlaminate intermediate, perhaps aquatic or desert adapted (Doyle and Hickey, 1976). Despite how morphologically and developmentally distinct the angiosperms may be however, this scenario does not seem necessary for a trait that has evolved so frequently among ferns. More recently, study of the ecophysiology of basal angiosperm lineages suggests that the angiosperms originally evolved in the understory of warm, wet environments (Feild *et al.,* 2004). It appears likely that the other lineages to independently evolve similar leaves all have evolved in comparable environments (Boyce, 2005).

---

with open dichotomous veins that converge toward their marginal endings, (D) reticulate veins with marginal endings, and (E) more complex angiosperm-like vein patterns involving vein reticulation and nonmarginal vein organization. Boxes in the drawing for leaf type E show some of the diverse vein patterns lumped into this grouping. Abbreviations on the timescale represent the geological periods of Silurian, Devonian, Mississippian, Pennsylvanian, Permian, Triassic, Jurassic, Cretaceous, Tertiary, and Quaternary. The Mississippian and Pennsylvanian together form the Carboniferous. No land plant macrofossils are available from the Cambrian and Ordovician periods of the Paleozoic, which are not included in the timescale (leaf drawings modified from Beck and Labandeira, 1998; Dobruskina, 1995; Zurakowski and Gifford, 1988; Tryon and Tryon, 1982; Boureau and Doubinger, 1975; Esau, 1953).

Recent physiological work indicates that an equitable supply of water to all parts of the angiosperm leaf lamina is ensured by a pressure drop between successive vein orders, owing to the presence of vessels in the midvein and secondaries, but not higher order veins, thereby preventing a disproportionate amount of water being lost to the most proximal areas of the finer vein network (in *Laurus*) (Zwieniecki *et al.,* 2002). It is notable that the other two known instances of vessel evolution among seed plants, the gigantopterids (Li and Taylor, 1999) and Gnetales, also have complex leaves with many vein orders that are reminiscent of angiosperm leaves. The cooccurrence of vessels and complex foliar venation, traits that are each relatively rare across the history of seed plants, may represent the coevolution of components of a hydraulic syndrome that assures equitable hydraulic supply across large laminate leaves (but see Chapter 24 in this volume for a plausible alternative interpretation of the evolution of vessels).

## Conclusions and Future Research

The repeated evolution of roots and leaves each reflect both the great adaptive significance of these structures and the finite range of developmental mechanisms available for evolutionary modification. For example, the traits held in common by diverse rooting structures are notable as examples of evolutionary convergence, but also as so many parallel departures from the development and morphology inherited from stems. Why were all these changes necessary and why were they so often the same? While root caps may have obvious utility, the functional necessity of exarchy seems a harder case to make. It is here suggested that many of these changes do not reflect adaptive advantages so much as passive consequences of root development. The root apex is always simple and its history involves nothing like the vast elaboration of apical lateral organs seen in stems. Furthermore, the auxin transport that is crucial to vascular patterning is toward the apical meristem in roots, rather than away from it. The relative homogeneity of roots may be a by-product of these developmental facts rather than the achievement of functional optima. The evolutionary history of leaves also consists of multiple independent origins and the repeated convergence of leaf morphologies. Though there also may be functional consequences, the origins of this morphological convergence may simply reflect the repeated evolution of the limited set of possible mechanisms of laminar growth. This pattern is particularly important in the context of this book, because even though the morphological and anatomical patterns described in this chapter surely affect physiological function, many of the patterns may be shaped more by developmental constraints than physiological implications.

The developmental sections of this review are obviously speculative, but largely testable. Phylogenetic tests are possible for some hypotheses of developmental evolution derived from the fossil record (Kenrick, 2002b). Hypotheses can also be tested with investigation of living plants directed by the context of the fossil record. For example, relationships between leaf venation patterns and mechanisms of growth that have been suggested by the fossil record can be used to predict developmental patterns in previously unstudied living plants (Boyce, unpublished data). Finally, a more extensive survey, placed in a phylogenetic context, of the details of morphological diversity that have been documented in fossil and living plants would serve both to better constrain evolutionary hypotheses and to determine what exceptions exist to the generalities that have been drawn here. This will be particularly important for roots, which have been understudied in both the living world and the fossil record. For example, the discussion concerning patterns of xylem maturation in roots is disproportionately influenced by the lycopods, but there is diversity within the lycopod lineage and even through the ontogeny of individual root systems (Pixley, 1968), the detailed phylogenetic survey of which will be crucial.

There has been little reference here to the physiological significance of the morphological evolution that has been discussed. The more basal vascular plant lineages are in general little studied. Though these plants may be peripheral to our modern angiosperm-dominated world, they are an essential part of its evolution and it is hoped that the context of the fossil record may foster study of these groups. There is a large reservoir of anatomical and developmental information concerning basal lineages in the older literature, but much of the work done in the first half of the twentieth century has been lost. In part, this loss is only because the older literature is rarely read. This information must also be reevaluated in light of the dramatic changes in our understanding of phylogenetic relationships and of the evolutionary and developmental processes, all of which, for better or worse, affect anatomical observations and interpretations of homology. As modern physiological and developmental interest broadens beyond model angiosperms, reinvestigation of basal taxa will have increasing importance for understanding the evolution of plant physiology and the development of the transport system.

## Acknowledgments

I thank T. Feild, N. M. Holbrook, A. Knoll, L. Sack, R. Sage, and M. Zwieniecki for providing helpful discussions or critiques of the manuscript. Some of the work described in this chapter was supported by the National Science Foundation (ERA 0106816 administered by A. Knoll and N. Holbrook).

# References

Algeo, T. J. and Scheckler, S. E. (1998) Terrestrial-marine teleconnections in the Devonian: Links between the evolution of land plants, weathering processes, and marine anoxic events. *Philos Trans R Soc London B* **353:** 113-130.

Beck, A. L. and Labandeira, C. C. (1998) Early Permian insect folivory on a gigantopterid-dominated riparian flora from north-central Texas. *Palaeogeo Palaeoclimatol Palaeoecol* **142:** 139-173.

Beck, C. B. (1967) *Eddya sullivanensis,* gen. et sp. nov., a plant of gymnospermic morphology from the Upper Devonian of New York. *Palaeontographica Abt B* **121:** 1-22.

Bell, A. D. (1993) *Plant Form.* Oxford University Press, Oxford.

Berleth, T. and Sachs, T. (2001). Plant morphogenesis: Long-distance coordination and local patterning. *Curr Opin Plant Biol* **4:** 57-62.

Bierhorst, D. W. (1971) *Morphology of Vascular Plants.* MacMillan, New York.

Boureau, E. and Doubinger, J. (1975) *Traité de Paléobotanique, Tome IV, Fascicle 2:Pteridophylla (Premiere Partie).* Masson et Cie., Paris.

Bowe, L. M., Coat, G. and dePamphilis, C. W. (2000) Phylogeny of seed plants based on all three genomic compartments: Extant gymnosperms are monophyletic and Gnetales' closest relatives are conifers. *Proc Natl Acad Sci* **97:** 4092-4097.

Bower, F. O. (1908) *The Origin of a Land Flora.* Macmillan and Co., London.

Boyce, C. K. and Knoll, A. H. (2002) Evolution of developmental potential and the multiple independent origins of leaves in Paleozoic vascular plants. *Paleobiology* **28:** 70-100.

Boyce, C. K. (2005) Patterns of segregation and convergence in the evolution of fern and seed plant leaf morphologies. *Paleobiology* **31:** 117-140.

Chaloner, W. G. (1986) Reassembling the whole fossil plant, and naming it. *SystematicsAsso Special Vol* **31:** 67-78.

Chaw, S.-M., Parkinson, C. L., Cheng, Y., Vincent, T. M. and Palmer, J. D. (2000) Seed plant phylogeny inferred from all three plant genomes: Monophyly of extant gymnosperms and origin of Gnetales from conifers. *Proc Natl Acad Sci* **97:** 4086-4091.

Crane, P. R. (1985) Phylogenetic analysis of seed plants and the origin of the angiosperms. *Ann Mo Botan Gardens* **72:** 716-793.

Damus, M., Peterson, R. L., Enstone, D. E. and Peterson, C. A. (1997) Modification of cortical cell walls in roots of seedless vascular plants. *Botan Acta* **110:** 190-195.

Dobruskina, I. A. (1995) Keuper (Triassic) Flora from Middle Asia (Madygen, Southern Fergana). *New Mexico Museum Nat History Sci Bull* **5:** 1-49.

Dolan, L. and Poethig, R. S. (1998) Clonal analysis of leaf development in cotton. *Am J Bot* **85:** 315-321.

Dolan, L. and Scheres, B. (1998) Root pattern: Shooting in the dark? *Semin Cell Dev Biol* **9:** 201-206.

Donoghue, M. J., Doyle, J. A., Gauthier, J., Kluge, A. G. and Rowe, T. (1989) The importance of fossils in phylogeny reconstruction. *Annu Rev Ecol Syst* **20:** 431-460.

Doyle, J. A. and Donoghue, M. J. (1992) Fossils and seed plant phylogeny reanalyzed. *Brittonia* **44:** 89-106.

Doyle, J. A. and Hickey, L. J. (1976) Pollen and leaves from the mid-Cretaceous Potomac Group and their bearing on early angiosperm evolution. In *Origin and Early Evolution of Angiosperms* (C. B. Beck, ed.) pp. 139-206. Columbia University Press, New York.

Driese, S. G. and Mora, C. I. (2001) Diversification of Siluro-Devonian plant traces inpaleosols and influence on estimates of paleoatmospheric $CO_2$ levels. In *Plants Invade the Land: Evolutionary and Environmental Perspectives* (P. G. Gensel and D. Edwards, eds.) pp. 237-253. Columbia University Press, New York.

Eames, A. J. (1936) *Morphology of Vascular Plants. Lower Groups (Psilophytales to Filicales).* McGraw Hill Book Company, New York.

Edwards, D. (1993) Cells and tissues in the vegetative sporophytes of early land plants. *New Phytol* **125**: 225-247.

Edwards, D. and Fanning, U. (1985) Evolution and environment in the late Silurian-early Devonian: The rise of the pteridophytes. *Philos Trans R Soc London B* **309**: 147-165.

Esau, K. (1953) *Plant Anatomy.* John Wiley and Sons, New York.

Fairon-Demaret, M. and Li, C.-S. (1993) Lorophyton goense gen. et sp. nov. from the Lower Givetian of Belgium and a discussion of the Middle Devonian *Cladoxylopsida. Rev Palaeobot Palynol* **77**: 1-22.

Feild, T. S., Arens, N. C., Doyle, J. A., Dawson, T. E. and Donoghue, M. J. (2004) Dark and disturbed: A new image of early angiosperm ecology. *Paleobiology* **30**: 82-107.

Galtier, J. and Phillips, T. L. (1996) Structure and evolutionary significance of Palaeozoic ferns. In *Pteridology in Perspective* (J. M. Camus, M. Gibby and R. J. Johns, eds.) pp. 417-433. Royal Botanic Gardens, Kew.

Gensel, P. G., Kotyk, M. E. and Basinger, J. F. (2001) Morphology of above- and below-ground structures in Early Devonian (Pragian-Emsian) plants. In *Plants Invade the Land: Evolutionary and Environmental Perspectives* (P. G. Gensel and D. Edwards, eds.) pp. 83-102. Columbia University Press, New York.

Gifford, E. M. and Foster, A. S. (1989) *Morphology and Evolution of Vascular Plants.* W. H. Freeman and Company, New York.

Grierson, J. D. and Bonamo, P. M. (1979) *Leclercqia complexa*: Earliest Ligulate Lycopod (Middle Devonian). *Am J Bot* **66**: 474-476.

Groff, P. A. and Kaplan, D. R. (1988). The relation of root systems to shoot systems in vascular plants. *Bot Rev* **54**: 387-422.

Hagemann, W. and Gleissberg, S. (1996) Organogenetic capacity of leaves: The significance of marginal blastozones in angiosperms. *Plant Syst Evol* **199**: 121-152.

Hao, S.-G. (1989) A new zosterophyll from the Lower Devonian (Siegenian) of Yunnan, China. *Rev Palaeobot Palynol* **57**: 155-171.

Hébant, C. (1977) *The Conducting Tissue of Bryophytes.* Cramer, Vaduz, Germany.

Kenrick, P. (2002a) The origin of roots. In *Plant Roots: The Hidden Half,* 3rd Ed. (Y. Waisel, A. Eshel and U. Kafkafi, eds.) pp. 1-13. Marcel Dekker, Inc., New York.

Kenrick, P. (2002b) The telome theory. In *Developmental Genetics and Plant Evolution* (Q. C. B. Cronk, R. M. Bateman and J. A. Hawkins, eds.) pp. 365-387. Taylor and Francis, London.

Kenrick, P. and Crane, P. R. (1997) *The Origin and Early Diversification of Land Plants.* Smithsonian Institution Press, Washington.

Kenrick, P., Remy, W. and Crane, P. R. (1991). The structure of water-conducting cells in the enigmatic early land plants *Stockmensella langii* Fairon-Demaret, *Huvenia kleui* Hass et Remy, and *Sciadophyton* sp. Remy *et al.* 1980. *Argumenta Palaeobot* **8**: 179-191.

Kidston, R. and Lang, W. H. (1917) On Old Red Sandstone plants showing structure, from the Rhynie Chert Bed, Aberdeenshire. Part 1. *Rhynia Gwynne-Vaughanii,* Kidston and Lang. *Trans R Soc Edinburgh* **51**: 761-784.

Kidston, R. and Lang, W. H. (1920a) On Old Red Sandstone plants showing structure, from the Rhynie Chert Bed, Aberdeenshire. Part 2. Additional note on *Rhynia Gwynne-Vaughanii,* Kidston and Lang; with descriptions of *Rhynia major,* n. sp., and *Hornea Lignieri* n. g., n. sp. *Trans R Soc Edinburgh* **52**: 603-627.

Kidston, R. and Lang, W. H. (1920b) On Old Red Sandstone plants showing structure, from the Rhynie Chert Bed, Aberdeenshire. Part III. *Asteroxylon mackiei,* Kidston and Lang. *Trans R Soc Edinburgh* **52**: 643-680.

Knoll, A. H. and Niklas K. J. (1987) Adaptation, plant evolution, and the fossil record. *Rev Palaeobot Palynol* **50**: 127-149.

Lersten, N. R. (1997) Occurrence of endodermis with a casparian strip in stem and leaf. *Bot Rev* **63:** 265-272.

Li, H. and Taylor, D. W. (1999) Vessel-bearing stems of *Vasovinea tianii* gen. et sp. nov. (Gigantopteridales) from the Upper Permian of Guizhou Province, China. *Am J Bot* **86:** 1563-1575.

Lu, P., and Jernstedt, J. A. (1996) Rhizophore and root development in Selaginella martensii: Meristem transitions and identity. *Int J Plant Sci* **157:** 180-194.

Ma, Y. and Steeves, T. A. (1992) Auxin effects on vascular differentiation in Ostrich Fern. *Ann Bot* **70:** 277-282.

Magallón, S., and Sanderson, M. J. (2002). Relationships among seed plants inferred from highly conserved genes: Sorting conflicting phylogenetic signals among ancient lineages. *Am. J. Bot.* **89:** 1991-2006.

Mapes, G. and Rothwell, G. W. (1988) Diversity among Hamilton conifers. In *Regional Geology and Paleontology of Upper Paleozoic Hamilton Quarry Area in Southeastern Kansas* (G. Mapes and R. H. Mapes, eds.) pp. 225-244. Kansas Geological Survey, Guidebook Series Vol. 6, Lawrence, Kansas.

Nixon, K. C., Crepet, W. L., Stevenson, D. and Friis, E. M. (1994) A reevaluation of seed plant phylogeny. *Ann Mo Bot Gardens* **81:** 484-533.

Ogura, Y. (1972) *Comparative Anatomy of Vegetative Organs of the Pteridophytes.* Hanbuch der pflanzenanatomie Vol. 7. Gebrüder Borntraeger, Berlin.

Pirozynski, K. A. and Malloch, D. W. (1975) The origin of land plants: A matter of mycotrophism. *BioSystems* **6:** 153-164.

Pixley, E. Y. (1968) A study of ontogeny of primary xylem in roots of *Lycopodium*. *Bot Gazette* **129:** 156-160.

Poethig, R. S. and Sussex, I. M. (1985). The cellular parameters of leaf development in tobacco: A clonal analysis. *Planta* **165:** 170-184.

Pray, T. R. (1955) Foliar venation of angiosperms. II. Histogenesis of the venation of *Liriodendron*. *Am J Bot* **42:** 18-27.

Pray, T. R. (1960) Ontogeny of the open dichotomous venation in the pinna of the fern *Nephrolepsis*. *Am J Bot* **47:** 319-328.

Pray, T. R. (1962) Ontogeny of the closed dichotomous venation of *Regnellidium*. *Am J Bot* **49:** 464-472.

Pryer, K. M., Schneider, H., Smith, A. R., Cranfill, R., Wolf, P. G., Hunt, J. S. and Sipes, S. D. (2001) Horsetails and ferns are a monophyletic group and the closest living relatives to seed plants. *Nature* **409:** 618-622.

Raven, J. A. and Edwards, D. (2001) Roots: Evolutionary origins and biogeochemical significance. *J Exp Bot* **52:** 381-401.

Read, D. J., Duckett, J. G., Francis, R., Ligrone, R. and Russell, A. (2000) Symbiotic fungal associations in 'lower' land plants. *Phil. Trans. Roy. Soc. London B* **355:** 815-831.

Redecker, D., Kodner, R. and Graham, L. E. (2000). Glomalean fungi from the Ordovician. *Science.* **289:** 1920-1921.

Reinhardt, D., Pesce, E.-R., Stieger, P., Mandel, T., Baltensperger, K., Bennett, M., Traas, J. and Friml, J. (2003). Regulation of phyllotaxis by polar auxin transport. *Nature* **426:** 255-260.

Remy, W., Gensel, P. G. and Hass, H. (1993) The gametophyte generation of some Early Devonian land plants. *Int J Plant Sci* **154:** 35-58.

Remy, W., Taylor, T. N., Hass, H. and Kerp, H. (1994) Four hundred-million-year-old vesicular arbuscular mycorrhizae. *Proc Natl Acad Sci U S A* **91:** 11841-11843.

Rothwell, G. W. (1984) The apex of Stigmaria (Lycopsida), rooting organ of Lepidodendrales. *Am J Bot* **71:** 1031-1034.

Rothwell, G. W. (1987) Origin of seed plants: An aneurophyte/seed-fern link elaborated. *Am J Bot* **74:** 970-973.

Rothwell, G. W. (1995) The fossil history of branching: Implications for the phylogeny of land plants. In *Experimental and Molecular Approaches to Plant Biosystematics* (P. C. Hoch and A. G. Stephenson, eds.) pp. 71-86. Missouri Botanical Garden, St. Louis.

Rothwell, G. W. and Erwin, D. M. (1985) The rhizomorph apex of *Paurodendron*; implication for homologies among the rooting organs of Lycopsida. *Am J Bot* **72:** 86-98.

Rothwell, G. W., and Serbet, R. (1994) Lignophyte phylogeny and the evolution of spermatophytes: A numerical cladistic analysis. *System Bot* **19:** 443-482.

Sabatini, S., Beis, D., Wolkenfelt, H., Murfett, J., Guilfoyle, T., Malamy, J., Benfey, P., Leyser, O., Bechtold, N., Weisbeek, P. and Scheres, B. (1999) An auxin-dependent distal organizer of pattern and polarity in the *Arabidopsis* root. *Cell* **99:** 463-472.

Sachs, T. (1991) *Pattern Formation in Plant Tissues.* Cambridge University Press, Cambridge.

Schneider, H. (1996) The root anatomy of ferns: A comparative study. In *Pteridology in Perspective* (J. M. Camus, M. Gibby and R. J. Johns, eds.) pp. 271-283. Royal Botanic Gardens, Kew.

Schneider, H., Pryer, K. M., Cranfill, R., Smith, A. R. and Wolf, P. G. (2002) Evolution of vascular plant body plans: A phylogenetic perspective. In *Developmental Genetics and Plant Evolution* (Q. C. B. Cronk, R. M. Bateman and J. A. Hawkins, eds.) pp. 330-364. Taylor and Francis, London.

Scott, D. H. (1909) *Studies in Fossil Botany.* Adam and Charles Black, London.

Stein, W. (1993) Modeling the evolution of stellar architecture in vascular plants. *Int J Plant Sci* **154:** 229-263.

Taylor, T. N. and Taylor, E. L. (1993) *The Biology and Evolution of Fossil Plants.* Prentice Hall, Englewood Cliffs, New Jersey.

Tryon, R. M. and Tryon, A. F. (1982) *Ferns and Allied Plants with Special Reference to Tropical America.* Springer-Verlag, New York.

Uggla, C., Moritz, T., Sanberg, G. and Sundberg, B. (1996) Auxin as a positional signal in pattern formation in plants. *Proc Natl Acad Sci* **93:** 9282-9286.

Wagner, W. H. (1979) Reticulate veins in the systematics of modern ferns. *Taxon* **28:** 87-95.

Watson, E. V. (1971) *The Structure and Life of Bryophytes.* Hutchinson, London.

Wing, S. L., Hickey, L. J. and Swisher, C. C. (1993) Implications of an exceptional fossil flora for Late Cretaceous vegetation. *Nature* **363:** 342-344.

Winter, K.-U., Becker, A., Muenster, T., Kim, J. T., Saedler, H. and Theissen, G. (1999) MADS-box genes reveal that gnetophytes are more closely related to conifers than to flowering plants. *Proc Natl Acad Sci* **96:** 7342-7347.

Wochok, Z. S. and Sussex, I. M. (1974) Morphogenesis in *Selaginella.* II. Auxin transport in the root (rhizophore). *Plant Physiol* **53:** 738-741.

Zimmerman, W. (1952) Main results of the "Telome Theory." *Palaeobot* **1:** 456-470.

Zurakowski, K. A. and Gifford, E. M. (1988) Quantitative studies of pinnule development in the ferns *Adiantum raddianum* and *Cheilanthes viridis. Am J Bot.* **75:** 1559-1570.

Zwieniecki, M. A., Melcher, P. J., Boyce, C. K., Sack, L. and Holbrook, N. M. (2002) Hydraulic architecture of leaf venation *Laurus nobilis* L. *Plant Cell Environ* **25:** 1445-1450.

# 24

## Are Vessels in Seed Plants Evolutionary Innovations to Similar Ecological Contexts?

*Taylor S. Feild*

Today, seed plants are represented by five major lineages. These lineages include angiosperms (or "flowering plants"), which, with 250,000 or so species, dominate the biomass and species composition of many modern terrestrial habitats—conifers (550 species), cycads (150 species), ginkgoes (1 species), and the Gnetales (70 species). Long-distance xylem hydraulic networks in seed plants consist of two principal cell types: tracheids and vessels (Zimmermann, 1983; Gifford and Foster, 1989; Doyle, 1998). Tracheids are dead, cytoplasm-free cells with lignified secondary cell walls. Tracheids connect to other cells through bordered pit-pairs, that are formed from overarching extensions of the secondary cell wall (Zimmermann, 1983). Pits, however, are not simply open holes but are sealed by primary cell wall materials (i.e., pit membranes) that are permeable to xylem sap. Consequently, tracheids are functionally unicellular.

In contrast, xylem vessels are continuous multicellular tubes (although not necessarily straight pipes, see André 1993 for descriptions of branched, zig-zag, and even circular vessels) that are formed by the differentiation and death of a file of individual cells (termed vessel elements or vessel members). Vessel elements are typically joined end-to-end with their end wall pits open such that sap flows freely through a vessel. Each element develops from a single cell (typically elongated) that before death deposits a lignified secondary cell wall. Contiguous vessel elements share a primary cell wall that remains mostly unlignified. The unlignified portions are hydrolyzed before cell death, leaving large openings that link the lumens of contiguous vessel elements. The basal and apical end walls of consecutive vessel elements are ornamented with barlike or, more often, circular pits that form a region referred to as a perforation plate. The hydraulic compartments defined by vessels, however, are finite in length, with the cells at the top and bottom retaining intact pit membranes (Zimmermann, 1983) but how this is co-ordinated is unknown. Vessel elements were proposed to arise from modifications of the developmental program producing tracheids, such that pit membranes were lost between end walls of adjacent cells (Frost, 1930). Correlated with cellular perforation (and thus vessel origin), tracheary cells become increasingly dimorphic as vessel elements are specialized for hydraulic conductance, through increasing diameter and thinner walls, and fibers for mechanical support, by increasing wall thickness and decreasing cell lumen area as well as pit apertures (Carlquist and Schneider, 2002).

Among the seed plants, xylem vessels are viewed as a quintessential feature of angiosperms, and for good reason—nearly all angiosperms possess vessels. Vessels in the wood also occur in the seed plant group Gnetales (Arber and Parkin, 1908; Thompson, 1918). More recently, xylem vessels were found in the woody stems of an extinct plant, *Vasovinea tianii* (Gigantopteridales), from Permian age rocks in China that was interpreted as a woody vine (Li *et al.*, 1996; Li and Taylor, 1999). However, most seed plant lineages, both living and extinct, rely solely on tracheid-based xylem for long-distance water movement. Vesselless wood occurs in conifers, cycads, ginkgoes, and several extinct lineages, including Bennettitales, Cordaitales, Corystosperms, Glossopterids, and *Pentoxylon* (Doyle, 1996).

What evolutionary processes bear on the transition(s) from a tracheid-to-vessel-dominated hydraulic transport system in seed plants? From time to time, botanists have viewed vessels as structures reflecting a common ancestry, or homologous characters, as in the case of angiosperm and gnetalian vessels (Arber and Parkin, 1908; Muhammad and Sattler, 1982). Other anatomists suggested that vessels were convergent traits, or repeatedly evolved structures that resemble each other in form but are misleading for evolutionary relatedness (Thompson, 1918; Bailey, 1944; Carlquist, 1996). This conclusion was based on observations that vessel perforation plates developed differently in Gnetales and angiosperms. In Gnetales, perforation plates are derived from slightly raised, circular bordered (foraminate) pits, while in angiosperms plates are based on flat, barlike (scalariform) pits. Gigantopterid plates differ from gnetalian and angiosperm designs in that they consist of numerous flattened circular pits that lack pit-borders, as in the vine *Vasovinea* (Li and Taylor, 1999). Convergent vessel origins are also indicated by recent molecular phylogenetic studies that place Gnetales close to vesselless conifers, or even derived from conifers, and that no other extant seed plant group is closely related to angiosperms, although the phylogenetic position of gigantopterids remains unknown (Bowe *et al.*, 2000; Chaw *et al.*, 2000; Magallon and Sanderson, 2002; see Donoghue and Doyle, 2000, and Doyle, 2001 for reviews).

If angiosperm, gigantopterid, and gnetalian vessels are indeed analogous, are vessels a functional motif that evolves whenever the "right" ecological and/or environmental conditions converge, or are the selective contexts for the origins of mega-porous tubes different in each lineage? Exploring these questions is the focus here. First, I summarize our understanding of tracheid and vessel hydraulics of seed plants. Then what is known about the extant and extinct ecophysiology of vessel-bearing seed plants is discussed. With this information in mind, some hypotheses on the paleohabitat context for vessel origins in angiosperms, Gnetales, and gigantopterids are developed. Significant areas where future work is necessary to supply convincing tests of these ideas are highlighted. Pteridophytes' vessels and their ecological consequences are not explored here, as it is presently unclear which groups possess vessels, and little functional information is currently available for these taxa. Recent work on the possible functional and habitat context involving vessel loss in possibly some vesselless angiosperms is also not considered here and is treated elsewhere (Feild *et al.*, 2002).

## Tracheid and Vessel Hydraulic Properties

Conduit size and whether the pit membranes are intact or lysed influence long-distance water transport characteristics of leaves, roots, and stems (Zimmermann, 1983; Tyree and Ewers, 1991; Comstock and Sperry, 2000; Sperry, 2003; and see Chapter 13 in this volume). Of the characters related to tracheary element size, conduit diameter exerts the greatest effect on xylem hydraulic conductance. If water movement through a tracheid or vessel is approximated as laminar hydraulic flow through an ideal capillary, it follows that an increase in conduit diameter results in an approximately four-power increase lumen conductivity (Zimmermann, 1983). Cross-sectional areas of tracheid lumens, approximated as maximum diameter, generally range from 5 μm to 60 μm, with an apparent upper limit of approximately 80 μm in extant seed plants (Bailey and Tupper, 1918; Carlquist, 1975; Lancashire and Ennos, 2002; Sperry, 2003). Seed plant vessels, however, exhibit a greater range in maximum conduit diameters, from 5 μm to as wide as 700 μm in some angiosperm monocot vines (Bailey and Tupper, 1918; Tyree and Ewers, 1991; Davis *et al.*, 1999).

Tracheid and vessel-based xylem also differ in the frequency that pit membranes are crossed during water flow, stemming from differences in effective conduit length. Tracheids form unicellular, relatively small hydraulic compartments varying from 980 to 10,500 μm long in seed plants (Bailey and Tupper, 1918; Zimmerman, 1983). Vessels are functionally multicellular, potentially consisting of several thousand cells (Zimmermann, 1983) and include compartments as short as 1 cm to up to 25 m long (Zimmermann and Jeje, 1981; Zimmermann, 1983; Fisher and Ewers, 1995; Martre and Durand, 2001; T. S. Feild, unpublished data 2001-2003).

Consequently, water transport through a tracheid-based network requires frequent crossing of pit membranes that possess resistive primary cell walls (porosities diameters 25 to 300 nm) (Sperry and Tyree, 1988; Feild *et al.*, 2000; Choat *et al.*, 2003).

Taking greater diameter and hydraulic volume into account, vessels permit greater potential hydraulic conductivity per unit conduit area than a tracheid-based transport system. Indeed, stems of vessel-bearing angiosperms from a variety of habitats generally possess greater xylem hydraulic capacity, expressed in terms of hydraulic flow relative to the amount of cross-sectional area of xylem invested ("sapwood specific hydraulic conductivity"), than conifers and vesselless angiosperms (Sperry *et al.*, 1994; Davis *et al.*, 1999; Brodribb and Feild, 2000; Field *et al.*, 2002; Sperry, 2003; T. Brodribb and T. Feild, unpublished data, 2001-2003). Although many tracheid-bearing species are capable achieving similar or greater stem or root xylem hydraulic capacities on a leaf area basis (i.e., similar leaf-specific hydraulic conductivities, $k_L$, see Brodribb and Feild, 2000), the advantage of vessels is in maintaining these conductivities with less total investment in wood required to support a given flow rate and leaf stomatal conductance (Sperry, 2003). In what sorts of habitats do we find early members of vessel-bearing seed plants? In the next section, existing evidence on the ancestral habitats and comparative (paleo-) ecophysiology of angiosperms, Gnetales, and gigantopterids are examined.

## Ancestral Habitats for Vessel Origin in Vessel-Bearing Seed Plants

### Angiosperms

*Extant "Basal" Angiosperm Biology*    Vessel-bearing angiosperms display a diverse array of growth habits, functional types, and extremely high species diversification into a variety of ecological habitats. To reconstruct the initial selective pressures bearing on the appearance of angiosperm vessels, however, a picture of early angiosperm biology and the sorts of habitats they occupied first is needed. A major step forward in this goal was provided by recent advances in molecular phylogenetics that have converged on a well-supported structure of extant flowering plant phylogeny. Based on several independent lines of phylogenetic evidence, the "basal" angiosperm lineage divergences (e.g., living representatives of nodes that split from the main angiosperm line early on, and which, if studied in a phylogenetic context, may allow resolution of ancestral ecophysiological states Feild *et al.*, 2000, 2001, 2003a, 2003b, 2004) involve woody tropical scandent shrubs, small trees, and straggling woody vines [*Amborella*, 1 species, Austrobaileyales (*Austrobaileya*, 1 species;

*Trimenia,* 9 species; *Illicium,* ~40 species; *Schisandra-Kadsura,* ~40 species), Chloranthaceae (*Ascarina,* 12 species; *Chloranthus,* 10 species; *Hedyosmum,* 40 species; *Sarcandra,* 2 species)], plus a few herbaceous aquatics (water lilies and *Ceratophyllum*), with the remainder of angiosperms grouped into one large clade (i.e., "core angiosperms"; (Mathews and Donoghue, 1999; Doyle and Endress, 2000; Zanis *et al.,* 2002).

***Habitat Context for Vessel Origin*** To develop an image of early angiosperm ecology, Feild *et al.* (2003a, 2004) mapped several ecophysiological traits (e.g., photosynthetic rate, leaf anatomical traits, growth forms) taken from representatives of each major basal lineage onto their phylogeny to reconstruct those present in their common ancestor. The phylogenetic distribution of ecophysiological traits suggested that flowering plants first occupied shaded, disturbed, and moist habitats. Examples of these habitats include the margins of understory watercourses and steep soil washouts in montane tropical cloud forests (Feild *et al.,* 2004). With few exceptions, *Amborella,* Austrobaileyales, and Chloranthaceae shared low and easily light saturated leaf photosynthesis, leaf anatomical features related to the capture of dim understory light (e.g., leaf cross-sections dominated by spongy parenchyma), small seeds, and heavy reliance on clonal reproduction (Feild *et al.,* 2003a, 2004). The aquatic habits of water lilies, although apparently appearing early on, were interpreted as derived; a similar case has been made for the extinct Early Cretaceous aquatic *Archaefructus* (Sun *et al.,* 2002; Friis *et al.,* 2003). Also, some Chloranthaceae (i.e., *Ascarina* and *Hedyosmum*) and Schisandraceae vines showed physiological specializations for higher light habitats (e.g., greater leaf specific xylem hydraulic conductivity, higher leaf photosynthetic rates, and palisade parenchyma cells). Current phylogenies also suggest that vessels were absent in the first angiosperms. *Amborella* is vesselless, and Nymphaeales lack wood and have primary xylem tracheary cells that appear intermediate between tracheids and vessel elements (Schneider *et al.,* 1995; Feild *et al.,* 2000; Carlquist and Schneider, 2002). Vessels, however, probably appeared early in angiosperm evolution and under shady, wet conditions, as the woods of Austrobaileyales and Chloranthaceae contain vessels of "primitive" structure (Carlquist and Schneider, 2002).

How does this ancestral image based on living plants compare to the fossil record of early angiosperms? Evaluating this question is vitally important because basal taxa, like *Amborella,* are equally as far from the angiosperm root in time as sunflowers and orchids. Thus, it is possible that the pattern of ecophysiological conformity that we see in extant basal angiosperms was produced by habitat-biased extinction or convergent evolution for wet, dark, and disturbed habitats among basal angiosperm lineages, even though this is not a parsimonious possibility (Feild *et al.,* 2003a, 2004). At present, the fossil record of early angiosperms (from the Early Cretaceous, 110 to 130 million years ago [Ma]) is inadequate to falsify any of these possibilities. Thus, future paleobotanical work will require a reevaluation of the Early Cretaceous fossil record with greater attention being paid to functional characters preserved in fossils that are linked to low-light, wet, and disturbed habitats (Feild *et al.*, 2003a, 2004).

Nevertheless, there are a few tantalizing clues from the fossil record that suggest extant basal angiosperms may be appropriate analogs for the first angiosperms (Doyle, 2001; Doyle *et al.*, 2003; Feild *et al.*, 2004). For example, some of the earliest fossil angiosperms' leaf cuticles known, such as *Eucalyptophyllum* and D.B. Leaf Type #1, from zone I of the Early Cretaceous Potomac Group sequence, although not linked directly to any extant basal angiosperm lineage, possess anatomical features that are shared only by extant basal angiosperm taxa. These features include presence of "chloranthoid" marginal leaf teeth (consisting of a triple-vein union at a hydathodal apex, Doyle, 2001) that are involved in the release of root-pressure driven guttation sap in extant Chloranthaceae (T. S. Feild and T. L. Sage, unpublished data, 2004), striated cuticles with radiostriate cells, and stomata with steep cuticle ledges and lamellar thickenings (Upchurch, 1984) that appear to help in keeping stomatal pores free of water infiltration or stomatal decoupled from flucuations in humidity during understory gas-exchange (Feild *et al.*, 2003b) (T. S. Feild, personal observations). Similar morphological correspondences between basal angiosperms and Early Cretaceous angiosperms are also seen in the flower and seed fossil record (Doyle, 2001; Doyle *et al.*, 2003; Feild *et al.*, 2004).

## Gnetales

***Extant Gnetalian Biology***   Gnetales are comprised of three major lineages, with *Ephedra* at the base followed by a clade of consisting of *Welwitschia* and *Gnetum* (Bowe *et al.*, 2000; Chaw *et al.*, 2000; Magallon and Sanderson, 2002). Although molecular phylogenies draw Gnetales as a coherent group, the three gnetalian lineages are wildly divergent in ecology and morphology (Gifford and Foster, 1989; Price, 1996).

*Ephedra* contains approximately 35 species of sun-loving and arid-adapted prostrate and profusely branched shrubs as well as a few species occur as scandent (vinelike) shrubs (Price, 1996; Lev-Yedan, 1999). *Ephedra* occurs in temperate dry and desert regions of Asia, Europe, northern Africa, western North America, and South America. Photosynthetic stems are the primary sites of carbon gain in many *Ephedra* species, but small scalelike leaves occur in a few species (Price, 1996).

*Welwitschia*, consisting of a single species, is also desert-dwelling, confined to dry, coastal deserts (0 to100 mm yr$^{-1}$ rainfall) in Angola and Namibia (Henschel and Seely, 2000). *Welwitschia*'s growth habit is, with no exaggeration, unparalleled among all living plants. It occurs as a short, woody, unbranched stem and a massive woody concave crown bearing two huge strap-shaped leaves that function as the permanent photosynthetic organs and last potentially for several centuries (Gifford and Foster, 1989).

*Gnetum* (35 species) inhabits a variety of humid, tropical lowland, riparian, and swamp rainforests (> 1000 m above sea level) of South East Asia, Papua New Guinea, Fiji, the Americas (i.e., South America and as far north as Costa Rica), and Africa (Croat, 1970; Markgraf, 1951, 1965, 1972). *Gnetum* bears remarkably angiosperm-like leaves, consisting of a broad, entire-margined lamina with pinnate-reticulate venation and multiple vein orders (Arber and Parkin, 1908; Markgraf, 1951; Rodin, 1966). The majority of *Gnetum* species are large woody climbers, producing xylem from multiple cambia (Carlquist, 1996). Some *Gnetum* vines ascend high into the canopies of dense riparian vegetation and lowland forest trees; other species occur as low scramblers in open, fire-burnt pastures and disturbed forest edges (Markgraf, 1951, 1965, 1972; T. S. Feild, unpublished field observations, 2002). Two *Gnetum* taxa (i.e., *G. gnemon*, widespread in the Indo-Pacific; *G. costatum*, from eastern Papua New Guinea and the Solomon Islands), that form sun-exposed medium-sized subcanopy trees in lowland rainforest and riverine gallery forests (typically 7 to 15 m, as high as 20 m) (Markgraf, 1951) possess some peculiar liana-like features. In addition to the occasional production of scandent branches, older trees (stems > 15 cm diameter at breast height) of *G. gnemon* develop additional anomalous cambia in the bark that are akin to multiple cambia of lianoid *Gnetum* species (Carlquist, 1996; T. S. Feild and L. Balun, unpublished data, 2002). Interestingly, molecular phylogenetic analyses indicate that tree-forming *Gnetum* species are well nested among *Gnetum* climbers, suggesting that arborescence is secondarily derived (Won and Renner, 2003).

***Habitat Context for Vessel Origin*** On first examination, the extant biology of Gnetales seems to indicate vessel origin in arid, high-light habitats (Carlquist, 1975). Unlike the case in basal angiosperms, however, the extreme morphological and ecological divergence of the living Gnetales suggests that ancestral states inferred from living taxa will likely be misleading (Donoghue and Doyle, 2000). Indeed, the fossil record suggests that the majority of gnetalian biology is extinct (Crane and Upchurch, 1987; Crane, 1996; Mohr *et al.*, 2003; Rydin *et al.*, 2003). This legacy of extinction is reflected by the molecular phylogenetics of extant taxa, as each of the major gnetalian lineages are separated from each other by long molecular phylogenetic branch lengths. This pattern does not appear to reflect ancient origins of each lineage followed by long ecological stasis and low speciation because species of *Ephedra* and *Gnetum* appear to be relatively recent in origin (see Huang and Price, 2003; Won and Renner, 2003).

Although dry habitats are implied by many fossil pollen records of Gnetales (Doyle *et al.*, 1982; Crane, 1996), there is growing evidence from the paleobiology of gnetalian macrofossils that Gnetales were diverse in wet, tropical habitats, which were similar to those occupied by *Gnetum* today. For example, Crane and Upchurch (1987) described *Drewria potomacensis*, a gnetalian from the Aptian (Zone I) Potomac Group sediments of Virginia that consisted of slender stems (potentially lacking wood) bearing opposite and decussate broad leaves. The venation of *Drewria* leaves was reticulate, but more similar to *Welwitschia* than *Gnetum. Drewria* fossils have been interpreted as small shrubs or herbs in mesic habitats similar to those described for some early angiosperms from the same stratigraphic sequence—disturbed floodplain areas or possibly disturbed understory habitats (Hickey and Doyle, 1977; Taylor and Hickey, 1996; Feild *et al.*, 2004). Given the slender stem habit and the extant representation of scandent habits in Gnetales, another possibility is that fossilized shoots were produced by vines climbing in the forest canopy or understory along stream margins. Furthermore, the lineage containing extant *Welwitschia* also appears to have contained a member that occupied much wetter conditions. *Cratonia cotyledon*, a spectacularly preserved intact seedling from the Early Cretaceous (late Aptian, to early Albian) of Brazil with broadleaves that is related to *Welwitschia* based on guard cell morphology and presence of a unique Y-type venation pattern (Rydin *et al.*, 2003), occurred in a lacustrine (lake) depositional environment. *Cratonia* is fossilized along with a diverse gymnosperm-dominated flora and some flowering plants, which indicates greater moisture availability than extant *Welwitschia* habitats, which are sparsely vegetated (Henschel and Seely, 2000; Rydin *et al.*, 2003). Other gnetophytes, awaiting taxonomic description, have also been found in the same sediments as *Cratonia* (Mohr *et al.*, 2003). Finally, although the ephedroid-pollen is usually taken as a faithful indicator of dry conditions, *Ephedra*-like pollen is associated with paleoclimate indicators of both dry (evaporites) and wet (coals) conditions in the Early Cretaceous, which suggests greater habitat breadth for the *Ephedra* lineage in the past (Doyle *et al.*, 1982). Thus, it is conceivable that Gnetales first radiated, and perhaps evolved vessels, in wet, tropical environments.

### Gigantopterids

*Gigantopterid Biology*     Gigantopterids are an extinct seed plant group, represented primarily by several large-leaved "morpho-species" that display a great diversity of reticulate venation patterns (Mamay *et al.*, 1988; Li and Taylor, 1998, 1999). Significantly, one of these gigantopterid leaf types was found in close association with vessel-bearing stems (Li *et al.*, 1996); although direct attachment was not demonstrated (Li and Taylor, 1999). This was taken as evidence that gigantopterids represent the oldest seed plant group with vessels (Li *et al.*, 1996).

During the Early to Late Permian (245 to 290 Ma), a great diversity of gigantopterids occurred in Southeast Asia, with some relic gigantopterids persisting until the Early Triassic (~240 Ma) (Li and Taylor, 1998, 1999). A smaller gigantopterid flora, but possibly unrelated to plants in China, occurred disjunctly in North America as represented by Early Permian rocks from Texas and Oklahoma (Mamay *et al.*, 1988; DiMichele *et al.*, 2000). Although the Permian was predominately a period of global drying (DiMichele and Hook, 1992) in southern China, gigantopterids were fossilized in sediments indicative of wet, tropical lowland swamp deposits. This conclusion is supported by the co-occurrence of gigantopterid leaves with well-known mesic-adapted taxa (lycopsid trees [Lepidodendrales], Calamitales, and *Psaronius*) and presence of thick coal seams in the same deposits (Li and Taylor, 1998). In North America, gigantoperids apparently occurred only in local wet pockets, including coastal and/or floodplain channel deposits, possibly in both fresh and brackish water, whereas the broader landscape may have supported a more drought-adapted flora based on the apparent xeromorphy of the represented leaf fossils (DiMichele *et al.*, 2000).

Sufficient fossil material, as well as close associations between particular leaf and stem fossil morpho-species, allows for partial reconstructions of the habit for two gigantopterids (i.e., *Aculeovinea yunguiensis* and *Vasovinea tianii*) (Li and Taylor, 1998, 1999). The stems of both species were slender, branchless for up to a meter, and apparently only bore a few large (400 cm$^2$) leaves, which were gigantic compared to the leaves of co-occurring plants (Li and Taylor, 1998, 1999). This suggests that gigantopterids were characterized by extremely low allocation of sapwood area to leaf areas; a morphological pattern exhibited by extant vessel-bearing angiosperm and gnetalian lianas (Tyree and Ewers, 1991; T. S. Feild and L. Balun, unpublished data, 2002). Consequently, *Acleovinea* and *Vasovinea* were interpreted as scrambling lianas in wet, arborescent lycopod-dominated forests. *Vasovinea* stems possessed bifurcated hooks extensions that were interpreted as climbing devices (Li and Taylor, 1999). However, the vasculatures of *Acleovinea* and *Vasovinea* were plumbed differently. The stems of *Acleovinea* were described as vesselless, bearing large tracheids (up to 80 μm in diameter) with abundant elliptical pits on their sidewalls (Li and Taylor, 1998). *Vasovinea*, in contrast, possessed large mega-porous vessels (up to 500 μm in diameter), with numerous open oval pits in the perforation plate region (Li *et al.*, 1996; Li and Taylor, 1999).

***Habitat Context for Vessel Origin***    The fossil of gigantopterids is again suggestive of vessel origin in wet habitats, possibly understory zones in lowland rainforest tropical habitats. Although, it is important to point out that the understories of lycopsid tree-dominated forests were likely to be relatively well lit given the "telephone-pole" growth habits of lycopsid trees (DiMichele, 2001).

## Why Vessels in Wet, Shaded Habitats

In the three lineages of vessel-bearing seed plants, there are indications that xylem vessels appear first in lineages that originated in wet tropical forested habitats characterized by high moisture and shade. Setting aside for the moment the great difficulties in ancestral habitat reconstruction, why might vessels be favored in wet, shady habitats? Because vessel-based vascular systems can maintain a given $k_L$ with less cross-sectional investment in wood than tracheids (Zimmerman, 1983; Brodribb and Feild, 2000; Sperry, 2003), one hypothesis is that vessels increase shade tolerance by decreasing leaf, stem, and root wood construction costs compared to ves-selless taxa, as less vascular plumbing is necessary. Thus, the initial selection pressures for vessels may have been increased vegetative growth efficiency under carbon-limited humid conditions (Feild *et al.*, 2004), rather than water stress in dry climates or under high transpiration load in full sun as has been previously suggested (Stebbins, 1974; Carlquist, 1975; Cronquist, 1988).

Vessels appear to be just one part of a whole plant-scale morphological "revolution" in angiosperms, Gnetales, and gigantopterids that is linked to radiation into wet, shady zones. For example, the evolution of net-veined, broad angiosperm-like leaves in seed plants and ferns appears to be associated with wet, warm understory habitats (see Chapter 22). In these habitats, reticulate venation may be functionally important in providing mechanical support, as well as an even allocation of water across large leaf blades that would be favored in mesic, carbon-limited environments (see Chapter 22). Another possibility is that sunfleck capture is increased by broad, entire photosynthetic surfaces. What is especially curious is that scandent growth habits, which experience relaxed mechanical constraints and increased needs for a high conductance long-distance hydraulic pathway to supply their leaves (Tyree and Ewers, 1991), are a recurring theme in basal angiosperms (Feild *et al.*, 2001, 2003a, 2003b, 2004), Gnetales (Fisher and Ewers, 1995; Carlquist, 1996; Lev-Yadun, 1999), and gigantopterids (Li and Taylor, 1998, 1999). Thus, an initial exploration of shady habitats as viney plants may have played a role in spurring vessel origin in seed plants, because hydraulic efficiency would have been placed under strong selection. Scandent habits may offer a new grade shade-tolerance by increasing the ability to use patchy resources in understory environments.

The coupling of increased hydraulic efficiency to scandent growth forms may have allowed more opportunistic exploitation of ephemeral water, nutrient, and light resources compared to co-occurring taxa relying on a single shoot axis.

## Conclusions

We are still a long way from understanding the ecological settings for vessel origins in seed plants. Nevertheless, the hypothesis that vessels arose independently and under the similar circumstances of wet, shady conditions deserves further attention. In particular, additional functional studies, both of the fossils themselves and the extant products of these ancient radiations, are needed. For instance, information on the comparative hydraulic properties of basal angiosperms, Gnetales, and gigantopterids is currently uninvestigated, which is necessary to test whether vessels in these groups are truly analogous. Using the excellent fossil record of gigantopterids, it may also be possible to make accurate inferences in physiologically important aspects of stem water transport (e.g., sapwood specific hydraulic conductivity and leaf area specific hydraulic conductivity). Comparative physiological work on basal angiosperms will be especially informative because the extant flora provides a continuum of xylem morphologies (from vesselless to large-diameter vessels with simple perforation plate) that occur in habitats differing primarily in light availability (Feild *et al.*, 2004). This fortunate condition may allow piecing together how different patterns of vessel development drive hydraulic function. For instance, some vessels in basal angiosperms are wide, but very short (1 to 3 mm), whereas other can be composed of narrow vessel elements that form long (up to 15 cm) and potentially wide hydraulic compartments (T. S. Feild, unpublished data, 2004). A blurred distinction between vessels and tracheids contrasts with vascular architectures of Gnetales and gigantopterids, which possess vessel elements that are well differentiated from tracheids (Carlquist, 1996; Li *et al.*, 1996).

Coupled to functional studies, advances in the phylogenetics and paleobiology of extinct seed plants will be needed to constrain the paleohabitat hypotheses for vessel origins in seed plants. A critical step for testing the "wet, shady" vessel origin hypothesis will be to fit important fossil taxa into a phylogenetic context that have come to light (e.g., *Aculeovinea, Cratonia, Drewria, Vasovinea*). By the same token, phylogenetic studies (including both molecules and fossils) directly addressing the sister-groups to angiosperms, Gnetales, and the gigantopterids, which remain enigmatic as ever (Doyle, 2001), are vital.

## Acknowledgments

I thank C. K. Boyce, T. Brodribb, and J. Sperry for helpful discussions on some of the ideas developed here. I also appreciate the help of L. Balun in the field for work on the physiological ecology of *Gnetum* in Papua New Guinea. I thank N. Arens, C. K. Boyce, U. Hacke, N. M. Holbrook, and M. A. Zwieniecki for comments on the manuscript. Finally, I dedicate this chapter to the memory of O. S. Feild and B. Feild. Support from this research came from funding from the Miller Institute for Basic Research and an NSERC discovery grant.

## References

André, J.-P. (1993) A study of the vascular organization of bamboos (Poaceae-Bambuseae) using a microcasting method. *IAWA J* **19:** 265-278.

Arber, E. A. and Parkin, J. (1908) Studies on the evolution of the angiosperms. The relationship of the angiosperms to the Gnetales. *Ann Bot* **22:** 489-515.

Bailey, I. W. (1944) The development of vessels in angiosperms and its significance in morphological research. *Am J Bot* **31:** 421-428.

Bailey, I. W. and Tupper, W. W. (1918) Size variation in tracheary cells. I. A comparison between the secondary xylems of vascular cryptograms, gymnosperms, and angiosperms. *Proc Am Arts Sci Soc* **54:** 149-204.

Bowe, L. M., Coat, G. and de Pamphilis, C. (2000) Phylogeny of seed plants based on all three genomic compartments: Extant gymnosperms are monophyletic and Gnetales' closest relatives are conifers. *Proc Natl Acad Sci U S A* **97:** 4092-4097.

Brodribb, T. J. and Feild, T. S. (2000) Stem hydraulic supply is linked to leaf photosynthetic capacity: Evidence from New Caledonian and Tasmanian rainforests. *Plant Cell Environ* **23:** 1381-1388.

Carlquist, S. (1975) *Ecological Strategies of Xylem Evolution.* University of California Press, Berkeley.

Carlquist, S. (1996) Wood, bark, and stem anatomy of Gnetales: a summary. *Int J Plant Sci* **157(Suppl.):** S58-S76.

Carlquist, S. and Schneider, E. L. (2002) The tracheid-vessel element transition in angiosperms involves multiple independent features: Cladistic consequences. *Am J Bot* **89:** 185-195.

Chaw, S.-M., Parkinson, C. L., Cheng, Y., Vincent, T. M. and Palmer, J. D. (2000) Seed plant phylogeny inferred from all three plant genomes: Monophyly of extant gymnosperms and origin of Gnetales from conifers. *Proc Natl Acad Sci USA* **97:** 4086-4091.

Choat, B., Ball, M., Luly J. and Holtum, J. (2003) Pit membrane porosity and water stress-induced cavitation in four co-existing dry rainforest tree species. *Plant Physiol* **131:** 41-48.

Comstock, J. P. and Sperry, J. S. (2000) Theoretical considerations of optimal conduit length for water transport in vascular plants. *New Phytol* **148:** 195-218.

Crane, P. R. (1996) The fossil history of Gnetales. *Int J Plant Sci* **157:** S50-S57.

Crane, P. R. and Upchurch, G. R. (1987) *Drewria potomacensis* gen. et sp. nov., an Early Cretaceous member of Gnetales from the Potomac group of Virginia. *Am J Bot* **74:** 1722-1736.

Croat, T. (1970) Gnetaceae. Flora of Panama. *Ann Mo Bot Garden* **57:** 1-4.

Cronquist, A. (1988) *The Evolution and Classification of Flowering Plants.* 2nd Ed. New York Botanical Garden, New York.

Davis, S. D., Sperry, J. S. and Hacke, U. (1999) The relationship between xylem conduit diameter and cavitation caused by freezing. *Am J Bot* **86:** 1367-1372.

DiMichele, W. A. (2001) Carboniferous coal-swamp forest. In *Palaeobiology II* (D. E. G. Briggs and P. R. Crowther, eds.) pp.79-81. Blackwell Science, London.

DiMichele, W. A., Chaney, D. S., Dixon, W. H., Nelson, W. J. and Hook, R. W. (2000) An Early Permian coastal flora from the Central Basin Platform of Gaines County, west Texas. *Palaios* **15:** 524-534.

DiMichele, W. A. and Hook, R. W. (1992) Paleozoic terrestrial ecosystems. In *Terrestrial Ecosystems Through Time* (A. K. Behrensmeyer, J. D. Damuth, W. A. DiMichele, R. Potts, H.-D. Sues and S. L. Wing, eds.) pp. 205-325. Chicago University Press, Chicago.

Donoghue, M. J. and Doyle, J. A. (2000) Seed plant phylogeny: Demise of the anthophyte hypothesis? *Curr Biol* **10:** R106-R109.

Doyle, J. A. (1996) Seed plant phylogeny and the relationships of Gnetales. *Int J Plant Sci* **157:** S3-S39.

Doyle, J. A. (1998) Phylogeny of vascular plants. *Annu Rev Ecol System* **29:** 567-599.

Doyle, J. A. (2001) Significance of molecular phylogenetic analyses for paleobotanical investigations on the origin of angiosperms. *Palaeobotanist* **50:** 167-188.

Doyle, J. A. and Endress, P. K. (2000) Morphological phylogenetic analysis of basal angiosperms: Comparison and combination with molecular data. *Int J Plant Sci* **161(Suppl.):** S121-S153.

Doyle, J. A., Eklund, H. and Herendeen, P. S. (2003) Floral evolution in Chloranthaceae: Implications of a morphological phylogenetic analysis. *Int J Plant Sci* **164(Suppl.):** S365-S382.

Doyle, J. A., Jardiné, S. and Doerenkamp, A. (1982) *Afropollis*, a new genus of early angiosperm pollen, with notes on the Cretaceous palynostratigraphy and paleoenvironments of Northern Gondwana. *Bulletin des Centres de Recherches Exploration-Production Elf-Aquaitane* **6:** 39-117.

Feild, T. S., Arens, N. C. and Dawson, T. E. (2003a) The ancestral ecology of angiosperms: Emerging perspectives from extant basal lineages. *Int J Plant Sci* **164:** S129-S142.

Feild, T. S., Arens, N. C., Doyle, J. A., Dawson, T. E. and Donoghue, M. J. (2004) Dark and disturbed: A new image of early angiosperm ecology. *Paleobiology* **30:** 82-107.

Feild, T. S., Brodribb, T. and Holbrook, N. M. (2002) Hardly a relict: Freezing and the evolution of vesselless wood in Winteraceae. *Evolution* **56:** 464-478.

Feild, T. S., Brodribb, T., Jaffré, T. and Holbrook, N. M. (2001) Acclimation of leaf anatomy, photosynthetic light use, and xylem hydraulics to light in *Amborella trichopoda* (Amborellaceae). *Int J Plant Sci* **162:** 999-1008.

Feild, T. S., Franks, P. J. and Sage, T. L. (2003b) Ecophysiological shade adaptation in the basal angiosperm, *Austrobaileya scandens* (Austrobaileyaceae). *Int J Plant Sci* **164:** 313-324.

Feild, T. S., Zwieniecki, M. A., Brodribb, T., Jaffré, T., Donoghue, M. J. and Holbrook, N. M. (2000) Structure and function of tracheary elements in *Amborella trichopoda*. *Int J Plant Sci* **161:** 705-712.

Fisher, J. B. and Ewers, F. W. (1995) Vessel dimensions in liana and tree species of *Gnetum* (Gnetales). *Am J Bot* **82:** 1350-1357.

Friis, E. M., Doyle, J. A., Endress, P. K. and Leng, Q. (2003) *Archaefructus*: Angiosperm precursor or specialized early angiosperm? *Trends Plant Sci* **8:** 369-373.

Frost, F. H. (1930) Specialization in secondary xylem in dicotyledons. I. Origin of vessels. *Bot Gazette* **90:** 67-94.

Gifford, E. M. and Foster, A. S. (1989) *Morphology and Evolution of Vascular Plants.* W. H. Freeman, New York.

Henschel J. R. and Seely, M. K. (2000) Long-term growth patterns of *Welwitschia mirabilis*, a long-lived plant of the Namib Desert (including a bibliography). *Plant Ecol* **150:** 7-26.

Hickey, L. J. and, Doyle, J. A. (1977) Early Cretaceous fossil evidence for angiosperm evolution. *Bot Rev* **43**: 1-104.

Huang, J. L. and Price, R. A. (2003) Estimation of the age of extant Ephedra using chloroplast rbcL sequence data. *Mol Biol Evol* **20**: 435-440.

Lancashire, J. R. and Ennos, A. R. (2002) Modeling the hydrodynamic resistance of bordered pits. *J Exp Bot* **53**: 1485-1493.

Lev-Yadun, S. (1999) Eccentric deposition of secondary xylem in stems of the climber *Ephedra campylopoda* (Gnetales). *IAWA J* **20**: 165-170.

Li, H. and Taylor, D. W. (1998) *Aculeovinea yunguiensis* gen. et sp. nov. (Gigantopteridales), a new taxon of gigantopterid stem from the Upper Permian of Guizhou Province, China. *Int J Plant Sci* **159**: 1023-133.

Li, H. and Taylor, D. W. (1999) Vessel-bearing stems of *Vasovinea tianii* gen. et sp. nov. (Gigantoperidales) from the Upper Permian of Guizhou Province, China. *Am J Bot* **86**: 1563-1575.

Li, H., Taylor, E. L. and Taylor, T. N. (1996) Permian vessel elements. *Science* **271**: 188-189.

Magallon, S. and Sanderson, M. J. (2002) Relationships among seed plants inferred from highly conserved genes: Sorting conflicting phylogenetic signals among ancient lineages. *Am J Bot* **89**: 1991-2006.

Mamay, S. H., Miller, J. M., Rohr, D. M. and Stein, W. E. (1988) Foliar morphology and anatomy of the gigantopterid plant *Delnortea abbottiae*, from the Lower Permian of west Texas. *Am J Bot* **75**: 1409-1433.

Markgraf, F. (1951) Gnetaceae. *Flora Malesiana Ser 1* **4**: 336-347.

Markgraf, F. (1965) New discoveries of Gnetum in tropical America. *Ann Mo Bot Garden* **52**: 379-386.

Markgraf, F. (1972) Gnetaceae. *Flora Malesiana Ser. 1* **6**: 944-949.

Mathews, S. and Donoghue, M. J. (1999) The root of angiosperm phylogeny inferred from duplicate phytochrome genes. *Science* **286**: 947-950.

Matre, P. and Durand, J.-L. (2001) Quantitative analysis of vasculature in the leaves of *Festuca arundinacea* (Poaceae): Implications for axial water transport. *Int J Plant Sci* **162**: 755-766.

Mohr, B., Rydin, C. and Friis, E.-M. (2003) Gnetalean diversity during the Early Cretaceous of Brazil. Botanical Society of America Meeting, Abstract 643.

Muhammad, A. F. and Sattler, R. (1982) Vessel structure of *Gnetum* and the origin of angiosperms. *Am J Bot* **69**: 1004-1021.

Price, R. A. (1996) The systematics of the Gnetales: A review of morphological and molecular evidence. *Int J Plant Sci* **157**: S40-S49.

Rodin, R. J. (1966) Leaf structure and evolution in American species of *Gnetum. Phytomorphol* **16**: 56-68.

Rydin, C., Mohr, B. and, Friis, E.-M. (2003) *Cratonia cotyledon* gen. et sp. nov.: A unique Cretaceous seedling related to *Welwitschia. Proc R Soc London, B (Suppl)* **270**: S29-S32.

Schneider, E. L., Carlquist, S., Beamer, K. and Kohn, A. (1995) Vessels in Nymphaeaceae: *Nuphar, Nymphaea,* and *Ondinea. Int J Plant Sci* **156**: 857-862.

Sperry, J. S. (2003) Evolution of water transport and xylem structure. *Int J Plant Sci* **164**: S115-S127.

Sperry, J. S., Nichols, K. L., Sullivan, J. E. M. and Eastlack S. E. (1994) Xylem embolism in ring-porous, diffuse-porous, and coniferous trees of northern Utah and interior Alaska. *Ecology* **75**: 1736-1752.

Sperry, J. S. and Tyree, M. T. (1988) Mechanism of water stress-induced xylem embolism. *Plant Physiol* **88**: 581-587.

Stebbins, G. L. (1974) *Flowering Plants: Evolution Above the Species Level.* Harvard University Press, Cambridge, Massachusetts.

Sun, G., Ji, Q., Dilcher, D. L., Zheng, S., Nixon, K. C. and Wang, X. (2002) Archaefructaceae, a new basal angiosperm family. *Science* **296**: 899-904.

Taylor, D. W. and Hickey, L.J. (1996) Evidence for and implications of an herbaceous origin for angiosperms. In *Flowering Plant Origin, Evolution, and Phylogeny* (D. W. Taylor and L. J. Hickey, eds.) pp. 232-266. Chapman and Hall, New York.

Thompson, W. P. (1918) Independent evolution of vessels in Gnetales and angiosperms. *Bot Gazette* **69:** 83-90.

Tyree, M. T. and Ewers, F. W. (1991) The hydraulic architecture of trees and other woody plants. *New Phytol* **119:** 354-360.

Upchurch, G. R. (1984) Cuticular anatomy of angiosperm leaves from the lower Cretaceous Potomac Group. I. Zone I leaves. *Am J Bot* **71:** 192-202.

Won, H. and Renner, S. S. (2003) Horizontal gene transfer from flowering plants to *Gnetum*. *Proc Natl Acad Sci U S A* **100:** 10824-10829.

Zanis, M. J., Soltis, D. E., Solits, P. S., Mathews, S. and Donoghue, M. J. (2002) The root of the angiosperms revisited. *Proc Natl Acad Sci U S A* **99:** 6848-6853.

Zimmermann, M. H. (1983) *Xylem Structure and the Ascent of Sap.* Springer-Verlag, Berlin.

Zimmermann, M. H. and Jeje, A. A. (1981) Vessel length distribution of some American woody plants. *Can J Bot* **59:** 1882-1892.

# 25

## Hydraulic Properties of the Xylem in Plants of Different Photosynthetic Pathways

*Ferit Kocacinar and Rowan F. Sage*

In terrestrial plants, large quantities of water evaporate from leaves during photosynthesis, an inevitable consequence of the need for open stomata to allow $CO_2$ entry into photosynthetic cells. Water lost during photosynthesis is replaced by soil water pulled up through the vascular conduits of the xylem. Because the structure of xylem tissue reflects water use and the photosynthetic capacity of the leaf canopy (Zimmermann, 1983; Tyree and Sperry, 1989; Hubbard *et al.*, 1999, 2001; Hacke and Sperry, 2001; Brodribb *et al.*, 2002; Sperry *et al.*, 2002), variation in water transport demands resulting from changes in humidity, atmospheric $CO_2$ content, allocation between roots and shoots, or intrinsic water use efficiency (WUE) should affect the evolution of xylem characteristics of plants.

In the angiosperms, xylem has undergone pronounced evolutionary specialization relative to older groups of plants (Willis and McElwain, 2002; Baas *et al.*, 2003). Early angiosperms evolved vessels, and during the past 100 million years, a wide range of vessel types appeared, leading to the highly specialized forms of wood that exist today. Major trends in xylem evolution include an increase in vessel width and length, opening of the perforation plate, and an increase in pit number and size (Wheeler and Baas, 1991; Willis and McElwain, 2002; Baas *et al.*, 2003). These traits increased xylem efficiency (flux capacity per unit of xylem), but are also associated with a reduction in xylem safety, where vulnerability to cavitation-induced failure of the xylem is high (Zimmermann, 1983). Climate change and evolutionary progression have been the primary explanations for xylem specialization through geological time; however, reductions in atmospheric $CO_2$ that occurred during the past 100 million years have been suggested to be a contributing factor in xylem evolution (Wheeler and Lehman, 2000; Willis and McElwain, 2002). Declining atmospheric $CO_2$ reduces the efficiency of water use during photosynthesis, and plants compensate by opening stomata to allow more $CO_2$ to enter the leaf, but at the cost of greater transpiration (Robinson, 1994; Beerling *et al.*, 2001; Hubbard *et al.*, 2001). If flow through the xylem cannot increase in step with transpiration, then tension on the water column will increase, potentially to the point where cavitation occurs. Therefore, reduced WUE associated with past reductions in atmospheric $CO_2$ should have selected for xylem characteristics that enhance the efficiency of water flux to leaves, namely longer, wider vessels. Although conceptually sound, the idea remains untested in part because of difficulties linking patterns in the fossil record with variation in atmospheric $CO_2$ content. Not only is the fossil record incomplete (Wheeler and Baas, 1991; Herendeen *et al.*, 1999), but multiple climatic factors covary with $CO_2$ content (Zachos *et al.*, 2001). In the absence of a direct test, linkages between atmospheric $CO_2$ and xylem evolution can be addressed by evaluating hydraulic properties in species differing in intrinsic water use efficiency. The most pronounced differences in intrinsic WUE are found between $C_3$, $C_4$, and Crassulacean Acid Metabolism (CAM) plant species.

In identical environments, WUE is 1.5- to 4-fold greater in $C_4$ than $C_3$ plants of similar ecological characteristics, whereas in CAM plants WUE is 3- to 10-fold greater than in $C_3$ species (Black, 1973; Osmond *et al.*, 1982). In a major summary of photosynthetic pathway differences, Larcher (1995) noted that $C_3$ WUE range between 1.4 and 3.6 g Dry Matter (DM) $kg^{-1}$ $H_2O$, whereas $C_4$ WUE range between 3 and 5 g DM $kg^{-1}$ $H_2O$. Ludlow (1976) showed that WUE of six $C_4$ grasses was twice that of 29 herbaceous $C_3$ species, including grasses and legumes. WUE of CAM plants generally ranges between 6 to 15 g DM $kg^{-1}$ water lost (Larcher, 1995; Zotz *et al.*, 1997). These differences in WUE substantially lower transpiration rates and water transport needs in $C_4$ and CAM plants relative to $C_3$ plants, assuming equivalent growth form and environmental conditions. As suggested for variation in $CO_2$, the evolution of the photosynthetic pathway may also alter the structure and function of xylem. More recently, a number of studies have evaluated hydraulic function in plants of varying photosynthetic pathways. Here, we review these studies and discuss the significance of the results in terms of xylem evolution in flowering plants.

## Conceptual Background: How Should WUE Changes Affect Xylem Hydraulics?

In plants optimally adapted to their environment, evolution is proposed to have selected for xylem properties that balance the conflicting demands of xylem safety and hydraulic efficiency (Tyree and Sperry, 1989). In resource rich environments, high competitive pressure favors efficient xylem that is able to support the transpiration demands of a larger leaf canopy with a high photosynthetic capacity. Efficient xylem is characterized by the presence of longer, wider vessels with numerous pits, open perforation plates between vessel elements, and pit membranes with relatively large pores (Tyree *et al.*, 1994; Tyree and Zimmermann, 2002). However, efficient xylem is prone to cavitation and hence would be selected against in arid habitats where high tension on the water column can cause catastrophic xylem failure (Tyree and Zimmermann, 2002). Safety features that reduce the risk of xylem cavitation include narrow, short vessels with relatively thick walls, fewer pits, and pit membranes with relatively smaller pores (Sperry *et al.*, 1991; Tyree *et al.*, 1994; Ewers *et al.*, 1997; Hacke *et al.*, 2001). An inevitable consequence of these safety features is increased resistance to rapid water transport in the xylem, and hence a reduction in the photosynthetic potential of the leaf canopy (Pockman and Sperry, 2000; Sperry, 2000).

Assuming there is an optimal balance between xylem safety and efficiency for any given environment, how would a change in WUE alter this balance? In resource-rich environments, features promoting rapid growth rate may be favored by natural selection. These generally include enhanced photosynthetic capacity, largely by increased leaf area allocation (Potter and Jones, 1977). By reducing water requirements corresponding to a given photosynthetic potential, conduits in a plant with higher WUE would be able to supply a larger leaf area than an identical set of conduits in a plant with low WUE. Thus, it can be predicted that the leaf area per xylem tissue would increase in mesic environments after WUE enhancements. In dry environments, by contrast, we predict that WUE advantage would allow for the selection of increased safety features. In the simplest case, plants of high WUE would maintain similar leaf area compared to less efficient plants, and the xylem supplying this area might evolve shorter, narrower, and stronger vessels with relatively high pit membrane resistance. In both situations, hydraulic conductivity per leaf area, $K_L$, is predicted to decrease as an evolutionary response to WUE increases.

## Initial Work on Xylem Function in $C_4$ and CAM Species

Until recently, only a few studies examined variation in hydraulic properties between $C_3$ and either CAM or $C_4$ species. In these studies, CAM and $C_4$ species were observed to have low indexes of hydraulic conductivity relative to most $C_3$ species, but the number of non-$C_3$ species was always low (one or two species) and hence any pattern was difficult to discern.

## CAM Plants

One of the most suggestive studies is that of Zotz and co-workers (Zotz *et al.*, 1994, 1997) who show that facultative CAM species *Clusia* have $K_L$ values that are lower than most $C_3$ species. *Clusia uvitana* exhibited $K_L$ values that were one third to one thirtieth of woody $C_3$ species from the same tropical forest on Barro Colorado Island, Republic of Panama. Most of the difference between *C. uvitana* and the $C_3$ species appeared to be in the maintenance of a much higher leaf area per stem in the CAM species, as the specific conductivity values were comparable between the species. *C. uvitana* exhibited 50% cavitation at a relatively high water potential (−1.3 MPa), indicating it had a relatively low safety factor in the xylem.

This study was not an ideal comparison, however, in that most of the $C_3$ species differed in life form from *C. uvitana*, which is a woody tropical hemiepiphyte. Hemiepiphytes in general have low hydraulic conductivities (Zotz *et al.*, 1997), demonstrating the problem of comparing different life forms. In a separate study addressing hydraulic conductivities in the hemiepiphyte life form, the $C_3$-CAM species *Clusia minor* exhibited the lowest $K_L$ values in a comparison group that included four $C_3$ species (Zotz *et al.*, 1997), thereby supporting the hypothesis that CAM photosynthesis enables plants to have a lower $K_L$ than $C_3$ species of similar life form. More work is needed to confirm the relationship between CAM and xylem hydraulics. *Clusia* could be the ideal genus for this work because it contains obligate $C_3$, obligate CAM, and $C_3$-CAM species where the degree of CAM expression varies (Lüttge, 1996).

## $C_4$ Plants

In a search of the literature, we found comparative analyses of wood function conducted on just two $C_4$ species, the desert shrubs *Atriplex confertifolia* and *Atriplex canescens* from arid rangelands of western North America. Relative to similar $C_3$ shrubs from the same arid zone, both *A. canescens* and *A. confertifolia* exhibited low cavitation pressures of stems and structural features that were similar to the most xeric adapted $C_3$ species in their vegetation community (Hacke *et al.*, 2000; Hacke and Sperry, 2001). *A. canescens* had the highest wood density of all the species in the study, and the wood density of *A. confertifolia* was equivalent to the densest wood in the $C_3$ species (Hacke *et al.*, 2000). Hydraulic conductivity data were not presented, limiting the ability to infer photosynthetic pathway effects. Given that wood density is correlated with hydraulic safety (Hacke *et al.*, 2001), however, the results are consistent with the possibility that $C_4$ photosynthesis may be associated with greater hydraulic safety in arid environments. The suggestive nature of the results highlights the need for a comprehensive evaluation of the influence that the $C_4$ photosynthetic pathway may have over xylem structure and function. This evaluation has recently become available (Kocacinar and Sage, 2003, 2004; Kocacinar, 2004).

# Comprehensive Surveys of Hydraulic Function in $C_3$ and $C_4$ Species

An inherent difficulty in evaluating the significance of photosynthetic pathway for xylem structure and function is that variation in xylem properties could be due to phylogenetic lineage and/or adaptation to local environments and not a response to the evolution of the $C_4$ or CAM pathways. To minimize these problems, research should focus on related species from a well-described phylogeny. This alone does not guarantee the resolution of a photosynthetic pathway effect on xylem function, because local ecological adaptation during species divergence could also alter xylem properties. Ideally, both close relationship and identical ecological habitat should be present across a lineage of photosynthetic pathway evolution.

While *Clusia* might be the model group for CAM evolution, the principal group for studying the effects of $C_4$ evolution has been *Flaveria*, a genus in the Asteraceae. *Flaveria* roughly fits the criteria for isolating photosynthetic pathway effects in that it is a closely related group of 21 $C_3$, $C_4$, and $C_3$-$C_4$ intermediate species that grow in semiarid or moderately saline habitats (Powell, 1978; Ku *et al.*, 1991; Monson, 1996). Two phylogenies of *Flaveria* are available. One compares morphological characteristics, ecogeographical data, and hybridization studies (Powell, 1978). The other is based on the gene encoding the H-protein in the glycine decarboxylase (Kopriva *et al.*, 1996). A comprehensive phylogeny of multiple molecular markers and morphological traits is not yet available. In *Flaveria*, there are two apparent $C_4$ lineages, one from $C_3$ herbaceous perennials to fully developed $C_4$ annual herbs, and a second from $C_3$ perennial herbs to $C_4$-like herbaceous perennials (Powell, 1978; Kopriva *et al.*, 1996). In the $C_4$-like perennials, between 80% and 95% of the initial photosynthetic products reside in $C_4$-acids (Monson *et al.*, 1988; Moore *et al.*, 1989). Presence of $C_4$-like and $C_4$ traits in *Flaveria* has increased WUE. WUE is 38% and 60% higher in the $C_4$-like *Flaveria brownii* and the $C_4$ *F. trinervia*, respectively, than the $C_3$ *F. cronquistii* (Monson *et al.*, 1987), whereas the $C_3$-$C_4$ intermediate *F. floridana* was estimated to have 83% higher WUE than the $C_3$ *F. cronquistii* (Schuster and Monson, 1990).

Using leaf-specific conductivity as an index for hydraulic effectiveness, Kocacinar (2004) compared hydraulic properties of two $C_3$ perennials, three $C_4$ annuals, and five herbaceous $C_3$-$C_4$ intermediates, including two $C_4$-like intermediates. The $C_3$ perennials had the highest $K_L$, whereas the $C_4$-like perennials had the lowest $K_L$ of the group, being some 80% less than the $C_3$ values. Fully developed $C_4$ plants did not have the lowest $K_L$ values. Instead they were 40% less than the $C_3$ species (see the *Flaveria* comparison in Fig. 25.1). This may reflect the presence of the annual habit in the $C_4$ species; annual species tend to have higher $K_L$ than perennial species, all else being equal (Kocacinar and Sage, 2003). The $C_3$-$C_4$ intermediates exhibited $K_L$ values that were less than the $C_3$, some threefold greater than the $C_4$-like perennials and equivalent to the $C_4$ annuals. These results show that the $C_4$ pathway is associated with substantial shifts in hydraulic conductivity per leaf area, which is consistent with the hypothesis that the higher WUE of the $C_4$ pathway and $C_3$-$C_4$ intermediates allow for a shift in the hydraulic system to either safer xylem or greater leaf area.

More than 40 independent evolutionary origins of $C_4$ photosynthesis in 18 taxonomic families are estimated to exist (Sage, 2004). Unfortunately, outside of *Flaveria*, phylogenetic analyses of the $C_4$ lineages are incomplete, if not altogether absent. To overcome this limitation, Kocacinar and Sage (2003, 2004) took a broad survey approach where hydraulic properties were compared in a wide range of $C_3$ and $C_4$ species that are either related by occurring in the same family or genus, or sharing the same ecological habitat, in the sense of having the same life form and leaf canopies that are often adjacent in their natural habitat. In a survey of 37 $C_3$ and $C_4$ species, $C_4$ species exhibited significantly lower $K_L$ values than $C_3$ species within the same functional or phylogenetic group (Fig. 25.1). When all $K_L$ values from both photosynthetic pathways were placed into a frequency diagram regardless of life form differences, most $C_4$ species had the lowest $K_L$ values measured (Fig. 25.2), and where there was overlap in $K_L$, it occurred between $C_4$ plants from a weedy or trailing herbaceous group and woody xerophytic $C_3$ species. In most cases, the anatomical properties reflected the differences in $K_L$ (Table 25.1). $C_3$ species within a comparison group tended to have longer and wider vessels (Fig. 25.3; see also Color Plate section) such that the Hagen-Poiseuille theoretical estimates of hydraulic flux and conduit efficiency values are greater in the $C_3$ species (Table 25.1). Xylem specific conductivity ($K_s$), determined as hydraulic conductivity per xylem-cross-sectional area, generally reflects differences in xylem safety. As safety increases, $K_s$ declines (Pockman and Sperry, 2000). $K_s$ was generally lower in the $C_4$ than $C_3$ species, particularly in the species from dry environments (Kocacinar and Sage, 2003, 2004). Associated with the $K_s$ differences, the xylem of the arid-zone $C_4$ species had smaller and shorter vessels.

In a number of cases, however, $C_3$ and $C_4$ species exhibited similar xylem structure and had identical $K_s$. Most overlap in $K_s$ occurred in fast growing herbaceous annuals, such as *Chenopodium album* ($C_3$) and *Amaranthus retroflexus* ($C_4$), *Atriplex hortensis* ($C_3$) and *Atriplex rosea* ($C_4$) (Kocacinar and Sage, 2003), and as shown in Fig. 25.4A (see also Color Plate section), *Sesuvium verrucosum* ($C_3$) and *Trianthema portulacastrum* ($C_4$). In each of these cases, the $C_4$ species had greater leaf area per unit xylem area, which caused the reduction in $K_L$. *T. portulacastrum*, for example, had nearly twice the leaf area per unit xylem area as *Sesuvium* (Fig. 25.4B; see also Color Plate section), leading to a $K_L$ that was half as great in *T. portulacastrum* as *S. verrucosum*

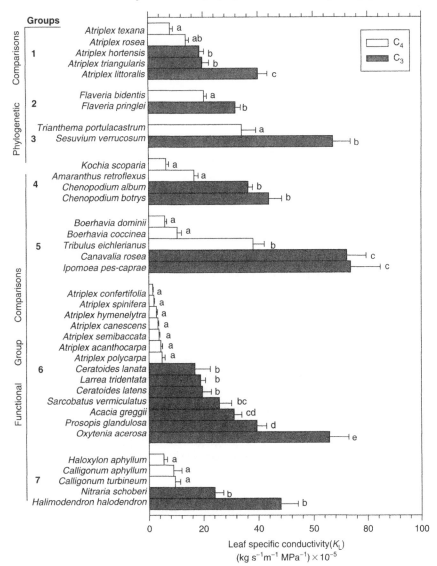

**Figure 25.1**    Leaf specific conductivity, $K_L$, of 37 $C_4$ and $C_3$ species segregated into seven groups based on either phylogenetic (first three groups) or ecological (last four groups) similarity. $K_L$ was determined as hydraulic conductivity divided by leaf area. Bars are mean (±SE) of 5 to 10 plants for each species. Different letters within each group represent significant difference ($P < 0.05$). (Compiled from data in Kocacinar, 2004; Kocacinar and Sage, 2003, 2004.)

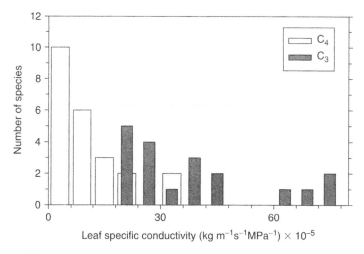

Leaf specific conductivity (kg m$^{-1}$s$^{-1}$MPa$^{-1}$) $\times$ 10$^{-5}$

**Figure 25.2**   A frequency diagram of leaf specific conductivity of 42 C$_4$ and C$_3$ species including the 37 species in Fig. 25.1 and 5 *Flaveria* species. (Adapted from Kocacinar, 2004).

(Fig. 25.4C; see also Color Plate section). This increase in leaf area per stem in the weedy C$_4$ species was visibly obvious in many cases, as shown in Fig. 25.4D (see also Color Plate section) for *T. portulacastrum* and *S. verrucosum*. The comparison between *T. portulacastrum* and *S. verrucosum* is noteworthy as it is phylogenetically and ecologically robust. C$_4$ photosynthesis arose in the Sesuvioideae tribe in the genus *Sesuvium* and diversified in the closely related genus *Trianthema* (Bittrich and Hartmann, 1988). Both species are weedy annuals growing in disturbed, often recently flooded habitats in hot, semiarid regions of the world. A high growth rate appears advantageous, as it allows these species to set seed before xeric conditions occur.

## Ecological Consequences of Photosynthetic Pathway on Xylem Function

The direct advantages of the C$_4$ pathway are higher photosynthetic capacity, light use efficiency at warmer temperatures, and water use efficiency (Pearcy and Ehleringer, 1984); these attributes have received most of the attention in studies addressing C$_4$ performance relative to C$_3$ plants. By reducing the water cost of carbon gain, C$_4$ plants also reduce hydraulic transport requirements, allowing for secondary responses that redistribute resources toward other critical environmental needs. (Similar phenomena occur in urban planning, where reduced water use through conservation reduces the need for municipalities to build expensive pipelines and aque-

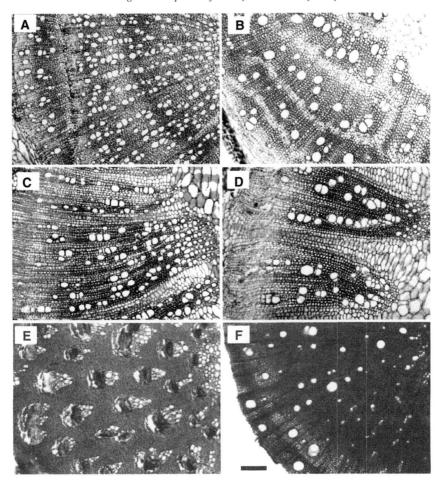

**Figure 25.3** Cross sections of stem xylem from representative pairs of $C_3$ and $C_4$ species of three functional groups. Left panels are $C_4$ and right panels are $C_3$. (A) *Kochia scoparia;* (B) *Chenopodium album;* (C) *Flaveria bidentis;* (D) *Flaveria pringlei;* (E) *Atriplex canescens;* (F) *Prosopis glandulosa.* The scale bar in F is 200 μm and applies to all panels. (See also Color Plate section.)

ducts.) Secondary evolutionary responses to changing water use efficiency represent an important yet poorly recognized phenomena, which could enhance the success of $C_4$ plants beyond the direct benefits of greater carbon gain and reduced water consumption. By adjusting the balance point between xylem safety and efficiency, $C_4$ photosynthesis increases the possible range of evolutionary solutions to other environmental challenges. This appears to be greater xylem safety in dry environments, allowing the $C_4$ species to maintain a given productive potential while reducing the risk

**Table 25.1** Hydraulic and anatomical parameters of $C_4$ and $C_3$ species.

| Functional group Species | PP | MVL (cm) | LVD (μm) | HMD (μm) | VF ($n$ mm$^{-2}$) | Hagen-Poiseulle (m$^4$s$^{-1}$MPa$^{-1}$)×10$^{-8}$ | CEV (m$^2$)×10$^{-14}$ |
|---|---|---|---|---|---|---|---|
| **Annuals** | | | | | | | |
| Kochia scoparia | $C_4$ | 13.2±0.9[a] | 57.2±1.1[a] | 38.3±1.4[a] | 177±13[a] | 10.7±2.0[a] | 1.5±0.3[a] |
| Amaranthus retroflexus | $C_4$ | 19.5±1.9[b] | 70.1±1.4[b] | 51.3±1.8[b] | 87±6.0[b] | 16.7±2.5[a] | 2.6±0.5[a] |
| Chenepodium album | $C_3$ | 26.8±1.4[c] | 93.1±3.3[c] | 65.1±2.4[c] | 43±3.0[c] | 28.3±3.0[b] | 6.8±0.6[b] |
| **Flaveria** | | | | | | | |
| Flaveria bidentis | $C_4$ | 8.6±0.7[a] | 66.1±1.2[a] | 43.0±0.5[a] | 154±10[a] | 21.4±1.5[a] | 3.9±0.2[a] |
| Flaveria pringlei | $C_3$ | 14.3±1.8[b] | 79.7±1.9[b] | 56.3±0.9[b] | 86±5.0[b] | 19.7±0.8[a] | 8.3±0.4[b] |
| **Desert shrubs** | | | | | | | |
| Atriplex canescens | $C_4$ | 7.3±0.4[a] | 42.9±1.7[a] | 28.1±0.9[a] | 188±9.0[a] | 0.68±0.1[a] | 0.7±0.1[a] |
| Atriplex acanthocarpa | $C_4$ | 10.5±2.0[a] | 44.6±0.6[a] | 28.3±1.8[a] | 204±29[a] | 0.60±0.2[a] | 0.8±0.1[a] |
| Prosopis glandulosa | $C_3$ | 27.6±1.9[b] | 71.2±1.6[b] | 51.6±1.1[c] | 47±3.0[c] | 2.09±0.3[b] | 4.9±0.9[c] |
| Acacia greggii | $C_3$ | 26.3±3.3[b] | 72.5±2.5[b] | 47.1±1.6[bc] | 50±3.0[c] | 1.16±0.1[a] | 3.1±0.4[b] |
| Oxytenia acerosa | $C_3$ | 23.2±2.8[b] | 71.0±5.0[b] | 44.6±1.6[b] | 119±8.0[b] | 3.48±0.3[c] | 5.0±0.3[c] |

PP = photosynthetic pathway; MVL = maximum vessel length; LVD = largest vessel diameter; HMD = hydraulic mean diameter; VF = vessel frequency (number of vessels per mm$^2$); Hagen-Poiseuille = estimate of hydraulic conductance by Hagen-Poiseuille equation; CEV = conduit efficiency value (sum of the fourth power of the radius of all conduits divided by the leaf area supplied by these conduits).

Values are the means (± SE) of 5 to 10 samples for each species. Different letters within each functional group represent significant difference at $P < 0.05$.

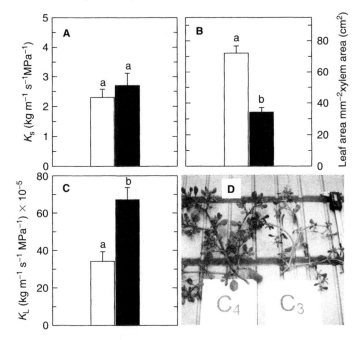

**Figure 25.4**    (A) Specific conductivity, $K_s$, (B) leaf area per unit xylem area, and (C) leaf-specific conductivity of *Trianthema portulacastrum* ($C_4$) (D, *left*) and *Sesuvium verrucosum* ($C_3$) (D, *right*). Bars are mean (± SE) of six plants. Different letters represent significant difference ($P < 0.05$). (From data presented in Kocacinar and Sage, 2003.) (See also Color Plate section.)

of injury or death from cavitation. In resource-rich environments, greater competitive potential appears to have been selected for as indicated by the greater leaf area per stem observed in the weedy $C_4$ species. This could explain in part why $C_4$ species such as the Amazonian floodplain grass *Echinochloa polystachya* exhibit the highest rates of primary productivity ever recorded (Long, 1999). In stochastic environments, a mix of safer xylem and a larger leaf area per stem could increase both safety and competitive potential.

These possibilities are important because competitive ability and catastrophic xylem failure are significant controls over species distribution (Tyree and Sperry, 1989; Pockman and Sperry, 2000; Davis *et al.*, 2002; Martínez-Vilalta *et al.*, 2002). For example, in the Mediterranean zone of northeastern Spain, vulnerability to embolism and drought tolerance are well correlated (Martínez-Vilalta *et al.*, 2002). Woody species from moist microhabitats have the most vulnerable xylem to water stress-induced embolism, and this limitation contributes to their scarcity on more arid soils in the region (Martínez-Vilalta *et al.*, 2002). Similarly, in the Sonoran Desert, riparian species have greater xylem conductivity and smaller safety margins than species from dry, upland sites (Pockman and Sperry, 2000). Based on this finding, Pockman and Sperry (2000) concluded that vulnerability to cavitation limits species distribution in drought-prone soils by determining minimum water potentials at which a plant can occur. By contrast, restrictions on growth imposed by safe xylem limits competitive potential of the upland species, and this inability to compete against riparian species apparently excludes them from the more resource-rich sites (Pockman and Sperry, 2000).

Air-seeding through pit membranes is a primary cause of xylem failure in response to drought (Zimmermann, 1983; Tyree *et al.*, 1994). This has led to much emphasis on the properties of the pit membrane as a key determinant of xylem safety (Tyree *et al.*, 1994; Jarbeau *et al.*, 1995; Choat *et al.*, 2003). In an effort to address this, we examined xylem vulnerability curves in several pairs of $C_3$ and $C_4$ species such as *Chenopodium album* ($C_3$) and *Kochia scoparia* ($C_4$), *Ipomoea pes-caprae* ($C_3$) and *Boerhavia coccinea* ($C_4$) (Kocacinar and Sage, 2003), and *Flaveria pringlei* ($C_3$) and *F. bidentis* (Kocacinar, 2004). In every case, the $C_4$ species exhibited a 50% loss of hydraulic conductivity at xylem tensions that were 0.5 to 2.8 MPa greater than observed in the $C_3$ species. Pit membranes were examined with a scanning electron microscope but showed no obvious difference in structure, and generally appeared to be intact once artifacts were taken into account.

While these vulnerability curves indicate greater safety in the xylem of $C_4$ plants in terms of air-seeding potential, the anatomical results demonstrate that evolution of xylem properties in response to changes in the photosynthetic pathway (and presumably WUE) is not simply changes in air-seeding potential but includes structural components of the hydraulic pathway. The response of conduit size and number to changes in WUE is important because it spreads the hydraulic load over the entire transport system, which should increase its resilience to drought and reduce the probability of the hydraulic system reaching the point of catastrophic xylem failure. For example, reduction in conduit size and associated increases in conduit number enhance redundancy in the transport pathway, thereby contributing to hydraulic safety (Tyree *et al.*, 1994). Furthermore, if links are to be made between variation in atmospheric $CO_2$ and wood evolution, it is important that conduit size and number respond to WUE variation in addition to the pit membrane structure. Xylem cells are usually preserved in fossils, whereas pit membranes are typically lost.

## Atmospheric $CO_2$ and the Evolution of Modern Wood

By associating improvements in WUE with differences in xylem structure and function, the $C_3$ to $C_4$ transition serves as a model linking reductions in atmospheric $CO_2$ during the past 100 million years with xylem evolution in plants. Atmospheric $CO_2$ levels are predicted to have been five to eight times the current level at the time angiosperms first appeared some 130 million years ago (Berner and Kothavala, 2001). Since then, atmospheric $CO_2$ levels have declined, approaching the current value of 370 ppm by about 7 million years ago, and then falling to as low as 180 ppm during the ice ages of the last million years (Fig. 25.5). The change in WUE corresponding to the $CO_2$ reduction over the past 100 million years can be estimated by comparing responses of modern plants grown at varying $CO_2$ levels. In rice, WUE increased more than fourfold as $CO_2$ content increased from ice age values to the levels present about 100 million years ago (Fig. 25.5, inset). This change in $CO_2$ corresponds to the upper range of the WUE differences observed between $C_3$ and $C_4$ species.

The relationship between photosynthetic pathway and xylem structure provides experimental support for the possibility that the efficient xylem of advanced angiosperm groups might be a specialization for low $CO_2$ conditions. Compared to species of the Cretaceous period, plants from the Tertiary are widely noted to have evolved xylem tissues that exhibit features indicative of enhanced conducting efficiency (Wheeler and Baas, 1991). The most obvious features are wider vessel elements with simple perforation plates. These changes are most likely reflecting changes at the leaf

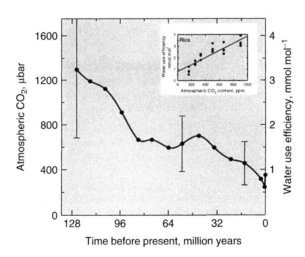

**Figure 25.5** The level of atmospheric $CO_2$ through geological time, based on modeled data from Sage (2004), using values in Berner (1994). Insert: the relationship between water use efficiency (WUE) in rice and growth $CO_2$. (From Baker *et al.*, 1990.) This relationship between WUE and growth $CO_2$ in rice was used to calibrate the relationship between WUE and atmospheric $CO_2$ through time assuming constant microclimatic conditions (right axis).

level that enhance water use. As $CO_2$ levels decline, greater stomatal conductance and larger leaf area are needed to compensate for inhibitory effects of low $CO_2$ availability (Robinson, 1994; Sage, 1994). Other factors, such as climate change, would have certainly played a contributing role. Between 40 million years ago and today, Earth's climate has deteriorated, resulting in widespread aridification (Zachos *et al.*, 2001); however, this would also have selected for safety features in the world flora. The implication that WUE differences may have driven xylem evolution in the past should have significance for the response of plants to future increases in atmospheric $CO_2$ caused by human activities. In the coming era of elevated atmospheric $CO_2$, there may be a shift in selection patterns such that plants with greater hydraulic safety may be favored over those with high xylem efficiency.

## Conclusion

Reductions in atmospheric $CO_2$ concentration during the past 100 million years substantially reduced WUE in land plants, whereas evolution of the $C_4$ photosynthetic pathway enhanced WUE. Because the structure of xylem reflects water use by the leaf canopy, variation in water transport requirements resulting from changes in WUE of the leaf canopy might affect the evolution of xylem characteristics. Recent studies of hydraulic conductivity and vascular anatomy between $C_3$, $C_4$, and CAM species show that $C_4$ and CAM plants consistently have lower hydraulic conductivity per unit leaf area than similar $C_3$ species. $C_4$ species produced shorter and narrower vessels and higher wood density, indicating the xylem of $C_4$ plants is less prone to cavitation. The results are consistent with a direct effect of WUE variation on patterns of xylem evolution, and indicate that other phenomena that affect WUE, such as variation in atmospheric $CO_2$, affect xylem structure and function and therefore may explain, in part, patterns of wood evolution in terrestrial plants.

## Acknowledgments

We thank Thomas Fung, Tammy Sage, and Emin Kocacinar for their assistance in the data collection used in this review. This review was prepared with financial assistance from the Ministry of Education of Turkey to F.K. and NSERC grant OGP0154273 to R.F.S.

# References

Baas, P., Ewers, F. W., Davis, S. D. and Wheeler, E. A. (2003) Evolution of xylem physiology. In: *The Evolution of Plant Physiology, Linnean Society symposium series Number 21,* (A. R. Hemsley and I. Poole, eds.) pp.277-299. Elsevier Academic Press, New York.

Baker, J. T., Allen, L. H., Jr., Boote, K. L., Jones, P. and Jones, J. W. (1990) Rice photosynthesis and evapotranspiration in subambient, ambient, and superambient carbon dioxide concentrations. *Agronomy J* **82:** 834-840.

Beerling, D. J., Osborne, C. P. and Chaloner, W. G. (2001) Evolution of leaf form in land plants linked to atmospheric $CO_2$ decline in the late Paleozoic era. *Nature* **410:** 352-354.

Berner, R. A. (1994) 3geocarb-II -a revised model of atmospheric $CO_2$ over Phanerozoic time. *Am J Sci* **294:** 56-91.

Berner, R. A. and Kothavala, Z. (2001) GEOCARB III: A revised model of atmospheric $CO_2$ over phanerozoic time. *Am J Sci* **301:** 182-204.

Bittrich, V. and Hartmann, H. E. K. (1988) The Aizoaceae-a new approach. *Bot J Linnean Soc* **97:** 239-254.

Black, C. J. (1973) Photosynthetic carbon fixation in relation to net $CO_2$ uptake. *Annu Rev Plant Physiol* **24:** 253-286.

Brodribb, T. J., Holbrook, N. M. and Gutierrez, M. V. (2002) Hydraulic and photosynthetic co-ordination in seasonally dry tropical forest trees. *Plant Cell Environ* **25:** 1435-1444.

Choat, B., Ball, M., Luly, J. and Holtum, J. (2003) Pit membrane porosity and water stress-induced cavitation in four co-existing dry rainforest tree species. *Plant Physiol* **131:** 41-48.

Davis, S. D., Ewers, F. W., Sperry, J. S., Portwood, K. A., Crocker, M. C. and Adams, G. C. (2002) Shoot dieback during prolonged drought in *Ceanothus* (Rhamnaceae) chaparral of California: A possible case of hydraulic failure. *Am J Bot* **89:** 820-828.

Ewers, F. W., Carlton, M. R., Fisher, J. B., Kolb, K. J. and Tyree, M. T. (1997) Vessel diameters in roots versus stems of tropical lianas and other growth forms. *IAWA J* **18:** 261-279.

Hacke, U. G. and Sperry, J. S. (2001). Functional and ecological xylem anatomy. *Perspect Plant Ecol Evol System* **4:** 97-115.

Hacke, U. G., Sperry, J. S. and Pittermann, J. (2000) Drought experience and cavitation resistance in six shrubs from the Great Basin, Utah. *Basic Appl Ecol* **1:** 31-41.

Hacke, U. G., Sperry, J. S., Pockman, W. T., Davis, S. D. and McCulloch, K. A. (2001) Trends in wood density and structure are linked to prevention of xylem implosion by negative pressure. *Oecologia* **126:** 457-461.

Herendeen, P. S., Wheeler, E. A. and Baas, P. (1999) Angiosperm wood evolution and the potential contribution of paleontological data. *Bot Rev* **65:** 278-300.

Hubbard, R. M., Bond, B. J. and Ryan, M. G. (1999) Evidence that hydraulic conductance limits photosynthesis in old *Pinus ponderosa* trees. *Tree Physiol* **19:** 165-172.

Hubbard, R. M., Ryan, M. G., Stiller, V. and Sperry, J. S. (2001) Stomatal conductance and photosynthesis vary linearly with plant hydraulic conductance in ponderosa pine. *Plant Cell Environ* **24:** 113-121.

Jarbeau, J. A., Ewers, F. W. and Davis, S. D. (1995) The mechanism of water-stress-induced embolism in two species of chaparral shrubs. *Plant Cell Environ* **18:** 189-196.

Kocacinar, F. (2004) Photosynthetic Pathway and Hydraulic Architecture in Higher Plants. Ph.D. Dissertation, University of Toronto, Toronto, Ontario.

Kocacinar, F. and Sage, R. F. (2003) Photosynthetic pathway alters xylem structure and hydraulic function in herbaceous plants. *Plant Cell Environ* **26:** 2015-2026.

Kocacinar, F. and Sage, R. F. (2004) Photosynthetic pathway alters xylem structure and hydraulic function in woody plants. *Oecologia* **139:** 214-223.

Kopriva, S., Chu, C. C. and Bauwe, H. (1996) Molecular phylogeny of *Flaveria* as deduced from the analysis of nucleotide sequences encoding the H-protein of the glycine cleavage system. *Plant Cell Environ* **19:** 1028-1036.

Ku, M. S. B., Wu, J. R., Dai, Z. Y., Scott, R. A., Chu, C. and Edwards, G. E. (1991) Photosynthetic and photorespiratory characteristics of *Flaveria* species. *Plant Physiol* **96:** 518-528.

Larcher, W. (1995) *Physiological Plant Ecology.* Springer-Verlag, New York.

Long, S. P. (1999) Environmental responses. In *C₄ Plant Biology* (R. F. Sage and R. K. Monson, eds.) pp. 215-249. Academic Press, San Diego.

Ludlow, M. M. (1976) Ecophysiology of C₄ grasses. In *Water and Plant Life: Problems and Modern Approaches* (O. L. Lange, L. Kappen and E.-D. Schulze, eds.), Vol. 19, pp. 364-386. Springer-Verlag, Berlin.

Lüttge, U. (1996). *Clusia*: Plasticity and diversity in a genus of C₃/CAM intermediate tropical trees. In *Crassulacean Acid Metabolism. Biochemistry, Ecophysiology and Evolution. Ecological Studies* (K. Winter and J. A. C. Smith, eds.), Vol. 114, pp. 296-311. Springer-Verlag, Berlin.

Martínez-Vilalta, J., Prat, E., Oliveras, I. and Piñol, J. (2002) Xylem hydraulic properties of roots and stems of nine Mediterranean woody species. *Oecologia* **133:** 19-29.

Monson, R. K. (1996) The use of phylogenetic perspective in comparative plant physiology and developmental biology. *Ann Mo Bot Garden* **83:** 3-16.

Monson, R. K., Schuster, W. S. and Ku, M. S. B. (1987) Photosynthesis in *Flaveria brownii* Powell, A.M. -a C₄-like C₃-C₄ intermediate. *Plant Physiol* **85:** 1063-1067.

Monson, R. K., Teeri, J. A., Ku, M. S. B., Gurevitch, J., Mets, L. J. and Dudley, S. (1988) Carbon-isotope discrimination by leaves of *Flaveria* species exhibiting different amounts of C₃-cycle and C₄-cycle co-function. *Planta* **174:** 145-151.

Moore, B. D., Ku, M. S. B. and Edwards, G. E. (1989) Expression of C₄-like photosynthesis in several species of *Flaveria. Plant Cell Environ* **12:** 541-549.

Osmond, C. B., Winter, K. and Ziegler, H. (1982) Functional significance of different pathways of CO₂ fixation in photosynthesis. In *Encyclopedia of Plant Physiology* (O. L. Lange, P. S. Nobel, C. B. Osmond and H. Ziegler, eds.), Vol. 12B. pp. 480-547, Springer-Verlag, Berlin.

Pearcy, R. W. and Ehleringer, J. (1984) Comparative ecophysiology of C₃ and C₄ plants. *Plant Cell Environ* **7:** 1-13.

Pockman, W. T. and Sperry, J. S. (2000) Vulnerability to xylem cavitation and the distribution of Sonoran desert vegetation. *Am J Bot* **87:** 1287-1299.

Potter, J. R. and Jones, J. W. (1977) Leaf area partitioning as an important factor in growth. *Plant Physiol* **59:** 10-14.

Powell, A. M. (1978) Systematics of *Flaveria* (Flaveriinae-Asteraceae). *Ann Mo Bot Garden* **65:** 590-636.

Robinson, J. (1994) Speculations on carbon dioxide starvation, late Tertiary evolution of stomatal regulation and floristic modernization. *Plant Cell Environ* **17:** 345-354.

Sage, R. F. (1994) Acclimation of photosynthesis to increasing atmospheric CO₂ -the gas-exchange perspective. *Photosynth Res* **39:** 351-368.

Sage, R. F. (2004) The evolution of C₄ photosynthesis. *New Phytol* **161:** 341-370.

Schuster, W. S. and Monson, R.K. (1990) An examination of the advantages of C3-C4 intermediate photosynthesis in warm environments. *Plant Cell Environ* **13:** 903-912.

Sperry, J. S. (2000) Hydraulic constraints on plant gas exchange. *Agric Forest Meteorol* **104:** 13-23.

Sperry, J. S., Hacke, U. G., Oren, R. and Comstock, J. P. (2002) Water deficits and hydraulic limits to leaf water supply. *Plant Cell Environ* **25:** 251-263.

Sperry, J. S., Perry, A. H. and Sullivan, J. E. M. (1991) Pit membrane degradation and air-embolism formation in aging xylem vessels of *Populus tremuloides* Michx. *J Exp Bot* **42:** 1399-1406.

Tyree, M. T., Davis, S. D. and Cochard, H. (1994) Biophysical perspectives of xylem evolution: Is there a tradeoff of hydraulic efficiency for vulnerability to dysfunction. *IAWA J* **15:** 335-360.

Tyree, M. T. and Sperry, J. S. (1989) Vulnerability of xylem to cavitation and embolism. *Annu Rev Plant Physiol Plant Mol Biol* **40:** 19-38.

Tyree, M. T. and Zimmermann, M. H., eds. (2002). *Xylem Structure and the Ascent of Sap.* 2nd Ed. Springer-Verlag, Berlin.

Wheeler, E. A. and Baas, P. (1991) A survey of the fossil record for dicotyledonous wood and its significance for evolutionary and ecological wood anatomy. *IAWA Bull* **12:** 275-332.

Wheeler, E. A. and Lehman, T. M. (2000) Late Cretaceous woody dicots from the Aguja and Javelina Formations, Big Bend National Park, Texas, USA. *IAWA J* **21:** 83-120.

Willis, K. J. and McElwain, J. C. (2002) *The Evolution of Plants.* Oxford University, Oxford.

Zachos, J., Pagani, M., Sloan, L., Thomas, E. and Billups, K. (2001) Trends, rhythms, and aberrations in global climate 65 Ma to present. *Science* **292:** 686-693.

Zimmermann, M. H. (1983) *Xylem Structure and the Ascent of Sap.* Springer-Verlag, Berlin.

Zotz, G., Patino, S. and Tyree, M. T. (1997) Water relations and hydraulic architecture of woody hemiepiphytes. *J Exp Bot* **48:** 1825-1833.

Zotz, G., Tyree, M. T. and Cochard, H. (1994) Hydraulic architecture, water relations and vulnerability to cavitation of *Clusia uvitana* Pittier: A $C_3$-CAM tropical hemiepiphyte. *New Phytol* **127:** 287-295.

# Part VII
## Synthesis

# 26

## Integration of Long Distance Transport Systems in Plants: Perspectives and Prospects for Future Research

*N. Michele Holbrook and Maciej A. Zwieniecki*

The modular and fundamentally decentralized body plan of plants suggests that their vascular tissues may play an important role in integrating information from distally located exchange surfaces. Although plant vascular tissues have long been known to be involved in the movement of chemical signals (Wilkinson and Davies, 2002), only recently have they been recognized as more than passive elements. The last decade has demonstrated that both the xylem and the phloem are involved in communicating and processing information (Holbrook *et al.*, 2002; van Bel, 2003a, 2003b). We anticipate that future work will fully establish the transport tissues as key integrating systems in plants in which the physiological properties of the conduits themselves play an active part in the distributed processes necessary for coordinated function. The ways in which the vascular tissues contribute to whole plant physiology forms the central theme of this book. Here we bring together many of the ideas raised in this volume, as well as draw attention to areas in need of further research.

### Xylem

Studies of xylem physiology have been dominated by the question of how this essential pipeline could provide reliable transport despite operating at pressures substantially below zero (Tyree and Zimmermann, 2002). Excellent correspondence between embolism formation resulting from external pressurization and that resulting from internal xylem tensions (Cochard *et al.*, 1992) provides strong evidence that meniscal failure is the dominant mode for xylem cavitation. This, in turn, has focused attention on pit membranes as the most likely site by which gas bubble can be "seeded" into water-filled conduits (Sperry and Tyree, 1988; Choat *et al.*, 2003; and see Chapter 16). Because the water column in the xylem is much more stable than what can be created in the laboratory (Smith, 1994), the question of how plants avoid internal nucleation should also be addressed. Because homogeneous nucleation within xylem conduits is predicted to be a rare event (Pickard, 1981), how the surface properties of xylem conduits minimize cavitation from occurring at water/solid interfaces (heterogeneous nucleation) is central to the issue of xylem transport. This is particularly an issue for species in which embolism repair occurs, as the presence of gases within these vessels should provide ample opportunity for the entrapment of gas nucleation sites along vessel walls (Alberti and DeSimone, 2005). There are practical reasons to pursue this line of research; understanding how plants stabilize water within xylem conduits will provide an important foundation for engineering microfluidic systems capable of operating under negative pressures (Stroock and Whitesides, 2003).

The ability to refill cavitated vessels contemporaneous with the transport of water under tension in neighboring vessels forms one of the most important and perhaps most challenging areas of xylem research. Although the repair of cavitated vessels has been demonstrated in a number of species (Salleo *et al.*, 1996; Zwieniecki and Holbrook, 1998; Zwieniecki *et al.*, 2000; Hacke and Sperry, 2003), both the costs of refilling and the associated physiological constraints remain largely unknown. A number of hypotheses have been proposed to explain refilling (see Chapter 18). These are based largely on indirect measurements that indicate that refilling is associated with the activities of living cells (Bucci *et al.*, 2003; Sakr *et al.*, 2003; Salleo *et al.*, 2004), but shed little light on the exact processes involved. Given the importance of refilling to xylem physiology, resolving this is critical and major investments should be directed toward understanding this phenomenon. Although we now know that the xylem exists in a state of dynamic equilibrium between cavitation and refilling, the real time scale for these events remains uncertain. It is possible, given the relatively coarse nature of current measurement techniques, that cavitation and repair occur at a much higher frequency that we can detect. Thus, *in vivo* studies that provide essentially continuous information on the status of individual vessels and that document not just the refilling of conduits with water but their return to a functional state (i.e., able to transport water under tension) are needed. Once the basic mechanisms of refilling are resolved, the next step will be to understand the signal transduction pathways by which xylem parenchyma sense and respond to the presence of cavitated conduits.

The xylem represents a major long-term investment, with tissues produced in any one year potentially being able to influence a plant's performance for years to come. Physiological processes controlling the development of this three dimensionally complex and long-lived tissue, however, are only partially understood (see Chapters 14 and 22). A number of studies have addressed the scaling laws governing optimal transport systems in relation to plant vascular systems (West *et al.*, 1999; McCulloh *et al.*, 2003). Although these studies provide valuable insights into the hydrodynamic design of such complex transport tissues, effort is also needed to understand the developmental processes that lead to these patterns in the first place (Cochard *et al.*, 1997). Aloni and Zimmerman's proposed linkage between auxin gradients and conduit size is an important step in this direction (Aloni and Zimmermann, 1983; Aloni, 1992). The idea of long-distance transport leads naturally to a focus on axial transport properties. Yet, because of the dangers of xylem dysfunction resulting from cavitation, water must also be able to move in both the radial and circumferential direction. The arrangement of xylem vessels is often quite species specific (Carlquist, 1975) and clearly of physiological importance (see Chapter 17), yet we know even less about the developmental processes that govern the spatial patterning of vessels, including the frequency with which individual conduits come into contact. Future work that bridges between molecular studies of cell differentiation and the large-scale vascular patterns governing water movement from soil to leaves is needed.

## Phloem

Studies of phloem transport have been beset by doubts in a manner reminiscent of the xylem, although, in this case, questions surrounding the Munch flow hypothesis have centered on the nature of the sieve plates (see Chapter 2). Although careful work has shown these pores to be open in undamaged sieve tubes, better resolution of the functional dimensions of sieve plate pores is needed to characterize the hydraulic properties of the phloem (Knoblauch and van Bel, 1998; Ehlers *et al.*, 2000). In general, our understanding of long-distance transport in the phloem has lagged behind characterization of the processes associated with both loading and unloading (see Chapters 3 and 8). A major factor contributing to this is the near absence of readily available tools for measuring the major parameters needed to characterize the hydraulic nature of the phloem. At present, techniques for measuring mass transport require access to short-lived radioisotopes (Minchin *et al.*, 1993), while direct turgor measurements require aphid collaborators (Wright and Fisher, 1980; Fisher and Cash-Clark, 2000). Thus, a major need is for new approaches that can be used to monitor sieve tube pressure and mass transport through the phloem. For example, pressure-sensitive molecular probes that allow one to combine measurements of transport and pressure would represent a major advance. However, conceptual advances that link processes operating at the level of the individual cells responsible for energizing the phloem with those relating to transport pathways that can be many tens of meters long are also needed.

Although we struggle with understanding phloem transport, plants continue to quietly transport sugars across distances between distal organs that can exceed 100 meters at a rate adequate to support their own growth and respiration, as well as feed rhizosphere microbial communities. The pressure drop in the phloem needed to transport sugar across the length of an average size tree, calculated based on our imperfect knowledge of both sieve tube dimensions and mass flow rates, significantly exceeds any experimentally determined pressure gradient. Furthermore, theoretical analyses indicate that both the transport of material and the transmission of information become inefficient when the axial pressure drop is large relative to the osmotic pressure of the sieve sap (Thompson and Holbrook, 2003b, 2004). Regulating such pressure gradients along the length of the plant might require that individual cells have a means of adjusting their physiology in relation to their position along the pathway. A simple structural solution to the problems attendant with large pressure gradients is to partition the flow path into discrete units. High rates of sugar reloading along the transport phloem (van Bel, 2003a) are consistent with the existence of symplastic discontinuities or "relays" in phloem transport (Lang, 1979). In addition to energizing transport at multiple points along the pathway, such relays might make an important contribution to transport security, as the redistribution of sugar at a local scale allows functional sieve tubes to bypass damaged regions. Studies of the functional anatomy of the phloem are needed to assess the classical idea of symplasmic continuity from source to sink.

Our understanding of the hydraulic architecture of the phloem lags well behind that of the xylem. Simple questions such as how the number and dimensions of sieve elements vary in relation to leaf area both within a single plant and between species with contrasting growth forms have received little attention (Schumacher, 1948). We believe that a system-level examination of the allometry of the phloem will yield insights comparable to those achieved through the study of xylem structure (Tyree and Zimmermann, 2002). The scaling relations governing the phloem, however, are likely to be quite different from those of the xylem. For example, the substantial hydraulic resistance of the sieve plates means that mass flow through sieve tubes will not follow the Hagen-Poisieulle law. Assuming that the hydraulic properties of the sieve plates themselves remain unchanged, there will be a diminishing effect of increased conduit diameter on pressure driven flow (Thompson and Holbrook, 2003a). System-level studies of the phloem, however, are hindered by the difficulty in identifying which cells are actually involved in long-distance transport. New anatomical approaches that allow sieve elements to be easily distinguished from surrounding phloem parenchyma cells are needed. In addition, the extent to which individual sieve tubes come into contact with one another should be quantified as a first step in understanding phloem "segmentation." The possibility of using transgenic plants to label sieve elements, followed by noninvasive techniques such as confocal microscopy to characterize the three-dimensional structure of the phloem should be explored.

## Xylem and Phloem Together: Whole Organism Integration

Xylem and phloem lie side-by-side throughout their entire path, and the exchange of energy and matter between the two has profound effects on their function. For example, phloem plays a critical, but as yet unspecified, role in the refilling of cavitated xylem vessels. Phloem girdling essentially eliminates refilling, although whether this is due to a cessation in energy input, a scarcity of solutes such as $K^+$, lack of auxin, or a general wounding response is not known. Because the xylem sets the water potential gradient along the stem, the phloem is also markedly influenced by the xylem. Modeling studies indicate that the phloem remains in osmotic equilibrium with the immediately adjacent apoplast (Thompson and Holbrook, 2003b). Thus, changes in xylem pressure gradients will influence turgor within sieve elements or require the active transport of solutes to regulate turgor. However, pressure gradients within the xylem may themselves be constrained by the need to maintain phloem transport. Because the typical direction of phloem transport runs counter to that of the xylem, the ability of the phloem to maintain suitable rates of mass transport may require that the water potential gradient in the xylem not exceed limitations imposed by the need to maintain phloem transport (Schulze, 1991).

Fundamental to the ways plants explore and exploit their environment is their ability to produce a highly bifurcated body. Although branching allows the major exchange surfaces (e.g., roots and leaves) to sample a spatially and temporally heterogeneous environment, it also means that the optimal supply networks between these distal regions will themselves vary with environmental conditions. One solution to this problem is to produce a transport system that is itself responsive to changes in resource availability at a local scale. Potassium is a key component of sieve tube osmotica, potentially playing the central role in the regulation of sieve tube turgor pressure. Potassium also influences xylem resistance through its effect on pit membrane hydrogels. Increased phloem export and thus higher rates of potassium recycling from leaves with currently high photosynthetic rates could, via enhanced potassium recycling to the xylem, increase xylem conductivity and thus water supply to those leaves. Similarly, new evidence suggests that the phloem is capable of flow control at a local scale owing to changes in path resistance (Knoblauch *et al.*, 2001, 2003) and/or through the local control of sieve tube turgor (Gould *et al.*, 2004). Local transport control is likely to extend to all tissues involved in the passage of materials in plants, including both roots and leaves. The challenge for future studies lies in testing if these local controls can be integrated at a larger scale by the transport system to allow the optimal redistribution of scarce resources (Fig. 26.1).

**Figure 26.1** Integrating role of transport systems in plants. In the classical view, information processing is restricted to distal organs. Recent work suggests that information is also locally understood by the transport system, with interactive control (IC) between xylem (x) and phloem (p) and autonomous control (AC) in both systems playing an important role in whole plant organization.

A new understanding of vascular tissues that includes their contributions to whole plant integration places on center-stage the question: To what extent are plants distributed systems in which processing of information at the local scale leads to self-organization at a larger scale? To date, studies of long-distance transport have focused on the effects manifested by both sources and sinks, the fact that plants must supply and therefore coordinate many sources and many sinks has been tackled in only the simplest of systems (Minchin *et al.*, 1993). There is no doubt that information is conveyed between the distal parts of the plant via material transport. Because this mode of communication might take days for a particular chemical messenger to travel between leaves and roots, it is likely that even sessile plants would benefit from a more rapid means of information transmission. Pressure waves (and in the phloem the associated concentration waves) (Thompson and Holbrook, 2004) form the most likely means of rapid communication across the plant, and it is precisely this form of communication that is most dependent on the structure and physiology of the transport tissues. Thus, understanding the hydraulic properties of the vascular system is central to the issue of whole plant function. The opportunity and challenge for the future will be to integrate across scales by manipulating physiological activities at the cellular level while monitoring mass flow and hydraulic parameters across the entire length of the plant. In particular, approaches that combine targeted manipulation of water and/or ion channels and membrane transporters in specific cell types, whether they be xylem parenchyma or sieve tube elements, are needed to unravel the complex interdependencies of the two vascular transport systems.

# References

Alberti, G. and DeSimone, A. (2005) Wetting of rough surfaces: A homogenization approach. *Proc R Soc London. Series A. Mathematical, Physical and Engineering Sciences* **461**: 79-97.

Aloni, R. (1992) The control of vascular differentiation. *Int J Plant Sci* **153**: S90-S92.

Aloni, R. and Zimmermann, M. H. (1983) The control of vessel size and density along the plant axis: A new hypothesis. *Differentiation* **24**: 203-208.

Bucci, S., Scholz, F., Goldstein, G., Meinzer, F. and Sternberg, L. (2003) Dynamic changes in hydraulic conductivity in petioles of two savanna tree species: Factors and mechanisms contributing to the refilling of embolized vessels. *Plant Cell Environ* **26**: 1633-1645.

Carlquist, S. (1975) *Ecological Strategies of Xylem Evolution.* University of California Press, Berkeley.

Choat, B., Ball, M., Luly, J. and Holtum, J. (2003) Pit membrane porosity and water stress-induced cavitation in four co-existing dry rainforest tree species. *Plant Physiol* **131**: 41-48.

Cochard, H., Croziat, P. and Tyree M. T. (1992) Use of positive pressure to establish vulnerability curves: Further support for the air-seeding hypothesis and implication for pressure-volume analysis. *Plant Physiol* **100**: 205-209.

Cochard, H., Peiffer, M., LeGall, K. and Granier, A. (1997) Developmental control of xylem hydraulic resistances and vulnerability to embolism in *Fraxinus excelsior* L: Impacts on water relations. *J Exp Bot* **48:** 655-663.

Ehlers, K., Knoblauch, M. and van Bel, A. J. E. (2000) Ultrastructural features of well-preserved and injured sieve elements: Minute clamps keep the phloem transport conduits free for mass flow. *Protoplasma* **214:** 80-92.

Fisher, D. B. and Cash-Clark, C. E. (2000) Sieve tube unloading and post-phloem transport of fluorescent tracers and proteins injected into sieve tubes via severed aphid stylets. *Plant Physiol* **123:** 125-137.

Gould, N., Minchin, P. E. H. and Thorpe, M. R. (2004) Direct measurements of sieve element hydrostatic pressure reveal strong regulation after pathway blockage. *Funct Plant Biol* **31:** 987-993.

Hacke, U. G. and Sperry, J. S. (2003) Limits to xylem refilling under negative pressure in *Laurus nobilis* and *Acer negundo*. *Plant Cell Environ* **26:** 303-311.

Holbrook, N. M., Zwieniecki, M. A. and Melcher, P. J. (2002) The dynamics of "dead" wood: Maintenance of water transport through plant stem. *Integrative Comp Biol* **42:** 492-496.

Knoblauch, M., Noll, G. A., Müller, T., Prüfer, D., Schneider-Hüther, I., Scharner, D., van Bel, A. J. E. and Peters, W. S. (2003) ATP-independent mechano-proteins that exert force in contraction and expansion. *Nat Mat* **2:** 600-603.

Knoblauch, M., Peters, W. S., Ehlers, K. and van Bel, A. J. E. (2001) Reversible calcium-regulated stopcocks in legume sieve tubes. *Plant Cell* **13:** 1221-1230.

Knoblauch, M. and van Bel, A. J. E. (1998) Sieve tubes in action. *Plant Cell* **10:** 35-50.

Lang, A. (1979) A relay mechanism for phloem translocation. *Ann Bot* **44:** 141-145.

McCulloh, K., Sperry, J. and Adler, F. (2003) Water transport in plants obeys Murray's law. *Nature* **421:** 939-942.

Minchin, P. E. H., Thorpe, M. R. and Farrar, J. F. (1993) A simple mechanistic model of phloem transport which explains sink priority. *J Exp Bot* **44:** 947-955.

Pickard, W. F. (1981) The ascent of sap in plants. *Prog Biophys Mol Biol* **37:** 181-229.

Sakr, S., Alves, G., Morillon, R. L., Maurel, K., Decourteix, M., Guilliot, A., Fleurat-Lessard, P., Julien, J. L. and Chrispeels, M. J. (2003) Plasma membrane aquaporins are involved in winter embolism recovery in walnut tree. *Plant Physiol* **133:** 630-641.

Salleo, S., Lo Gullo, M. A., De Paoli, D. and Zippo, M. (1996) Xylem recovery from cavitation-induced embolism in young plants of *Laurus nobilis*: A possible mechanism. *New Phytol* **132:** 47-56.

Salleo, S., Lo Gullo, M.A., Trifilo, P. and Nardini, A. (2004) New evidence for a role of vessel-associated cells and phloem in the rapid xylem refilling of cavitated stems of *Laurus nobilis* L. *Plant Cell Environ* **27:** 1065-1076.

Schulze, E. D. (1991) Water and nutrient interactions with plant stress. In *Response of Plants to Multiple Stresses* (H. A. Mooney, W. E. Winner and E. J. Pell, eds.) pp. 89-103. Academic Press, San Diego.

Schumacher, A. (1948) Beitrag zur Kenntnis des Stofftransportes in dem Sichbröhrensgotem höherer Pflanzen. *Planta* **35:** 642-700.

Smith, A. M. (1994) Xylem transport and the negative pressure sustainable by water. *Ann Bot* **74:** 647-651.

Sperry, J. S. and Tyree, M. T. (1988) Mechanism of water stress-induced xylem embolism. *Plant Physiol* **88:** 581-587.

Stroock, A. D. and Whitesides, G. M. (2003) Controlling flows in microchannels with patterned surface charge and topography. *Accounts Chem Res* **36:** 595-604.

Thompson, M. V. and Holbrook, N. M. (2003a) Application of a single-solute non-steady-state phloem model in the study of long-distance assimilate transport. *J Theoret Biol* **220:** 419-455.

Thompson, M. V. and Holbrook, N. M. (2003b) Scaling phloem transport: Water potential equilibrium and osmoregulatory flow. *Plant Cell Environ* **26:** 1561-1577.

Thompson, M. V. and Holbrook, N. M. (2004) Scaling phloem transport: Information transmission. *Plant Cell Environ* **24:** 509-519.

Tyree, M. T. and Zimmermann, M. H. (2002) *Xylem Structure and the Ascent of Sap.* 2nd Ed. Springer-Verlag, Berlin.

West, G., Brown, J. and Enquist, B. (1999) A general model for the structure and allometry of plant vascular systems. *Nature* **400:** 664-667.

van Bel, A. J. E. (2003a) Transport phloem: Low profile, high impact. *Plant Physiol* **131:** 1509-1510.

van Bel, A. J. E. (2003b) The phloem, a miracle of ingenuity. *Plant Cell Environ* **26:** 125-150.

Wilkinson, S. and Davies, W. J. (2002) ABA-based chemical signalling: The coordination of responses to stress in plants. *Plant Cell Environ* **25:** 195-210.

Wright, J. P. and Fisher, D. B. (1980) Direct measurements of sieve tube turgor pressure using severed aphid stylets. *Plant Physiol* **65:** 1133-1135.

Zwieniecki, M. A. and Holbrook, N. M. (1998) Diurnal variation in xylem hydraulic conductivity in white ash (*Fraxinus americana* L.), red maple (*Acer rubrum* L.) and red spruce (*Picea rubens* Sarg.). *Plant Cell Environ* **21:** 1173-1180.

Zwieniecki, M. A., Hutyra, L., Thompson, M. V. and Holbrook, N. M. (2000) Dynamic changes in petiole specific conductivity in red maple (*Acer rubrum* L.), tulip tree (*Liriodendron tulipifera* L.) and northern fox grape (*Vitis labrusca* L.). *Plant Cell Environ* **23:** 407-414.

# Index

# Physiological Ecology

## A Series of Monographs, Texts, and Treatises

### Series Editor
**Harold A. Mooney**
*Stanford University, Stanford, California*

### Editorial Board
Fakhri A. Bazzaz    F. Stuart Chapin    James R. Ehleringer
Robert W. Pearcy    Martyn M. Caldwell    E.-D. Schulze

T. T. KOZLOWSKI. Growth and Development of Trees, Volumes I and II, 1971

D. HILLEL. Soil and Water: Physical Principles and Processes, 1971

V. B. YOUNGER and C. M. McKELL (Eds.). The Biology and Utilization of Grasses, 1972

J. B. MUDD and T. T. KOZLOWSKI (Eds.). Responses of Plants to Air Pollution, 1975

R. DAUBENMIRE. Plant Geography, 1978

J. LEVITT. Responses of Plants to Environmental Stresses, Second Edition
Volume I: Chilling, Freezing, and High Temperature Stresses, 1980
Volume II: Water, Radiation, Salt, and Other Stresses, 1980

J. A. LARSEN (Ed.). The Boreal Ecosystem, 1980

S. A. GAUTHREAUX, JR. (Ed.). Animal Migration, Orientation, and Navigation, 1981

F. J. VERNBERG and W. B. VERNBERG (Eds.). Functional Adaptations of Marine Organisms, 1981

R. D. DURBIN (Ed.). Toxins in Plant Disease, 1981

C. P. LYMAN, J. S. WILLIS, A. MALAN, and L. C. H. WANG. Hibernation and Torpor in Mammals and Birds, 1982

T. T. KOZLOWSKI (Ed.). Flooding and Plant Growth, 1984

E. L. RICE. Allelopathy, Second Edition, 1984

M. L. CODY (Ed.). Habitat Selection in Birds, 1985

R. J. HAYNES, K. C. CAMERON, K. M. GOH, and R. R. SHERLOCK (Eds.). Mineral Nitrogen in the Plant–Soil System, 1986

T. T. KOZLOWSKI, P. J. KRAMER, and S. G. PALLARDY. The Physiological Ecology of Woody Plants, 1991

H. A. MOONEY, W. E. WINNER, and E. J. PELL (Eds.). Response of Plants to Multiple Stresses, 1991

F. S. CHAPIN III, R. L. JEFFERIES, J. F. REYNOLDS, G. R. SHAVER, and J. SVOBODA (Eds.). Arctic Ecosystems in a Changing Climate: An Ecophysiological Perspective, 1991

T. D. SHARKEY, E. A. HOLLAND, and H. A. MOONEY (Eds.). Trace Gas Emissions by Plants, 1991

U. SEELIGER (Ed.). Coastal Plant Communities of Latin America, 1992

JAMES R. EHLERINGER and CHRISTOPHER B. FIELD (Eds.). Scaling Physiological Processes: Leaf to Globe, 1993

JAMES R. EHLERINGER, ANTHONY E. HALL, and GRAHAM D. FARQUHAR (Eds.). Stable Isotopes and Plant Carbon—Water Relations, 1993

E.-D. SCHULZE (Ed.). Flux Control in Biological Systems, 1993

MARTYN M. CALDWELL and ROBERT W. PEARCY (Eds.). Exploitation of Environmental Heterogeneity by Plants: Ecophysiological Processes Above- and Belowground, 1994

WILLIAM K. SMITH and THOMAS M. HINCKLEY (Eds.). Resource Physiology of Conifers: Acquisition, Allocation, and Utilization, 1995

WILLIAM K. SMITH and THOMAS M. HINCKLEY (Eds.). Ecophysiology of Coniferous Forests, 1995.

MARGARET D. LOWMAN and NALINI M. NADKHARNI (Eds.). Forest Canopies, 1995

BARBARA L. GARTNER (Ed.). Plant Stems: Physiology and Functional Morphology, 1995

GEORGE W. KOCH and HAROLD A. MOONEY (Eds.). Carbon Dioxide and Terrestrial Ecosystems, 1996

CHRISTIAN KÖRNER and FAKHRI A. BAZZAZ (Eds.). Carbon Dioxide, Populations, and Communities, 1996

FAKHRI A. BAZZAZ and JOHN GRACE (Eds.). Plant Resource Allocation, 1997

J. J. LANDSBERG and S.T. GOWER. Application of Physiological Ecology to Forest Management, 1997

THEODORE T. KOZLOWSKI and STEPHEN G. PALLARDY. Growth Control in Woody Plants, 1997

LOUISE E. JACKSON (Ed.). Ecology in Agriculture, 1997

ROWAN F. SAGE and RUSSELL K. MONSON (Eds.). $C_4$ Plant Biology, 1999

YIGI LUO and HAROLD MOONEY. Carbon Dioxide and Environmental Stress, 1999

JACQUES ROY, BERNARD SAUGIER, and HAROLD A. MOONEY (Eds.). Terrestrial Global Productivity, 2001

L.B. FLANAGAN, J.R. EHLERINGER and D.E. PATAKI. Stable Isotopes and Biosphere-Atmosphere Interactions: Processes and Biological Controls, 2005

Printed and bound by CPI Group (UK) Ltd, Croydon, CR0 4YY

08/05/2025

01864787-0001